Springer
Berlin
Heidelberg
New York
Barcelona
Budapest
Hong Kong
London
Milan
Paris
Tokyo

NATO ASI Series

Advanced Science Institutes Series

A series presenting the results of activities sponsored by the NATO Science Committee, which aims at the dissemination of advanced scientific and technological knowledge, with a view to strengthening links between scientific communities.

The Series is published by an international board of publishers in conjunction with the NATO Scientific Affairs Division

A Life Sciences B Physics	Plenum Publishing Corporation London and New York
C Mathematical and Physical Sciences D Behavioural and Social Sciences E Applied Sciences	Kluwer Academic Publishers Dordrecht, Boston and London
F Computer and Systems Sciences G Ecological Sciences H Cell Biology I Global Environmental Change	Springer-Verlag Berlin Heidelberg New York London Paris Tokyo Hong Kong Barcelona Budapest

PARTNERSHIP SUB-SERIES

1. Disarmament Technologies	Kluwer Academic Publishers
2. Environment	Springer-Verlag
3. High Technology	Kluwer Academic Publishers
4. Science and Technology Policy	Kluwer Academic Publishers
5. Computer Networking	Kluwer Academic Publishers

The Partnership Sub-Series incorporates activities undertaken in collaboration with NATO's Cooperation Partners, the countries of the CIS and Central and Eastern Europe, in Priority Areas of concern to those countries.

NATO-PCO DATABASE

The electronic index to the NATO ASI Series provides full bibliographical references (with keywords and/or abstracts) to about 50000 contributions from international scientists published in all sections of the NATO ASI Series. Access to the NATO-PCO DATABASE compiled by the NATO Publication Coordination Office is possible in two ways:

- via online FILE 128 (NATO-PCO DATABASE) hosted by ESRIN, Via Galileo Galilei, I-00044 Frascati, Italy.
- via CD-ROM "NATO Science & Technology Disk" with user-friendly retrieval software in English, French and German (© WTV GmbH and DATAWARE Technologies Inc. 1992).

The CD-ROM can be ordered through any member of the Board of Publishers or through NATO-PCO, Overijse, Belgium.

Series G: Ecological Sciences, Vol. 37

Azospirillum VI and Related Microorganisms

Genetics – Physiology – Ecology

Edited by

Istvàn Fendrik

Institute of Biophysics, University of Hannover
Herrenhäuser Straße 2
D-30419 Hannover, Germany

Maddalena del Gallo

Department of Food Technology, Environmental and Microbiological Sciences, University of Molise
Via Tiberio 21
I-86100 Campobasso, Italy

Jos Vanderleyden

F. A. Janssens Laboratory of Genetics
Department of Applied Plant Sciences, K. U. Leuven
Willem de Croylaan 42
B-3001 Heverlee, Belgium

Miklos de Zamaroczy

Unité de Physiologie Cellulaire, CNRS URA 1300
Département des Biotechnologies, Institut Pasteur
F-75724 Paris Cedex 15, France

Springer

Proceedings of the NATO Advanced Research Workshop on Azospirillum VI and Related Microorganisms, held at Sárvár, Hungary, September 4-7, 1994

Library of Congress Cataloging-in-Publication Data

Azospirillum VI and related microorganisms : genetics, physiology, ecology / edited by István Fendrik ... [et al.].
p. cm. -- (NATO ASI series. Series G, Ecological sciences ; no. 37)
Includes bibliographical references and index.
ISBN-13:978-3-642-79908-2 (hard)
1. Azospirillum. I. Fendrik, I. II. Series.
QR82.A9A9725 1995
589.9'5--dc20 95-33021
CIP

Additional material to this book can be downloaded from http://extra.springer.com.

ISBN-13:978-3-642-79908-2 e-ISBN-13:978-3-642-79906-8
DOI: 10.1007/978-3-642-79906-8

Softcover reprint of the hardcover 1st edititon 1995

Typesetting: Camera ready by authors/editors
Printed on acid-free paper
SPIN 10477097 31/3136 - 5 4 3 2 1 0

pathway might be dependent on changes in the proton motive force. Although the role of indole-3-acetic acid (IAA) biosynthesis in *Azospirillum* is still not clear, significant progress was reported on the genetics and biochemistry of IAA biosynthesis.

In the "Ecology session", new and intriguing research fields were introduced: the growth of *Azotobacter* inside algal cells proves the possibility to produce stable and functional artificial symbiosis; the formation of stable coaggregates of *Azospirillum* and *Rhizobium* cells was reported as an interesting model for delivery beneficial bacteria with multiple applications. Using new methodologies, such as the use of ribosomal sequences, taxonomic and ecologic studies of diazotrophic bacteria were presented; soil parameters which affect *Azospirillum* survival were for the first time systematically analysed and lead to the conclusion that this bacterium cannot survive, or at least cannot survive in a high number in the soil, without the associated plant. As a logical consequence, tactic response appears to be one of the most important capability of *Azospirillum* to occupy its ecological niche.

The key lecture on Field Application evaluated the physiological and agronomical aspects of *Azospirillum* inoculations. A survey of worldwide field experiments over the past 20 years revealed that in 60 to 70% of the trials, an increase in crop yield could be observed. These increases compensated the inoculation costs and replaced generally about 30% N-fertilizer. A better coordination of field experiments among the researchers and more clarification of the basic mechanisms of plant growth promotion will be necessary in the future.

The workshop demonstrated how important it is to encourage contacts between scientists of NATO and Central and East European Countries in particular - to evaluate the impact of research on environmental friendly and low energetically demanding agriculture and food prodction. Such meetings are especially useful in eliminating differences between research activities of laboratories and in initiating co-operation in future projects. The exchange of ideas and results are improving efficiency both in basic and applied sciences.

The Organizing Committee wants to express its deepest gratitude for the very generous support of the Workshop given by the North Atlantic Treaty Organization. The Organization Committee also wishes to thank the following sponsoring Organizations: Hannoversche Hochschulgemeinschaft, Agricultural University of Gödöllö, University of Hannover, ENEA, Deutsche Forschungsgemeinschaft, University of Molise, Hungarian Academy of Sciences. We gratefully acknowledge the Township of Sárvár, especially Dr. M. Radics (Manager of the City) and I. Szabó (Major of Town). Our special thanks go to all members of the Local Organizing Committee (Peter Gaál, Eva Kárpáti), who were able to overcome all difficulties and are fully responsible for the success of the meeting. Many thanks must go to Roswitha Fendrik for her excellent management of the organization before and during the Workshop.

I. Fendrik	Director
M. Del Gallo	Co-Director
T. Sik	Co-Director
J. Vanderleyden	Co-Director

Preface

Three years ago the 5th *Azospirillum* Workshop was held in Hannover. Since 1991, *Azospirillum* research has made significant progress and the NATO A.R.W. in Sárvár/Hungary was very important to consolidate the present interest and to discuss novel aspects. Among these novel aspects, major emphasis was given to bacteria that live inside the plants - endophytes - and their mode of infection. Fungi are presently considered as good candidates for infection needles. This means that the ecosystem to consider becomes more complex and that the experimental settings to study so-called associative bacteria-plant interactions, might have to be adjusted.

The Workshop covered in an interdisciplinary way, genetical aspects, interactions with other microorganisms, physiology and molecular biology, ecology and field applications.

Among the large number of diazotrophs isolated from the rhizosphere of cereal crops, *Azospirillum* appears as a model system to study root colonization during association. However, in this type of association, which occurs in the absence of root differentiated structures, it is difficult to analyse what limits nitrogen transfer to the host plant. Therefore new tools had to be developed, and are now successfully applied in *Azospirillum* and other associative diazotrophs.

An area which has progressed rapidly in the last few years deals with plant-endophytic nitrogen fixing organisms. Several genera of bacteria are involved and these and their various hosts were reviewed by J. Döbereiner. Also J. Sprent developed this theme, discussing first, ways in which endophytic bacteria enter host plants and second, cells/intercellular spaces/vessels as an environment of supporting nitrogen fixation. This led to the formulation of key questions for further study in the near future. Endophytic bacteria are present in high numbers in all plants analysed (wild and cultivated species). This is known for a long time but was not considered in the right perspective. Now we are aware that endophytes are part of the plant ecosystem and it should therefore be possible to reconstruct a more equilibrated crop-environment through inoculation with selected endophytes and VAM-fungi.

Meanwhile the study of Rhizobia as possible endophytes for non-leguminous plants continues. The group of E. Cocking studies the structure of pseudo-nodules induced in roots of wheat and rice by *Rhizobium* sp.

Using the most advanced tools in molecular genetics, several groups reported on genome analysis, gene identification and gene regulation in *Azospirillum*. The new findings with respect to *nif* and *gln* gene regulation are of particular interest, since *Azospirillum* holds also for other plant root associated diazotrophs. It could also be observed from the various presentations at the Workshop, that statements, concerning the genetics and the physiology of nitrogen fixation and association with plant roots, are slowly merging, as was previously observed in the field of symbiotic nitrogen-fixing bacteria.

In the session "Physiology and Molecular Biology" several papers focussed on the Wheat-*Azospirillum* interactions. Colonization of wheat roots causes alterations in the metabolism of both the plant and bacterial partner. In colonized plant roots, accumulation of hydroxy-proline rich proteins and activation of glycosidases occurs. The role of bacterial polysaccharides and plant lectins in the interaction *Azospirillum*-plant roots was addressed by several groups. New ideas for the lectin-recognition hypothesis have been forwarded and appear worthwhile to pursue in the future. It should be noticed that the lectin-hypothesis gains also renewed interest in the symbiotic *Rhizobium*-legumes interaction. With respect to sensing between *Azospirillum* and plant roots, nice studies on chemo- and aerotaxis of *Azospirillum* were presented. At least for *Azospirillum brasilense*, the main chemosensing

TABLE OF CONTENTS

SECTION IV - PHYSIOLOGY AND MOLECULAR BIOLOGY

SECTION V - ECOLOGY

SECTION I:

PLENARY LECTURES

Endophytic Occurrence of Diazotrophic Bacteria in Non-Leguminous Crops.

Johanna Döbereiner , Vera L. D. Baldani and Veronica M. Reis.

Centro Nacional de Pesquisa em Agrobiologia/EMBRAPA.

23851-970, Seropédica, Rio de Janeiro,

Brazil.

Abstract

The recent discovery of obligate endophytic diazotrophs, like *Acetobacter diazotrophicus*, *Herbaspirillum seropedicae* and *H. rubrisubalbicans* colonizing sugar cane roots, stems and leaves, may finally explain the high contributions of BNF to this crop. *A. diazotrophicus* so far has only been isolated from sugar cane, sweet potatoe and Cameroon grass, all three with high sugar contents and propagated by stem cuttings. *Azoarcus* spp., another obligate endophytic genus, isolated from Kallar grass (*Leptochloa fusca*), has been found to establish within wetland rice roots when inoculated. *Herbaspirillum seropedicae* has been isolated from many graminaceous plants, forage grasses as well as from maize, sorghum and rice, but it does not seem to occur in other plant families. *H. rubrisubalbicans* so far has only been isolated from sugar cane roots, stems and leaves. These two species, even though clearly identified as two distinct species by DNA homology, are extremely similar and can only be distinguished by growth on n-acetyl-glucosamine (*H. seropedicae*) and that on meso-erythritol (*H. rubrisubalbicans*). Transmission of H. seropedicae may occur within seeds or via graminaceous weed plants because even high numbers of this organism inoculated into soil, do not survive for more than one month. Inoculation of endophytic diazotrophs, especially *A. diazotrophicus* is better achieved by the use of VA mycorrhizae together with bacterial cultures or with spores containing the bacteria. Spores of various VAM species were found to contain this organism and also *A. lipoferum* and *Klebsiella* sp. The infection of the root cap zone in micropropagated sugar cane plants with *A.diazotrophicus* and its colonization of xylem, has been confirmed by optical and electron microscope observations with immunogold-silver-enhanced techniques. Also *Herbaspirillum* spp. have been localized in the metaxylem after artificial inoculation of sugar cane and sorghum, using the same technique.

Introduction

Colonization of plant roots by endophytic bacteria until recently, has been thought to be restricted to plant pathogens. Several recent reports however have shown that such colonizations can have beneficial effects on plant growth (Kloepper and Beauchamp, 1992). Meletsus et al. (1993) introduced the phytopathogenic bacterium *Clavibacter michiganensis*

NATO ASI Series, Vol. G 37
Azospirillum VI and Related Microorganisms
Edited by I. Fendrik et al.

after genetic manipulation to eliminate its pathogenicity and observed that it continued to colonize maize plants endophytically. Indigenous plant-growth promoting endophytic bacteria were found in sweet corn and cotton, suggesting that internal plant habitats are exploited by several different bacteria (McInroy and Kloepper 1991). Based on this, Kloepper et al. (1992) suggested the replacement of the term "endorhizosphere" so far used for the site of root interiors, by "endophytic colonization of roots".

Colonization of the root interior by *Azospirillum* spp. was already observed in the seventies (Umali- Garcia et al. 1978; Patriquin and Döbereiner, 1978; Schank et al. 1979). Patriquin et al. (1983) suggested that localization of *Azospirillum* spp. within the roots could be expected to facilitate exchange of substrates and products between host and bacteria. The low pO_2 and the presence of malate in the xylem, the favored C source of *Azospirillum*, might be the key to active root colonization of C_4 plants. The more efficient effects of inoculation with strains of *Azospirillum* spp which have been isolated from surface sterilized roots rather than with soil isolates, has been confirmed in maize, sorghum and wheat experiments (Pereira et al. 1988; Baldani et al. 1986 a). These last authors confirmed the importance of plant infection even with wheat, which is a C_3 plant, by the highly significant correlation (r=0.92**) of numbers of *A. brasilense* SP245 in surface sterilized roots with the total N incorporation by the plant, while the correlation with root surface isolates was not significant (r=0.S2). Recently, Schloter et al. (1993) confirmed the specific localization of the strain SP245 in wheat roots using strain specific monoclonal antibodies. The immunogold-label showed that under axenic cultures, strain SP245 infected the xylem and phloem two weeks after inoculation while the strains SP7 and WA3 could only be detected on the root surface. Even though these first results were published more than 10 years ago and new techniques are confirming them, most papers on associations with diazotrophs still call them rhizosphere associations or rhizocoenoses.To a certain extend this is understandable as azospirilla are typical rhizosphere bacteria and only a few strains are able to infect plants efficiently. In the present paper we will review more recent findings on the identification of three predominantly or obligate endophytic diazotrophs isolated from non-leguminous plants.

Identification and Occurrence of *Herbaspirillum* spp. in Graminaceous Plants

Herbaspirillum seropedicae was originally isolated from cereal roots and rhizosphere soil (Baldani et al. 1986 b) but more recently it was observed that it survives poorly in soil and that the rhizosphere soil samples analyzed by Baldani et al. (1986 b) probably contained root hairs or small root pieces (Olivares et al. 1995). These authors showed that even large numbers of *H. seropedicae* inoculated into soil could not be reisolated 30 days later (Baldani et al. 1992), except when sorghum was planted in this soil, where it multiplied again within the roots. This diazotroph has been isolated from many samples of roots, stems, leaves and seeds of rice, maize and sorghum, from stems and leaves of Cameroon grass (*Pennisetum purpureum*), sugar cane and several graminaceous weed species, and from one sample of a legume weed species found growing in a maize field (Olivares et al. 1995). Preliminar results from our laboratory indicate that it might also occur in various palm trees which are also monocots. It has however not been found in a large number of samples of plant species from dicotyledoneous plants (Table 1).

The recent discovery of a very close relationship of the misclassified "Pseudomonas" *rubrisubalbicans* with *H. seropedicae* (Gillis et al. 1991) observed by DNA/rRNA hybridization which has been confirmed by auxanographic tests, also supports the endophytic nature of *Herbaspirillum* spp.. *H. rubrisubalbicans*, known as a mild sugar cane pathogen causing mottled stripe disease in some cane genotypes but not in the Brazilian cultivars, was also shown to be able to fix N_2 in rates comparable to those of *H. seropedicae* (Table 1), which were confirmed by the incorporation of $^{15}N_2$ (Boddey and Döbereiner 1994). The occurrence of this diazotroph seems restricted to sugar cane. However, the Brazilian varieties, which show no symptoms of mottled stripe disease, also contain the bacteria in high numbers in roots, stems, and leaves (unpublished data from our laboratory). Artificial inoculation by injection into leaves can also introduce these diazotrophs into sorghum and *Pennisetum* plants (Pimentel et al. 1991).

The inclusion of "P." *rubrisubalbicans* within the genus *Herbaspirillum* is now being definitively proposed by Baldani et al. (in preparation) as a second species which had been already identified by Gillis et al. (1991) by clearly differentiated DNA/DNA homologies. Additional differences between these two species are given in Table 1.

Table 1. Characteristics and occurrence of *H. seropedicae* and *H. rubrisubalbicans* (Pimentel et al. 1991, Boddey & Döbereiner 1994, Olivares et al. 1995)

	H. seropedicae	H. rubrisubalbicans
Growth in N-free semi-solid JNFB medium	white aerotactic pellicle	white aerotactic pellicle
ARA after 44h growth in JNFB medium (n moles c_2H_4 h^{-1} 5 ml^{-1}). Means of 3 strains	187	262
Total N accumulation (mg 5ml^{-1}) in N-free medium	0.442	0.371
$^{15}N_2$ incorporation (%)	0.746	0.427
Growth on n-acetyl-glucosamine	yes	no
Growth on meso-erythritol with NH_4	no	yes
Optimal growth temperature	34°C	30°C
Spontaneous occurrence in the field in roots, stems and leaves of		
• sugar cane	yes	yes
• forage grasses and cereals	yes	n.i.
• dicotyledoneous plants	n.i.	n.i.
• soil	n.i.	n.i.

n.i.- not isolated.

Electron microscopy studies have shown the colonization of vascular tissues of sorghum, rice and sugar cane by *H. seropedicae* and *H. rubrisubalbicans*. Analyses of the sugar cane leaf tissues by optical microscopy showed that *H. rubrisubalbicans* infected the stomata of the variety B 4362 which is sensitive to mottled stripe disease, colonized the intercellular space of mesophyll cells, and completely filled the protoxylem and metaxylem vessels (Fig. 1). On the other hand, the variety resistant to mottled stripe disease (SP 70-114 3) showed the bacteria in the xylem vessels forming structures of encapsulated bacteria within membranes, probably of plant origin (Fig. 2). The mottled stripe disease is probabily caused by the blocking of the vessels by *H. rubrisubalbicans* in the sensitive variety, which does not occur in the resistant variety.

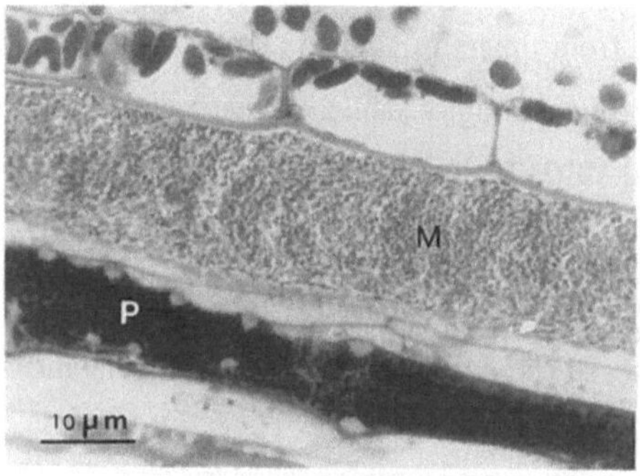

Fig 1. Light micrograph showing the protoxylem (P) and metaxylem (M) vessels of a sensitive variety of sugar cane completely filled with *H. rubrisubalbicans.*

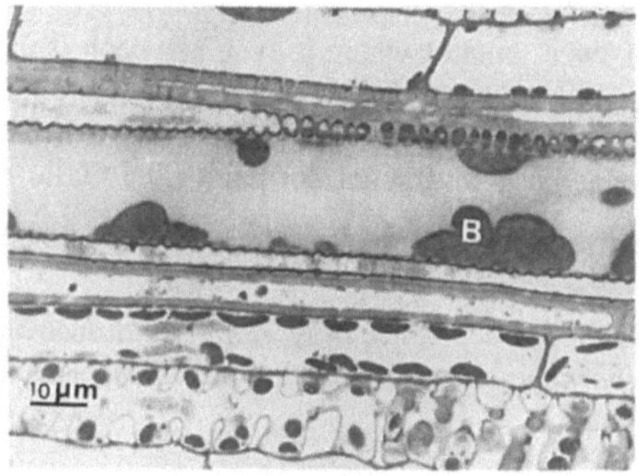

Fig 2. Light micrograph showing *H. rubrisubalbicans* encapsulated (B) inside the metaxylem vessels of sugar cane

Endophytic Occurrence of *Acetobacter diazotrophicus*.

This new *Acetobacter species* (Cavalcante and Döbereiner, 1988; Gillis et al. 1989) seems well adapted to the conditions prevailing within sugar cane stems, as it shows best growth with 10% sugar at pH 5.5 or even less. Under these conditions about half of the fixed N can be excreted and liberated to the plant as shown in mixed cultures with an yeast (Cojho et al. 1993). *Acetobacter diazotrophicus* also does not possess a nitrate reductase, being able to fix N_2 in the presence of high concentrations of NO_3^- (Cavalcante and Döbereiner, 1988). In the presence of 10% sucrose, the NH_4^+ assimilation by this bacteria is reduced by 65% (Boddey et al. 1991). These characteristics may enable the bacteria to fix N_2 in complementation to the N assimilation by the plant from the soil.

A. diazotrophicus, in contrast to *H. seropedicae*, so far has only been isolated from sugar cane, sweet potato and Cameroon grass, which are all propagated vegetatively and contain high sugar concentrations. This indicates that its habitat and spontaneous distribution is limited to such plants (Paula et al. 1993; Reis et al. 1994). It colonizes roots, stems and leaves of sugar cane and Cameroon grass in numbers up to 10^6 cells per g fresh weight or more and of sweet potato, where it even has been isolated from tubers in numbers up to 10^5 (Table 2). Artificial inoculation with the aid of VA mycorrhizal fungi established the bacteria also in sugar beets and sugar sorghum (Del Gallo et al.1992; Paula et al. 1991). For this, *Glomus clarum* spores containing the bacteria have been shown to be especially efficient (Table 3).

Table 2. Occurrence of *Acetobacter diazotrophicus* in stems and pulp of sweet potato plants (*Ipornoea batatas*) collected at various sites in Brazil (N^0 of cells x 10^4 per g fresh wheight) (Paula et al. 1993).

Site	Stems	Pulp
Rio de Janeiro		
Itaguai	7.3	17.0
Santa Cruz	2.0	4.0
Seropedica	4.7	16.0
Pernambuco		
Alto Bonito	n.a.	2.5
Itaperuna	n.a.	17.5

n.a. not analyzed

Table 3. Establishment of inoculated *A. diazotrophicus*, with or without *Glomus clarum*, in micropropagated plantlets of sugar cane and sweet potato (Paula et al. 1991) (N^0 of cells per g fresh weight x 10^4)

Inoculant	Sugar cane		Sweet potato	
	Roots	Stems	Roots	Stems
Controll	0	0	0	0
VAM spores	0	0	0	0
A. diazotrophicus	0.8	0.5	2	0
VAM + *A. diazotrophicus*	1.6	0.1	10	15
VAM spores containing *A.diazotrophicus*	0.1	3.0	13	30

Infection of micropropagated sugar cane plantlets in "in-vitro" systems has been shown by optical and electron microscopy (James et al. 1994) and confirmed by the immuno-gold silver enhancement technique. External colonization by the bacteria was seen especially in cavities of lateral root junctions (Fig. 3). The loose cells of the root cap at root tips were a site of entry of the bacteria into root tissues. There, the bacteria can be seen in enlarged, but intact epidermal cells. After 15 days, the bacteria were present in xylem vessels at the base of the stem (Fig. 4) connected via mucus to the spiral secondary thickenings. This indicates that the bacteria are distributed throughout the plant via the xylem vessels in the transpiration stream. This suggests that the xylem vessels within the dense shoots of mature plants are the most probable site of BNF as they provide the low pO_2 and energy necessary for nitrogenase activity (James et al. 1994).

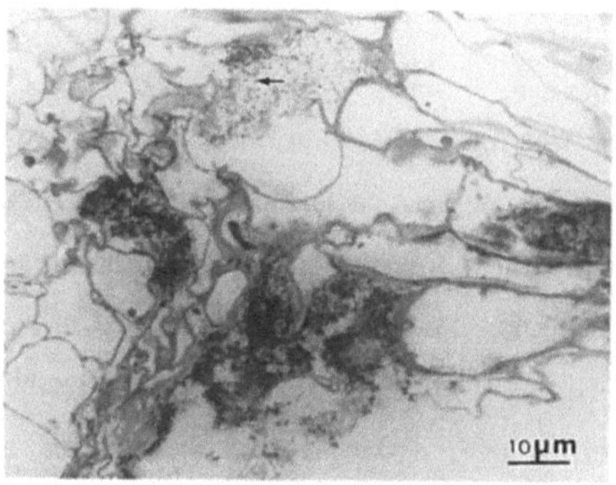

Fig. 3. Light micrograph of an silver enhanced immuno-gold lateral root junction of sugar cane showing bacteria accumulating in intercellular cavities. Some bacteria are within cells (arrow).

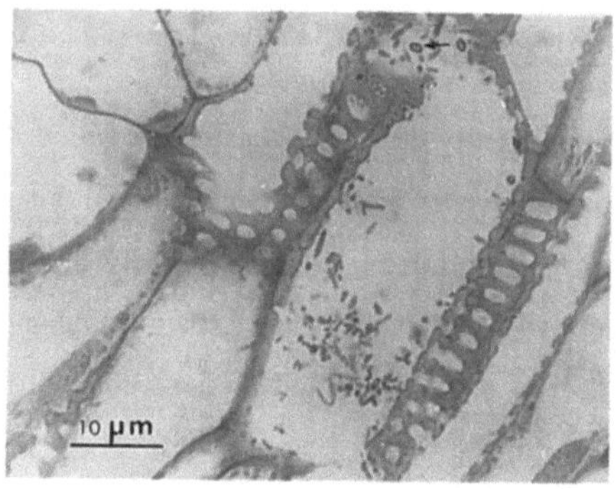

Fig. 4. Light micrograph of xylem vessels in the lower stem of sugar cane plantlets 15 days after inoculation, showing bacteria within the vessels and in intimate association with the spiral secondary thickening (arrow).

James et al. (1994) showed that infection of sterile micropropagated sugar cane plantlets by *A. diazotrophicus* could be achieved when the plants were grown in axenic conditions in a modified plant-growth medium (MS- Murashige and Skoog, 1962) containing 2 g sucrose l^{-1}. However, this bacterium does not survive in soil, or similar substrates low in sugars, and probably for this reason successful inoculation under these conditions is very difficult. The most successful inoculation of such plantlets so far has been that with mixed inoculants of VA mycorrhizae (Table 3).

The natural distribution of *A. diazotrophicus* seems restricted to sugar rich plants which propagate vegetatively, and where the bacteria are transfered from one field to another within the stem cuttings used for planting these crops in the field. In addition *A. diazotrophicus* has been isolated from pink mealybugs (*Saccharococcus sacchari*) which suck sugar cane juice (Ashbolt and Inkerman, 1990) and therefore could also be responsible for local distribution of this diazotroph. The very limited habitat *A. diazotrophicus* could be an explanation for the surprisingly low variability of the genome of different strains of this species reported recently by Caballero-Mellado and Martinez-Romero (1994).

Occurrence of *Azoarcus* spp. in Kallar grass (*Leptochloa fusca*) in Salin-sodicc Soils in Pakistan (Reinhold-Hurek et al. 1993 a,b).

This recently described diazotroph called *Azoarcus* was repeatedly found inside Kallar grass roots, however was only isolated from this grass in Pakistan. This bacterium is strictly aerobic but fixes nitrogen microaerobically and grows only on a few amino acids. Colonies are yellowish especially when grown on ethanol and peptone. This new genus is divided in two species, *A. indigens* and *A. communis* (Reinhold-Hurek et al. 1993 b). In contrast to other diazotrophs which colonize this grass in the rhizosphere, *Azoarcus* spp. were found to colonize the root interior where the authors suggest more efficient transfer of metabolites between host and diazotrophs. Axenic Kallar grass plants showed no pathogenic reaction when inoculated with this bacterium (Reinhold-Hurek and Hurek 1993 a). These authors showed that the bacteria invaded roots of Kallar grass and rice using the reporter gene Tn5-gus constitutively expressed by transposon mutagenesis. The colonization of the root is usually observed in the elongation zone but not in the more mature parts, forming colonies inter and intracellularly mainly in the cortex region. The bacteria entered the xylem vessels allowing systemic spreading and increased yields of rice by a process, other than N_2 fixation (Reinhold-Hurek and Hurek 1993 a). In addition these authors reported close interactions of *Azoarcus* spp. with rhizosphere fungi and that in mixed culture there is an induction of tubular arrays of internal membrane stacks which they call "diazosomes". Immunogold electron microscopy using the nitrogenase serum showed that the nitrogenase changes its localization from the bacterial cytoplasm to this membrane.

Physiological Characteristics of Endophytic Diazotrophs.

These new findings reviewed in this paper completely change the classical concepts of rhizosphere associations of N_2-fixing bacteria with Gramineae. No data are as yet available to explain why certain plant genotypes obtain much higher contributions from BNF than others and therefore better understanding of the localization of the different bacteria and the physiology of various plant genotypes is needed. It seems that regulation of nitrogenase activity in the endophytic diazotrophs is different, permitting N_2 fixation at low levels of NH_4^+ and in the case of *A. diazotrophicus* even with high levels of NO_3. In addition, about half of the N_2 fixed by this organism can be excreted into the nutrient solution (Cojho et al. 1993) which means that it can go directly to the plant. Inhibition of nitrogenase by NH_4^+ in *H. seropedicae* is only partial, even at 20 mM concentration (Fu & Burris 1989). Treatments with NH_4^+ also gave no detectable change in the electrophoretic pattern of the nitrogenase reductase indicating that there is no evidence that ADP ribosylation plays a part in turning off nitrogenase in this organism (Burris et al. 1991). These authors also report that there is no evidence of a switch-off-on process in *A. diazotrophicus* indicating that this organism is not subject to control by covalent modification of its dinitrogenase reductase. In addition high sugar concentrations as found within sugar cane stems, limit the assimilation of NH_4^+ by this diazotroph (Boddey et al. 1991). It is also unclear which effect low pH-values may have on the NH_4^+ assimilation prosses in these bacteria. In contrast to *A. brasilense* and *A. lipoferum*, which harbor the switch off/on enzymes DRAG and DRAT for dinitrogenase, the acid-tolerant *A. amazonense* is lacking this regulatory system (Hartmann et al. 1986; Fu and Burris 1989). Better understanding of the role of differences in the protein membrane patterns among *Azospirillum* spp. and *Herbaspirillum seropedicae* (Dianese et al. 1989) might also contribute to the characterization of endophytic diazotrophs.

All these characteristics indicate an adaptation of the endophytic diazotrophs to the internal plant environment where plant photosynthates are available to the bacteria and where nitrogen fixed by the bacteria is much more available to the plant than in rhizosphere associations. These characteristics explain much easier the high contributions sugar cane obtains from associations with this kind of diazotrophs (Urquiaga et al. 1992).

References

Ashbolt NJ, Inkerman PE (1990) Acetic acid bacterial biota of the pink sugar cane mealybug *Saccharococcus sacchari* and its environs. Appl Environ Microbiol 56:707-712.

Baldani VLD, Alvarez MA de B, Baldani JI, Döbereiner J (1986 a) Establishment of inoculated *Azospirillum* spp. in the rhizosphere and in roots of field grown wheat and sorghum. Plant Soil 90:35-46

Baldani VLD, Baldani JI, Olivares F, Döbereiner J.(1992) Identification and ecology of *Herbaspirillum seropedicae* and the closely related *Pseudomonas rubrisubalbicans.* Symbiosis 13: 65-73.

Baldani JI, Baldani VLD, Seldin L, Döbereiner J (1986 b) Characterization of *Herbaspirillum seropedicae* gen. nov., sp. nov., a root-associated nitrogen-fixing bacterium. Int J of Syst Bacteriol 36: 86-93.

Boddey RM, Urquiaga S, Reis V, Döbereiner J (1991) Biological nitrogen fixation associated with sugar cane. In: Polsinelli M Materrassi R, Vincenzini M (eds), Nitrogen Fixation, Kluwer Academic Publishers, Dordrecht, pp.105-111.

Boddey RM, Döbereiner J (1994) Biological nitrogen fixation associated with gramineous plants. In: Okon Y (ed) *Azospirillum*/Plant associations, CRC Press, Boca Raton, London, pp. 119-135.

Burris RH, Hartmann A, Zhang Y, Fu H (1991) Control of nitrogenase in *Azospirillum* sp.Plant Soil 137: 127-134.

Cavalcante V, Döbereiner J (1988) A new acid tolerant bacterium associated with sugar cane. Plant Soil 108: 23-31.

Caballero-Mellado J, Martinez-Romero E (1994) Limited genetic diversity in the endophytic sugarcane bacterium *Acetobacter diazotrophicus.* Appl Environ Microbiol 60: 1532-1537.

Cojho EH, Reis VM, Schenberg ACG, Döbereiner J (1993) Interactions of *Acetobacter diazotrophicus* with an amylolytic yeast in nitrogen free batch culture. FEMS Microbiol Lett 106: 341-346.

Del Gallo M, Fabbri P, Marini F, Döbereiner J (1992) First results of inoculation of sugar beet (*Beta vulgaris*) with *Acetobacter diazotrophicus* via V.A. mycorrhizae. 6th. Int Symp Microb Ecol (Abstract pg: 03-05).

Dianese JC, Döbereiner J, Dos Santos LTP (1989) Membrane protein patterns of three *Azozpirillum* species and *Herbaspirillum seropedicae.* An Acad Bras Ci 61: 223-230.

Fu H, Burris RH (1989) Ammonium inhibition of nitrogenase activity in *Herbaspirillum seropedicae.* J Bacteriol 171: 3168-3175.

Gillis M, Döbereiner J, Pot B, Goor M, Falsen E, Hoste B, Reinhold B, Kersters K (1991) Taxonomic relationships between (*Pseudomonas*) *rubrisulbalbicans*, some clinical isolates (EF Group 1), *Herbaspirillum seropedicae* and (*Aquaspirillum*) *autotrophicus.* In Polsinelli M, Materassi R, Vincenzini M, (eds), Nitrogen Fixation, Kluwer Academic Publishers, Dordrecht, pp. 293-298.

Gillis M, Kertsers K, Hoste B, Janssens D, Kroppenstedt RM, Stephan MP, Teixeira KR, Döbereiner J, De Ley J (1989) *Acetobacter diazotrophicus* sp. nov., a nitrogen fixing acetic acid bacterium associated with sugar cane. Int J Syst Bacteriol 39:361-364.

Hartmann A, Fu H, Burris RH (1986) Regulation of nitrogenase activity by ammonium chloride in *Azospirillum* spp. J Bacteriol 165: 864-870.

James EK, Reis VM, Olivares FL, Baldani JI, Döbereiner J (1994) Infection of sugar cane by the nitrogen fixing bacterium *Acetobacter diazotrophicus*. J Exp Bot 45: 757-766.

Kloepper Jw, Beauchamp CJ (1992) A review of isues related to measuring colonization of plant roots by bacteria. Can J Microbiol 38: 1219-1232.

Kloepper Jw, Schippers B, Bakker PAHM (1992) Proposed elimination of the term endorhizosphere. Phytopathology 82: 726-727.

McInroy JA, Kloepper (1991) Analysis of population densities and identification of endophytic bacteria in maize and cotton in the field. In: Keel C, Koller B, Défago G (eds), Plant-growth-promoting rhizobacteria/progress and prospects, IOBC/WPRS Bulletin XIV 8: 328-331

Meletzus D, Bermpohl A, Dreier J, Eichenlaub R (1993) Evidence for plasmid-encoded virulence factors in the phytopathogenic bacterium *Clavibacter michiganensis* subsp. *michiganensis* NCPPB382. J Bacteriol 175: 2131-2136.

Murashige T, Skoog F A (1962)Revised medium for rapid growth and bioassays with Tobacco tissue culture. Physiol Plantarum 15: 473-497.

Olivares FL, Baldani VL, Reis VM, Baldani JI, Döbereiner J (1995) Occurrence of the endophytic diazotrophs *Herbaspirillum* spp. in roots, stems and leaves predominantly of Gramineae. Biol Fert of Soil (in press).

Patriquin DG, Döbereiner J (1978) Light microscopy observations of tetrazolium reducing bacteria in the endorhizosphere of maize and other grasses in Brasil. Can J Microbiol 24: 734-742.

Patriquin DG, Döbereiner J, Jain DK (1983) Sites and processes of association between diazotrophs and grasses. Can J Microbiol 29: 900-915.

Paula MA, Reis VM, Döbereiner J (1991) Interactions of *Glomus clarum* with *Acetobacter diazotrophicus* in infection of sweet potato (*Ipomoea batatas*), sugarcane (*Saccharum* spp.) and sweet sorghum (*Sorghum vulgare*). Biol Fert Soils 11: 111-115.

Paula MA, Siqueira JO, Döbereiner J (1993) Ocorrência de fungos micorrízicos vesículo-arubusculares e de bactérias diazotróficas em batata-doce (*Ipomoea batatas* L. Lam). Rev Bras Ciên Solo 17: 349-356.

Pereira JAR, Cavalcante VA, Baldani JI, Döbereiner J (1988) Field inoculation of sorghum and rice with *Azospirillum* spp. and *Herbaspirillum* seropedicae. Plant Soil 110: 269-274.

Pimentel JP, Olivares F, Pitard RM, Urquiaga S, Akiba F, Döbereiner J (1991) Dinitrogen fixation and infection of grass leaves by *Pseudomonas rubrisubalbicans* and *Herbaspirillum seropedicae*. Plant Soil 137: 61-65.

Reinhold-Hurek B, Hurek T (1993 a) Capacities of *Azoarcus* sp., a new genus of grass-associated diazotrophs. In: Palacios R, Mora J, Newton WE (eds) New horizons in Nitrogen Fixation, pp.691-694.

Reinhold-Hurek B, Hurek T, Gillis M, Hoste B, Vancanneyt M, Kersters K, De Ley J (1993b) *Azoarcus* gen. nov., nitrogen fixing Proteobacteria associated with roots of Kallar grass (*Leptochloa fusca* (L.) Kunth) and description of two species *Azoarcus indigens* sp. nov. and *Azoarcus communis* sp. nov. Int J Syst Bacteriol 43: 574-588.

Reis VM, Olivares F, Döbereiner J (1994) Improved methodology for isolation of *Acetobacter diazotrophicus* and confirmation of its endophytic habitat. World J of Appl Microbiol and Biotech. 10: 101-104

Schank SC, Smith RL, Weiser GC, Zuberer DA, Bouton JH, Quessenbery KH, Tyler ME, Milan JR, Littell RC (1979) Fluorescent antibody technique to identify *Azospirillum brasilense* associated with roots of grasses. Soil Biol Biochem 11: 287-295.

Schloter M, Kirchhof G, Gundish C and Hartmann A (1993) Localization, quantification and characterization of *Azospirillum brasilense* strains in inoculated wheat using strain specific monoclonal antibodies (1993). In: Hegazi NA, Fayez M and Monib M (eds), Sixth International Symposium on Nitrogen Fixation with Non-Legumes, Ismailia, Egypt, Abstract Book pg. 36.

Umali Garcia M, Hubbell DH, Gaskins MH (1978) Process of infection of *Panicum maximum* by *Spirillum lipoferum.* In Granhall U. (ed), Environmental Role of Nitrogen-fixing Blue-green Algae and Asymbiotic bacteria, Swedish Natural Science Research Council, Uppsala, pp. 373-379 (Ecological Bulletins, 26).

Urquiaga S, Cruz KHS, Boddey RM (1992) Contribution of nitrogen fixation to sugar cane: Nitrogen-15 and nitrogen-balance estimates. Soil Sci Soc Amer J 56: 105-114

N_2-fixation by endophytic bacteria: questions of entry and operation

Janet I. Sprent and Euan K. James[1]
Department of Biological Sciences
University of Dundee
Dundee DD1 4HN
Scotland, UK

Abstract

Endophytic N_2-fixing bacteria from several genera can enter plants through natural openings (stomata, hydathodes), natural and artificial wounds (points of emergence of roots, via vegetative propagation techniques) and also intact plant tissues. They may penetrate to xylem vessels or intercellular spaces, but there is no evidence that they can survive in healthy plant cells. The nutrition of endophytic bacteria in these various locations is discussed. It is concluded that subtle interactions between the C and N metabolisms of both host and endophyte may lead to significant N_2 fixation without causing undue burdens on the host. There are good prospects for improving these associations by interdisciplinary laboratory and field studies involving both microbiologists and plant scientists.

Introduction

The range of microorganisms known to fix nitrogen has widened greatly in recent years. Indeed Young (1993) has suggested that *nif* is an ancient gene, possibly possessed by all early bacteria and since lost by many. The evidence that extant rhizobia are related to, not only plant pathogens such as *Agrobacterium*, but also to animal pathogens such as *Brucella*, is consistent with this suggestion. It is perhaps not surprising therefore that of the newly described endophytic bacteria, some are closely related to pathogens and may even act as pathogens in some hosts (Pimental *et al.*, 1991, Baldani *et al.*, 1992). Even with an effective legume-*Rhizobium* symbiosis the balance between the two symbionts is very delicately poised. For example, Vasse *et al.* (1993) recently studied abortive infections in alfalfa (lucerne). These were

[1]Present address: Department of Biology and Preclinical Medicine, University of St. Andrews, St. Andrews, Fife KY16 9AJ, Scotland UK.

NATO ASI Series, Vol. G 37
Azospirillum VI and Related Microorganisms
Edited by I. Fendrik et al.

characterised by necrosis of both plant and bacterial cells and were thought to result from a hypersensitive reaction. In this way plants can regulate the number of nodules produced. The balance may also be modified by stress: in soybean nodules treated with NaCl, renewed host meristematic activity, proliferation of infection threads and also intercellular bacteria embedded in a glycoprotein matrix were found (James *et al.*, 1993). When lipopolysaccharide mutants of *R. leguminosarum* infected pea, abnormalities were observed in both symbionts: these included hypertrophy and death of host cells and presence of rhizobia surrounded by large quantities of glycoprotein normally associated with the infection thread matrix (Perotto *et al.*, 1993). Some of the structures illustrated in all three papers just cited are very similar to those published for N_2-fixing endophytes in non-legume cells (e.g. James *et al.*, 1994, Spencer *et al.*, 1994). However, we should point out that in wound infections in some legumes, such as *Stylosanthes* host cell hypertrophy, preceding intercellular movement of rhizobia is a normal part of the infection process (Chandler *et al.*, 1982).

In this paper we assess the structural and functional relations between some endophytic bacteria and their hosts. For a general review of the occurrence of these associations see Döbereiner *et al.* (1994).

Entry of bacteria into plant tissues

Kluepfel (1993) has summarised the way in which bacteria generally obtain entry into plants. He points out that invasion is usually through natural openings and wounds. Natural openings include points of emergence of lateral roots (which we have tended to consider a natural wound) and stomata (including hydathodes). Wounds include abrasion of roots by soil and damage of any part by pests and pathogens and harvesting of perennial crops (Gagné *et al.*, 1987). In the case of pathogenic bacteria further movement of cells is accompanied by cell damage and/or death. The success of rhizobia in forming symbiotic associations has been associated with a refinement of such a pathogenic situation (Djordjevic *et al.*, 1987). One feature of this refinement is the very localised entry into root hairs (or in some cases between epidermal cells) which was considered in an earlier meeting of this group

(Sprent and Faria, 1988). We suggested that, rather than large-scale production of wall-degrading enzymes a more targeted process is required. Recent work by Mateos *et al.* (1992) has shown that *Rhizobium leguminosarum* bv trifolii produces multiple enzymes that cleave glycosidic bonds, but that these are bound to the bacterial cells and not readily detectable by assays developed for extracellular enzymes. A similar suggestion was made for the entry of the pathogen *Pseudomonas* sp. into canola (rape) seedling roots (Campbell *et al.*, 1987).

Evidence from endophytic N_2-fixing bacteria suggests that most also obtain entry through natural openings or wounds. The better documented cases of entry through intact surfaces, are *Azoarcus* sp., *Acetobacter diazotrophicus*, *Alcaligenes faecalis*, and *Pantoea agglomerans* all of which associate with grasses, and in some cases, other plants (Hurek *et al.*, 1994, James *et al.*, 1994, You *et al.*, 1990, Ruppel *et al.*, 1992, respectively). In all these, cells were labelled with molecular markers in order to be certain of their identity: in all, bacteria penetrated to the root xylem. The work with *Acetobacter* (James *et al.*, 1994) clearly showed entry into sugar cane both at emergent lateral roots and also in the root cap/apical meristem region. The latter have only the early primary wall layers, a condition suggested to be necessary by Sprent and Faria (1988), based on work with rhizobial systems. Localised entry may be an essential part of ensuring a pure culture of endophyte within its host.

Thus the evidence that endophytic bacteria can enter intact as well as wounded tissue is now strong. However, there is no convincing evidence to date that entry progresses beyond the cell wall without causing severe cell damage, probably death. Hurek *et al.* (1994) stated "plant cells were apparently destroyed (by *Azoarcus*) after bacteria had penetrated the cell wall". However, they found no signs of phytopathogenicity such as visible necrosis. Examination of all published light and transmission electron micrographs leads to the conclusion that plant cells containing bacteria are unhealthy (in the case of scanning electron microscope studies it is not possible to say whether cells are healthy or not, e.g. Christiansen-Weniger and Vanderleyden, 1994). Indeed they look very like potato cells in which we have recently established infection by rhizobia (Spencer *et al.*, 1994). In a legume nodule cell, rhizobia are confined by membrane vesicles and a full suite of eukaryotic organelles can be seen, in addition to various nodulins such as leghaemoglobin. The confinement of rhizobia by the peri-bacteroid membrane is an essential feature of the functional symbiosis. It enables the rhizobia to be kept in an environment low in organic N and O_2, conditions which favour induction of nitrogenase.

Movement of endophytic bacteria across the root to the xylem occurs between cells and through dead cells: exactly how entry into vessels is achieved is not yet clear. At the other extreme, some endophytic bacteria can also enter stomata (Döbereiner *et al.*, 1993).

Once inside tissue, bacteria can be transmitted by vegetative propagation of the host plant, as is the normal practice in cultivation of sugar cane and sweet potato. Indeed *Acetobacter diazotrophicus* and *Herbaspirillum seropedicae* are apparently obligate endotrophs, not found living freely in soil (Döbereiner *et al*, 1993). This does not preclude their occurrence on damaged or senescent roots and stems, from where new infections might originate. *H. seropedicae* can also, apparently, be seed transmitted in rice (Baldani *et al.*, 1993).

Plant tissues as an environment for nitrogen fixation

In order for cells to multiply, and fix N_2, conditions inside the host must be appropriate. Each known location of endophytic bacteria will be examined in turn and the likely conditions within them discussed.

Parenchyma and other living cells. The root cortex is largely composed of parenchyma cells and it is these where endophytic bacteria can often be seen. However, they are never confined by membrane vesicles and are thus exposed to all metabolites of the plant protoplast. At the time of entry, this probably consists of cytoplasm and vacuole. The cytoplasm is likely to have a C:N ratio of 4-5, with relatively high levels of amino compounds: it is not likely to permit induction of nitrogenase. Once the cell becomes moribund, the vacuolar membrane is disrupted (evidence from published micrographs) and thus the vacuolar contents will be added to those of the cytoplasm. Contents of vacuoles vary greatly; in mesophytes the total osmotic concentration is usually in the range of 300-350 mosmol kg^{-1}. Many bacteria, including rhizobia and *Azospirillum brasilense* (Altabe *et al.*, 1994) produce cyclic or linear glucans; production may be reduced under high osmolarity. Contributions to the total osmolarity vary greatly with species and/or growth conditions. Nitrate may be absent or in concentrations of 30-40 mmol m^{-3}. Organic anions (particularly malate) may be major components. The principle cation is usually K^+, but in halophytes high concentrations of Na^+ (and Cl^-) may be found (Wyn-Jones, 1981). Both total osmolarity and the nature of the osmolytes are thus likely to affect bacterial metabolism, including nitrogen fixation. If an infected cell dies and assuming this prevents further transport of solutes into the cell, then the total pool of metabolites available to support N_2 fixation is strictly limited.

Bacteria inside cells may appear healthy and may be seen to be dividing (e.g. Spencer *et al.*, 1994) or they may appear senescent. In the case of newly-formed cells of the root-tip meristem, bacteria may become surrounded by exopolysaccharide-like substances of host origin and sometimes appear to be digested within lytic vacuoles (James *et al.*, 1994). Thus there appear to be two extreme situations (1) bacteria enter host cells causing their disruption and death; (2) bacteria enter host cells and are digested by them. An intermediate stage where the two components live together has not yet been reported.

Intercellular spaces. This is the most widely reported location for endophytic bacteria. In all cases where sufficient detail can be seen, bacteria are embedded within a more-or-less fibrillar matrix (Plate 1 B), sometimes as in sorghum, in discrete globules (Döbereiner *et al.*, 1993). This may react with antibodies which also react with components of the legume infection thread matrix. In some cases the resemblance to infection threads is quite striking (James *et al.*, 1994, Spencer *et al.*, 1994). In that infection threads of legumes have been described as an extension of the intercellular space system (VandenBosch *et al.*, 1989), we have a close parallel between endophytic bacteria and rhizobia. The volume of intercellular space available for colonisation varies greatly: in the case of Kallar grass, *Leptochloa fusca*, which grows in waterlogged soils, extensive aerenchyma is present and this may contain large numbers of cells of *Azoarcus* sp. (Reinhold *et al.*, 1987). What do such bacterial cells use for growth and metabolism? It is known that the glycoprotein found in intercellular spaces of the cortex of some legume nodules (sometimes surrounding rhizobia) is synthesised inside the cell and transported across the wall to the intercellular space system (de Lorenzo *et al.*, 1993). If the material surrounding the bacteria in intercellular spaces is of a similar nature, it could act as a major source of C and a minor source of N. Is this sufficient, or are other metabolites supplied by the host in an analogous way? It is possible that the bacteria have adapted to use products of host defence reaction for their own metabolism: a similar activation could obtain in the infection threads of legume nodules where rhizobia multiply and thus need nutrients, and also in the ineffective pea nodules described by Perotto *et al.* (1994).

A further likely role of the matrix is that it acts to reduce pO_2, so facilitating induction of nitrogenase synthesis by the bacteria. This would be analogous to the role of glycoprotein in the operation of the O_2 diffusion barrier in legumes (de Lorenzo *et al.*, 1993 and references therein). The importance of pO_2 cannot be overemphasised, as not only can it affect nitrogenase induction (Alexander and Zuberer, 1989), but also efficiency of N_2 fixation. For example *Azoarcus* sp. when incubated under low pO_2 can adapt its respiratory pathway and fix N_2 at very high efficiencies, possibly an order of magnitude higher than bacteroids of *Bradyrhizobium japonicum* (Hurek *et al.*, 1994). If this observation holds true for cells inside host grasses and for other endophytic bacteria, then the case for significant N_2 fixation by these systems is greatly strengthened.

Bacteria entering via stomata may be found in large numbers in substomatal cavities (Olivares *et al.*, 1993; Plate 1 A,B). Nothing is known of their metabolism in this location.

Xylem vessels. Evidence is now accumulating that nitrogen-fixing bacteria are frequent residents of the xylem of various members of the Gramineae, both C_4 (sugar cane, maize, kallar grass, sorghum) and C_3 (rice, wheat) species (Plate 1 C,D): we know of no convincing evidence that they are present in either phloem or xylem parenchyma. Whilst xylem vessels are an obvious way of transporting bacteria from

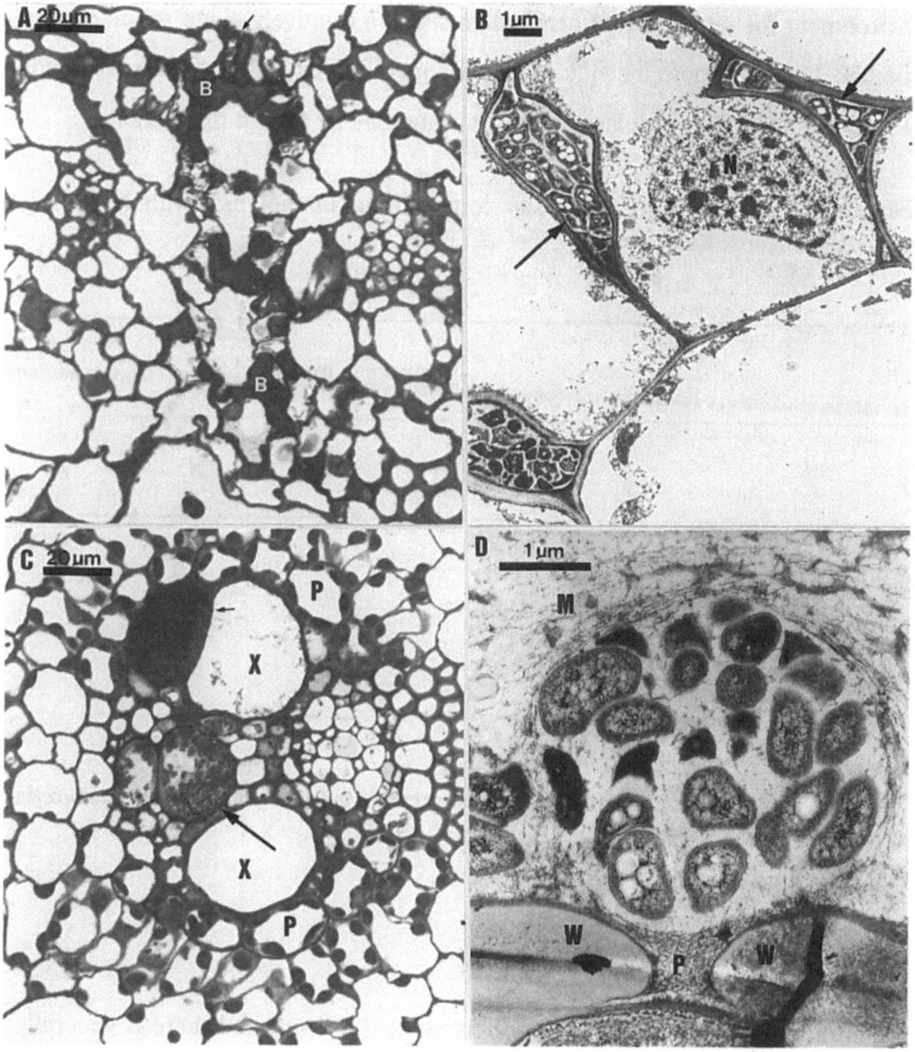

Plate 1. Infection of grasses by *Herbaspirillum rubrisubalbicans*. A. Transverse section of sugar cane leaf showing stoma (arrow), and sub-stomatal cavities filled with bacteria (B). B. Transmission electron micrograph (TEM) of intercellular bacteria in sugar cane leaf. Host cells appear healthy with intact nuclei (N) and bacteria are embedded in a matrix (arrows). C. Transverse section of vascular tissue in a sorghum leaf. The protoxylem is filled with bacteria (arrow), some in aggregates (large arrow), the metaxylem (X) is relatively empty. The surrounding host parenchyma (P) is healthy in appearance. D. TEM of sorghum xylem vessel showing a discrete aggregate of bacteria adjacent to a pit (P) in the thickened vessel wall (W). The aggregate is surrounded by a fibrillar matrix (M).

the root system to stems and leaves, they should also be considered as a potential environment for nitrogen fixation. Xylem sap is a relatively dilute solution, usually at a pH ~5.5, but whose individual components vary greatly with species and amount (Tables 1 - 3). Like vacuolar contents, nitrate may be absent or in high

Table 1 Variation of xylem sap composition (in mol m^{-3}) with N source in *Phaseolus vulgaris*: data from Allen *et al* (1988).

	N source 1 mol m^{-3}		
	NH_4^+	**NO_3^-0**	**N_2**
carboxylic acid[1]	3.4	7.6	9.6
amino acids[2]	71.92	22.9	25.9
inorganic cations[3]	19.3	19.7	24.1
inorganic anions	9.5	2.5[3]	6.3

[1] of which malate was the major components; independent analysis of carboxylate suggested that the values given are minimal
[2] of which most was asparagine
[3] of which 1.3 was nitrate

concentrations. Amino acids vary in number and amount and there is generally a significant organate component (usually malate). Minor components include hormones and these may be supplemented by products of endophytes (e.g. IAA by *Acetobacter diazotrophicus*, Fuentes-Ramirez *et al.*, 1993). A number of recent observations on xylem constituents suggest that the ability to support nitrogen fixing

Table 2 Composition of xylem sap of *Ricinus communis* grown on different N sources: data of Allen and Smith (1986).

	N source mol m^{-3}		
	12 NO_3^-	8 NH_4^+	1 NH_4^+
organic N	12.6	41.3	38.0
inorganic cations[1]	33.2	43.1	26.8
inorganic anions	24.5[2]	22.2[3]	22.5[3]
malate	6.0	0.2	3.0

[1] mainly K^+ [2] of which NO_3^- 22.3 [3] mainly Cl^-

bacteria may vary greatly with species and with genotype within species. Surprisingly, current evidence suggests that maize (Canny and McCully, 1988) but <u>not</u> sugar cane (Welbaum *et al.*, 1992) xylem sap contains sugar. The apoplast of xylem parenchyma

Table 3 Composition of xylem sap of field grown maize: data from Canny and McCully (1988).

Component	Concentration/mol m^{-3}	Comment
amino acids	1.8 ± 1.1	major components; asn, asp, ala, glu, gln, ser
nitrate	1.8 ± 1.0	
carboxylic acids	18.3 ± 13.0	major component; malate
sucrose	3.8 ± 2.2	
glucose	0.8 ± 0.6	
fructose	0.8 ± 0.6	
inorganic ions	not given	

of sugar cane may have 700 mol m^{-3} of sucrose, but (unless endophytic bacteria cause leakage into vessels) this would not be available to support N_2 fixation in mature internodes as xylem is thought to be isolated from storage parenchyma (Jacobsen *et al.*, 1992). Because sugar-cane has been shown unequivocally both to be associated with significant N_2 fixation (Urquiaga *et al.*, 1992) and to house *Acetobacter diazotrophicus* in its xylem (James *et al.*, 1994), we shall concentrate on this host species: similar arguments can be applied to other xylem-residing N_2-fixing bacteria e.g. sorghum or sugar cane with *Herbaspirillum* (Olivares *et al*, 1993), rice with *Azospirillum brasilense* (Baldani *et al.*, 1993). Sugar cane can utilize both nitrate and ammonium, but the extent to which it does so varies with genotype (Abellan *et al.*, 1994). Further, nitrate metabolism is linked to carbon metabolism, including sucrose synthesis. Abellan *et al.* (1994) found one cultivar to have high rates of sucrose synthesis that were drastically reduced by nitrate, whereas another had a lower rate of sucrose synthesis that was stimulated by nitrate (Table 4). In both cultivars, the main effect of nitrate upon phosphoenolpyruvate carboxylase was to reduce its sensitivity to malate: the latter is likely to be a major substrate for some cane endophytes, including *Herbaspirillum seropedicae* (Baldani *et al.*, 1992) although *A. diazotrophicus* is normally grown in high sucrose media (Boddey *et al.*, 1991). A possible consequence of this is that sugar-cane grown in the absence of nitrate but

Table 4 Effect of nitrate on sucrose synthesis (μmol mg chl^{-1}) of two sugar cane cultivars: data from Abellan *et al* (1994).

	NO_3^- given/mol m^{-3}		
Cultivar	0	40	80
B 63-118	1.9	2.7	3.4
Ja 60-5	2.8	2.0	0.4

containing endophytes metabolising malate, could have a higher rate of photosynthesis than those without endophytes. Even in the presence of nitrate, malate-utilizing endophytes with active nitrate reductase could also enhance photosynthesis. Since *A. diazotrophicus* has no nitrate reductase and can fix N_2 in 25 mol m^{-3} NO_3^- (Boddey *et al.*, 1991) it may be a better endophyte for some sugar genotypes and *H. seropedicae*, which has nitrate reductase (Baldani *et al.*, 1986) for others. These complex interactions between C and N metabolism are entirely consistent with observed effects of genotype where highest yield of *Saccharum officinarum* genotypes was correlated with high N_2 fixation (Urquiaga *et al.*, 1992) and suggest considerable potential for improvement by host plant selection. Indeed Brazilian cultivars of sugar cane which grow with low N input may have been unwittingly selected for N_2 fixation (Boddey, 1993).

Other effects of host genotype can be involved. In rice, for example, xylem infections by *Herbaspirillum* may lead to blockage of vessels and disease symptoms, or be confined to the sides of vessels and be beneficial (Döbereiner *et al.*, 1993). In such cases, bacteria are frequently seen in aggregates, surrounded by a fibrillar matrix (see section on intercellular spaces) and located close to the pits between xylem thickenings (Plate 1 D). In this location they might be able to affect exchange of nutrient materials between vessels and neighbouring parenchyma. A more extreme situation is seen in some pathogenic systems (Bretschneider *et al.*, 1989).

A further general point to note is that the normal osmolarity of xylem sap is much less than that of mesophytic parenchyma cells and is unlikely to reduce glucan synthesis.

Conclusions and Suggestions for Future Work

We believe that there is now a good case that significant nitrogen fixation can occur in a variety of situations involving endophytic bacteria and in the absence of specialised structures (nodules). Great advances have been made in the last five years. However, there are still many gaps in our knowledge which need to be filled before the systems are well-enough understood to be properly evaluated and exploited. Many of these, as Döbereiner and Urquiaga (1992) have pointed out are physiological. In particular:

1. More information is needed on how bacteria enter xylem vessels: does the process involve wall changes that could lead to leakage of material (e.g. sucrose) from xylem parenchyma to support N_2 fixation? Can vegetative propagation avoid this problem (see Döbereiner *et al.*, 1994)?
2. Quantification of bacteria inside plants to see if they can account for observed rates of N_2 fixation. Kloepper and Beauchamp (1992) have reviewed the problems of measuring colonisation of plant roots by bacteria. An additional problem for endophytes that is not discussed in detail is that the quantification of bacterial numbers depends on appropriate culture media. To take one example, *Acetobacter diazotrophicus* in pure culture grows best on high sucrose, low pH media (Boddey *et al.*, 1991), which may be found in cane mealybug honeydew (Ashbolt and Inkermen, 1990): it is by no means clear that this what is obtained from host xylem. Cultures should be grown on synthetic xylem sap as close as possible in composition to the natural fluid.
3. Multidisciplinary studies in which metabolism of both host and endophyte are examined together. This could compare different host genotypes with different endophytes. A further hypothesis which could be tested is that bacterial respiration lowers the pO_2 in xylem vessels sufficiently to allow "reversal" of the proton pump by which xylem constituents may be abstracted by contiguous parenchyma cells (de Boer *et al.*, 1985). The intriguing possibility of using a model system involving a yeast and *Acetobacter* has been put forward by Cojho *et al.* (1993).

Where possible, laboratory and field studies should be closely compared since, for example, the differentiation of xylem vessels may vary between the two. In field-grown maize metaxylem elements remain living for longer than suspected from studies in artificial culture media (St Aubin *et al.*, 1986).

Acknowledgements

We are grateful to our many colleagues in Brazil and Dundee for helpful discussion and to the Inter-American Institute for Cooperation in Agriculture (IICA) and NATO for financial support. We give especial thanks to Fabio Olivares for supplying material used for microscopy.

References

Abellan IP, Armas Urguiza de R, Valadier M-H, Champigny M-L (1994) Short-term effect of nitrate on carbon metabolism of two sugar cane cultivars differing in their biomass production. Phytochem 36:819-823.

Alexander DB, Zuberer DA (1989) $^{15}N_2$ fixation by bacteria associated with maize roots at a low partial O_2 pressure. Appl Environ Microbiol 55:1748-1753.

Allen S, Smith JAC (1986) Ammonium Nutrition in *Ricinus communis*: Its Effect on Plant Growth and the Chemical Composition of the Whole Plant, Xylem and Phloem Saps. J Exp Bot 37:1599-1610.

Allen S, Raven JA, Sprent JI (1988) The role of long-distance transport in intracellular pH regulation in *Phaseolus vulgaris* grown in ammonium or nitrate as nitrogen source, or nodulated. J Exp Bot 39:513-528.

Altabe SG, Iñón de Iannino N, Mendoza de D, Ugalde RA (1994) New osmoregulated ß(1-3),ß(1-6) Glucosyltransferase(s) in *Azospirillum brasilense.* J Bacteriol 176:4890-4898.

Ashbolt NJ, Inkerman PA (1990) Acetic acid bacterial biota of the pink sugar cane mealybug, *Saccharococus sacchari*, and its environs. Appl Environ Microbiol 56: 707-712.

Baldani JI, Baldani VLD, Seldin L, Döbereiner J (1986) Characterization of *Herbaspirillum seropedicae* gen nov sp nov a Root-associated Nitrogen-Fixing Bacterium. Int J Syst Bacteriol pp 86-93.

Baldani VLD, Baldani JI, Olivares F, Döbereiner J (1992) Identification and ecology of *Herbaspirillum seropedicae* and the closely related *Pseudomonas rubrisubalbicans*. Symbiosis 13:65-73.

Baldani VLD, James EK, Baldani JI, Döbereiner J (1993) Colonization of rice by the nitrogen-fixing bacteria *Herbaspirillum* spp. and *Azospirillum brasilense*. In: R Palacios, J Mora & WE Newton (eds). *New Horizons in Nitrogen Fixation* p705. Kluwer Academic Publishers, Dordrecht.

Boddey R (1993) 'Green' energy from sugar cane. Chemistry & Industry 17 May, pp 355-358.

Boddey RM, Urquiaga S, Reis V, Döbereiner J (1991) Biological nitrogen fixation associated with sugar cane. Plant and Soil 137:111-117.

Brentschneider KE, Gonella MP, Robeson DJ (1989) A comparative light and electron microscopical study of compatible and incompatible interactions between *Xanthomonas campestris* pv. *campestris* and cabbage (*Brassica oleracea*). Physiol Mol Plant Path 34:285-297.

Canny MJ, McCully ME (1988) The xylem sap of maize roots: its collection, composition and formation. Aust J Plant Physiol 15:557-566.

Campbell JN, Cass DD, Peteya DJ (1987) Colonization and penetration of intact canola seedling roots by an opportunistic fluorescent *Pseudomonas* sp. and the response of host tissue. Phytopath 77:1166-1173.

Chandler MR, Date RA, Roughley RJ (1982) Infection and root nodule development in *Stylosanthes* species by *Rhizobium*. J Exp Bot 33:47-57.

Christiansen-Weniger C, Vanderleyden J (1994) Ammonium-excreting *Azospirillum* sp. become intracellularly established in maize (*Zea mays*) para-nodules. Biol Fert Soils 17:1-8.

Cojho EII, Reis VM, Schenberg ACG, Döbereiner J (1993) Interactions of *Acetobacter diazotrophicus* with an amylolytic yeast in nitrogen-free batch culture. FEMS Microb Lett 106:341-346.

de Boer AH, Katou K, Mizuno A, Kojima H, Okamoto H (1985) The role of electrogenic xylem pumps in K^+ absorption from the xylem of *Vigna unguiculata*: the effects of auxin and fusicoccin. Plant, Cell and Environ 8:579-586.

de Lorenzo C, Iannetta PPM, Fernandez-Pascual M, James EK, Lucas MM, Sprent JI, Witty JF, Minchin FR, de Felipe MR (1993) Oxygen diffusion in lupin nodules. J Exp Bot 44:1469-1474.

Djordjevic MA, Gabriel DW, Rolfe BG (1987) *Rhizobium*-the refined parasite of legumes. Annu Rev Phytopathol 25:145-168.

Döbereiner J, Urquiaga S (1992) Soil biology and sustainable agriculture. An Acad Bras Ci 64:127-133.

Döbereiner J, Reis VM, Lazarini AC (1988) New N_2-fixing bacteria in association with cereals and sugar cane. In: Bothe H, De Bruijn FJ, Newton WE, eds. *Nitrogen fixation: hundred years after*. Stuttgart: Gustav Fischer, 717-722.

Döbereiner J, Reis VM, Paula MA, Olivares F (1993) Endophytic diazotrophs in sugar cane, cereals and tuber plants. In: R Palacios, J Mora & WE Newton (eds). *New Horizons in Nitrogen Fixation* pp 671-676. Kluwer Academic Publishers, Dordrecht.

Döbereiner J, Baldani VLD, Olivares F, Reis VM (1994) Endophytic diazotrophs: the key to BNF in gramineous plants. Proceedings of VIth International Symposium on BNF in non-legumes, Ismailia, Egypt. In press.

Fuentes-Ramirez LE, Jimenez-Salgado T, Abarca-Ocampo IR, Caballero-Mellado J (1993) *Acetobacter diazotrophicus*, an indoleacetic acid producing bacterium isolated from sugarcane cultivars of Mexico. Plant and Soil 154:145-150.

Gagné S, Claude R, Rosseau H, Antoun H (1987) Xylem-residing bacteria in alfalfa roots. Can J Microbiol 33:996-1000.

Hurek T, Reinhold-Hurek B, van Montagu M, Kellenberger E (1994) Root colonization and systemic spreading of *Azoarcus* sp. strain BH72 in grasses. J Bacteriol 176:1913-1923.

Hurek T, Reinhold-Hurek B, Turner GL, Bergersen FJ (1994) Augmented rates of respiration and efficient nitrogen fixation at nanomolar concentrations of dissolved O_2 in hyperinduced *Azoarcus* sp. strain BH72. J Bacteriol 176:4726-4733.

Jacobsen KR, Fisher DG, Maretzki A, Moore PH (1992) Developmental changes in the anatomy of the sugarcane stem in relation to phloem unloading and sucrose storage. Bot Acta 105:70-80.

James EK, Sprent JI, Hay GT, Minchin FR (1993) The effect of irradiance on the recovery of soybean nodules from sodium chloride-induced senescence. J Exp Bot 44:997-1005.

James EK, Reis VM, Olivares FL, Baldani JI, Döbereiner J (1994) Infection of sugar cane by the nitrogen-fixing bacterium *Acetobacter diazotrophicus*. J Exp Bot 45:757-766.

Kloepper JW, Beauchamp CJ, (1992) A review of issues related to measuring colonization of plant roots by bacteria. Can J Microbiol 38:1219-1232.

Kluepfel DA (1993) The behaviour and tracking of bacteria in the rhizosphere. Annu Rev Phytopathol 31:441-472.

Mateos PF, Jimenez-Zurdo JI, Chen J, Squartini AS, Haack SK, Martinez-Molina E, Hubbell, DH, Dazzo FB (1992) Cell-associated pectinolytic and cellulolytic enzymes in *Rhizobium leguminosarum* biovar trifolii. Appl Environ Microbiol 588:1816-1822.

Olivares FL, James EK, Reis VM, Baldani VLD, Baldani, JI, Döbereiner J (1993) Colonization of vascular tissue by *Herbaspirillum* spp. in sorghum and sugar cane. In: *Proceedings of the Twenty Third Brazilian Congress of Plant Pathology*.

Perotto S, Brewin NJ, Kannenberg EL (1994) Cytological evidence for a host defense response that reduces cell and tissue invasion in pea nodules by lipopolysaccharide-defective mutants of *Rhizobium leguminosarum* strain 3841. Mol Plant-Microbe Interact 7:99-112.

Pimentel JP, Olivares F, Pitard RM, Urquiaga S, Akiba F, Döbereiner J (1991) Dinitrogen fixation and infection of grass leaves by *Pseudomonas rubrisubalbicans* and *Herbaspirillum seropedicae*. Plant and Soil 137:61-65.

Reinhold B, Hurek T, Fendrik I (1987) Cross-reaction of predominant nitrogen-fixing bacteria with enveloped, round bodies in the root interior of Kallar grass. Appl Environ Microbiol 53:889-891.

Ruppel S, Hecht-Buchholz C, Remus R, Ortmann U, Schmelzer R (1992) Settlement of the diazotrophic, phytoeffective bacterial strain *Pantoea agglomerans* on and within winter wheat: An investigation using ELISA and transmission electron microscopy. Plant and Soil 145:261-273.

Spencer D, James EK, Ellis GJ, Shaw JE, Sprent JI (1994) Interaction between rhizobia and potato tissues. J Exp Bot 45: 1475-1482.

Sprent JI, de Faria SM (1988) Mechanisms of infection of plants by nitrogen fixing organisms. Plant and Soil 110:157-165.

St Aubin G, Canny MJ, McCully ME (1986) Living vessel elements in the late metaxylem of sheathed maize roots. Ann Bot 58: 577-588.

Urquiaga S, Cruz, KHS, Boddey RM (1992) Contribution of nitrogen fixation to sugar cane: Nitrogen-15 and nitrogen-balance estimates. Soil Sci Soc Amer J 56: 105-514.

VandenBosch KA, Bradley DJ, Knox JP, Perotto, S, Butcher GW, Brewin NJ (1989)

Common components of the infection thread matrix and the intercellular space identified by immunocytochemical analysis of pea nodules and uninfected roots. EMBO J 8:335-342.

Vasse J, de Billy F, Truchet G (1993) Abortion of infection during the *Rhizobium meliloti*-alfalfa symbiotic interaction is accompanied by a hypersensitive reaction. Plant J 4:555-566.

Welbaum GE, Meinzer FC, Grayson RL, Thornham KT (1992) Evidence for and consequences of a barrier to solute diffusion between the apoplast and vascular bundles in sugarcane stalk tissue. Aust J Plant Physiol 19:611-623.

Wyn-Jones RG (1981) Salt Tolerance. In: *Physiological Processes Limiting Plant Productivity* CB Johnston (ed) pp 271-292 Butterworths, London.

Young JPW (1993) Molecular phylogeny of rhizobia and their relatives. In: R Palacios, J Mora & WE Newton (eds). *New Horizons in Nitrogen Fixation* pp 587-592 Kluwer Academic Publishers, Dordrecht.

Achievements in the Genetics of Nitrogen Fixation in Hungary

Tibor Sík
Department of Biotechnology
University of Agricultural Sciences
H-2103 Gödöllő
Hungary

Abstract

The research development of microbial and molecular genetics is followed by the results on the genetics of *Rhizobium meliloti* and its phages. Two temperate and one virulent *Rhizobium phage* was characterized. By the established, original phage genetic system genetic and physical maps were constructed, genes identified as well as molecular mechanisms in temperate regulation, phage-bacterium interactions clarified. Transduction was worked out for mapping and for transposon suicide transfer into defective lysogenic recipients. Transformation and transfection was carried out with *R. meliloti*. The bacterial genome was mapped by transduction and conjugation. The *nod* and *nif* genes were localized, characterized, the genetic communication between Rhizobium meliloti and alfalfa revealed with molecular genetic techniques. Plant host specific chemotaxis of Rhizobia was demonstrated and genes involved identified. For practical application and strain improvement a competition test was elaborated including the identification of reisolated symbionts by phage superinfection.

Introduction

This lecture is a review of the development of microbial and molecular genetics in Hungary. It is in fact the history of research on the genetics of symbiotic nitrogen fixing *Rhizobium meliloti* and its phages. The experimental systems worked out with these were producing plenty of valuable results of general interest. It will certainly be useful in our further research on associative nitrogen fixation and genetics of *Azospirillum*, which is just about to start in cooperation with the organizers of this Workshop.

More than 40 years ago in Budapest at the Institute of Genetics the director and renowned Hungarian geneticist Barna Győrffy suggested to choose *Rhizobium* as a subject of genetic studies. This was the time when *E. coli* was world-wide studied with

NATO ASI Series, Vol. G 37
Azospirillum VI and Related Microorganisms
Edited by I. Fendrik et al.

great success. Since by science policy we were isolated from these advanced laboratories, the genetics of symbiotic nitrogen fixation was a promising field of research with both theoretical and practical importance. The review will point out milestones in the research with results of meaningful impacts.

Rhizobium transformation

In the early fifties, attempting genetic information transfers, *Rhizobium lupini* was transformed (Balassa, 1956). The experiments claiming transformation of host specificity are hard to explain but the transfer of Streptomycin resistance and the transformation of auxotrophs was well worked out and convincing (Balassa, 1960). Streptomycin dependents and resistants were distiguished within the mapped gene region (Balassa and Gábor, 1964). It was characteristic of that time to use in the first papers the term adaptation instead of recombination and variant instead of mutant.

Later *Rhizobium meliloti* phage sensitive cells were transformed with DNA from lysogenic cells and superinfection immunity could be detected in the transformants (Szende et al. 1961). Transfection was also successful especially with a helper phage (Kondorosi et al. 1974). *R. meliloti* becomes poorly competent for transformation, but Ca^{++} and cold treatment facilitated the uptake of plasmids or R-primes (Kiss and Kálmán 1982).

Rhizobium phage genetics

A really sophisticated genetic system was worked out with the isolated *Rhizobium phages* (Szende and Ördögh, 1960; Szekeres, unpublished). The two tempertate phages 16-3 and 11 as well as the virulent 32 were characterized physiologically, morphologically and their DNA with molecular methods (Table 1).

The temperate phages were isolated in different parts of Hungary. Phage 16-3 is forming turbid plaques, stable lysogenes which are inducible. The plaques of phage 11 are less turbid indicating less stable lysogenes and mutants can frequently be detected among the progeny.

Table 1. Characteristics of *Rhizobium meliloti* phages.

Name	Phage			DNA			
	Head (nm)	Tail (nm)	Density (g/cm^{-3})	Molecular weight (MD)	Density (g/cm^{-3})	Tm (°C)	Base composition (GC)
16-3	64x58	127	1.558	42	1.717	77.8	0.58
11	64x58	127	1.558	42	1.717	77.7	0.58
32	82x78	-	1.534	59	1.707	72.4	0.47
Error	S=3.5	S=2.3	±0.005	±5%	±0.002	±0.1	±0.01

Dimensions of phage particles from electron microscopy.
Density of phages and DNA-s by CsCl gradient centrifugation.
Molecular weight by sucrose gradient centrifugation.
Tm: thermal denaturation in 0,1 SSC. (Sík 1983. unpublished)

According to morphological and DNA data phage 16-3 and 11 seems to be identical. A deletion could be detected with electronmicroscopy in the regulatory DNA region of phage 11 by hybridizing the corresponding DNA fragments of the two phages (Mink et al. 1982).

For genetic crosses plack morphology, ti, ts, lysis and host range mutants were isolated, arranged by complementation and the genetic map constructed by recombination frequences as shown in Figure 1 (Sík and Orosz, 1971; Orosz et al. 1973).

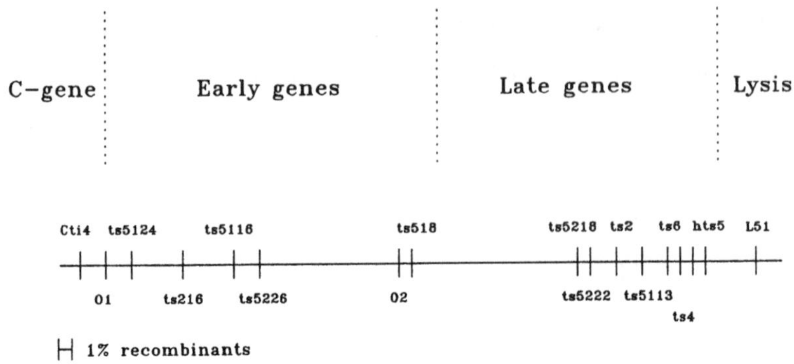

Figure 1 Genetic map of *Rhizobium phage* 16-3.
Mutant C: clear plaque; ti: thermo inducible; ts: thermo sensitive; o: operator; h: host specificity; Ant: antigen character; L: lysis.

The research of *Rhizobium* phages was partly continued in the Department of Genetics of the Attila József University. The physical map of the phage 16-3 DNA was determined with several endonucleases (Dallmann et al. 1979). After mapping the genetic fine structure of the c-cistron, the regulation of phage 16-3 was studied in detail. Reversion studies detected frequent spontaneous deletions in the regulatory region (Orosz, 1980). Due to poor mismatch repair in *R. meliloti,* heteroduplexes are thought to be resolved by replication rather than repair (Orosz et al. 1980a). Among progeny recombinants 5-8% heterozygotes were found and the results revealed how branch migration participates in recombination (Orosz et al. 1980b).

Also Rec^- *R. meliloti* mutants were isolated as non inducible lysogenes (Olasz et al. 1983). These were useful to study phage determined recombination and genetic complementation of gene clones by using them as recipients.

Molecular genetic analysis of the temperate phage 16-3 revealed homology between helix-turn-helix repressor motifs of phage 16-3 and λ_{434} and between their operator consensus structures (Dallmann et al. 1987). It is also of general interest that the 51 bp long att target site of phage 16-3 was identified as the 3' end of proline-tRNA and functional interpretation was given for the prophage integration (Papp et al. 1993).

In the Institute of Genetics transduction was carried out with *Rhizobium* phages. The temperate phage 16-3 induced from lysogenic cells was performing specialized transduction of Cys prototrophy (Sváb et al. 1978). Generalized transduction was worked out with the phage 11 and the constructed partial genetic map of *R. meliloti* is shown in Figure 2 (Sík at al. 1980).

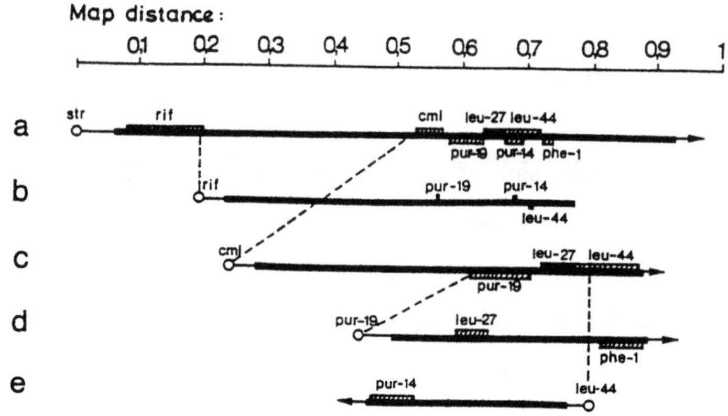

Figure 2 Partial genetic map of *Rhizobium meliloti* 41 by transduction.
a, b and c linkage between auxotrophic and resistance genes
d and e linkage of auxotrophic gene sites.

Phage particles performing generalized transduction have been shown to contain bacterial DNA as a result of packaging error. This is probably due to the less temperate growth cycle of phage 11 carrying the deletion localized in its regulatory region (Mink et al. 1982).

With a clear plaque mutant of phage 11 several defective lysogenes could be isolated with various superinfection sensitivity and exclusive for the heteroimmune phage 32. These defective lysogenes were excellent recipients, forming a suicide system for transposon transduction. (Hutvagner and Sík, unpublished).

There is no direct relation between nitrogen fixation and the phages in *R. meliloti*. Besides genetic information transfer and genetic manipulations, phages with certain limits are useful in the identification of strains especially when modified in phage sensitivity or superinfection specificity. The phages isolated in our laboratory did not infect other *R. meliloti* strains suggesting that phage receptor sites differ from recognition sites involved in binding to roots.

Genetics of *Rhizobium meliloti*

It was difficult to start genetics since mutants in the most interesting symbiotic genes could not be selected with free-living bacteria. Inspired by the hypothesis that nitrogenase and nitrate reductase enzymes share a common subunit, nitrate reductase mutants were isolated which were also defective in nitrogen fixation (Kondorosi et al. 1973). Although pleiotropy was confirmed, detailed analysis proved, that the mutants are utilising nitrate, but do not form nitrate reductase like the wild type, when grown on protein hydrolisate with strong aeration. This induction of nitrate reductase in the wild type bacteria was found to be due to ammonia accumulation by deamination of amino acids to produce carbon source for growth. The mutants were defective in a metabolic regulation or in detoxification of ammonia which is essential in nitrogen fixation (Sík et al. 1976; Sík and Barabás, 1977). Efficient nitrogen fixation of the mutants could be restored by the addition of adenine, glutamic and aspartic acids, resembling phenotypic reversion (Barabás and Sík, 1979). The common subunit theory was disproved but an interesting regulatory factor, involved in the nitrogen fixation was found.

The low frequency of transformation and the limited resolution of transduction was unsatisfactory for genetic mapping of *Rhizobia*. The real genetics started in Salamanca at the Symposium on Nitrogen Fixation in 1976, when the general use of conjugative R-plasmids was published. Simultaneously with other laboratories antibiotic resistant and polyauxotrophic mutants were isolated and conjugative crosses were carried out with the plasmid R 6845. The resulting circular linkage map of *R. meliloti* 41 is shown in Figure 3 (Kondorosi et al. 1977).

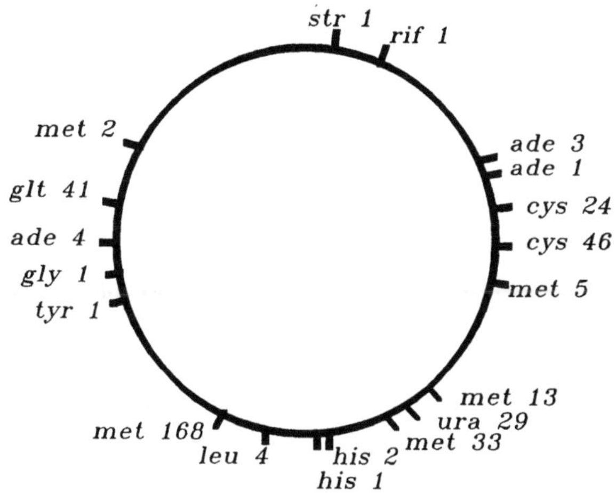

Figure 3 Genetic map of *Rhizobium meliloti* 41.

From the exconjugants an R-prime was isolated containing the *cys* region. Phage 16-3 integrated into this R-prime, thus the linkage between *cys* gene and $att_{16\text{-}3}$, as concluded from the specialized transduction, was confirmed (Kiss et al. 1980).
Inspite of very careful crossing experiments no symbiotic *nod* or *nif* genes could be localized on the chromosome. Nod-mutants were found to harbour smaller pRme41b plasmids, indicating deletions of various sizes on that megaplasmid. The *nifHDK* probe hybridized with the same *nod*-plasmid, proving the clustering of symbiotic genes in *R. meliloti* (Bánfalvi et al. 1981).

From Tn*5* tagged symbiotic mutants R-primes were isolated with the plasmid pJB3JI (Km^s), a derivative of R 6845. The overlapping 80 kb inserts were representing the *nif-nod* region. By complementation with Nod and Nif mutants and hybridization with *nif*-DNA probes a 120-240 kb region of the sym-plasmid was identified and characterized (Bánfalvi et al. 1983).

A thorough analysis was carried out with gene cloning, transposon tagging, physical mapping and complementation with known mutants to delimit cistrons of *nif* and *nod* genes, to determine the direction of transcription, the regulatory connections and functions. The identified nodulation genes were distinguished as common *nod* and host specific *hsn* genes (Kondorosi et al. 1984). The priority of *nodD* gene in the activation of symbiotic genes was confirmed, *nifH* and *nod* genes were sequenced (Török and Kondorosi, 1981; Török et al. 1984). With these molecular genetic approaches several details of the communication between plants and bacteria have been revealed. Three functionally different copies of *nodD* gene were identified. The gene products of these copies are binding signal molecules of different host plants by slightly altered specificity to activate symbiotic genes (Györgypál et al. 1988). *Nod* genes were shown to be involved in the control of nitrogen metabolism and NH_3 assimilation. A distinctive role of nodD products was also found in this crucial way of regulation in nitrogen fixation (Dusha et al. 1989; Dusha et al. 1993).

Chemotaxis of *Rhizobia*

The chemotaxis experiments with *R. meliloti* were started in the Institute of Genetics, but most of the work was carried out in the Department of Biotechnology in Gödöllő. The first result was the demonstration of plant-host specific chemotaxis distinguished from metabolic chemotaxis of the bacteria. (Kárpáti et al. 1989). Host plant signal molecules were shown to have high specific attractant activity, detectable at very low concentration, in diluted (10^{-4} or less) root exudates. Host specificity was confirmed by comparing chemotaxis of *R. meliloti, R. leguminosarum* and *R. trifolii* towards exudates from alfalfa, pea, trifolium (Figure 4) or wheat and corn seedlings (Figure 5).

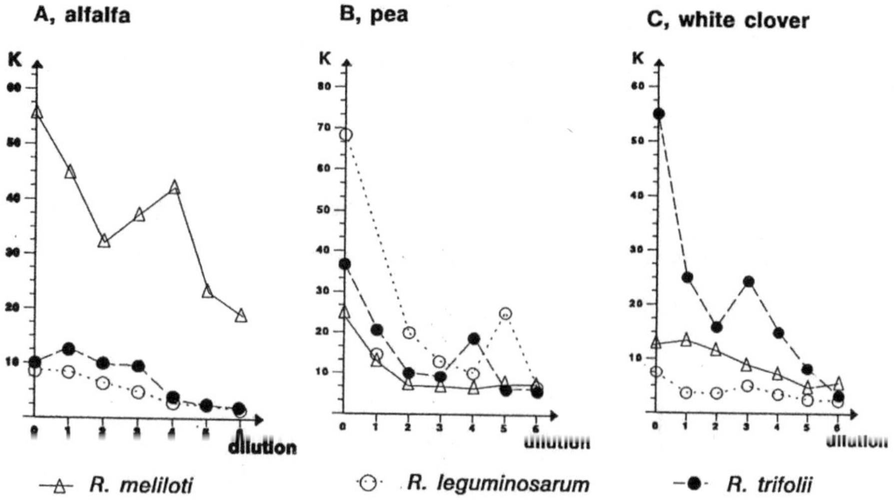

Figure 4 Host plant specific chemotaxis *Rhizobium* spp.
Root exudates of 2 days old plants (10mg dry weight/ml).
Values are chemotaxis coefficients (K) by modified capillary method, control: K_{tyr}:42.6 (*R. m.*); 46.9 (*R. l.*); 59.7 (*R. t.*).

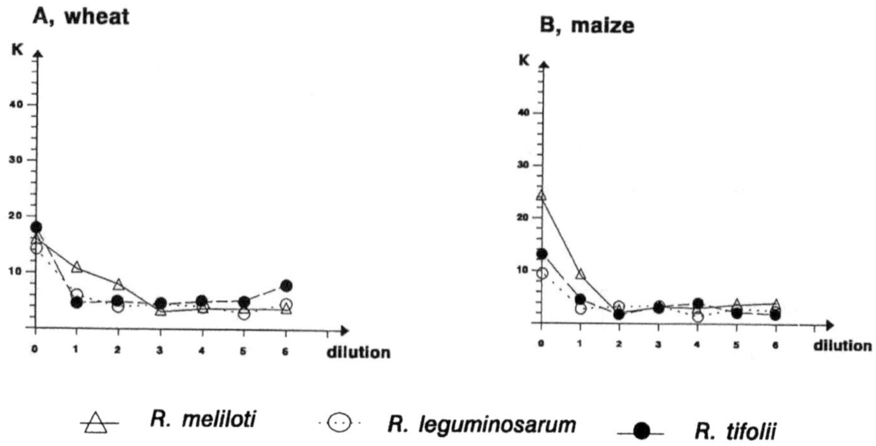

Figure 5 Attractant activity of wheat and maize exudates.
Root exudates of 5 days old plants (10mg dry weight/ml).
Chemotaxis and control values as in Figure 4.

Each *Rhizobium* strain reacted specifically with the exudate of its own host-plant seedlings as seen on the second peak at 10^{-4} dilution. No specific chemotaxis could be detected with the non leguminous plants.

The exudate of alfalfa was fractionated by molecular sieving and the fractions with highest attractant activity were inducing specific chemotaxis with *R. meliloti* as shown in Figure 6 (Kárpáti et al. 1989).

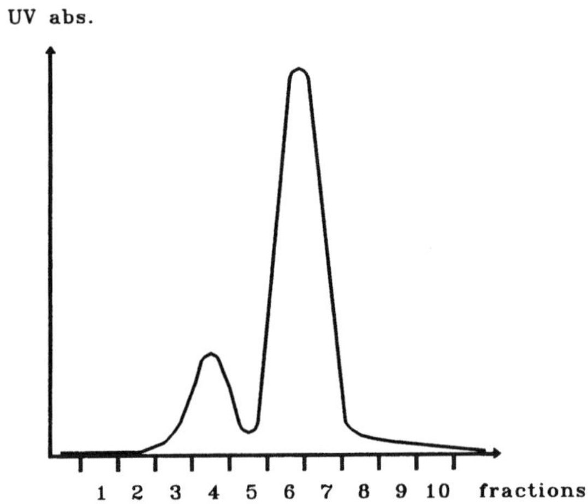

Fractionation on Sephadex G-50

Bacterium	Chemotaxis activity of fractions			
	2-3	4-5	6-7	8-10
R. meliloti 41	2	13	22	3
R. meliloti 18	5	9	19	3
R. leguminosarum	-	4	6	-
R. trifolii	-	7	4	-

Figure 6 Attractant activity of fractionated alfalfa root exudate. Chemotaxis coefficient (K) of unfractionated exudate was 16 and 18 with *R. meliloti* 41 and *R. meliloti* 18 respectively.

In the exudates besides flavonoids also amino acids are inducing chemotaxis, but with lower specific activity (Ponyi, unpublished). The genes involved in the host specific and metabolic chemotaxis were distinguished by transposon tagging and characterized by complementation and physical mapping of the insertion sites (Kárpáti et al., unpublished).

Practical applications

Our research was not directly oriented to the practical applications of *Rhizobia* as inoculants. Isolated strains, their mutants, defective lysogenes and other derivatives were compared in symbiotic efficiency. Some of them were significantly more active both in laboratory and field experiments, suitable for inoculation.

For field tests defective lysogens were used instead of antibiotic resistant strains. It was found a selectable genetic marker to identify reisolated inocula by phage superinfection. The defective lysogens were constructed and the superinfecting mutant phages characterized in our laboratory.

An easy test was developed to measure competition. The strains to be tested were mixed for the inoculation with a Nod^{+} Nif^{-} mutant in different proportions. The symbiosis was carried out in test tubes and the nitrogenase activity measured after six weeks. The lowest proportion of the tested strain which produced detectable nitrogenase activity was the measure of relative competitiveness. Field experiments confirmed the estimated competition with the use of defective lysogenes, reisolated and identified.

The reviewed results on microbial and molecular genetics of *R. meliloti* and its *phages* did inspire research in several Institutes, Departments and Laboratories in Hungary and attracted international collaborations. The experience and knowledge collected during the past decades fruitfully inspired research projects with various other bacteria and their phages. It will certainly be a useful scientific background of our studies concerning *Azospirillum* and associative nitrogen fixation.

References:

Balassa R (1956) Durch Desoxyribonukleinsäuren induzierte Veränderungen an Rhizobien. Naturwissenschaften 43:133-134

Balassa R (1960) Transformation of a strain of *Rhizobium lupini*. Nature 188: 246-247

Balassa R, Gábor M (1964) Transformation of Streptomycin markers in rough strains of *Rhizobium lupini*. Acta Microbiologica Acad. Sci. Hu. 11:329-339

Bánfalvi Z, Randhava GS, Kondorosi E, Kiss A, Kondorosi A (1983) Construction and characterization of R-prime plasmids carrying symbiotic genes of Rhizobium. Mol. Gen. Genet. 189:129-135

Bánfalvi Z, Sakanyan V, Koncz Cs, Kiss A, Dusha I, Kondorosi A (1981) Location of nodulation and nitrogen fixation genes on a high molecular weight plasmid of *Rhizobium meliloti*. Mol. Gen. Genet. 184:318-325

Barabás I, Sík T (1979) Phenotypic reversion of nitrogenase in pleiotropic mutants of *Rhizobium meliloti*. Can. J. Microbiol. 25:298-301

Dallmann G, Papp P, Orosz L (1987) Related repressor specificity of unrelated phages. Nature 330: 298-401

Dallmann G, Orosz L, Sain B (1979) Restriction mapping of DNA of temperate *Rhizobium meliloti phage 16-3*: Comparison of genetic and physical maps indicates a long genetically silent chromosomal arm. Mol. Gen. Genet. 176:439-448

Dusha I, Bakos Á, Kondorosi A, de Bruijn FJ, Schell J (1989) The *Rhizobium meliloti* early nodulation genes (*nodABC*) are nitrogen-regulated: Isolation of a mutant strain with efficient nodulation capacity on alfalfa in the presence of ammonium. Mol. Gen. Genet. 219:89-96

Dusha I, Kondorosi A (1993) Genes at different regulatory levels are required for the ammonia control of nodulation in *Rhizobium meliloti*. Mol. Gen. Genet. 240:435-444

Dusha I, Kovalenko S, Bánfalvi Z, Kondorosi A (1987) *Rhizobium meliloti* insertion element ISRm2 and its use for identification of the *fixX* gene. J. Bacteriol. 169:1403-1409

Györgypál Z, Narayan Iyer, Kondorosi A (1988) Three regulatory *nodD* alleles of diverged flavonoid-specificity are involved in host-dependent nodulation by *Rhizobium meliloti*. Mol. Gen. Genet. 212:85-92

Kárpáti É, Ponyi T, Sík T (1989) Host-specific chemotaxis of *Rhizobium meliloti* induced by alfalfa. Bull. Univ. Agr. Sci. Gödöllő 15-19

Kiss GB, Dobo K, Dusha I, Breznoivits Á, Orosz L, Vincze E, Kondorosi A (1980) Isolation and characterization of an R-prime plasmid in *Rhizobium meliloti*. J. Bacteriol. 141:121-128

Kiss GB, Kálmán Z (1982) Transformation of *Rhizobium meliloti* 41 with plasmid DNA. J. Bacteriol. 150:465-470

Kondorosi A, Barabás I, Sváb Z, Orosz L, Sík T, Hotchkiss RD (1973) Evidence for common genetic determinants of nitrogenase and nitrate reductase in *Rhizobium meliloti*. Nature, New Biol. 246:153-154

Kondorosi A, Kiss GB, Forrai T, Vincze E, Bánfalvi Z (1977) Circular linkage map of *Rhizobium meliloti* chromosome. Nature 268:525-527

Kondorosi A, Orosz L, Sváb Z, Sík T (1974) Genetic studies on *Rhizobiophage 16-3*. II. Helper-induced Transfection. Mol. Gen. Genet. 132:153-163

Kondorosi E, Bánfalvi Z, Kondorosi A (1984) Physical and genetic analysis of a symbiotic region of *Rhizobium meliloti*: Identification of nodulation genes. Mol. Gen. Genet. 193:445-452

Mink M, Orosz L, Sík T (1982) Specialized and generalized transducing *Rhizobiophage 16-3* and *11* are closely related. FEMS Microbiol. Lett. 13:383-387

Olasz F, Dorgai L, Páy A, Orosz L (1983) Recombination Deficient Mutants of *Rhizobium meliloti* 41. Mol. Gen. Genet. 191:393-396

Orosz L (1980) Methods for analysis of the C cistron of temperate *phage 16-3* of *Rhizobium meliloti*. Genetics 94:265-276

Orosz L, Páy A, Dallmann G (1980) Heterozygosis of phage *16-3* of *Rhizobium meliloti*: Moderate level of mismatch repair or gene conversion. Mol. Gen. Genet. 179:163-167

Orosz L, Rostás K, Hotchkiss RD (1980) A comparison of two-point, three-point and deletion mapping in the *C* cistron of *Rhizobiophage 16-3*, with an explanation for the recombination pattern. Genetics 94:249-263

Orosz L, Sváb Z, Kondorosi A, Sík T (1973) Genetic studies on *Rhizobiophage 16-3*. I. Genes and functions on the chromosome. Mol. Gen. Genet. 125:341-350

Papp I,Dorgai L, Papp P, Jónás E, Olasz F, Orosz L (1993) The bacterial attachment site of temperate *Rhizobium phage 16-3* overlaps the 3' end of a putative proline tRNA gene. Mol. Gen. Genet. 240:258-264

Putnoki P, Petrovics G, Kereszt A, Grosskopf E, Dang thi Cam Ha, Bánfalvi Z, Kondorosi A (1990) *Rhizobium meliloti* lipopolysaccharide and exopolysaccharid can have the same function in the plant-bacterium interaction. J. Bacteriol. 172:5450-5458

Sík T, Barabás I (1977) The correlation of nitrate reductase and nitrogenase in *Rhizobium* symbiosis. In: Recent Developments in nitrogen fixation (Eds. Newton W, Postgate JR, Rodriguez-Barrueco C) Academic Press Inc. pp.365-373

Sík T, Horváth J, Chatterjee S (1980) Generalized transduction in *Rhizobium meliloti*. Mol. Gen. Genet. 178:511-516

Sík T, Kondorosi A, Barabás I, Sváb Z (1976) Nitrate reductase and effectiveness in *Rhizobium*. In: Proceedings of the 1st International Symposium on nitrogen fixation (Eds. Newton WE, Nyman CJ) ashington State University Press Vol.2 pp.374-382

Sík T, Orosz L (1971) Chemistry and genetics of *Rrizobium meliloti phage 16-3*. Plant and Soil Sp.Vol.: 57-62

Sváb Z, Kondorosi A, Orosz L (1978) Specialized transduction of cysteine marker by *Rhizobium meliloti phage 16-3*. J. Gen. Microbiol. 106:321-327

Szende K, Ördögh F (1960) Die Lysogenie von *Rhizobium meliloti*. Naturwissenschaften 47:404-405

Szende K, Sík T, Ördögh F, Győrffy B (1961) Transfer of immunity by nucleic acids of a lysogenic *Rhizobium* strain. Biochim. Biophys. Acta 47:215-217

Török I, Kondorosi A (1981) Sequence of the *Rhizobium meliloti* nitrogen reductase (nif H) gene. Nucl. Acids Res. 9:5711-5723

Török I, Kondorosi E, Stepkowski T, Pósfai J, Kondorosi A (1984) Nucleotide sequence of *Rhizobium meliloti* nodulation genes. Nucleic Acids Res. 12:9509-9524

SECTION II:

GENETICS

Azospirillum-Cereals: An Intriguing Partnership

Jos Vanderleyden, My Ali Bekri, Antonia Costacurta, Jos Desair, Paul De Troch, Sofie Dobbelaere, Veerle Keijers, Kris Michiels, Anne Milcamps[1], Sara Moens, Daniel Petersen[2], Els Prinsen[3], Els Van Bastelaere, Ann Vande Broek, Anne Van Dommelen, August Van Gool, Harry Van Onckelen[3]
F.A. Janssens Laboratory of Genetics
Department of Applied Plant Sciences, K.U. Leuven
Willem de Croylaan 42
3001 Heverlee
Belgium

Abstract

Azospirillum is probably the best studied example of beneficial plant rhizosphere bacteria. Studies in our laboratory focus on the identification of bacterial genes and gene products that are of importance in the physical and metabolic interaction of *Azospirillum brasilense* with plant roots. Here we report for *Azospirillum brasilense*, flagellation, motility, the physical interaction with plant roots, the synthesis of indole-3-acetic acid, the expression of *nif* genes in plant-root associated bacteria, and the induction of gene expression with plant root exudates.

Introduction

Historically, the genus *Azospirillum* was first studied for its nitrogen-fixation capacity and occurrence in the rhizosphere of cereals and grasses (von Bulow and Döbereiner, 1975). Today there is still no clear evidence that significant amounts of nitrogen are fixed by cereals like rice or wheat, when inoculated with *Azospirillum*. Nevertheless overwhelming knowledge has been acquired over the past years on the genetics, biochemistry and physiology of *Azospirillum*. These studies revealed that *Azospirillum* has a number of traits, besides the potential to fix nitrogen, that might be of importance for its plant growth promoting effect.

---

[1] MSU-DOE, Plant Research Laboratory, East Lansing, Michigan, USA
[2] Department of Plant Science, University of British Columbia, Vancouver, Canada
[3] Department of Biology, Universitaire Instelling Antwerpen, Wilrijk, Belgium

NATO ASI Series, Vol. G 37
Azospirillum VI and Related Microorganisms
Edited by I. Fendrik et al.

Flagellation of *Azospirillum brasilense*

A. brasilense possesses two types of flagella: a single polar flagellum that is always present, and several lateral flagella that are only synthesized in semi-solid or solid growth medium (Tarrand *et al.*, 1978). Recently we have isolated the structural gene that codes for the flagellin of the lateral flagella (Moens *et al.*, submitted *b*). The gene, designated as *laf1*, has considerable homology with the *flaA*, *flaB* and *flaC* genes of *Agrobacterium tumefaciens* (Shaw, GenBank), and the *flaA* and *flaB* genes of *Rhizobium meliloti* (Pleier and Schmitt, 1990). In *R. meliloti*, the *flaA* and *flaB* genes code for different subunits of the flagellin. These subunits, although they show considerable homology at protein level, have a different function in the assembly of the flagella. In *A. brasilense*, there appears to be only one gene, based on DNA hybridization data, and the phenotype of the mutant constructed by insertion inactivation of the cloned gene.

Partial amino acid sequencing of the protein isolated from the lateral flagellum does not reveal homology with any known protein in databanks. Attempts to isolate the structural gene have not been successful so far. However, different mutants that no longer make the polar flagellum have been isolated from a Tn*5* mutant bank. Surprisingly, these mutants no longer make the lateral flagella as well. The genes that are mutated in these strains are presently under study. Recently, it was also shown (Milcamps *et al.*, submitted) that an *A. brasilense ntrA* mutant is devoid of both the polar and the lateral flagella. NtrA control of flagella synthesis was also reported for *Pseudomonas aeruginosa* and *P. putida* (Inouye *et al.*, 1990; Totten *et al.*, 1990), *Caulobacter crescentus* (Ramakrishnan *et al.*, 1994) and *Campylobacter coli* (Guerry *et al.*, 1991).

Based on immunological data (Schloter *et al.*, 1994) and DNA hybridization studies with the structural gene for the lateral flagella, it is obvious that proteins (genes) of the two types of flagella are quite different. Genes that are needed for the synthesis of the polar flagellum are localized on the chromosome and on the pRhico (90 MDa plasmid). Structural analysis of the polar flagellum demonstrated that the major protein is glycosylated (Moens *et al.*, submitted *a*). This evidence is based on specific carbohydrate staining of protein gels, the effect of glycosylation inhibitors, chemical deglycosylation and the use of sugar specific monoclonal antibodies. Moreover mutations that affect the synthesis of specific sugar nucleotides (e.g. a mutation in a gene that encodes phosphomannomutase activity) in *A. brasilense* also change the size of the polar flagellin on a denaturing polyacrylamide gel.

An *A. brasilense laf1* mutant shows wild-type root colonization patterns. However, the polar flagellum of *A. brasilense* is important for plant root attachment (Croes *et al.*, 1993) and plant root colonization (Vande Broek, 1994). This was demonstrated in several ways. Mutants that no longer synthesize the polar flagellum are drastically reduced in plant root colonization capacity and total protein extracts of plant roots inoculated with *A. brasilense* wildtype show the presence of the polar flagellin while the lateral flagellin is absent. The role of the polar flagellum, in particular the importance of the sugar moiety, in plant root interaction is presently studied.

Indole-3-acetic (IAA) synthesis in *Azospirillum brasilense*

A. brasilense produces considerable amounts of IAA in the presence of L-tryptophan (Reynders and Vlassak 1979; Tien *et al.*, 1979). Attempts to isolate IAA negative mutants were not successful and this led to the speculation that more than one pathway might be involved in *Azospirillum* IAA biosynthesis (Abdel-Salam and Klingmüller, 1987). The best documented IAA biosynthesis pathways in microorganisms and plants are illustrated in Figure 1. For reviews, see also Arshad and Frankenberger, 1991; Costacurta and Vanderleyden, 1995. Prinsen *et al.* (1993) provided biochemical evidence for the existence of three IAA biosynthesis pathways in *A. brasilense*, two tryptophan-dependent and one tryptophan-independent pathway. One of the tryptophan-dependent pathways was identified as the indole-3-acetamide pathway, based on the following observations: (I) IAM is detected in culture supernatants; (ii) IAM can be converted to IAA by *A. brasilense*; (iii) tryptophan-2-monooxygenase activity was detected in crude cell extracts of *A. brasilense* (Bar and Okon, 1993). The second tryptophan-dependent pathway was recently identified as the indole-3-pyruvic acid pathway, based on cloning and mutagenesis of the gene encoding indole-3-pyruvic decarboxylase (Costacurta *et al.*, 1994). This pathway is extensively studied in *Enterobacter cloacae* (Koga *et al.*; 1991; 1992) and seems to be widespread in other *Enterobacter* species

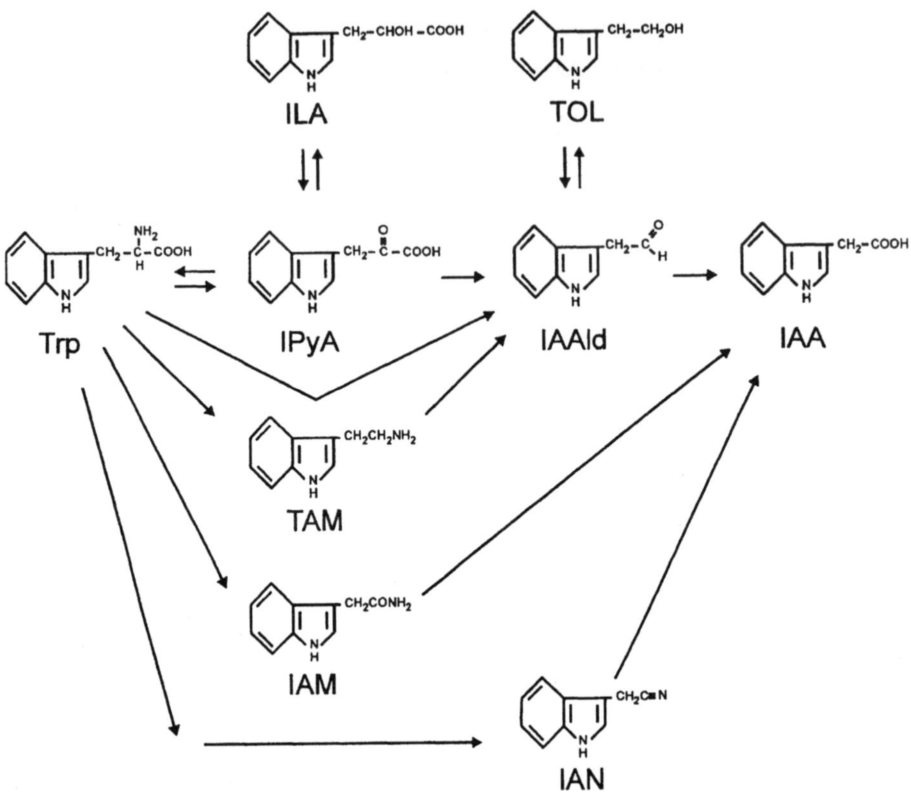

Figure 1. Biosynthetic pathways for the conversion of tryptophan to indole-3-acetic acid. Abbreviations: TRP, tryptophan; IPyA, indole-3-pyruvic acid; IAAld, indole-3-acetaldehyde; IAM, indole-3-acetamide; IAN, indole-3-acetonitrile; TAM, tryptamine; IAA, indole-3-acetic acid. Recently, a new pathway for conversion of indole-3-acetonitrile to indole-3-acetamide was discovered in *Rhizobium* and *Agrobacterium* (Kobayashi *et al.*, 1995)

as well (see Zimmer *et al.*, this volume). The identity of the tryptophan-independent pathway is not known yet.

Concerning the contribution of each of these pathways in the overall IAA production by *A. brasilense*, it is important to discriminate between experiments in which trypthophan is added and experiments in which no external tryptophan is added (Prinsen *et al.*, 1993). In the latter case, 90 % of the overall IAA production is mediated through the tryptophan-independent pathway. With exogenously added L-tryptophan, most IAA synthesis is believed to occur through the indole-3-pyruvic acid pathway. This is based on the observation that a mutation in the indole-3-pyruvic acid decarboxylase gene causes a reduction of approximately 95 % of IAA produced. However, at present it cannot be excluded that precursors other than tryptophan can be converted to IAA via the IpyA pathway.

In order to study the regulation of the indole-3-pyruvic acid decarboxylase gene, we have constructed an *ipdC:gusA* gene fusion (Costacurta, 1995). It was observed that IAA and salt stress result in 3-fold induction of the gene fusion. Downstream of the *ipdC* gene, a partial ORF that shows homology to a putative sigma factor of *Escherichia coli* was identified. Insertion mutagenesis of this ORF drastically reduces expression of the *ipdc:gusA* fusion.

Expression of *A. brasilense nifH* in plant root associated bacteria

Studies on the regulation of *nif* genes in *Azospirillum* reveal both common features with other diazotrophic bacteria as well as unique characteristics (see de Zamaroczy, this volume). However, little is known about the expression of *nif* genes in plant root colonizing Azospirilla. It was shown that the *gusA* reporter gene is a suitable system to visualize and to quantify plant root colonizing Azospirilla and to monitor expression of bacterial genes in *Azospirillum*-plant root associations (Vande Broek *et al.*, 1992; 1993). Using a *nifH-gusA* fusion on an IncP plasmid, which is stably maintained in *A. brasilense* in absence of antibiotic selection pressure, environmental conditions that determine transcription of the *nifHDK* operon in wheat root colonizing *A. brasilense* Sp245 cells could be determined. Under the experimental conditions of the plant test system used (see Vande Broek *et al.*, 1993), highest expression is observed at an initial oxygen concentration of 3 % in the head space, and in the presence of an extra carbon source, such as malate or fructose, in the plant nutrition medium (Table 1). Under aerobic conditions, that is in test tubes where the oxygen concentration is in equilibrium with the ambient atmosphere, expression levels are 4 to 5 times lower. The observed *nifH* expression patterns coincide very nicely with nitrogen fixation activity, as measured by the acetylene reduction activity (ARA).

Table 1. Effect of the carbon and nitrogen content on the plant nutrient solution and the oxygen tension on the expression of the *A. brasilense nifH* gene during wheat root association

	Associative *nifH* induction [a]	
Nutrient solution [b]	Aerobic	Microaerobic[c]
MPCL - N	7(8)	32(13)
MPCL - N + 0.1 % malate	61(20)	258(118)
MPCL - N + 0.1 % fructose	85(38)	328(112)
MPCL	0	ND
MCL + 0.1 % malate	0	ND
MPCL + 0.1 % fructose	0	ND
MPCL - N (no seedling)	0	0

[a] Bacteria were reisolated from the roots and the sand 16 days after inoculation, as described by Vande Broek *et al.* (1993). β-glucuronidase activity is expressed as nanomoles *p*-nitrophenol produced per minute per 10^8 cells. Standard deviations are given within parentheses
[b] Seedlings were grown in sand cultures with MPCL or MPCL-N medium. MPCL medium is a plant-growth medium described by Lavigne (1987). MPCL-N is nitrogen free MPCL medium.
[c] The oxygen concentration in the test tubes was lowered to 3 %.

From these and other experiments, it can be concluded that available carbon source and prevailing O_2 tension at the root surface are limiting factors for significant nitogenase activity to occur. Therefore plant cultivars with higher root exudation are worthwhile to investigate in a screening program. Also engineering strategies for *A. brasilense* to lower the intracellular oxygen concentration have to be considered.

Induction of *Azospirillum* genes with plant root exudates

The question as to whether plant root colonizing *Azospirillum* respond by differential gene expression to signals exuded from the plant root, was addressed by two different approaches (Van Bastelaere *et al.*, in preparation). Using the promoterprobe vector Tn*5*-B30 (Simon *et al.*, 1989) for constructing an *A. brasilense* Sp7 mutant bank, an insertion mutant in which the neomycin fosfotransferase gene was specifically induced in the presence of wheat root exudates, was isolated. The Tn*5*-B30 in this mutant was found to be inserted in a gene that is very likely part of an operon (Figure 2).

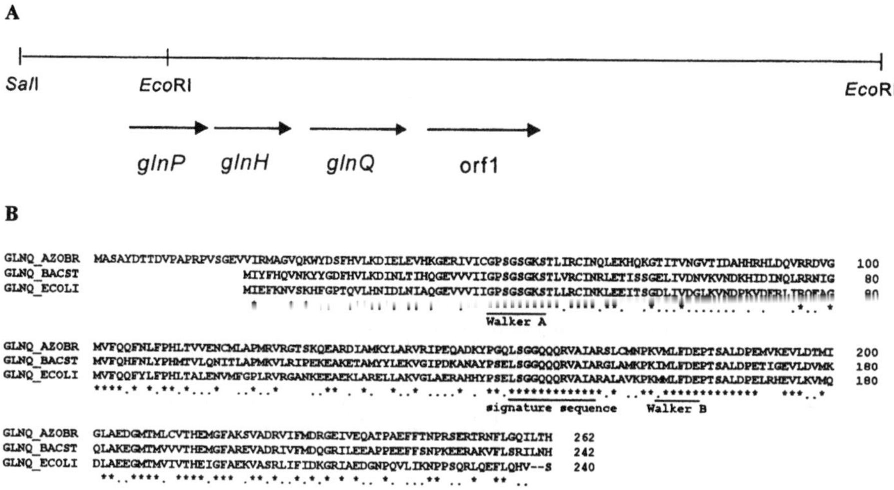

Figure 2. Physical and genetic map of an *A. brasilense* operon (A) that shows high homology with the *E. coli gln HPQ* operon. Amino acid alignment is given for GlnQ (B).

Sequence analysis of this operon shows significant homology with the *E. coli glnHPQ* operon (Nohno *et al.*, 1986) and to a lesser extent with the *his* operon. These operons are expressed under nitrogen-limiting conditions and encode high-affinity transport systems for glutamine and histidine respectively, and belong to the class of ATP-Binding-Cassette (ABC) transporters. Presently a promoter *gusA* fusion and mutants with an antibiotic resistance cassette in various positions in the operon are being analyzed in monoxenic plant test systems.

When comparing 2D-protein profiles (De Mot and Vanderleyden, 1989) of *A. brasilense* Sp7 cells grown in presence and absence of wheat root exudates, a number of quantitative and qualitative differences can be observed (Van Bastelaere *et al.*, 1993). A protein that is specifically induced in presence of wheat root exudates and that is not observed when cells are grown in normal medium, like minimal MMAB medium (Vanstockem *et al.*, 1987) or LB* medium (Luria-Bertani broth supplemented with 2.5 mM $CaCl_2$ and 2.5 mM $MgSO_4$) was further studied. Partial amino acid sequence revealed significant homology with the

Agrobacterium tumefaciens sugar-binding protein ChvE. This was confirmed by DNA sequence analysis of the complete gene (Van Bastelaere *et al.*, in preparation). Organization of the *A. brasilense chvE* homolog is given in Figure 3.

A

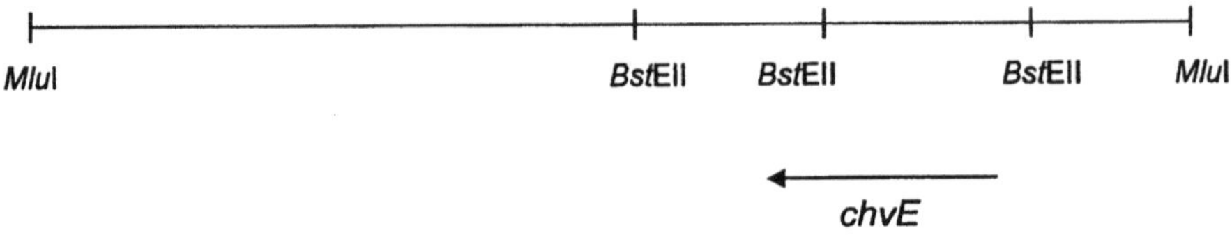

B

```
CHVE_AGRTU- MKSIISLMAACAIGAASFAAPAFAQDKGSVGIAMPTKSSARWIDDGNNIVKQLQEAGYKTDLQYADDDIPNQLSQIENMVTKGVKVLVIASIDGTTLSDV -100
CHVE_AZOBR-                            GPRXRHQSDSAQDKPTVGIAMPTKSSARWIDDGNNMVKQFQAKGYKTDLQYAEDDIPNQLAQIETMVAKNSKVLVIAAIDGTTLTDV -87

CHVE_AGRTU- LKQAGEQGIKVIAYDRLIRNSGDVSYYATFDNFQVGVLQATSITDKLGLKDGKGPFNIELFGGSPDDNNAFFFYDGAMSVLKPYIDSGKLVVKSGQMGMD -200
CHVE_AZOBR- LQQAKDRGVKVIAYDRLIRGSENVDYYATFDNFQVGVLQGSYIVDALGLKDGKGPFNIELFGGSPDDNNAYFFYNGAMSVLQPYIDSGKLTVGSGQVGMD -187

CHVE_AGRTU- KVGTLRWDPATAQARMDNLLSAYYTDAKVDAVLSPYDGLSIGIISSLKGVGYGTKDQPLPVVSGQDAEVPSVKSIIAGEQYSTIFKDTRELAKVTVNMVN -300
CHVE_AZOBR- KVSTLRWDGATAQARMDNLLSAFYGNRRVDAVLSPYDGISIGIISSLKGVGYGSPSQPMPVVTGQDAEVPSIKSILAGEQXXTVFKDTRELARITVEMVD -287

CHVE_AGRTU- AVMEGKEPEVNDTKTYENGVKVVPSYLLKPVAVTKENYKQVLVDGGYYKEDQLK -354
CHVE_AZOBR- XVLX-------------------------------------------------- -291
```

Figure 3. Physical and genetic map (A) of the *A. brasilense* DNA region that contains homology with the *A. tumefaciens chvE* gene (B).

The deduced amino acid sequence of *chvE* is similar to that of the galactose-binding protein (GBP) of *E. coli.* In *A. tumefaciens,* ChvE is required for *vir* induction by sugars (Lafferty Doty *et al.*, 1993; Shimoda *et al.*, 1993). In *A. brasilense* cells, the protein is also found to be induced when galactose, fructose or arabinose, in stead of wheat root exudates, are added to minimal MMAB medium. Using an *A. brasilense chvE-gusA* fusion, the same sugars that were found to induce an *A. tumefaciens chvE-lacZ* fusion, are able to induce the expression of the *A. brasilense chvE* gene. It remains to be answered whether chvE in *A. brasilense* is part of a signal transduction pathway, as observed in *A. tumefaciens.* It was observed that *A. brasilense* Sp7 cells show strong chemotactic response to galactose, fructose and arabinose (I. Zhulin, personal communication), and this chemotactic response is inducible. Therefore, it is worthwhile to investigate the role of *chvE* in chemotaxis.

Concluding remarks

At this stage, it is still not clear how *Azospirillum* can promote plant growth under certain conditions. Nevertheless, it appears that *Azospirillum* has some properties that are an advantage to proliferate in the rhizosphere plants: strong chemotaxis, firm attachment to the roots, synthesis of auxins. Auxins stimulate root proliferation and increase the number of lateral roots, thereby increasing the root surface for colonization and stimulating root exudation for bacterial proliferation. For some of these properties, differences exist among *Azospirillum* species. *A. irakense* produces little or no auxins but on the other hand is the only species with significant cell wall hydrolytic activity. A better understanding of *Azospirillum* plant root interactions will depend on studies in which changes in bacterial gene expression and potential changes in plant root metabolism during the interaction are evaluated and quantified.

Acknowledgements

This work was supported by the FGWO (No. 3.0095.93) and Geconcerteerde Onderzoeksacties (GOA93 Vanderleyden).

References

Abdel-Salam MS, Klingmüller W (1987) Transposon Tn*5* mutagenesis in *Azospirillum lipoferum*: isolation of indole-3-acetic acid mutants. Mol Gen Genet 210: 165-170

Arshad M, Frankenberger JWT (1991) Microbial production of plant hormones. Plant Soil 133: 1-8

Bar T, Okon Y (1993) Tryptophan conversion to indole-3-acetic acid via indole-3-acetamide in *Azospirillum brasilense* Sp7. Can J Microbiol 39: 81-86

Costacurta A (1995) Genetic studies on the auxin hypothesis in the *Azospirillum*/plant interaction. Ph.D. Thesis, Katholieke Universiteit Leuven

Costacurta A, Keijers V, Michiels K, Vanderleyden J (1994) Molecular cloning and sequence analysis of an *Azospirillum brasilense* indole-3-pyruvate decarboxylase. Mol Gen Genet 243: 463-472

Costacurta A, Vanderleyden J (1995) Synthesis of phytohormones by plant-associated bacteria. Crit Rev Microbiol: 21, 1-18

Croes CL, Moens S, Van Bastelaere E, Vanderleyden J, Michiels KW (1993) The polar flagellum mediates *Azospirillum brasilense* adsorption to wheat roots. Gen Microbiol 139: 960-967

De Mot R, Vanderleyden J (1989) Application of two-dimensional protein analysis for strain fingerprinting and mutant analysis of *Azospirillum* species. Can J Microbiol 35: 960-967

Guerry P, Alm RA, Power ME, Logan SM, Trust TJ (1991) Role of two flagellin genes in *Compylobacter* motility. J Bacteriol 173: 4757-4764

Inouye S, Kimoto M, Nakazawa A, Nakazawa T (1990) Presence of *Pseudomonas putida* is dependent on *ntrA (rpoN)* gene. Mol Gen Genet 221: 295-298

Kobayashi M, Suzuki T, Fujita T, Masuda M, Shimizu S (1995) Occurrence of enzymes involved in biosynthesis of indole-3-acetic acid from indole-3-acetonitrile in plant-associated bacteria, *Agrobacterium* and *Rhizobium*. Proc Natl Acad Sci USA 92: 714-718

Koga J, Adachi T, Hidaka H (1991) Molecular cloning of the gene for indole pyruvate decarboxylase from *Enterobacter cloacae*. Mol Gen Genet 226: 10-16

Koga J, Adachi T, Hidaka H (1992) Purification and characterization of indolepyruvate decarboxylase. J Biol Chem 26: 15823-15828

Lafferty Doty S, Chang M, Nester EW (1993) The chromosomal virulence gene, *chvE*, of *Agrobacterium tumefaciens* is regulated by a LysR family member. J Bacteriol 175: 7880-7886

Lavigne C (1987) Contribution à l'étude du système racinaire du bananier. Mise au point de rhizotrons et premiers resultats. Fruits 42: 265-271

Milcamps A, Van Dommelen A, Stigter J, Vanderleyden J, de Bruijn FJ (1995) The *Azospirillum brasilense ntrA* gene is involved in nitrogen fixation, nitrate assimilation, ammonium uptake and flagella biosynthesis. Submitted

Moens S, Michiels K, Vanderleyden J (1995) Glycosylation of the flagellin of the polar flagellum of *Azospirillum brasilense*, a Gram-negative nitrogen-fixing bacterium. Submitted *a*

Moens S, Michiels K, Keijers V, Van Leuven F, Vanderleyden J (1995) Cloning, sequencing and phenotypic analysis of *laf1*, encoding the flagellum of the lateral flagella of *Azospirillum brasilense* Sp7. Submitted *b*

Nohno T, Saito T, Hog J (1986) Cloning and complete nucleotide sequence of the *Escherichia coli* glutamine permease operon (*glnHPQ*). Mol Gen Genet 205: 260-269

Pleier E, Schmitt R (1991) Expression of two *Rhizobium meliloti* flagellin genes and their contribution to the complex filament structure. J Bacteriol 173: 2077-2085

Prinsen E, Costacurta A, Michiels K, Vanderleyden J, Van Onckelen H (1993) *Azospirillum brasilense* indole-3-acetic acid biosynthesis: evidence for a non-tryptophan dependent pathway. Mol Plant-Microbe Interact 6: 609-615

Ramakrishnan G, Zhao JL, Newton A (1994) Multiple structural proteins are required for both transcriptional activation and negative autoregulation of *Caulobacter crescentus* flagellar genes. J Bacteriol 176: 7587-7600

Reynders L, Vlassak K (1979) Conversion of tryptophan to indole acetic acid by *Azospirillum* sp. Soil Biol Biochem 11: 547-548

Shimoda N, Toyoda-Yamamoto A, Aoki S, Machida Y (1993) Genetic evidence for an interaction between the VirA sensor protein and the ChvE sugar-binding protein of *Agrobacterium*. J Biol Chem 268: 26552-26558

Simon R, Quandt J, Klipp W (1989) New derivatives of transposon Tn*5* suitable for mobilization of replicons, generation of operon fusions and induction of genes in Gram-negative bacteria. Gene 80: 161-169

Tarrand JJ, Krieg NR, Döbereiner J (1978) A taxonomic study of the *Spirillum lipoferum* group, with description of a new genus, *Azospirillum* gen. nov., and two species *Azospirillum lipoferum* (Beijerinck) sp. nov. and *Azospirillum lipoferum* sp. nov. Can J Microbiol 24: 967-980

Tien TM, Gaskins MH, Hubbell DH (1979) Plant growth substances produced by *Azospirillum brasilense* and their effect on the growth of pearl millet (*Pennisetum americanum* L.). Appl Environm Microbiol 37: 1016-1024

Totten PA, Cano Laro J, Lory S (1990) The *rpoN* gene product of *Pseudomonas aeruginosa* is required for expression of diverse genes, including the flagellin gene. J Bacteriol 172: 389-396

Van Bastelaere E, De Mot R, Michiels K, Vanderleyden J (1993) Differential gene expresion in *Azospirillum* spp. by plant root exudates: analysis of protein profiles by two-dimensional polyacrylamide gel electrophoresis. FEMS Microbiol lett 112: 335-342

Vande Broek A (1994) Histochemical and genetic analysis of the *Azospirillum brasilense*-wheat root association. Ph.D. Thesis, Katholieke Universiteit Leuven

Vande Broek A, Michiels J, de Faria SM, Milcamps A, Vanderleyden J (1992) Transcription of the *Azospirillum brasilense nifH* gene is positively regulated by NifA and NtrA and is negatively controlled by the cellular nitrogen status. Mol Gen Genet 232: 279-283

Vande Broek A, Michiels J, Van gool A, Vanderleyden J (1993) Spatial-temporal colonization patterns of *Azospirillum brasilense* on the wheat root surface and expression of the bacterial *nifH* gene during the association. Mol Plant-Microb Interact 6: 592-600

Vanstockem M, Michiels K, Vanderleyden J, Van Gool A (1987) Transposon mutagenesis of *Azospirillum brasilense* and *Azospirillum lipoferum*: Physiological analsyis of Tn*5* and Tn*5*-*mob* insertional mutants. Appl Environm Microbiol 53: 410-415

von Bulow JGD, Döbereinder J (1975) Potential for nitrogen fixation in maize genotypes in Brazil. Proc Natl Acad Sci USA 72:2389-2393

Phylogenetic Studies of the Genus *Azospirillum*

Renato Fani, Claudio Bandi[1], Marco Bazzicalupo, Giuseppe Damiani[2], Francescopaolo Di Cello, Silvia Fancelli, Enzo Gallori, Letizia Gerace, Annamaria Grifoni, Pietro Liò and Elena Mori
Department of Animal Biology and Genetics "Leo Pardi"
University of Florence
Via Romana 17
I-50125 Firenze
Italy

Abstract

The 16S rDNA of seventeen *Azospirillum* strains, fourteen of which assigned to one of known *A. amazonense, A. brasilense, A. halopraeferens, A. irakense* and *A. lipoferum* species, and other three of uncertain taxonomic collocation, was sequenced and analysed after polymerase chain reaction amplification, in order to investigate the phylogenetic relationships at the intra-generic and super-generic level. Moreover the use of histidine biosynthetic genes and of anonymous sequences generated by RAPD technique to infer phylogenetic relationships among *Azospirillum* strains is also described. The relationships among the five species, as revealed by DNA sequence data, appeared to be congruent with other molecular and phenotypic results, showing that the genus *Azospirillum* constitutes a phylogenetically separate entity in which *A. brasilense* and *A. lipoferum* are very close species, whereas the species *A. amazonense* is the most divergent.

Introduction

Bacteria of the genus *Azospirillum* are Gram-negative, nitrogen-fixing rods forming different kinds of association with non-leguminous plants (cereals and grasses) from various geographical origins.

----

1) Istituto di Patologia Generale Veterinaria and 2) I.D.V.G.A. - CNR, Via Celoria 10, 20133 Milano, Italy

NATO ASI Series, Vol. G 37
Azospirillum VI and Related Microorganisms
Edited by I. Fendrik et al.

Since 1976, when Dobereiner and Day reisolated *Azospirillum lipoferum*, (originally *Spirillum lipoferum*) from the roots of tropical grasses, a number of taxonomic studies were carried out using phenotypic and physiological characters, such as:biochemical tests, G+C base composition, DNA homology (Tarrand et al. 1978), serological tests (De Polli et al. 1980), chemical analysis of lipopolysaccharides (Choma et al. 1987). These studies, although weakened by the particular phenotypic instability of this genus, showed that azospirilla could be divided into two distinct groups: *A. brasilense* and *A. lipoferum*. During the last 15 years, a wide variety of molecular techniques were developed to perform taxonomic and phylogenetic studies. In particular nowadays the comparison of rDNA sequences plays a key role in the description, classification, and identification of microbial taxa: these techniques were also applied to the genus *Azospirillum* (reviewed by Gillis and Reinhold-Hurek, 1994).

This body of molecular studies supports the idea that the *Azospirillum* genus constitutes a phylogenetically separate entity in which *A. amazonense* and *A. halopraeferens* (and probably also *A. irakense*) are divergent from *A. brasilense* and *A. lipoferum*; nevertheless up to now it is not clear whether *A. irakense* is genotypically closer to *A. amazonense* and *A. halopraeferens* than to *A. brasilense* and *A. lipoferum*. Moreover, the relationships between the azospirilla and the other diazotrofic Proteobacteria should be supported by further studies.

Although not all molecular chronometers are equally useful for reconstructing phylogenies, and no method for phylogenetic inference in use today is completely satisfactory, the most commonly used tool is the analysis of small subunit ribosomal rRNA-rDNA sequences (Woese, 1987). In fact ribosomal RNA is universal in distribution; its function was established at an early stage in the evolutionary process, and appears extremely constant and highly constrained. Being large molecules, the ribosomal rRNAs contain a great deal of information. For this and other reasons rRNAs are less erratic than smaller molecules (Woese, 1987); moreover they do not appear to be subject to lateral gene transfer. Nevertheless the fact that the whole microbial superstructure rests almost solely on this molecular species is disquieting (Woese, 1991). Indeed some authors agree that evolutionary trees inferred from ribosomal RNA sequences could be sometimes misleading when there are large differences in base composition among the species (Loomis and Smith, 1990). Moreover the 16S rDNA is often not able to distinguish among strains belonging to the same species. Therefore phylogenies produced from rRNAs need to be tested with that obtained with other molecular approaches (Woese, 1991).

In this paper we report an exhaustive phylogenetic analysis of the genus *Azospirillum* carried out by determining and comparing the nucleotide sequence of the 16S rDNA gene of seventeen strains belonging to the five *Azospirillum* species. Moreover we describe the preliminary results that we have obtained with two other different approaches; the first one involves genes with known function(s) (the histidine biosynthetic genes) and the second the anonymous sequences generated by RAPD technique.

We focused our attention on the histidine biosynthetic genes for many reasons. First of all there are several independent indications of the *antiquity* of the histidine biosynthesis pathway. Although how this route originated is still an open question, the apparently universal phylogenetic distribution of the *his* genes (Fani et al. 1993a; Fani et al. 1995) suggests that histidine synthesis was already part of the metabolic abilities of the last common ancestor of the three extant cell lineages (Lazcano et al. 1992) and that the function of histidine biosynthetic genes was established at an early stage in the evolutionary process. Secondly, the phylogenetic trees constructed for the available sequence dataset appear to be robust, i.e., they give the same general topology showing similar interdomain relationships (Fani et al. 1995).

The second approach we have used takes advantage from the random amplified polymorphic DNA (RAPD) methodology (Williams et al. 1990; Welsh and McClelland, 1990), which has been extensively used in the last years for many purposes, including DNA fingerprinting of microorganisms (Hadrys et al. 1992; Williams et al. 1993) and for generating specific DNA probes for microorganisms (Fani et al. 1993b). The RAPD methodology is based on a modification of the polymerase chain reaction (PCR) and its basic principle is the possibility to amplify unknown sequences of a given genome by using a single oligonucleotide as primer. A

RAPD experiment generates patterns of bands which can be easily analyzed by agarose gel electrophoresis. A bacterial RAPD pattern generally contains both polymorphic and monomorphic bands. The greater the genetic similarity among bacterial strains, the greater the similarity of RAPD patterns (i.e. the number of monomorphic bands). Although strains belonging to different species often show only polymorphic bands, it is possible, by changing the reaction conditions, to obtain RAPD pattern containing monomorphic bands also for these strains.

The working hypothesis is that the comparison of nucleotide sequence of monomorphic bands can be used for the study of phylogenetic relationships among bacterial strains. This is based on the assumption that the amplification bands with the same molecular weight, obtained in two or more samples amplified with the same primer, have related sequence structure as suggested by previous hybridization experiments (Welsh and McClelland, 1990; Williams et al. 1993; Fani et al. 1993b).

Materials and Methods

Bacterial strains and growth conditions

The *Azospirillum* strains used are listed in Table 1. *A. halopraeferens* was grown at 41°C in Nutrient Broth supplemented with 0.25% NaCl; all other *Azospirillum* strains were grown in LB medium at 35°C.

DNA manipulations

All DNA manipulations, including DNA purification, restriction digestions and Southern hybridization, were carried out as described by Fani et al. (1993b).

Amplification and sequencing of 16S rDNA

The 16S rDNA amplification and sequencing reactions were performed according to Bandi et al. (1994). The primers used were designed on the basis of the conserved sequences of the eubacterial 16S rDNA and were named 27f, 561r, 704f, 765r, 930f and 1495r using the number of the *E. coli* rRNA positions to which the 3' end of the primer anneals in forward (f) or reverse (r) orientation. The primers were synthesized, deprotected, dried, dissolved in TE (10 mM TRIS-HCl pH 8.0, 1 mM EDTA) and used without any further purification. The primers 27f and 1495r, located respectively at the extreme 5' and 3' ends of the rDNA sequence, allowed the amplification of nearly the entire gene.

Table 1. *Azospirillum* strains used in this work and relative EMBLGenBank accesion number for 16S rDNA sequences.

Species	Strain	EMBLGenBank accession number	Obtained from
A. brasilense	Cdr	X79732	This laboratory
	Sp7	X79739	DSM1690
	SpF94	X79740	Fani et al. (1988)
	DSM1859	X79733	DSM
	DSM2298	X79734	DSM
	DSM2287		DSM
Azospirillum spp	129/1	X79726	C. Scotti
	129/2	X79727	C. Scotti
	129/3		C. Scotti
	137/5	X79728	C. Scotti
	137/6		C. Scotti
A. lipoferum	ATCC29707		ATCC
	ATCC29708	X79729	ATCC
	ATCC29731	X79730	ATCC
	F	X79736	Washutza et al.(1992)
	W03	X79741	Washutza et al.(1992)
	PA1	X79738	L. Giovannetti
A. amazonense	Y1	X79735	DSM2787
	Y2	X79742	A. Hartman
A. irakense	103312	X79737	A. Hartman
A. halopraeferens	Au4	X79731	A. Hartman

Analysis of sequence data

Sequences were checked for secondary structural constraints, manually aligned with groups of prealigned sequences (Neefs et al. 1993) and analysed using the neighbour joining method according to the models of Jukes and Cantor and Kimura with TREECON 2.2 (see references in Van de Peer and De Wachter, 1993). Parsimony analysis was effected using the DNAPARS option in PHYLIP 3.5 (Felsenstein, 1989) after selecting the informative sites with the SELECT option in TREECON. The robustness of the inferred tree was evaluated with the SEQBOOT option in PHYLIP 3.5 by applying 500 bootstrap resampling of the data. Data sets, excluding those stretches that could not be reliably aligned, were also analysed in order to assess the influence of the 16S rDNA variable regions on the inferred trees.

Nucleotide sequence accession number.

The 16S rDNA nucleotide sequence of the seventeen *Azospirillum* strains were submitted to the EMBLGenBank and they have been assigned the accession numbers reported in Table 1.

Generation of RAPD patterns and cycle sequencing of RAPD markers

The overall strategy utilized for obtaining and sequencing of RAPD marker was carried out as follows:

1. *Generation of RAPD patterns with different primers*. Amplification reactions were performed in 25 μl final volume containing 1 ng of genomic DNA as described (Fani et al. 1993b). The reaction mixtures, after an incubation at 90°C for 1 minute and at 95°C for 1' 30", were cycled 45 times through the following temperature profile: 95°C for 30", Ta for 1' and 75°C for 2'. Ta was 36°C or 45°C for 10-nt or 17-nt primers, respectively.

2. *Reamplification of RAPD band(s)*. The bands were excised from the agarose gel and reamplified as described (Williams et al. 1993).

3. *Digestion of the RAPD band with different restriction endonucleases* in order to find an enzyme that recognises a single site within the fragment. In this way two fragments

were generated, each containing one single annealing site for the primer. The restriction reactions were performed by digesting 2-μl aliquots of the reamplified RAPD marker with 3 units of restriction endonuclease in a 25-μl reaction volume.

4. *Purification of the products of restriction digestion from low melting temperature agarose gel.* The products of digestion were loaded on 1.2 % low melting temperature agarose gel and electrophoresed at 10 V per cm for 2 hours. After a short analysis under a long-wave UV light, two small agarose slices, containing the digestion products of one of the enzymes that cut the marker only once, were excised from the gel. The agarose was placed in a 1.5 ml tube, frozen at -20°C for 5 minutes and centrifuged 10' at 12000 rpm. 8 μl of the supernatant, that usually contains at least 30 ng of DNA, was used as such in the sequencing reactions.

5. *Sequencing via PCR of the RAPD marker* using the same primer employed for the generation of RAPD pattern. The nucleotide sequencing of the fragments was carried out *via* PCR using the same primer used for the generation of the RAPD pattern. The reactions were performed for 25 cycles according to the instruction manual of the CircumVent^{TM} DNA sequencing kit with the Vent$_R$^{TM} (exo$^-$) DNA polymerase (New England Biolabs Inc. Usa). The annealing temperature (Ta) was 50°C for 17-mer or longer primers, whereas a 25°C Ta was used 10-mer primers.

6. *Design of specific primer(s) on the sequence obtained.* The sequence obtained was then to design a new specific oligonucleotide primer for direct sequencing of the remaining part of the marker, carried out on the undigested DNA fragment.

7. *Nucleotide sequence via PCR of the undigested RAPD band using the new primer.*

Results and Discussion

Phylogeny of the genus Azospirillum *based on 16S rDNA sequence*

Universal eubacterial primers were used to amplify almost the whole 16S rDNA by means of PCR. 16S rDNA amplifications were obtained from seventeen *Azospirillum* strains, fourteen of which were previously assigned to the species *A.*

amazonense, A. brasilense, A. halopraeferens, A. lipoferum or *A. irakense*. In addition, three Italian isolates were tested, two of which (129/1 and 129/2) were tentatively assigned to the species *A. brasilense*, and one (137/5) to the species *A. lipoferum*. Nearly all the 16S rDNA (about 95%) of the seventeen *Azospirillum* strains was sequenced.

Phylogenetic analysis (Van de Peer and De Wachter, 1993) unambiguously placed the seventeen strains into a coherent cluster of the Proteobacteria in total agreement with previous data (De Smedt et al. 1980; Woese et al. 1984) (Figure 1). Within the genus all of the species were recovered on separate lines of descent, indicating that they represent distinct taxa. Both the node defining the genus and those defining the different species were supported by high bootstrap values. Moreover, all tested procedures for tree construction resulted in dendrograms showing similar topologies.

Among the Proteobacterial rDNA sequences so far available, the nearest relatives of the azospirilla belong to magnetospirilla (alias aquaspirilla) and to phototrophic rhodospirilla. The relationship between *Azospirillum* and the other *alpha* Proteobacteria was in general agreement with the results of Woese et al. (1982), based on DNA:rRNA hybridization, in which *Azospirillum* was near to *R. rubrum* and to an *Aquaspirillum* species.

Concerning the intrageneric relationships of the *Azospirillum* strains, it is clear that all the strains previously assigned to the *A. brasilense* species were collected in a rather homogeneous cluster, where the two unidentified strains 129/1 and 129/2 were also placed, according to physiological traits (C. Scotti, personal communication). The species *A. brasilense* appears to be closely related to the species *A. lipoferum*, in agreement with other taxonomic approaches (reviewed in Gillis and Reinhold-Hucker, 1994).

The species *A. lipoferum* appears to be more heterogeneous than the species *A. brasilense* as suggested by the length of the branches in Figure 1; this result is in agreement with data obtained by Tarrand et al. (1978) based on DNA homology studies, with data obtained by restriction endonuclease analysis of total DNA

(Giovannetti et al. 1990) and random amplified polymorphic DNA analysis (Fani et al. 1993c) and with physiological traits which showed a great heterogenity of strains belonging to the species *A. lipoferum*. The unassigned strain 137/5 also joined this cluster.

Although the *A. irakense* species shows some phenotypic peculiarities (pectinolytic activity and a low G + C content), *A. irakense* strain 103312 was

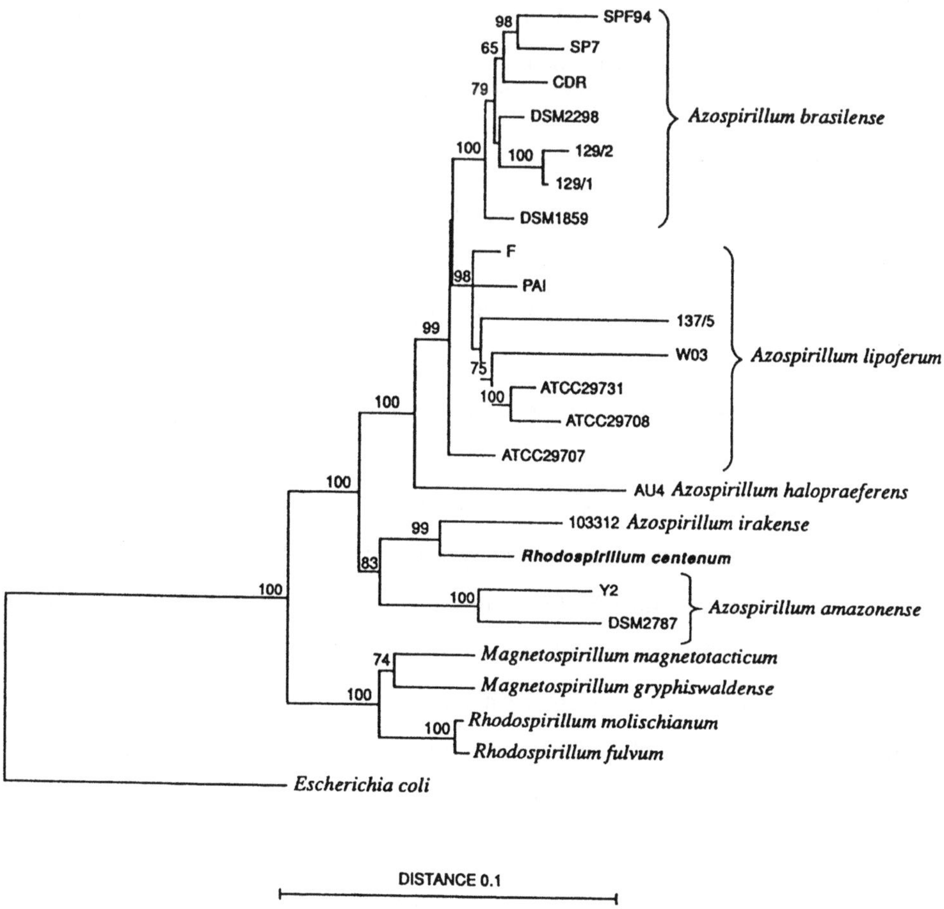

Figure 1.16S rDNA-based tree showing the relationships between the seventeen analysed strains of *Azospirillum* spp. and some representatives of closely related genera. Numbers at the nodes are the bootstrap confidence values obtained after 500 replicates. The scale bar indicates the distance in substitutions per nucleotide (Jukes & Cantor correction). The reported tree is based on an alignment which included the variable regions. Outgroup: *E. coli*.

unambiguously placed among the other *Azospirillum* strains. Moreover this species appears to be closer to *A. amazonense* and *A. halopraeferens* than to *A. brasilense* and *A. lipoferum*.

A. halopraeferens appears phylogenetically closer to *A. brasilense* and *A. lipoferum* than to the other two species *A. amazonense* and *A. irakense* that constituted a second cluster where the nitrogen-fixing, purple photosynthetic bacterium *Rhodospirillum centenum* (Favinger et al. 1989) is included. Although *R. centenum* looks like azospirilla in certain phenotypic characters (such as nitrogen fixation, the spiral shape and the intracellular production of poly-β-hydroxybutyrate), it also shows some differences (photosynthesis, growth at high temperatures, and the production of refractile intracellular structures, named "R" bodies). The placement of this microorganism in the *Azospirillum* cluster gives rise to a new question on whether *R. centenum* has to be reclassified within the *Azospirillum* genus or if rhodospirilla do not constitute a real taxon but have a polyphyletic origin.

Use of histidine biosynthetic genes to study phylogenetic relationships among Azospirillum *strain*

In *A. brasilense* a *his* operon (contained in a 7.7 Kb *Pst*I-*Pst*I fragment) consisting of five genes encoding proteins whose activity is known (*his Bd H A F E*) and two ORF, ORF168 and ORF122 with unknown function, has been extensively studied (Fani et al. 1993a). Our attention was particularly focused on the four genes *hisBdHAF* since in all of those microorganisms where the *his* genes are clustered they are always contiguous and share the same relative order (Fani et al. 1995). It has been postulated that it could be due to the fact that they encode enzymes involved in the central consequential steps of the histidine biosynthetic pathway, which cross-connect it to the nitrogen metabolism and to the *de novo* synthesis of purines and that they represent the "core" of the pathway which evolved in the early stages of molecular evolution and which has evolved as a unique block (Fani et al. 1995). For this reason these genes could be used for establishing phylogenetic relationships among bacterial species.

The final aim of this work is to determine the nucleotide sequence of the *hisBdHAF* genes from different *Azospirillum* strains and to compare the phylogenetic trees constructed by using them with that produced by 16S rDNA.

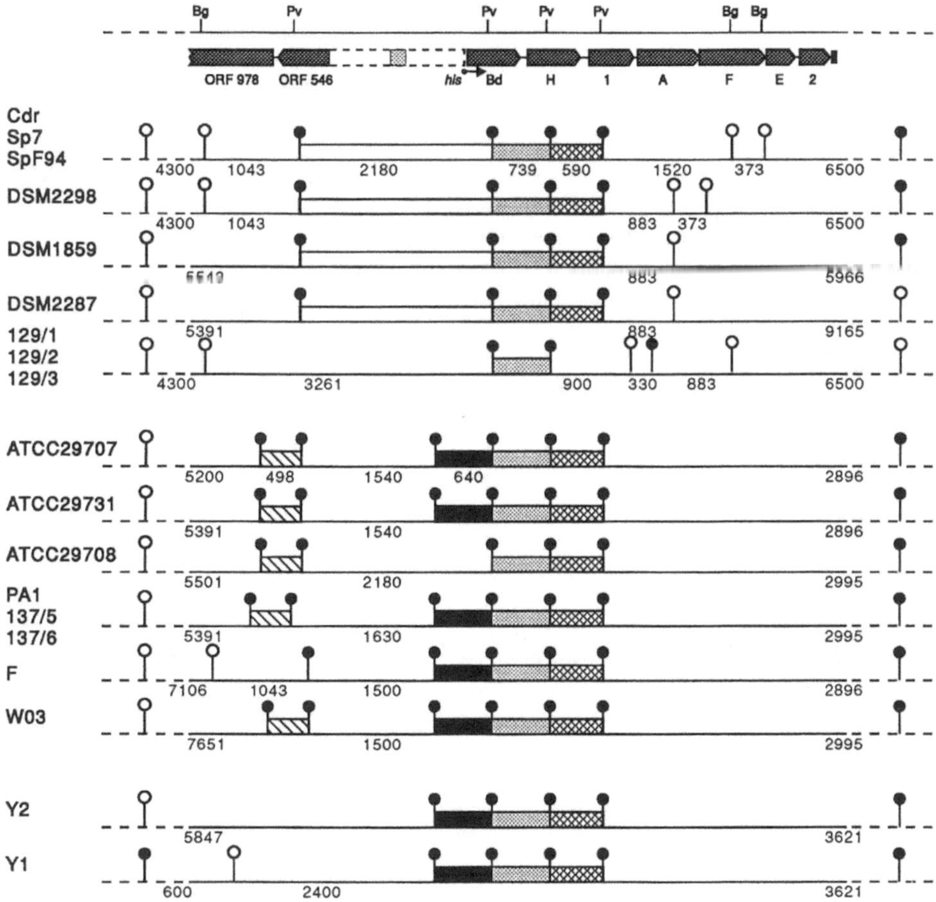

Figure 2. Restriction map of the *his* operon and the upstream region of nineteen *Azospirillum* strains. Symbols: Bg (⚲), Pv (⚲): recognition sites for the restriction endonucleases *Bgl*II, *Pvu*II; Bd, H, 1, A, F, E, 2: genes belonging to the histidine cluster *hisBd*, *hisH*, ORF1, *hisA*, *hisF*, *hisE*, ORF2; black box: transcription terminator; shaded box indicates a putative iteron; arrow with closed circle indicates the promoter and direction of the trascription of *his* genes. ORF978 and ORF546 represents two putative ORF homologous to the *Escherichia coli hslUV* genes. Numbers indicate the length of restriction fragments (bp).

For this purpose the 7.7 Kb *A. brasilense* DNA region (reported in the upper part of figure 2) and parts thereof have been used as probe in hybridization experiments with the DNA extracted from nineteen *Azospirillum* strains and digested with the restriction endonucleases *Pvu*II and/or *Bgl*II.

						***Pvu* II**			
A. brasilense	GAC	CAC	ATG	CTG	GAG	**CAG**	**CTG**	TCG	
E. coli	* * T	* * T	* * *	* * *	* * T	* * *	A* C	G* T	
S. typhimurium	* * *	* * T	* * *	* * C	* * T	* * A	A* C	G* C	***hisBd***
S. coelicolor	T* *	* * *	* * *	* * C	* * C	* * *	* * C	GGC	
L. lactis	* * *	* * *	* * *	* * *	ACA	* TT	* * C	A* C	
S. cerevisiae	* * *	* * T	* * *	A* A	C* T	GCT	* * *	G* C	

A. brasilense	D	H	M	L	E	Q	L	S	
E. coli	*	*	*	*	D	*	I	A	
S. typhimurium	*	*	*	*	D	*	I	A	**HisBd**
S. coelicolor	*	*	*	*	D	*	*	G	
L. lactis	*	*	*	*	T	L	*	T	
S. cerevisiae	*	*	*	I	H	A	*	A	

			***Pvu* II**						
A. brasilense	GGC	ATG	**CAG**	**CTG**	ATC	GGC	AGG	GCA	
E. coli	* *G	* * *	* * A	* * *	C *G	* * G	C * *	CGC	
S. typhimurium	* * G	* * *	* * A	* * *	C *G	* * T	C * T	CGC	***hisH***
S. coelicolor	* * *	* * *	* * *	A* C	C * *	TT *	TC *	CGC	
L. lactis	* * G	* * *	* * A	G* *	C *T	TTT	GAA	AA*	
S. cerevisiae	* * G	C *A	* * A	GC *	C * *	TTT	GCC	* GT	

A. brasilense	G	M	Q	L	M	A	E	G	
E. coli	*	*	*	*	L	G	R	R	
S. typhimurium	*	*	*	*	L	G	R	R	**HisH**
S. coelicolor	*	*	*	I	L	F	S	R	
L. lactis	*	*	*	V	*	*	*	K	
S. cerevisiae	*	L	*	A	L	F	A	*	

Figure 3. Comparison of nucleotide and predicted aminoacid sequence of two DNA regions of *hisBd* (upper) and *hisH* (lower) genes of *A. brasilense, E. coli, S. typhimurium, L. lactis, S. coelicolor* and *S. cerevisiae*. Nucleotides or aminoacids identical to those of the *A. brasilense* sequences are indicated by asterisks. The various sequences were retrieved from the EMBLGenBank.

Results obtained allowed to construct a restriction map of the *his* region for nineteen strains on the basis of the comparison of hybridization patterns obtained with different *his* probes (Figure 2).

Three restriction sites are strongly conserved among the different strains; they delimit two DNA fragments: the former, of 590 bp, common to all strains but 129/1, 129/2 and 129/3 corresponds to a *Pvu*II fragment comprising the 3' region of the *hisH* gene and the 5' region of ORF1. The latter, of 739 bp, present in all strains, corresponds to a *Pvu*II fragment including the 3' portion of *hisBd* and the 5' region of *hisH* (Figure2). The strong conservation of two of these sites is expected since they lie in regions of the *hisBd* and *hisH* genes coding for highly conserved aminoacids (Fani et al. 1989; Fani et al. 1993a) as shown in Figure 3.

Therefore the *hisBd* and *hisH* genes (and probably ORF1) appear to be organized in the same order in the nineteen *Azospirillum* strains tested. This is in agreement with the fact that in all those microorganisms where the *his* genes are clustered they are always contiguous and share the same relative order (Fani et al. 1995).

The analysis of the similarities of restriction maps supported the relationships among the *Azospirillum* strains that were postulated from rDNA sequences.

Use of RAPD monomorphic markers to infer phylogenetic relationships among bacterial strains

If the RAPD monomorphic bands are to be used in reconstructing phylogeny one must be sure that differences among homologous sequences actually reflect differences in the original genome and that they are not to be ascribed to error(s) produced by *Taq* Polymerase during the RAPD reaction. An useful approach is the direct sequencing of the PCR product(s), since only a small amount of misincorporated nucleotides are present at each position in a pool of the amplified molecules. Neverthless the use of direct sequencing of RAPD markers *via* PCR requires the solving of two questions. The former concerns the structure of the RAPD markers. In fact because in the RAPD the amplification of DNA sequences is primed by the same oligonucleotide at both ends, each amplified fragment contains the same tails, the primer sequences. Therefore it is not possible to sequence the RAPD bands directly *via* PCR using the same oligonucleotides employed in the generation of RAPD pattern. The latter regards the length of the primer employed in RAPD patterns. Indeed nucleotide sequence is tipically achieved using 17-nt primers, but the oligonucleotides employed in RAPD technique are 10-nt long. For this purpose we have developed a method (whose details are reported in the section Materials and Methods) to obtain (*via* PCR) the nucleotide sequence of RAPD markers from microorganisms, obtained with primers of different length (ranging from 10 to 24 nucleotides).

The working hypothesis described in the Introduction was tested by determining the nucleotide sequence of a monomorphic band isolated from the RAPD patterns of two *A. brasilense* strains, Cdr and 129/1, obtained with primer 1253 (5'GTTTCCGCCC3') (Figure 4). In the region analyzed the two sequences show more than 99% of sequence homology out of 450 nucleotides.

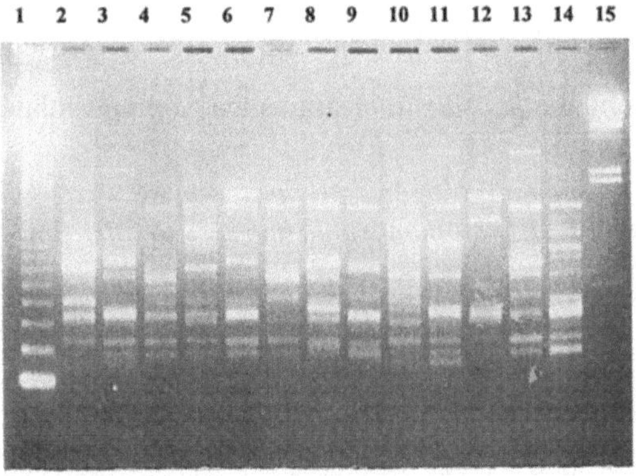

Figure 4. RAPD patterns of genomic DNA from thirteen different *Azospirillum* strains. Lanes 1 and 15; molecular weight markers. Lanes: 2-9) *A. lipoferum* F, W03, PA1, ATCC29708, ATCC29731, ATCC29707, 137/5; lane 10) *A. amazonense* Y2; lanes 11-14) *A. brasilense* SP7, SPF94, Cdr and 129/1. The arrow indicates the bands chosen for sequencing.

This result clearly indicates that monomorphic bands have related sequence structure as previously suggested on the basis of hybridization experiments (Kersulyte et al. 1992; Fani et al. 1993b; Williams et al. 1993) and that the same could be selected in order to study microevolution and phylogeny among strains belonging to the same genus and/or to the same species.

As already pointed out, none of the molecular techniques described so far, alone can support reliable phylogenetic inferences. For this reason we have chosen to test, on the genus *Azospirillum* a multiple approach based on highly conserved rDNA genes, on conserved genes with known functions, as the *his* biosynthetic genes, and on completely unknown sequences generated by RAPDs.

We hope, exploiting these analysis, to produce a really complete picture of the evolution of genomes at species and genus level.

References

Bandi C, Damiani G, Magrassi L, Grigolo A, Fani R, and Sacchi L (1994) *Flavobacteria* as intracellular symbiont in cockroaches. Proc R Soc Lond B 257: 43-48.

Choma A, Russa R, Mayer H and Lorklewicz Z (1987) Chemical analysis of *Azospirillum* lipopolysaccharides. Arch Microbiol 146: 341-345.

De-Polli H, Bohlool BB and Dobereiner J (1980) Serological differentiation of *Azospirillum* species belonging to different host plant specificity groups. Arch Microbiol 126: 217-222.

De Smedt J, Bauwens M, Tytgat R and De Ley J (1980) Intra- and intergeneric similarities of ribosomal ribonucleic acid cistrons of free-livig, nitrogen-fixing bacteria. Int J Syst Bact 30: 106-122.

Dobereiner J and Day JM (1976) Associative symbiosis in tropical grasses: characterization of microorganisms and dinitrogen fixing sites, *in* "Proceedings of the 1st International Symposium on Nitrogen Fixation" (W.E. Newton & J.C. Nyman) (pp 518-538). Pullman, Washington State University Press.

Fani R, Bazzicalupo M, Ricci F, Schipani C and Polsinelli M (1988) A plasmid vector for the selection and study of transcription promoters in *Azospirillum brasilense*. FEMS Microbiol Lett 50: 271-276.

Fani R, Bazzicalupo M, Damiani G, Bianchi A, Schipani C, Sgaramella V and Polsinelli M (1989) Cloning of the histidine genes of *Azospirillum brasilense*: organization of the *ABFH* gene cluster and nucleotide sequence of the *hisB* gene. Mol gen Genet 216: 224-229.

Fani R, Alifano P, Allotta G, Bazzicalupo M, Carlomagno MS, Gallori E, Rivellini F and Polsinelli M (1993a) The histidine operon of *Azospirillum brasilense*: organization, nucleotide sequence and functional analysis. Res Microbiol 144: 187-200.

Fani R, Damiani G, Di Serio C, Gallori E, Grifoni A and Bazzicalupo M (1993b) Use of random amplified polymorphic DNA (RAPD) for generating specific DNA probes for microorganisms. Molecular Ecology 2: 243-250.

Fani R, Bandi C, Bardin MG, Comincini S, Damiani G, Grifoni A and Bazzicalupo M (1993c) RAPD fingerprinting is useful for identification of *Azospirillum* strains. Microbial Releases 1: 217-221.

Fani R, Liò P and Lazcano A (1995) Evolution of the histidine biosynthetic pathway: a preliminary analysis. J Mol Evol (submitted).

Favinger J, Stadtwald R and Gest H (1989) *Rhodospirillum centenum* sp. nov., a thermotolerant cyst-forming anoxygenic photosyntetic bacterium. Antonie van Leewenhoek 55: 291-296.

Felsenstein J (1989) PHYLIP-phylogeny inference package (version 3.2). Cladistic 5: 164-166.

Gillis M and Reinhold-Hurek B (1994) Taxonomy in *Azospirillum*, in "*Azospirillum*-Plant Associations" (Y. Okon) (pp. 1-14) C.R.C. press, Boca Raton.

Giovannetti L, Ventura S, Bazzicalupo M, Fani R and Materassi R (1990) DNA restriction fingerprint analysis of the soil bacterium *Azospirillum*. J Gen Microbiol 136: 1161-1166.

Hadrys H, Balick M and Schierwater B (1992) Applications of random amplified polymorphic DNA (RAPD) in molecular ecology. Molecular Ecology 1: 55-63.

Kersulyte D, Woods JP, Keath EJ, Goldman WE and Berg DE (1992) Diversity among clinical isolates of *Histoplasma capsulatum* detected by polymerase chain reaction with arbitrary primers. J Bact 174: 7075-7079.

Lazcano A, Fox GE and Oro J (1992) Life before DNA: the origin and evoution of early Archean cells. In RP Mortlock (ed) *The Evolution of Metabolic Function* (CRC Press, Boca Raton FL), pp. 237-339.

Loomis WF and Smith DW (1990) Molecular phylogeny of *Dictiostelium discoideum* by protein sequence comparison. Proc Natl Acad Sci USA 87: 9093-9097.

Neefs J-M, Van de Peer Y, De Rijk P, Chapelle S and De Wachter R (1993) Compilation of small ribosomal subunit RNA structures. Nucl Acids Res 21: 3025-3049.

Tarrand JJ, Krieg NR and Dobereiner J (1978) A taxonomic study of the *Spirillum lipoferum* group, with descriptions of a new genus, *Azospirillum* gen. nov. and two species, *Azospirillum lipoferum* (Beijerink) comb. nov. and *Azospirillum brasilense* sp.nov. Can J Microbiol 24: 967-980.

Van der Peer Y and De Wachter R (1993) TREECON: a software package for the construction and drawing of evolutionary trees. Comput Appl Biosci 9: 177-182.

Washutza S, Hofmann N, Niemann EG and Fendrik I (1992) Investigations on root exudates of Korean rice. Symbiosis 13: 181-189.

Welsh J and McClelland M (1990) Fingerprinting genomes using PCR with arbitrary primers. Nucl Acids Res 18: 7213-7218.

Williams JGK, Kubelik AR, Livak KJ, Rafalski JA and Tingey SV (1990) DNA polymorphisms amplified by arbitrary primers are useful as genetic markers. Nucl Acids Res 18: 6531-6535.

Williams JGK, Rafalski JA and Tingey SV (1993) Genetic analysis using RAPD markers. Methods in Enzymology 218: 704-740.

Woese CR, Blanz P, Hespell RB and Hahan CM (1982) Phylogenetic relationships among various helical bacteria. Curr Microbiol 7: 119-124.

Woese CR, Stackebrandt E, Weisburg WG, Paster BJ, Madigan MT, Fowler VJ, Hann CM, Blanz P, Gupta R, Nealson KH and Fox GE (1984) The phylogeny of purple bacteria: the alpha subdivision. Syst Appl Microbiol 5: 315.

Woese CR (1987) Bacterial evolution. Microbiol Rev 51: 221-271.

Woese CR (1991) The use of ribosomal RNA in reconstructing Evolutionary Relationships among bacteria. In "Evolution at Molecular Level", Selander RK, Clark AG, Whittam TS (eds), pp 1-24.

Genetic control of nitrogen assimilation and nitrogen fixation in free living *Azospirillum brasilense* : A review

Miklos de Zamaroczy

Unité de Physiologie Cellulaire, CNRS URA 1300
Département des Biotechnologies
Institut Pasteur
25, rue du Dr. Roux 75724 Paris Cedex 15
France

Abstract

Among plant rhizospheric bacteria, genetics and molecular biology of the regulatory pathways, that control the nitrogen metabolism, have been studied, over the past 10 years, mostly in *Azospirillum brasilense*. To date, more and more changes, as compared to the canonical reference system of *Klebsiella pneumoniae* have been discovered in *A. brasilense*, with regard to : (i) the qualitative modifications of physiological functions of some key regulators; (ii) the identification of novel types of promoter and new regulatory proteins. I will try here to give an overview of our current knowledge on this subject, with a particular reference to recent work.

Introduction

The regulation of nitrogen fixation and nitrogen metabolism has been the subject of investigations for several years in *Azospirillum brasilense*, a bacterium that fixes nitrogen in microaerobiosis and associates with the root of grasses (see Elmerich et al. 1992). *Azospirillum* bacteria are Gram negative, aerobic chemoorganotrophs and belong to the alpha subgroup of the proteobacteria.

The nitrogen fixation (*nif*) genes are largely conserved among diazotrophs. As a general rule, *nif* gene expression in proteobacteria requires σ^{54} factor, encoded by *rpoN*, and the transcriptional activator NifA, encoded by *nifA* (see Merrick, 1992). However, a variety of regulatory pathways for the overall regulation of nitrogen metabolism have evolved in nitrogen fixing bacteria.

In the model system of *Klebsiella pneumoniae*, *ntrBC* (encoding a two-component regulatory system) and *glnA*, that codes for glutamine synthetase (GS) map close together on the chromosome. The signal transduction protein P_{II} encoded by *glnB*, which maps elsewhere on the chromosome. This protein plays a central role by co-ordinating the response of the Ntr system to the intracellular nitrogen status (Son and Rhee, 1987; Holtel and Merrick, 1988; Vasudevan et al. 1991). Hexagonal crystals of the unuridylylated form of *E. coli* P_{II} have

NATO ASI Series, Vol. G 37
Azospirillum VI and Related Microorganisms
Edited by I. Fendrik et al.

been recently analysed, strongly suggesting that the native protein is a homotrimer (Vasudevan et al. 1994).

Under conditions of nitrogen limitation, P_{II} is uridylylated on a tyrosine residue (Adler et al. 1975). NtrB, a histidine kinase, phosphorylates NtrC (see Magasanik, 1988; Stock et al. 1989). NtrC-P, in turn, activates transcription of the *glnA-ntrBC* cluster. This results in the synthesis of large amounts of GS in response to a decrease in glutamine availability. NtrC-P also activates the transcription of the *nifLA* operon and NifA, in turn, activates the other *nif* operons. P_{II} is not absolutely required for nitrogen fixation, since a *glnB* null mutant of *K. pneumoniae* displays a Nif^{+} phenotype (Holtel and Merrick, 1989). Under conditions of nitrogen excess, P_{II} is not uridylylated and NtrC is not phosphorylated, preventing activation of transcription of both the *glnA-ntrBC* cluster and the *nifLA* genes. Indeed, *ntrC* null mutant strains of *K. pneumoniae* are impaired in nitrogen fixation and glutamine synthesis. They are also impaired in other phenotypic properties under the *ntr* control, such as the utilization of nitrate and several amino acids as nitrogen sources (see Elmerich, 1991).

This review article describes the current knowledge of the regulatory pathways governing nitrogen assimilation and nitrogen fixation in *A. brasilense*. It appears that the structural and functional organization of several nitrogen-regulated genes in *A. brasilense*, including *glnA, glnB, ntrBC* and *nifA* differs greatly from the *Klebsiella* model and that *A. brasilense* possesses features which are unique among diazotrophs.

Nitrogen assimilation in *A. brasilense*

Key enzymes for glutamine and glutamate synthesis

A. brasilense possesses a GS whose activity is modulated by reversible adenylylation as in enterobacteria in response to the cellular N-status. The GS holoenzyme is composed of twelve identical subunits of 52 kDa (Colombo Pirola et al. 1992). The adenylylation level of GS depends on the fixed-nitrogen concentration in the cells. GS is maintained in its non-adenylylated form under nitrogen limiting conditions, whereas the adenylylation level increases in cells growing under conditions of ammonia excess (Gauthier and Elmerich, 1977; Westby et al. 1987). It should be noted, that the adenylylation cascade of GS in *E. coli* is regulated by modulation of the activities of UTase and ATase, rather than by changes in the expression of their genes (van Heeswijk et al. 1993). Homologous enzymes have not yet been identified in *A. brasilense*.

The second key enzyme, glutamate synthase (GOGAT or GltS), has also been characterized in *A. brasilense*. It is composed of two different subunits of 135 kDa and 50 kDa, respectively and possesses a *E. coli* type, $\alpha 4$-$\beta 4$ quaternary structural organization (Ratti et al. 1985). The tandemly organized *gltD-gltB* structural genes encoding GltS, were characterized and the deduced amino acid sequences are similar to those of the homologous ironsulfur flavoprotein complex of *E. coli* (Pelanda et al. 1993).

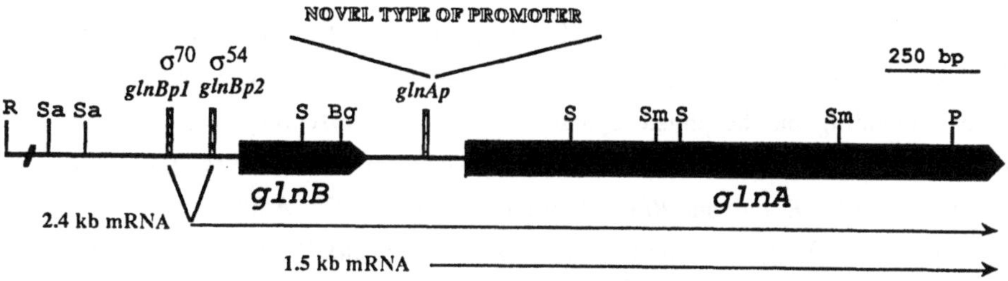

glnBp1 : TTGGCACG-N4-GTGCTT(-9)

glnBp2 : TGCGC-N24-ATATT(-8)

glnAp : CTTTTTGTGCTGTCACGTTGTGTGCTTGGGAAAGGCATGACATAA(-1)

FIGURE 1. Organization of *glnB-glnA* gene cluster in *A. brasilense*. The location of the three different N-regulated promoters is indicated upstream of the *glnB* and *glnA* coding sequences. Restriction sites : Bg: *Bgl*II; R: *Eco*RI; P: *Pst*I; S: *Sal*I; Sm: *Sma*I. Data from de Zamaroczy et al. 1990; 1993/A.

Structural organization of the *glnB-glnA* cluster

Structural genes for glutamine synthetase (*glnA*), and P_{II} protein (*glnB*) have been characterized (Bozouklian and Elmerich, 1986; Bozouklian et al. 1986; de Zamaroczy et al. 1990). The *glnB* gene is located upstream of *glnA*. This clustered structural organization (Figure 1) resembles that found in some *Rhizobiaceae* and in *Rhodobacter capsulatus* (Colonna-Romano et al. 1987; Martin et al. 1989; Kranz et al. 1990), but is completely different from the well-documented *glnA-ntrBC* gene cluster, in enteric bacteria (see Magasanik, 1988; Stock et al. 1989).

Construction and phenotypic properties of *glnB* null mutant strains

A *glnB* null mutant strain was constructed by homologous recombination of a kanamycin cartridge, at the unique 3'-end proximal *Bgl*II site of the coding sequence, with the same transcriptional orientation as *glnB* (de Zamaroczy et al. 1993/A). It was verified by Southern hybridization that the kanamycin cartridge had recombined at the expected location. A second *glnB::kan* mutant strain was also constructed (data not shown), in which the kanamycin cartridge was inserted at the central *Sal*I site of the *glnB* coding sequence (Figure 1). Both *glnB* mutant strains were glutamine prototrophs and displayed a Nif⁻ phenotype (see below).

Characterization of P_{II}, the *Azospirillum glnB* gene product

Total proteins of the wild type strain grown under conditions of ammonia excess or nitrogen limitation were labelled *in vivo*, using uracil-(^{3}H). A single band corresponding to a polypeptide of 13 kDa (in agreement with the predicted molecular weight of *glnB* polypeptide) was detected by SDS-PAGE, under nitrogen limitation (data not shown). No band was detectable in the presence of ammonia excess. Thus, P_{II}, whose regulatory roles differ depending on the physiological conditions, is post-translationally modified by uridylylation in response to N-limitation. Uridylylation of *A. brasilense* P_{II} is comparable to that described for *E. coli* and *Rhizobium leguminosarum* P_{II} (Adler et al. 1975; Colonna-Romano et al. 1993), but differs from the *Synechococcus* P_{II}, where P_{II} is modified by phosphorylation of a serine residue, depending on both the N-status and the spectral light quality (Forchhammer and Tandeau de Marsac, 1994).

Surprisingly, the two *glnB* null mutant strains of *A. brasilense* labelled with uracil-(^{3}H) also displayed a band corresponding to a 13 kDa polypeptide. Western blot analysis, using a specific anti *A. brasilense* P_{II}-antiserum, led to the identification of P_{II} polypeptide with the same migration in SDS-PAGE as the tritiated polypeptide, both in the wild type and *glnB*-mutant strains (de Zamaroczy et al. unpublished data). These results strongly suggest that the 13 kDa polypeptide band is a doublet and consequently a second P_{II}-like protein may be produced by *A. brasilense.*

Modulation of the GS activity by ammonia does not require P_{II}

The GS activity was assayed by monitoring the production of γ-glutamyl hydroxamate (γGT) (Bender et al. 1977), in both wild type and *glnB* null mutant strains. The proportion of non-adenylylated GS was similar in both stains under all physiological conditions tested (de Zamaroczy et al. unpublished data). This indicates that the reversible adenylylation / deadenylylation process is not regulated by the *glnB* gene product in *A. brasilense.* In the other words, P_{II} seems not to be required for the nitrogen control of GS activity. An "alternative P_{II}" (possibly the P_{II}-like protein mentioned above) may play a regulatory role in the modulation of GS activity.

Regulation of P_{II} and GS synthesis

Two different nitrogen-regulated promoters are required for the *glnB*- *glnA* transcription

Northern blot analysis showed that *glnA* can be co-transcribed with *glnB* as a 2.4 kb mRNA or transcribed from its own promoter as a 1.5 kb mRNA, depending on the N-status of the cells. Transcription start site analysis revealed the presence of two tandemly organized promoters upstream of *glnB* (Figure 1). The *glnB-A* mRNA was transcribed selectively from either one : Promoter *glnBp1*, resembling *E. coli* σ^{70}-dependent promoters, is used in ammonia containing media, while *glnBp2*, resembling σ^{54}-dependent promoters, is used essentially under conditions of nitrogen limitation (de Zamaroczy et al. 1990). This transcriptional regulation of *glnB-A* in *A. brasilense* is different from that described in other species with a similar clustered organization of the two genes (see above).

Mapping of the *glnA* promoter region

The minimal part of the *glnA* promoter region essential for transcription of *glnA* alone was defined using plasmid-borne *glnA-lacZ* fusions carrying various 5'-deleted or mutated *glnAp* fragments. *glnAp* was located within a 45 bp fragment upstream of the mRNA start site (Figure 1) that contains no similarity with the consensus of any already described σ factor (de Zamaroczy et al. 1993/A). Under conditions of nitrogen fixation transcription from the *glnAp* was very low, whereas in media containing ammonia transcription proceeded mainly from this promoter. For these reasons the *glnA* promoter appears to be a novel type of N-regulated promoter, which has no equivalent in *E. coli.*

Involvement of P_{II} and GS in their own transcription

Analysis of the promoter activities of the *glnB-glnA* cluster was investigated under different physiological conditions with the aid of translational *gln-lacZ* fusions. Promoter-containing fragments from *glnB* and *glnA* were amplified by PCR and ligated in frame with the 8th codon of the *lacZ* gene, carried by the broad host range vector pGD926 (de Zamaroczy *et al.* 1993/A). Expression of the *glnB-* and *glnA-lacZ* fusions was monitored in wild type and mutant strains, including *glnB* and *glnA* mutants. The pattern of expression of the fusions in the wild type strain confirmed the opposite responses of the *glnB* and *glnA* promoters to ammonia availability, in agreement with the Northern blot analysis.
Other relevant results can be summarized as follows : (i) P_{II} and GS are required for full expression from *glnAp* under conditions of nitrogen excess; (ii) P_{II} is involved in the repression of its own synthesis under conditions of nitrogen excess; (iii) GS is required for nitrogen regulation of *glnBp1p2*. GS acts probably indirectly on the transcriptional regulation of *glnB-glnA* cluster, by modulating the intracellular nitrogen status (de Zamaroczy *et al.* 1993/A).
However, it should be stressed that a functional P_{II} is not essential for ammonia assimilation. The *glnB* null mutant strains do not require glutamine for growth in minimal medium containing ammonia as sole nitrogen source, a property in agreement with transcription of

glnA from its own promoter and with the absence of involvement of P_{II} in adenylylation / deadenylylation of the GS.

Control of nitrogen fixation

Structural organization of the *A. brasilense nif* and *fix* genes

The major *nif* cluster covers about 30 kb of chromosomal DNA region and contains the nitrogenase structural genes *nifHDK-Y* and other *nif* (*E*, *US*, *W*) and *fix* (*ABCX*) genes (Galimand et al. 1989; Passaglia et al. 1991; Milcamps et al. 1993). Northern blot analysis showed that the *nifHDK* genes are organized in a single transcription unit and suggested that nitrogenase expression is regulated at both the initiation and termination of transcription (de Zamaroczy et al. 1989). *nifA* and *nifB* are located elsewhere on the chromosome and are transcribed separately (Liang et al. 1991/A). The isolation of another *nifA*-like gene has been reported (Singh et al. 1989), but its identification has not been fully investigated.
Null mutant strains obtained by directed or random mutagenesis of all characterized *nif* and *fix* genes exhibit a Nif^- phenotype, although strains carrying mutations in the *fixABCX* region and in the *nifUS* region retain a residual nitrogenase activity (Galimand et al. 1989).

Regulation of the nitrogenase switch-off by ammonia

The phenomenon of "ammonia switch-off/on", a reversible post-translational modification of nitrogenase Fe-protein, has been described in *A. brasilense* and *A. lipoferum*, but has not been found in *A. amazonense* (Hartman et al. 1986; Hartman and Burris, 1987). In presence of ammonia the inactivation is catalyzed by the enzyme DRAT, an ADP-ribosyl transferase (Fu et al. 1989). An ADP-ribose group from NAD is covalently linked to an arginine residue at position 101, as in *Rhodospirillum rubrum* (Ludden and Roberts, 1989; Liang et al. 1991/B). Reactivation results from the activity of DRAG, a membrane-bound glycohydrolase, under conditions of nitrogen limitation.

The structural genes, *draT-draG* have been characterized in *A. lipoferum* and *A. brasilense* (Fu et al. 1990; Zhang et al. 1992). They belong to the 3' distal part of the major *nif* cluster (see above), and map next to *nifH*. The two genes are co-transcribed, but in an orientation opposite to *nifHDK*. Both genes appear to be constitutively expressed, suggesting selective post-translational activation / inactivation of DRAT and DRAG, in response to variations of the N-status.

Characterization and role of NtrB and NtrC in *Azospirillum*

The structural genes in *A. brasilense*, *ntrB-ntrC*, encoding the two component sensor-transactivator regulatory system NtrB / NtrC (Keener and Kustu, 1988; see Parkinson, 1993) have been characterized (Liang et al. 1993). They are probably co-transcribed with an unidentified ORF, located upstream of *ntrB* (Arsène et al. 1993), that exhibits similarity with a *nif* regulatory protein, NifR3, of *Rhodobacter capsulatus* (Foster-Hartnett et al. 1993) and with the ORF1 that precedes *ntrB-ntrC* in *R. leguminosarum* (Patriarca et al. 1993.) In *R. capsulatus nifR3-ntrB-ntrC* are organized in a single operon. In contrast, the possibility that *ntrC* can be transcribed from a promoter located within the *ntrB* coding sequence was shown in *A. brasilense* (Liang et al. 1993).
Azospirillum NtrC is a functional homologue of *Azorhizobium caulinodans* NtrC : It can efficiently complement the Nif$^-$ phenotype of an *Azorhizobium ntrC*::Tn*5* mutant strain. However, *Azospirillum ntrC* null mutant strains exhibit a Nif$^+$ phenotype, indicating that NtrB and NtrC are not required for the regulation of the nitrogen fixation process in this species (Liang et al. 1993), as also shown in another free living diazotroph, *Azotobacter vinelandii* (Toukdarian and Kennedy, 1986). In agreement with this conclusion, no sequence resembling an NtrC-binding site consensus (UAS) was found in the promoter region of *nifA* (Liang et al. 1991/A). However, NtrC has been shown to be involved in nitrate utilization as a nitrogen source in *A. brasilense*.

Common regulatory mechanisms for *nif* transcription

σ^{54}-type promoter consensus sequences (TGT-N_{10}-ACA) have been found upstream of *Azospirillum nif* genes, namely *nifH*, *nifB* and *nifW* (de Zamaroczy et al. 1989; Fani et al. 1989; Liang et al. 1992, Milcamps et al. 1993). No such promoter consensus is present upstream of *nifA*. In addition, the *nifA* promoter region does not contain sequence elements similar to any known promoter consensus (Liang et al. 1991/A). Characterization of *A. brasilense nif* operons showed that they were transcribed under conditions of nitrogen limitation, from σ^{54}-dependent promoters controlled by NifA, as generally observed among diazotrophs.
The *nifH* promoter is preceded by two elements homologous to the upstream activator sequences (UAS) required for the classical NifA-mediated activation of *nif* genes. Two mRNA start sites were identified upstream of *nifHDK* (de Zamaroczy et al. 1989). One is located at position -11, in agreement with the usually observed -12 position of *nif* promoter consensus.
A *rpoN*-like gene in *A. brasilense* encoding σ^{54} factor has been recently cloned (Vanderleyden personal communication). Moreover, it was shown that expression of an *A.*

brasilense nifH-gusA fusion required functional *rpoN* and *nifA* products in *R. meliloti* (Vande Broek et al. 1992).

The biosynthesis and the activity of the *A. brasilense* NifA activator are modulated by both the cellular N-status and the oxygen tension

The regulation of *nifA* expression in *A. brasilense* is unusual. The *nifA* gene is highly expressed under conditions of nitrogen and oxygen limitation, but a significant level of expression is also observed under conditions non-compatible with nitrogen fixation, i.e. in the presence of ammonia and oxygen (Liang et al. 1992; de Zamaroczy et al. 1993/B). Since in these conditions *nifH* transcription is abolished, NifA is expected to be in an inactive form under conditions incompatible with nitrogen fixation.

NifA contains the four conserved cysteine residues likely involved in oxygen control, as suggested in the case of *Bradyrhizobium japonicum* NifA (Fischer et al. 1988). This suggests a comparable post-translational regulation of NifA by oxygen in *Azospirillum.*

P_{II} is a new element required for the regulation of nitrogen fixation in *A. brasilense*

The *glnB::kan* mutant strains displayed a strict Nif$^-$ phenotype, and no nitrogenase activity was detectable under conditions of nitrogen and oxygen limitation (de Zamaroczy et al. 1993/A). The ß-galactosidase activity of a *nifA-lacZ* fusion was not affected in *glnB::kan* strains, but the expression of *nifH-* and *nifB-lacZ* fusions was abolished. This indicates that the P_{II} protein is absolutely required for nitrogen fixation, and that in the *glnB* mutant NifA was inactive under conditions of nitrogen fixation.

Moreover, the nitrogenase activity in a *glnB* null mutant strain was partially restored (about 35% of the wild-type level) by *K. pneumoniae nifA* but not by a constitutively expressed *Azospirillum nifA*. Since, P_{II} does not participate in *nifA* transcription, it is most probably involved in the activation of NifA, a mechanism that appears to be specific to *A. brasilense.*

Nitrogen assimilation and nitrogen fixation : A model for the regulation of *glnB-A* and *nif* genes

A tentative model of regulation is depicted in Figure 2. It can be summarized as follows. The central regulatory element in this model is the P_{II} protein whose activity is modulated by uridylylation in response to the N-status. P_{II} plays a role in regulation of the *glnBA* transcription and also in modulating NifA activity.

It has been established that the *glnBp2* promoter is of the σ^{54}-dependent type. This implies the existence of a specific activator such as NtrC or NifA. However, expression of the *glnB-*

lacZ fusion was not impaired either in the *ntrC*, the *ntrB*, nor in the *nifA* null mutant strains. Thus, there is presumably another transcriptional activator, referred as regulatory protein X. By switching between an active and inactive form, depending on the nitrogen status, X could promote *glnB* transcription under conditions of nitrogen limitation.

Under conditions of nitrogen excess, non-uridylylated P_{II} is involved in the transcriptional regulation of the *glnB-A* cluster, since disruption of the *glnB* gene led to a high expression of *glnB*. Thus, P_{II} could promote the inactivation of the X regulator. This allows the initiation of transcription from the canonical σ^{70}-type *glnBp1*.

The transcriptional activator for *nif* genes expression is NifA, which activates σ^{54}-dependent promoters. NifA is synthesized both under conditions of nitrogen limitation and excess. When P_{II} is under its non-uridylylated form, NifA is inactive and the *nif* structural genes are repressed. Uridylylated P_{II} is required for nitrogen fixation, through the modulation of NifA activity.

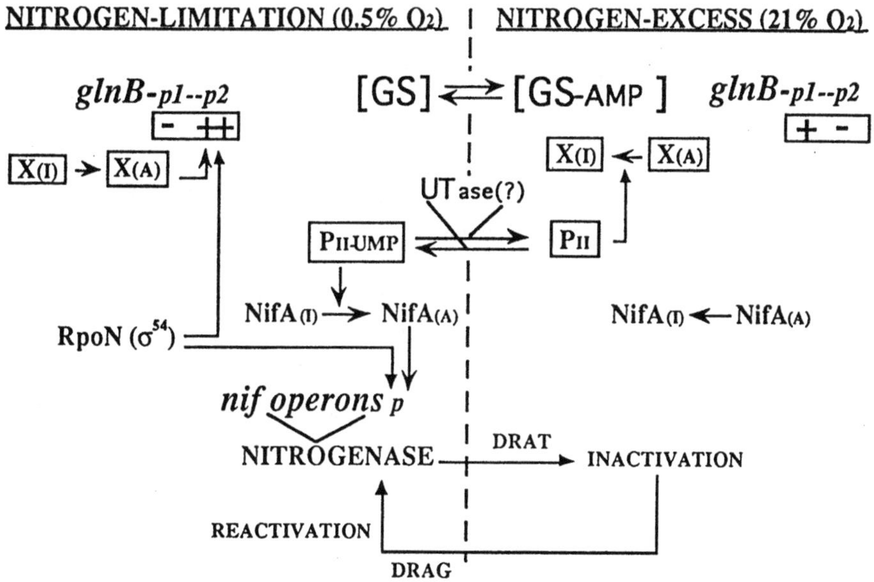

FIGURE 2. Model of regulation of nitrogen fixation and ammonia assimilation in *A. brasilense*. $X_{(A)}$, $NifA_{(A)}$ and $X_{(I)}$, $NifA_{(I)}$correspond to the inactive and active forms of an unidentified regulatory protein X and NifA, respectively. + and - indicate the relative efficiency of transcription from the *glnB* promoters.

The nitrogenase complex consists of dinitrogenase (MoFe protein), which contains the active site for dinitrogen reduction, and dinitrogenase reductase (Fe protein), which donates electrons to dinitrogenase. Biological nitrogen fixation is an energy-demanding process and is therefore subject to elaborate regulatory controls at the level of both :

(i) the expression of the *nif* genes in response to fixed nitrogen and oxygen;.and

(ii) the post-translational regulation of nitrogenase activity, which has also been termed short-term inhibition or ammonia switch-off/on of nitrogenase.

The *in vivo* regulation of nitrogenase activity by NH_4Cl or glutamine is correlated with reversible covalent modification of dinitrogenase reductase. Two enzymes are responsible for this regulation. DRAT transfers the ADP-ribose to the Arg^{101} residue of one subunit of the dinitrogenase reductase dimer, resulting in inactivation of that enzyme. Under conditions of nitrogen fixation, DRAG removes the ADP-ribose from the covalently modified enzyme, restoring its activity.

Concluding remarks

The regulation of ammonia assimilation in *A. brasilense* displays unusual features:

- (i) The transcription of *glnB-A* depends on three different nitrogen-regulated promoters. Expression from the *glnA* promoter is high under conditions of nitrogen excess;
- (ii) NtrBC are not involved in the regulation of *glnB-A* expression;
- (iii) P_{II} is involved in the repression of its own synthesis in response to N excess. The glutamine synthetase is also required for the transcriptional regulation of *glnB-A*. It acts probably indirectly by modulating the N status.

In *A. brasilense* the regulation of *nif* transcription shares many features with those described in the great majority of N_2-fixing organisms. In particular, the archetypal *nif* gene promoter structure with its specific requirements for an alternative RNA polymerase, containing an *rpoN*-encoded σ^{54} factor, and NifA as activator protein, constitute the central part of the common regulatory organization. However, N- regulation of both the expression and activity of NifA in *A. brasilense* differs considerably from that described in the other free living diazotrophs :

- (i) The *nifA* gene, which is not controlled by a σ^{54}-dependent promoter, is transcribed both under conditions compatible and non-compatible with nitrogen fixation;
- (ii) NtrB and NtrC are not essential for nitrogen fixation, but they are required for nitrate utilization as a nitrogen source. It should also be noted that P_{II} is probably not involved in the regulation of NtrC activity;
- (iii) P_{II} is essential for the regulation of nitrogen fixation, by activating the NifA protein, in response to ammonia limitation.

In conclusion, *A. brasilense* shows an unexpected complexity of regulatory pathways controlling both nitrogen assimilation and nitrogen fixation in the free living state. It provides a unique example, among nitrogen fixing bacteria studied so far, where *glnB* is an essential regulatory gene for nitrogen fixation. On the other hand, the post-translationally reversible adenylylation of glutamine synthetase, in response to ammonia availability in the cells, does not require the presence of a functional *glnB* gene product. The unusual properties of the product of the *glnB* gene suggest, the existence of a second P_{II}-like encoding gene in *A. brasilense*. Further analysis should lead to a better understanding of the specific signal transduction processes involved in nitrogen metabolism of *A. brasilense*. Such studies should be also taken into consideration for engineering strains, with improved in nitrogen fixation properties in association with cereals.

Acknowledgements

The contributions of C. Elmerich to all phases of this work are gratefully acknowledged. The author wishes to thank Prof. J.-P. Aubert for critical reading of the manuscript. This work was supported by research funds from the University of Paris 7.

References

Adler, S. P., Purich, D. and Stadtman, E. R. (1975) Cascade control of *E. coli* glutamine syhthetase : Properties of the P_{II} regulatory protein and the uridylyltransferase-uridylylremoving enzyme. J. Biol. Chem. 250, 6264-6272.

Arsène, F., Liang, Y. Y., Kaminski, P.-A., Michel-Reydellet, N., Vieille, C., Isnard, A.-D. and Elmerich, C. (1994) Functional organization of the ORFX*ntrBC* locus of *Azospirillum brasilense* SP7 and study of gene regulation by use of *lacZ* fusions; *In* Proceedings of the 6th International Symposium on Nitrogen Fixation with Non-Legumes, (N. A. Hegazi et al. Eds.) Kluwer Academic Publishers, in press.

Bender, R. A., Janssen, K. A., Resnick, A. D., Blumenberg, M., Foor, F. and Magasanik, B. (1977) Biochemical Parameters of glutamine synthetase from *Klebsiella aerogenes*. J. Bacteriol. 129, 1001-1009.

Bozouklian, H. and Elmerich, C. (1986) Nucleotide sequence of the *Azospirillum brasilense* Sp7 glutamine synthetase structural gene. Biochimie 68, 1181-1187.

Bozouklian, H., Fogher, C. and Elmerich, C. (1986) Cloning and characterization of the *glnA* gene of *Azospirillum brasilense* Sp7. Ann. Inst. Pasteur/Microbiol. 137B, 3-18.

Colombo Pirola, M., Monopoli, R., Aliverti, A. and Zanetti, G. (1992) Isolation and characterization of glutamine synthetase from the diazotroph *Azospirillum brasilense*. Int. J. Biochem. 24, 1749-1754.

Colonna-Romano, S., Riccio, A., Guida, M., Defez, R., Lamberti, A., Iaccarino, M., Arnold, W., Priefer, U. and Pühler, A. (1987) Tight linkage of *glnA* and a putative regulatory gene in *Rhizobium leguminosarum*. Nucleic Acids Res. 15, 1951-1964.

Colonna-Romano, S., Patriarca, E. J., Amar, M., Bernard, P., Manco, G., Lamberti, A., Iaccarino, M. and Defez, R. (1993) Uridylylation of the P_{II} protein in *Rhizobium leguminosarum*. FEBS Lett. 330, 95-98.

Elmerich, C. (1991) Genetics and regulation of Mo-nitrogenase; *In* Biology and Biochemistry of Nitrogen Fixation (M. J. Dilworth and A. R. Glenn Eds.), Elsevier, Amsterdam, pp. 103-141.

Elmerich, C., Zimmer, W. and Vieille, C. (1992) Associative nitrogen-fixing bacteria; *In* Biological Nitrogen Fixation (G. Stacey, R. H. Burris, and H. J. Evans Eds.), Chapman and Hall, New-York, pp. 212-258.

Fani, R., Allotta, G., Bazzicalupo, M., Ricci, F., Schipani, C. and Polsinelli, M. (1989) Nucleotide sequence of the gene encoding the nitrogenase iron protein (*nifH*) of *Azospirillum brasilense* and identification of a region controlling *nifH* transcription. Mol. Gen. Gent. 220, 81-87.

Fischer, H. M., Bruderer, T. and Hennecke, H. (1988) Essential and non-essential domains in the *Bradyrhizobium japonicum* NifA protein : Identification of indispensable cysteine residues potentially involved in redox reactivity and/or metal binding. Nucl Acids Res. 16, 2207-2224.

Forchhammer, K. and Tandeau de Marsac, N. (1994) The P_{II} protein in the Cyanobacterium *Synechococcus* sp. PCC 7942 is modified by serine phosphorylation and signals the cellular N-status. J. Bacteriol. 176, 84-91.

Foster-Hartnett, D., Cullen, P. J., Gabbert, K. K. and Kranz, R. G. (1993) Sequence, genetic, and *lacZ* fusion analyses of a *nifR3-ntrB-ntrC* operon in *Rhodobacter capsulatus*. Mol. Microbiol. 8, 903-914.

Fu, H., Hartmann, A., Lowery, R. G., Fitzmaurice, W. P., Roberts, G. P. and Burris, R. H. (1989) Posttranslational regulatory system for nitrogenase activity in *Azospirillum* spp. J. Bacteriol. 171, 4679-4689.

Fu, H. A., Fitzmaurice, W. P., Roberts, G. P. and Burris, R. H. (1990) Cloning and expression of *draTG* genes from *Azospirillum lipoferum*. Gene 86, 95-98.

Galimand, M., Perroud, B., Delorme, F., Paquelin, A., Vieille, C., Bozouklian, H. and Elmerich, C. (1989) Identification of DNA regions homologous to nitrogen fixation genes *nifE*, *nifUS* and *fixABC* in *Azospirillum brasilense* Sp7. J. Gen. Microbiol. 135, 1047-1059.

Gauthier, D. and Elmerich, C. (1977) Relationship between glutamine synthetase and nitrogenase in *Spirillum lipoferum* . FEMS Microbiol. Lett. 2, 101-104.

Hartmann, A. and Burris, R. H. (1987) Regulation of nitrogenase activity by oxygen in *Azospirillum brasilense* and *Azospirillum lipoferum*. J. Bacteriol. 169, 944-948.

Hartmann, A., Fu, H. and Burris, R. H. (1986) Regulation of nitrogenase activity by ammonium chloride in *Azospirillum* spp. J. Bacteriol. 165, 864-870.

van Heeswijk, W.C., Rabenberg, M., Westerhoff, H. V. and Kahn, D. (1993) The genes of the glutamine synthetase adenylylation cascade are not regulated by nitrogen in *E. coli*. Mol. Microbiol. 9, 443-457.

Holtel, A. and Merrick, M. (1988) Identification of the *Klebsiella pneumoniae glnB* gene : Nucleotide sequence of wild-type and mutant alleles. Mol. Gen. Genet. 215, 134-138.

Holtel, A. and Merrick, M. J. (1989) The *Klebsiella pneumoniae* P_{II} protein (*glnB* gene product) is not absolutely required for nitrogen regulation and is not involved in NifL-mediated *nif* gene regulation. Mol. Gen. Genet. 217, 474-480.

Keener, J. and Kustu, S. (1988) Protein kinase and phosphoprotein phosphatase activities of nitrogen regulatory proteins NTRB and NTRC of enteric bacteria : Roles of the conserved amino-terminal domain of NTRC. Proc. Natl. Acad. Sci. USA 85, 4976-4980.

Kranz, R. G., Pace, V. M. and Caldicott, I. M. (1990) Inactivation, sequence, and *lacZ* fusion analysis of a regulatory locus required for repression of nitrogen fixation genes in *Rhodobacter capsulatus*. J. Bacteriol. 172, 53-62.

Liang, Y. Y., Kaminski, P. A. and Elmerich, C. (1991/A) Identification of a *nifA*-like regulatory gene of *Azospirillum brasilense* Sp7 expressed under conditions of nitrogen fixation and in the presence of air and ammonia. Mol. Microbiol. 5, 2735-2744.

Liang, J., Nielsen, G. M., Lies, D. P., Burris, R. H., Roberts, G. P. and Ludden, P. W. (1991/B) Mutations in the *draT* and *draG* genes of *Rhodospirillum rubrum* result in loss of regulation of nitrogenase by reversible ADP-ribosylation. J. Bacteriol. 173, 6903-6909.

Liang, Y. Y., de Zamaroczy, M., Arsène, F., Paquelin, A. and Elmerich, C. (1992) Regulation of nitrogen fixation in *Azospirillum brasilense* Sp7: involvement of *nifA*, *glnA* and *glnB* gene products. FEMS Microbiol. Lett. 100, 113-120.

Liang, Y. Y., Arsène, F. and Elmerich, C. (1993) Characterization of the *ntrBC* genes of *Azospirillum brasilense* Sp7: their involvement in the regulation of nitrogenase synthesis and activity. Mol. Gen. Genet. 240, 188-196.

Ludden, P.W. and Roberts, G. P. (1989) Regulation of nitrogenase activity by reversible ADP-ribosylation. Curr. Topics Cell. Regul. 30, 23-55.

Magasanik, B. (1988) Reversible phophorylation of an enhancer binding protein regulates the transcription of bacterial nitrogen utilization genes. TIBS 13, 475-479.

Martin, G. B., Thomashow, M. F. and Chelm, B. K. (1989) *Bradyrhizobium japonicum glnB*, a putative nitrogen regulatory gene, is regulated by NtrC at tandem promoters. J. Bacteriol. 171, 5638-5645.

Merrick, M. J. (1992) Regulation of nitrogen fixation genes in free-living and symbiotic bacteria;*In* Biological Nitrogen Fixation (G. Stacey, R. H. Burris, and H. J. Evans Eds.), Chapman and Hall, New-York, pp. 835-876.

Milcamps, A., Keyers, V. and Vanderleyden, J. (1993) Identification of a *nifW*-like gene in *Azospirillum brasilense*. Biochim. Biophys. Acta 1173, 237-238.

Parkinson, J. S. (1993) Signal transduction shemes of bacrteria. Cell 73, 857-871.

Passaglia, L. M., Numes, C. L., Zaha, C. P. and Schrank, I. S. (1991) The *nifHDK* operon in the free-living nitrogen fixing bacteria *Azospirillum brasilense* comprise sequentialy genes *H, D, K* and 353 bp ORF and gene *nifY*. Braz. J. Med. Biol. Res. 24, 649-675.

Patriarca, E. J., Riccio, A., Taté, R., Colonna-Romano, S., Iaccarino, M. and Defez, R. (1993) The *ntrBC* genes of *Rhizobium leguminosarum* are part of a complex operon subject to negative regulation. Mol. Microbiol. 9, 569-577.

Pelanda, R., Vanoni, M. A., Perego, M., Puibelli, L., Galizzi, A., Curti, B. and Zanetti, G. (1993) Glutamate synthase genes of the diazotroph *Azospirillum brasilense*. J. Biol. Chem. 268, 3099-3106.

Ratti, S., Curti, B., Zanetti, G. and Galli, E. (1985) Purification and characterization of glutamate synthase from *Azospirillum brasilense*. J. Bacteriol. 163, 724-729.

Singh, M., Tripathi, A. K. and Klingmüller, W. (1989) Identification of a regulatory *nifA* type gene and physical mapping of cloned new *nif* regions of *Azospirillum brasilense*. Mol. Gen. Genet. 219, 235-240.

Son, H. S. and Rhee, S. G. (1987) Cascade control of *Escherichia coli* glutamine synthetase. Purification and properties of P_{II} protein and nucleotide sequence of its structural gene. J. Biol. Chem. 262, 8690-8695.

Stock, J. B., Ninfa, A. J and. Stock, A. M (1989) Protein phosphorylation and regulation of adaptative responses in bacteria. Microbiol. Rev. 53, 450-490.

Toukdarian, A. and Kennedy, C. (1986) Regulation of nitrogen metabolism in *Azotobacter vinelandii* : Isolation of *ntr* and *glnA* genes and construction of *ntr* mutants. EMBO J. 5, 399-407.

Vande Broek, A., Michiels, J., de Faria, S. M., Milcamps, A. and Vanderleyden, J. (1992) Transcription of the *Azospirillum brasilense nifH* gene is positively regulated by NifA and NtrA and is negatively controlled by the cellular nitrogen status. Mol. Gen. Genet. 232, 279-283.

Vasudevan, S. G., Armarego, W. L. F., Shaw, D. C., Lilley, P. E. Dixon, N. E. and Poole, R. K. (1991) Isolation and nucleotide sequence of the *hmp* gene that encodes a haemoglobin-like protein in *E. coli* K-12. Mol. Gen. Genet. 226, 49-58.

Vasudevan, S. G., Gedye, C., Dixon, N. E., Cheah, E., Carr, P. D., Suffolk, P. M., Jeffrey, P. D. and Ollis, D. L. (1994) *E. coli* P_{II} protein : purification, crystallization and oligomeric structure. FEBS Lett. 337, 255-258.

Wetsby, C. A., Enderlin, C. S., Steinberg, N. A., Joseph, C. M. and Meeks, J. C. (1987) Assimilation of $^{13}NH_4^+$ by *Azospirillum brasilense* grown under nitrogen limitation and excess. J. Bacteriol. 169, 4211-4214.

de Zamaroczy, M., Delorme, F. and Elmerich, C. (1989) Regulation of transcription and promoter mapping of the structural genes for nitrogenase (*nifHDK*) of *Azospirillum brasilense* Sp7. Mol. Gen. Genet. 220, 88-94.

de Zamaroczy, M., Delorme, F. and Elmerich, C. (1990) Characterization of three different nitrogen-regulated promoter regions for the expression of *glnB* and *glnA* in *Azospirillum brasilense*. Mol. Gen. Genet. 224, 421-430.

de Zamaroczy, M., Paquelin, A. and Elmerich, C. (1993/A) Functional organization of the *glnB-glnA* cluster of *Azospirillum brasilense*. J. Bacteriol. 175, 2507-2515.

de Zamaroczy, M., Liang, Y.Y., Kaminski, A., Arsène, F. and Elmerich, C. (1993/B) Regulation of NifA synthesis and activity in *Azospirillum brasilense* Sp7; *In* New Horizons in Nitrogen Fixation (R. Palacios, J. Mora and W. E. Newton Eds.), Kluwer Academic Publishers, Dordrecht, pp. 423-427.

Zhang, Y., Burris, R. H. and Roberts, G. P. (1992) Cloning, sequencing, mutagenesis, and functional characterization of *draT* and *draG* from *Azospirillum brasilense*. J. Bacteriol. 174, 3364-3369.

Involvement of *nifA* and *glnB* genes in expression of nitrogen fixation genes in *Azospirillum brasilense*.

Florence Arsène and Claudine Elmerich.
Unité de Physiologie Cellulaire and URA 1300 CNRS
Département des Biotechnologies
Institut Pasteur
25 rue du Dr. Roux
75724 Paris Cedex 15
France

Abstract

Using promoter fusions to *lacZ* as reporter gene, regulation of *nif* gene expression was studied in *Azospirillum brasilense* Sp7, in free-living state and during association of *Azospirillum brasilense* with wheat. In both conditions, *nif* gene expression requires NifA and the *glnB* product. The same protocol was successfully used to show that auxotrophic mutants (*glnA*, *trpC*, *trpD*) were impaired in the colonisation efficiency as compared to the wild type, whereas a *nifA*-Tn*5* mutant was not.

Introduction

The ecological distribution of *Azospirillum* is extremely wide. Strains have been found in association with monocotyledons, including maize, rice, surgarcane, sorghum, forage grasses and also with dicotyledons. These bacteria are Gram-negative, aerobic, non-fermentative chemoorganotrophs. Expression of *nif* genes has been studied at the free-living state and in association with wheat.

Use of *lacZ* fusions to study the regulation of nitrogen fixation genes at the free-living state.

Azospirillum brasilense strains are able to fix nitrogen under free-living conditions, in microaerobiosis and in absence of ammonia (c.a. nitrogen fixation conditions). In *A. brasilense* Sp7, the characterization of DNA regions containing *nif* , *ntrBC*, *gln* genes was previously reported (Galimand *et al.* 1989, Liang *et al.* 1991, 1992, 1993, de Zamaroczy *et al.* 1993, this symposium). Upstream of *glnA*, an ORF was found to have a high similarity with *glnB*, the structural gene for the P_{II} protein. Tn*5* insertions in *nifHDK*, *nifA* and *nifB*, and cartridge insertion in *glnB* led to a Nif$^-$ phenotype. The nucleotide sequence of *nifH* was established. Under conditions of nitrogen fixation, transcription proceeds from a σ^{54}-dependent promoter, that is preceded by two upstream elements (UAS) typical of genes controlled by *nifA* (Fani *et*

NATO ASI Series, Vol. G 37
Azospirillum VI and Related Microorganisms
Edited by I. Fendrik et al.

al. 1989, de Zamaroczy *et al.* 1989, Vande Broek *et al.* 1992). The *nifA* gene from strain Sp7 was cloned and sequenced (Liang *et al.* 1991). The coding sequence is not preceded by a promoter resembling σ^{54}-dependent promoters, suggesting that *nifA* regulation in *Azospirillum* differs from the *K. pneumoniae* model system. In order to determine precise involvement of NifA, and P_{II} in the regulation of *nif* gene expression, *lacZ* fusions to *nifA* and *nifH* have been used to monitor the expression of their promoters under conditions of nitrogen fixation and ammonia assimilation.

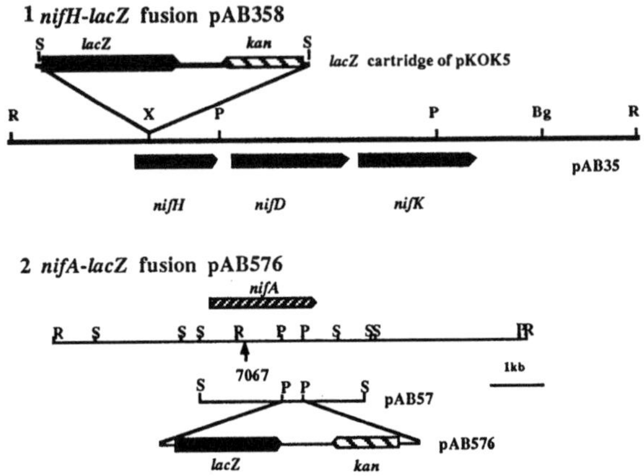

Figure 1: Construction of the *nif-lacZ* fusions. The physical map of the *nifHDK* and *nifA* regions is shown. Details of construction of pAB358 and pAB576 were reported by Liang *et al.* 1991. Restriction site: Bg: *Bgl*II; P: *Pst*I; R: *Eco*RI; S: *Sal*I; X: *Xho*I.

Figure 1 shows the physical maps of the *nifH*- and *nifA-lacZ* fusions. Both fusions are carried by broad host range plasmids and can be introduced into *Azospirillum* wild type and mutant strains by conjugation. It can be seen in Figure 2 that *nifH* is expressed only under nitrogen fixation conditions in the wild type (2A) while no expression is detected in a *nifA* background (7067) (2B). Similar data have been observed with a *nifB-lacZ* fusion. This strongly suggests that the *nifHDK* operon as well as *nifB* expression is under the control of the NifA transcriptional activator. By contrast, the *nifA* gene is expressed even in presence of air or ammonia both in the wild type and in the *nifA*-mutated background (2A/B). This suggests, that the *nifA* product is synthesized in an inactive form in the presence of oxygen and ammonia. Data reported in Figure 2C show that expression of *nifA* is not impaired in a *glnB-km* (7606), while expression of *nifH* is totally abolished. This is an indication that P_{II} might be involved in the modulation of NifA activity in *Azospirillum* (Liang *et al.* 1992).

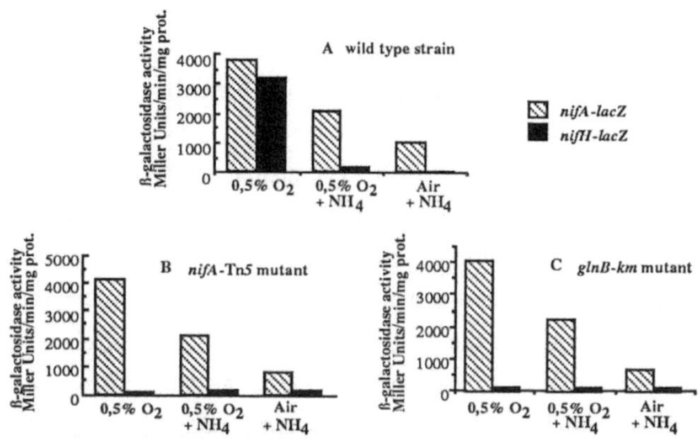

Figure 2: Expression of the *nif-lacZ* fusions in the wild type and mutant strains. ß-galactosidase activity is expressed in Miller Units/min/mg prot. Overnight culture are derepressed as previously reported by Liang *et al.* 1991. Details of contruction of *nifA*-Tn*5* (7067) and *glnB-km* (7606) mutants are described by Liang *et al.* 1991 and de Zamaroczy *et al.* 1993.

The model of regulation of *nifH* expression is schematized in Figure 3.This model is valid for *nifB* transcription and most probably for the others *nif* operons, excluding *nifA*. It is worth to recall that the requirement of *ntrBC* genes in *nifA* expression was ruled out in a previous work (Liang *et al.* 1993). This was based on the fact that *ntrBC*-Tn*5* insertion mutants were Nif$^+$.

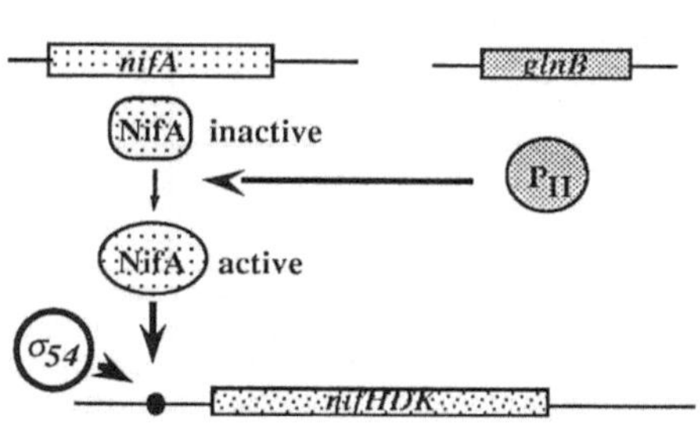

Figure 3: model of regulation of *nifHDK* genes.

Use of *lacZ* fusions to study the *Azospirillum* association with cereal crops.

A simple protocol, based on the use of promoter fusions to *lacZ* as reporter gene, has been established to study association of *Azospirillum brasilense* with wheat (Arsène *et al.* 1994). Wheat seeds are sterilized, germinated and inoculated under sterile conditions in test tubes containing a mineral solution devoid of carbon and nitrogen sources. After one week growth, the plantlets are inoculated with *Azospirillum* wild type or mutants containing or not plasmids carrying *lacZ* fusions. After ten days, the colonization is studied by *in situ* staining of the bacteria using 5-bromo-4-chloro-3-indolyl-ß-D-galactoside (X-gal) as a chromogenic substrate and examination by light microscopy. Determination of ß-galactosidase activity associated with plant extract can be easily performed.

Colonization of wild type or mutant strains was compared using totally or partially constitutive *lacZ* fusions (*km-lacZ* or *nifA-lacZ* respectively). It was verified that the *km-lacZ* fusion was expressed at the same rate in the different strains at the free-living state. Similarly, the *nifA-lacZ* fusion was expressed at the same rate under nitrogen fixing conditions in these strains. For these reasons, it is assumed that the ß-galactosidase activity is correlated to the number of bacteria associated to the root system and hence to the colonization efficiency.

Table 1 shows that auxotrophic mutants (*glnA*, *trpC*, *trpD*) displayed a reduced ß-galactosidase activity as compared to the wild type, whereas *nifA*-Tn*5* did not. This lower ß-galactosidase activity was correlated with a lower number of bacteria associated to the roots. After X-gal staining of the root inoculated with the wild type strain or the Nif$^-$ mutant, observation under microscope revealed blue stained bacteria, located on the surface of the lower part of the root. In the case of the auxotrophic mutants, the bacteria were preferentially found at the emergence of secondary roots.

Table 1: Expression of the *nifA*- and *km-lacZ* fusions in various strains, in association with wheat.

Strains	*km-lacZ*	*nifA-lacZ*
WT	1268 ± 508	490 ± 206
7067 (*nifA*, nif-)	ND	408 ± 235
7028 (*glnA*, Nif^C)	293 ±212	46 ± 25
7853 (*trpD*)	677 ± 349	ND
7854 (*trpC*)	587 ± 429	ND

ß-galactosidase activity is expressed in Miller Units/min/mg plant protein The background level obtained with strains carrying no fusion was about 20 Miller Units/min/mg prot. *glnA* mutant was obtained by chemical mutagenesis (Gauthier *et al* 1977). *trpD* (7853) and *trpC* (7854) mutants were obtained by Tn*5* mutagenesis (Zimmer *et al.* 1991). Results are average of at least 10 samples. Data from Arsène *et al.* 1994.

Expression of the plasmid-borne *nifA*- and *nifH-lacZ* fusions was then analysed. As shown in Table 2, the fusions were significantly expressed in the wild type or in the *ntrBC* mutants, as compared to the basal level obtained with bacteria lacking the fusions or with the *nifA*-Tn*5* and *glnB-km* mutants carrying the *nifH-lacZ* fusion. Similar results were obtained with the *nifB-lacZ* fusion (Data not shown). ß-galactosidase activity rates was well correlated with the intensity of staining of the samples with X-gal (Data not shown).

Table 2: Expression of the *nifA*- and *nifH-lacZ* fusions in various strains, in association with wheat.

Strains	*nifH-lacZ*	*nifA-lacZ*
WT	220 ± 112	421 ± 209
7067 (*nifA*, Nif$^-$)	26 ± 17	408 ± 235
7606 (*glnB*, Nif$^-$)	9 ± 4	194 ± 176
7148 (*ntrC*, Nif$^+$)	140 ± 47	254 ± 202
7194 (*ntrB*, Nif$^+$)	161 ± 57	312 ± 173
7028 (*glnA*, Nif^c)	67 ± 61	46 ± 25

ß-galactosidase activity is expressed in Miller Units/min/mg plant protein The background level obtained with strains carrying no fusion was about 20 Miller Units/min/mg prot. Strains 7148 (*ntrC*) and 7194 (*ntrB*) were obtained by Tn*5* insertion (Liang *et al.* 1993). Results are average of at least 10 samples. Data from Arsène *et al.* 1994.

Significant activity observed in the wild type carrying the *nifH-lacZ* fusions led to the conclusion that NifA is under its active form during the association. The absence of expression of *nifH* in the *nifA* and *glnB* mutants suggests that, in association with wheat, *nif* promoters are activated by NifA and P_{II} is involved in the activation of NifA. In the case of the *glnA* mutant (7028) the low ß-galactosidase activity was correlated with a poor colonization efficiency of this strain (see table 1).

Colonization of other cereal crops.

The protocol optimized with wheat was also successfully used with sorgho and rice, and significant expression of *nifA* and *nifH* was detected. Colonization of the lower part of the root was similar to what observed with wheat. Bacteria were located on the surface or in the first layer of the cortex. Colonization efficiency was also tested with various wild type strains of *A. brasilense* or *A. lipoferum* (Katupitiya *et al.* 1994). In fact, this protocol can be used with any gene, for wich a *lacZ* fusions is available and also with other plant or bacteria species.

References

Arsène, F., Katupitiya, S., Kennedy, I.R., and Elmerich, C. 1994. Use of *lacZ* fusions to study the expression of *nif* genes of *Azospirillum brasilense* in association with plants. Mol. Plant-Microbe Interact. in press.

Fani, R., Allotta, G., Bazzicalupo, M., Ricci, F., Schipani, C., and Polsinelli, M. 1989. Nucleotide sequence of the gene encoding the nitrogenase iron protein (*nifH*) of *Azospirillum brasilense* and identification of a region controlling *nifH* transcription. Mol. Gen. Genet. **220**:81-87.

Galimand, M., Perroud, B., Delorme, F., Paquelin, A., Vieille, C., Bozouklian, H., and Elmerich, C. 1989. Identification of DNA regions homologous to nitrogen fixation genes *nifE*, *nifUS* and *fixABC* in *Azospirillum brasilense* Sp7. J. Gen. Microbiol. **135**:1047-1059.

Katupitiya, S., New, P.B., Elmerich, C., and Kennedy, I.R. 1994. Improved nitrogen fixation in 2,4-D treated wheat roots associated with *Azospirillum lipoferum*: studies of colonization using reporter genes. Soil. Biol. Biochem. In press.

Liang, Y. Y., Kaminski, P. A., and Elmerich, C. 1991. Identification of a *nifA*-like regulatory gene of *Azospirillum brasilense* Sp7 expressed under conditions of nitrogen fixation and in the presence of air and ammonia. Molecular Microbiol. **5**:2735-2744.

Liang, Y. Y., de Zamaroczy, M., Arsène, F., Paquelin, A., and Elmerich, C. 1992. Regulation of nitrogen fixation in *Azospirillum brasilense* Sp7: involvement of *nifA*, *glnA* and *glnB* gene products. FEMS Microbiol. Lett. **100**:113-120.

Liang, Y. Y., Arsène, F., and Elmerich, C. 1993. Characterization of the *ntrBC* genes of *Azospirillum brasilense* Sp7: Their involvement in the regulation of nitrogenase synthesis and activity. Mol. Gen. Genet. **240**:188-196.

Vande Broek, A., Michiels, J., de Faria, S.M., Milcamps, A., and Vanderleyden J. 1992. Transcription of the *Azospirillum brasilense nifH* gene is positively regulated by NifA and NtrA and is negatively controlled by the cellular nitrogen status. Mol. Gen. Genet. **232**:279-283.

de Zamaroczy, M., Delorme, F., and Elmerich, C. 1989. Regulation of transcription and promoter mapping of the structural genes for nitrogenase (*nifHDK*) of *Azospirillum brasilense* Sp7. Mol. Gen. Genet. **220**:88-94.

de Zamaroczy, M., Paquelin, A., and Elmerich, C. 1993. Functional organization of the *glnB-glnA* cluster of *Azospirillum brasilense*. J. Bacteriol. **175**:2507-2515.

Polysaccharide synthesis in *Azospirillum brasilense*

Paul De Troch, Daniel J. Petersen and Jos Vanderleyden

F.A. Janssens Laboratory of Genetics, Department of Applied Plant Sciences, Catholic University Leuven
W. de Croylaan 42, B-3001 Leuven
Belgium

Abstract

The surface of bacteria consists of polysaccharides, like the exopolysaccharides and lipopolysaccharides and, as it is the case for *Azospirillum brasilense*, glycoproteins like glycosylated flagella. The precursors for the synthesis of polysaccharides are sugarnucleotides. Several genes, coding for enzymes involved in the synthesis and in the modification of sugarnucleotides, have been characterised is *Azospirillum brasilense* Sp7.
Two copies of *exoB* genes, encoding UDP-glucose 4'-epimerase, have been isolated. *exoB1*, localised on the p90 megaplasmid is responsible for 20% of UDP-glucose 4'-epimerase activity, and *exoB2*, localised on the chromosome is responsible for the remaining 80% of the activity. *Azospirillum* is the first micro-organism which has two totally functional copies of this important enzyme. *exoC* encodes a phosphomannomutase, involved in the synthesis of mannose-1-phosphate, a precursor for the synthesis of GDP-mannose. The upstream localised *exoD* encodes a GDP-mannose-dehydrogenase, which catalyzes the synthesis of GDP-mannuronic acid. The *exoD* gene may be cotranscribed with *exoC*.
In view of these data, it can be understood why mutations in some of these genes have a pleiotropic effect, whereas for others no apparent phenotype is observed.

Introduction

The surface of bacteria is important for interactions with the surroundings. The surface of bacteria consists of exopolysaccharides. *Azospirillum* is a Gram-negative bacterium, and therefore the surface consists of lipopolysaccharides and exopolysaccharides. In the case of *Azospirillum brasilense* Sp7, even the polar flagellum is glycosylated (Moens S., unpublished data). Previously, it had been shown that the polar flagellum and the exopolysaccharides are involved in the attachment to wheat roots (Michiels *et al.*, 1991). The synthesis of polysaccharides is a complex process and requires sugarnucleotides, such as UDP-glucose and UDP-galactose. These sugarnucleotides are the donors of sugars for incorporation into polysaccharides. In this study, several genes, involved in the synthesis of sugarnucleotides, have been characterized in *Azospirillum brasilense*.

NATO ASI Series, Vol. G 37
Azospirillum VI and Related Microorganisms
Edited by I. Fendrik et al.

```
ECOLI           MRVLVTGGSGYIGSHTCVQLLQNGHD-VIILDNLCNSKRSVLPVIERLGGKHPTFVEGDIRNEALMTEILHD   71
KLEPN           MKVLVTGGSGYIGSHTCVQLLQRGHE-VVILDNLCNSKRRMLPVIERLGGKEATFIEGDIRNEARMTEILHD   71
SALTY           MRVLVTGGSGYIGSHTCVQLLQNGHD-VVILDNLCNSKRSVLPVIERLGGKHPTFVEGDIRNEALITEILHD   71
HAEIN           MAILVTGGAGYIGSHTVVELLNVGKE-VVVLDNLCNSSPKSLERVKQITGKEAKFYEGDILDRALLQKIFAE   71
EAMYL           MSILVTGGAGYIGSHTVLSLLQRGDD-VVILDNLSNASRESINRVEKLTGKTATFFEGDLLDRSCLRSVFSA   71
NEIGO           MTVLITGGTGFIGSHTAVSLVQSGYD-AVILDNLCNSSAAVLPRLRQITGRNIPFYQGDIRDCQILRQIFSE   71
NEIME          MKKILVTGGTGFIGSHTVVSLLKSGHQ-VVILDNLCNSSINILPRLKTITGQEIPFYQGDIRDREILRRIFAE   72
CORDI           MKLLVTGGAGYVGSVCSTVLLEQGHE-VTIVDNLTTGNRDAVPL-------GATFVEGDIKDVAENVLSSDS   64
STRLI         MSGKYLVTGGAGYVGSVVAQHLVEAGNE-VVVLHNLSTGFREVVPA-------GASFVEGDIRDRAQ-VRGRLS   65
STRTR           MAILVLGGAGYIGSHMVDRLVEKGQEKVVVVDSLVTGHRAAVHP-------DAIFYQGDLSDQDFMRKVFKE   65
ABRA2       MNTPHTVLVTGGAGYVGSHCVAELLDRGFR-VVVLDNLRQGHRVAVPS-------DAAFVEADLADMDALTRVFAE   68
ABRA1   MTDQTAASPRVLVTGGAGYIGSHVLHALTDAGIP-AVTIDDLSAGRREAIPA-------AVPLVEGDIGSAELLDRVMRD   72
RHIME        MQNNNILVVGGAGYIGSHTCLQLAAKGYQ-PVVYDNLSNGHEEFVKW-------GV-LEKGDIRDRQRLDEVLAR   66
                   *. **.*..**     *   *    .    *       .               **.

ECOLI   HA-IDTVIHFAGLKAVGESVQKPLEYYDNNVNGTLRLISAMRAANVKNFIFSSSATVYGDQPKIPYVESFPTGTPQSPYG   150
KLEPN   HA-IEAVIHFAGLKAVGESVAKPLEYYDNNVTGTLKLVSAMRAAGVKNFIFSSSATVYGDQPKIPYVES·           110
SALTY   HA-IDTVIHFAGLKAVGNSVARPLEYYDNNVNGTLRLVSAMRAANVKNLIFSSSATVYGDQPKIPYVESFPTGAPQSPTG   150
HAEIN   NE-INSVIHFAGLKAVGESVQKPTEYYMNNVAGTLVLIQEMKKAGVWNFVFSSSATVYGDPKIIPITEDCEVGGTTNPYG   150
EAMYL   HR-ISAVIHFAGLKAVGASTRKPLEYYQNNVTGTLVLLEEMRSAGVNQFIFSSSATVYGADAPVPYVETTPIGGTTSPYG   150
NEIGO   HE-IESVIHFAGLKAVGESVAEPTKYYGNNVYGSLVLAEEMARAGVLKIVFSSSATVYGDAEKVPYTEDMRPGDTANPYG   150
NEIME   NR-IDSVIHFAGLKAVGESVAEPMKYYDNNVSGSLVLAEEMARAGVFSIVFSSSATVYGDPGKVPYTEDMPPGDTTSPYG   151
CORDI   ---FDAVLHFAARSLVGESVEKPDEYWQHNMVTTLALLDAMKRNNVRNIVFSSTAATYGEPETVPITEDAPT-HPTNPYG   150
STRLI   ---FDGVLHFAAFSQVGESVVKPEKYWDNNVGGTMALLEAMRGAGVRRLVFSSTAATYGEPEQVPIVESAPT-RPTNPYG   141
STRTR   NPDVDAVIHFAAYSLVGESMEKPLKYFDNNTAGMVKLLEVMNECGVKYIVFSSTAATYGIPEEIPILETTPQ-NPINPYG   144
ABRA2   WR-FDAVFHFAALSLVGESMRDPHLYLHGNTVNSLNLIRAATTAGVRKMVFSSTANLFGTPERLPIDEETTI-DPGSPYG   146
ABRA1   HRV-DAVMHFAGSIVVPESVVKPLDYYRNNTANSLTLLGACLRAGIDKVVFSSTAAVYGAPESVPIREDAPT-VPINPYG   150
RHIME   HKP-RAILHFAAMIEVGESVKDPAAFYDNNVIGTLTLLSAALAAGIDAFVFSSTCATYGLPDSVPMDESHKQ-APINPYG   144
                .***     *  *  *      *    .*         .   .***.   .*    .*  *         * *

ECOLI   KSKLMVEQILTDLQKAQPDWSIALLRYFNPVGAHPSGDMGEDPQGIPNNLMPYIAQVAVGRRDSLAIFGNDYPTEDGTGV   230
SALTY   KSKLMVEQILTDLQKAQPEWSIALLRYFNPVGAHPSGDMGEDPQGIPNNLMPYIAQVAVGRRESLAVFGNDYPTEDGTGV   230
HAEIN   TSKYMVEQILRDTAKAEPKFSMTILRYFNPVGAHESGLIGEDPNGIPNNLLPYISQVAIGKLAQLSVFGSDYDTHDGTGV   230
EAMYL   TSKLMVEQILRDYAKANPEFKTIALRYFNPVGAHESGQMGEDPNGIPNNLLPYIAQVAIGRLEKLGIFGDDYPTEDGTGV   230
NEIGO   ASKAMVERMLTDIQKADPRWSVILLRYFNPIGAHESGLIGEQPNGVPNNLLPYICQVASGRLPQLSVFGGDYPTPDGTGM   230
NEIME   ASKSMVERILTDIQKADPRWSMILLRYFNPIGAHESGLIGEQPNGIPNNLLPYICQVAAGKLPQLAVFGDDYPTPDGTGM   231
CORDI   ATKLSIDYAITSYAHAYGFAATS-LRYFNVAGAY--GLVGEN-REIETHLIPLVLQVALGHRDKIFMFGDDWPTEDGTPI   236
STRLI   ASKLAVDHMITGEAAAHGLGAG--LRYFNVAGAY--GEYGER-HDPESHLIPLVLQVAQGRREAISVYGDDYPTPDGTCV   236
STRTR   ESKLMMETIMKWSDQAYGIKYVP-LRYFNVAGANLMVRLVRT-RS-ETHLLPIILQVAQGVREKIMIFGDDYNTPDGTNV   241
ABRA2   ESKFFIERALHWADRCHGLRSAC-LRYFNAAGAHPNGTLGED-HAPETHLIPLVLDAPAAGRSHIEIFGDDYPTADGTCV   244
ABRA1   ASKLMTEQMLRDAGAAHGLRSVI-LRYFNVAGADPAGRTGQA-TPVATHLIKVACQALLGRRPPLAIFGTDYDTPDGTCI   248
RHIME   RTKWICEQALKDYGLYKGLRSVI-LRYFNAAGADFEGRIGEW-HEPETHAIPLAIDAALGRREGFKVFGTDYDTRDGTCV   242
         .*   .  .              *****  **                 .               ..* *. * *** .

ECOLI   RDYIHVMDLADGHVVAMEKLANKPGVHIYNLGAGVGNSVLDVVNAFSKACGKPVNYHFAPRREGDLPAYWADASKADREL   310
SALTY   RDYIHVMDLADGHVVAMEKLADKSGVHIYNLGAGVGSSVLDVVNAFSKACGKPINYHFAPR-DGDLPAYWADASKADREL   309
HAEIN   RDYIHVVDLAVGHLKALQRHENDAGLHIYNLGTGHGYSVLDMVKAFEKANNITIPYKLVERRSGDIATCYSDPSLAAKEL   310
EAMYL   RDYIHVMDLAEGHLKALDHLSAIEGYKAYNLGAGKGYSVLEMVKAFEKASGGTVAYQISPRRDGDLAAFWADATLADKEL   310
NEIGO   RDYIHVMDLAEGHIAAMKAKGGVAGVHLFNLGSGRAYSVLEIIRAFEAASGLHIPYRIQPRRAGDLACSYADPSHTKQQT   310
NEIME   RDYIHVMDLAEGHVAAMQAKSNVAGTHLLNLGSGRASSVLEIIRAFEAASGLTIPYEVKPRRAGDLACFYADPSYTKAQI   311
CORDI   RDYIHIRDLADAHILALQ-SNVEGSHRIFNLGSGEGYSVKQVIDTCREVTGHPIPAEVAPRRAGDPAVLIASSAKAQSEL   295
STRLI   RDYIHVADLAEAHLLAVR--RRPGESLICNLGNGNGFSVREVVETVRRVTGHPIPEIMAPRRGRDPAVLVASAGTAREKL   294
STRTR   RDYVHPFDLADAHLLAVEYLRKGNESTAFNLGSSTGFSNLQILEAARKVTGKEIPAEKADRRPGDPDILIASSEKARTVL   311
ABRA2   RDYIHVCDLADAHLRVLPAL---ETRSRFNLGNGTGYSVREVIAATERVTGLTVPVRVGPRRAGDPAVLIASSERIRGEL   311
ABRA1   RDYIHVSDLADAHVLALLHLRRGGGSLLMNCGYGRGASVREVVRTLEEVSGEQVPATFADRRPGDPPQLVAGADRIREQL   318
RHIME   RDYIHVLDLADAHVRAVDYLLEGGESVALNLGTGTGTTVKELLDAIEKVAKRPFNIGYAERREGDSTTLVANNDKARQVL   312
        ***.*  ***  *. .            * *     .   .. .                 *   *     .

ECOLI   NWRVTRT-LDEMAQDTWHWQSRHPQGYPD                                                    338
SALTY   NWRVTRT-LDEMAQDTWHWQSRHPQAYPD                                                    337
HAEIN   GWVAERG-LEKMMQDTWNWQKNNSKGYRD                                                    338
EAMYL   NWRVSRG-IDEMMRDTWNWQSQNPQG-DS                                                    337
NEIGO   GWETKRG-LQQMMEDSWRWVSRNPGRYGD                                                    338
NEIME   GWQTQRD-LTQMMEDSWRWVSNNPNGYDD                                                    339
CORDI   GWKPQRTDLHTIVSDAWAFTSQLGDKAHSASRG                                                328
STRLI   GWNPSRADLA-IVSDAWELPQRRAGQ                                                       318
STRTR   GWKPQFDNIEKIIASAWAWHSSHPKGYDDRG                                                  332
ABRA2   GWSPRFPDLDSIIGSAWEWRRRHPGGYRGPAVAQAAE                                            338
ABRA1   GWVPKHDRLDGIVRSALSWERSLEQSVGQAGAPGGSASRS                                         348
RHIME   GWEPQYD-LAAITESAWNWHSRRNQGG                                                      327
         *      .  .   .
```

Figure 1 Amino acid sequence alignment, using the method of Higgings and Sharp (Clustal, PCGene) of UDP-glucose 4'-epimerase of *Escherichia coli* (ECOLI) (Lemaire and Müller-Hill, 1986), *Klebsiella pneumoniae* (KLEPN) (amino terminal part) (Peng *et al.*, 1992), Salmonella typhimurium (SALTY) (Houng *et al.*, 1990), *Haemophilus influenzae* (HAEIN) (Maskell *et al.*, 1991), *Erwinia amylovora* (EAMYL) (Metzger *et al.*, 1994), *Neisseria gonorrhoeae* (NEIGO) (Robertson *et al.*, 1993), *Neisseria meningitidis* (NEIME) (Jennings *et al.*, 1993), *Corynebacterium diphtheriae* (CORDI) (Boyd *et al.*, 1990), *Streptomyces lividans* (STRLI) (corrected from Adams *et al.*, 1988), *Azospirillum brasilense exoB1* (ABRA1), *Azospirillum brasilense exoB2* (ABRA2), *Rhizobium meliloti exoB* (RHIME) (Buendia *et al.*, 1991), and *Streptococcus thermophilus* (STRTR) (Poolman *et al.*, 1990). Stars indicate conservative changes (S-T-A, L-V-I-M, K-R, D-E, Q-N, F-Y-W). Glycine residues, conserved in the fingerprint sequence GXGXXG, which is characteristic for the ADP-binding ßaß-fold domain of NAD-binding proteins are denoted as '' (Wieringa *et al.*, 1986). The substrate binding site (LRYFNXXGA) is indicated in bold.

A. brasilense contains two functional *exoB* genes

The *A. brasilense exoB* genes have been isolated by the genetic complementation of a *Rhizobium meliloti exoB* mutant (Michiels *et al.*, 1988). The *R. Meliloti exoB* gene encodes a UDP-glucose 4'-epimerase, which interconverts UDP-glucose and UDP-galactose (Buendia *et al.*, 1991). Two different genes were isolated, *exoB1*, which is located on pRhico, a 90 MDa megaplasmid, and *exoB2*, which is located on the chromosome (Michiels *et al.*, 1989; Croes *et al.*, 1991). The nucleotide sequence of both genes has been determined, and both gene products show strong homology to other prokaryotic UDP-glucose 4'-epimerases (EC 5.1.3.2) (Fig. 1.) (De Troch *et al.*, 1994). UDP-glucose 4'-epimerases form dimers. A putative substrate binding site has been identified, LRYFNXXGA, at position 170 of the *E. coli* GalE. The *exoB1* gene product is responsible for 20% for the UDP-glucose 4'-epimerase activity in *A. brasilense* Sp7. ExoB2 is responsible of the other 80% (Table 1). The construction of a *exoB1 exoB2* double mutant was unsuccessful, as the lack of UDP-glucose 4'-epimerase activity might be lethal for *Azospirillum*.

Table 1 UDP-glucose 4'-epimerase activity of *A. brasilense* Sp7 wild type, *exoB1* and *exoB2* mutants.

Strain or mutant	enzymatic activity[a]
Sp7	3.2
exoB1	2.5
exoB2	0.7

a one unit corresponds to the conversion of 1 mmol UDP-galactose to UDP-glucose per mg protein.

The *A. brasilense exoC* gene encodes a phosphomannomutase

The *A. brasilense exoC* gene has been isolated by the genetic complementation of the *R. meliloti exoC* mutant (Michiels *et al.*, 1988). The *R. meliloti exoC* gene has been suggested to encode a phosphoglucomutase (Uttaro *et al.*, 1990). Phosphoglucomutase (EC 5.4.2.2) is responsible for the interconversion of glucose-1-phosphate and glucose-6-phosphate. DNA sequence analysis of the *A. brasilense exoC* locus revealed the presence of a 1407 bp long open reading frame, encoding a polypeptide, showing homology with several genes, encoding phosphoglucomutase/phosphomannomutase (Fig. 2). The highest homology (40.7 % identity) was detected with AlgC of *Pseudomonas aeruginosa*. This latter enzyme has both phosphomannomutase (conversion of mannose-1-phosphate to mannose-6-phosphate, EC 5.4.2.8) and phosphoglucomutase activity (Coyne *et al.*, 1994). The *A. brasilense exoC* gene seems to encode mainly phosphomannomutase activity, with some phosphoglucomutase

```
PSALGC  MSTVKAPTLPASIFRAYDIRRVVGDTLTAETAYWIGRAIGSESLARGEPCVAVGRDGRLSGPELVKQLIQGLVDCGCQVS   80
NMEPGM       MASIARDIFKAYDIRGIVGKTLTDEAAYLIGKAIAAKAAEKGITRIALGRDGRLSGPELMEHIRRGFTDSGINVL   75
NGOPGM       MASITRDIFKAYDIRGIVGKTLTDDAAYFIGRAIAAKAAEKGIARIALGRDGRLSGPELMEHIQRGLTDSGISVL   75
ABEXOC    MSEAHTFHPTVLREYDIRGIVGSTLTAADARAVGRLRHRGRPAAVRKTVCVGYDGRLSSPELEAAMVDGLVACGLHVL   78
ECCPSG         MKKLTCFKAYDIRGKLGEELNEDIAWRIGRAYGEFLKPK---TIVLGGDVRLTSETLKLALAKGLQDAGVDVL   70
ECRFBK           MLTCFKAYDIRGKLGEELNEDIAWRIGRAYGEFLKPK---TIVLGGDVRLTSETLKLALAKGLQDAGVDVL   68
SECPSG         MTKLTCFKAYDIRGRLGEELNEDIAWRIGRAYGEYLKPK---TVVLGGDVRLTSEALNVALAKGLQDAGVDVL   70
SERFBK         MNNLTCFKAYDIRGRLGEELNEDIAWRIGRAYGEYLKPK---TIVLGGDVRLTSEALKLALAKGLQDAGVDVL   70
ECRFB2         MTQLTCFKAYDIRGELGEELNEDIAYRIGRAYGEFLKPG---KIVVGGDVRLTSESLKLALARGLMDAGTDVL   70
XCXANA          MTLPAFKAYDIRGRVPDELNEDLARRIGVALAAQLDQG---PVVLGHDVRLASPALQEALSAGLRASGREVI   69
                   ...****  . ..*... *  .*                ..* * **.. .*  .  *. . * *

PSALGC  DVGMVPTPVLYYAANVLEGKSGVMLTGSHNPPDYNGFKIVVAGE-TLANEQ-IQALRERIEKNDLASGV----GSVEQVD  154
NMEPGM  NVGMVATPMLYFAAVNECGGSGVMITGSHNPPDYNGFKMMLGGD-TLAGEA-IQELLSIIEKDGFARAG--KQGSVTEKD  151
NGOPGM  NVGMVTTPMLYFAAVNECGGSGVMITGSHNPPDYNGFKMMLGGD-TLAGEA-IQELLAIVEKDGFVAAD--KQGSVTEKD  151
ABEXOC  RIGLGPTPMLYFATRDREAAAGIMITGSHNPPDYNGIKMMLGKG-PVYGRQ-ILDIGAIASKADYVSG----EGSSEQLD  152
ECCPSG  DIGMSGTEEIYFATFHLGVDGGIEVTASHNPMDYNGMKLVREGARPISGDTGLRDVQRLAEANDFPPVDETKRGRYQQIN  150
ECRFBK  DIGMSGTEEIYFATFHLGVDGGIEVTASHNPMDYNGMKLVREGARPISGDTGLRDVQRLAEANDFPPVDETKRGRYQQIN  148
SECPSG  DIGMSGTEEIYFATFHLGVDGGIEVTASHNPMDYNGMKLVREGARPISGDTGLRDVQRLAEAGDFPPVNEAARGSYRQIS  150
SERFBK  DIGMSGTEEIYFATFHLGVDGGIEVTASHNPMDYNGMKLVREGARPISGDTGLRDVQRLAEAGDFPPVNDAARGSYRQIS  150
ECRFB2  DIGLSGTEEIYFATFHLGVDGGIEVTASHNPMNYNGMKLVRENAKPISGDTGLRDIQRLAEENQFPPVDPARRGTLRQIS  150
XCXANA  DIGLCGTEEVYFQTDHLKAAGGVMVTASHNPMDYNGMKLVREQARPISSDTGLFAI-RDTVAADTAAAGEPTAAEHSRTD  148
         .*.  *  .*...     .*. .*.**** .***.*.. . . . .. .   . ..          .      .      .

PSALGC  ILPRYFKQIRDDIAMA---KPMKVVVDCGNGVAGVIAPQLIE-----ALGCSVIPLYCEVDGNFPNHHPDPGKPENLKDL  226
NMEPGM  ISGEYLKHITGHIRLK---RPMNIAIDAGNGVGGAFAGKLYK-----GLGNEVTELFCDVDGTFPNHHPDPSKPKNLQDL  223
NGOPGM  ISGAYHDHIVGHVKLK---RPINIAIDAGNGVGGAFAGKLYK-----GLGNEVTELFCEVDGNFPNHHPDPSKPENLQDL  223
ABEXOC  IKDAYVERLLRDDDGT---RDLTIAWDAGNGASGEDPAPPDR-----EVPGKHVLLFDEIDGNFPNHHPDPTVEKNLVDL  224
ECCPSG  LRDAYVDHLFGYINVKNLT-PLKLVINSGNGAAGPVVDAIEARFKALGAPVELIKVHNTPDGNFPNGIPNPLLPECRDDT  229
ECRFBK  LRDAYVDHLFGYINVKNLT-PLKLVINSGNGAAGPVVDAIEARFKALGAPVELIKVHNTPDGNFPNGIPNPLLPECRDDT  227
SECPSG  LRDAYIDHLLGYISVNNLT-PLKLVFNAGNGAAGPVIDAIEARLKALGAPVEFIKIHNTPDGTFPNGIPNPLLPECRDDT  229
SERFBK  LRDAYIDHLLAYISVNNLT-PLKLVVNSGNGAAGPVIDAIEARLKALGAPVEFIKIHNTPDGTFPNGIPNPLLPECRDDT  229
ECRFB2  VLKEYVDHLMGYVDLANFTRPLKLVVNSGNGAAGHVIDEVEKRFAAAGVPVTFIKVHHQPDGHFPNGIPNPLLPECRQDT  230
XCXANA  -KTAYLEHLLSYVDRSTL-KPLKLVVNAGNGGAGLIVDLLAPHL-----PFEFVRVFHEPDGNFPNGIPNPLLQENRDAT  221
            *  ...            .... .  ***  .*                  . . .   ** ***  *.*   .   .

PSALGC  IAKVKAENADLGLAF [DGDGDR] VGVVTNTGTIIYPDRLLMLFAKDVVSRNPGADIIFDVKCTRRLIALISGYGGRPVMWKT  306
NMEPGM  IAALKNGDAEIGLAF [DGDADR] LGVVTKDGNIIYPDRQLMLFAQDVLNRNPGAKAIFDVESTRLVAPWIKEHGGKAIMEKT  303
NGOPGM  IAALKNGDAEIGLAF [DGDADR] LGVVTKDGNIIYPDRQLMLFAQDVLNRNPGAKVIFDVKSTRLLAPWIKEHGGEAIMEKT  303
ABEXOC  KAAVAEHGCDIGIGF [DGDGDR] IGAIDHLGRVVWGDQLVAIYAADVLKSHPGATIIADVKASQTLFDEIARLGGNPLMWKT  304
ECCPSG  RNAVIKHGADMGIAF [DGDFDR] CFLFDEKGQFIEGYYIVGLLAEAFLEKNPGAKIIHDPRLSWNTVDVVTAAGGTPVMSKT  309
ECRFBK  RNAVIKHGADMGIAF [DGDFDR] CFLFDEKGQFIEGYYIVGLLAEAFLEKNPGAKIIHDPRLSWNTVDVVTAAG-TPVMSKT  306
SECPSG  RKAVIEHGADMGIAF [DGDFDR] CFLFDEKGQFIEGYYIVGLLAEAFLEKHPGAKIIHDPRLTWNTEAVVTAAGGTPVMSKT  309
SERFBK  RKAVIEHGADMGIAF [DGDFDR] CFLFDEKGQFIEGYYIVGLLAEAFLEKHPGAKIIHDPRLTWNTEAVVTAAGGTPVMSKT  309
ECRFB2  ADAVREHQADMGIAF [DGDFDR] CFLFDDEASFIEGYYIVGLLAEAFLQKQPGAKIIHDPRLTWNTVDIVTRNGGQPVMSKT  310
XCXANA  AKAVKEHGADFGIAW [DGDFDR] CFFFDHTGRFIEGYYLVGLLAQAILAKQPGGKVVHDPRLTWNTVEMVEDAGGIPVLCKS  281
         .      ..*..    *** **    .. .    . . * . . . ..**.. . *  .   .    *   ... *.

PSALGC  GHSLIKKKMKETGALLAGEMSGHVFFKERWFGF-DDGIYSAARLLEILSQDQRDSEHVF-SAFPSDISTPEINITVTEDS  384
NMEPGM  GHSFIKSAMKETGAPVAGEMSGHIFFKERWFGF-DDGLYAGARLLEILSASDNPTE-VL-NNLPQSISTPELNIALPEGS  380
NGOPGM  GHSFIKSAMKKTGALVAGEMSGHVFFKERWFGF-DDGLYAGARLLEILSASDNPSE-VL-DNLPQSISTPELNISLPEGS  380
ABEXOC  GHAFIKERMRKEDAIYGGEMSAHHYFRD--YGF-DDALYCAVRLIGLVSKLNQPLSELR-DRLPDVVNTPETRFQVSEER  382
ECCPSG  GHAFIKERMRKEDAIYGGEMSAHHYFRD--FAYCDSGMIPWLLVAELVCLKDKTLGELVRDRMAAFPASGEINSKLAQPV  387
ECRFBK  GHAFIKERMRKEDAIYGGEMSAHHYFRD--FAYCDTGMIPWLLVAELVCLKGKTLGELVRDRMAAFPASGEINSKLAHPV  384
SECPSG  GHAFIKERMRTEDAIYGGEMSAHHYFRD--FAYCDSGMIPWLLVAELVCLKRQSLGELVRDRMAAFPASGEINSRLAEPA  387
SERFBK  GHAFIKERMRTEDAIYGGEMSAHHYFRD--FAYCDSGMIPWLLVAELVCLKGQSLGELVRDRMAAFPASGEINSRLAEPA  387
ECRFB2  GHAFIKERMRQEDAIYGGEMSAHHYFRD--FAYCDSGMIPWLLVAELLCLKNSSLKSLVADRQAAFPASGEINRKLGNAA  388
XCXANA  GHAFIKEKMRSENAVYGGEMSAHHYFRE--FAYADSGMIPWLLIAELVSQSGRSLADLVEARMQKFPCSGEINFKVDDAK  379
        **...*  *   ...   .****.* .* .  ... *...     . ....        .  .  .   . *  .  . .

PSALGC  K-FAIIEALQRDAQ--W-GEGNITTLDGVRVDYPKGWGLVRASNTTPVLVLRFEADPEEELERIKTVFRNQLKAVDSSLPVPF      463
NMEPGM  NGHQVIDELAAKAE--FEGATEIITIDGLRVEFPDGFGLMRASNTTPILVLRFEADTQEAIERIQNQFKAVIESNPNLIWPL      460
NGOPGM  NGHQVIEELAAKAE--FEGATEIITIDGLRVEFPDGFGLMRASNTTPILVLRFEADTQAAIERIQNRFKAVIESNPHLIWPL      460
ABEXOC  K-FQVVQEVEGRSSRLMAEGADVNDIDGVRVKDADGWWLLRASNTQDVLVARAESGTRRSWERLKGMVVAHWKPPASRPFLRGRRQLH 469
ECCPSG  EAINRVEQ------HFSREALAVDRTDGISMTFADWRFNLRTSNTEPVV--RLNVESRGDVPLMEARTRTLLTLLNE            456
ECRFBK  EAINRVEQ------HFSREALAVDRTDGISMTFADWRFNLRLLNTEPVV--RLNVESRGDVPLMEEKTKLILELLNK            453
SECPSG  AAIARVEA------HFAEEAQAVDRTDGLSMSFADWRFNLRSSNTEPVV--RLNVESRGDIPLMEARTRTLLALLNQ            456
SERFBK  AAIARVEA------HFAEEAQAVDRTDGLSMSFADWRFNLRSSNTEPVV--RLNVESRGDIPLMEARTKEILQLLNS            456
ECRFB2  EAIRRIRA------QYEPAAAHIDTTDGISIEYPEWRFNLRTSNTEPVV--RLNVESRADVALMNEKTTELLNLLKEELL         460
XCXANA  AAVARVMA------HYGDQSPELDYTDGISADFGQWRFNLRSSNTEPLL--RLNVETRGDAALLETRTQEISNLLRG            448
         . .           .  .    **.. .   .  .*  **  ..  *. .. .  .  ...
```

Figure 2 Amino acid sequence alignment, using the method of Higgings and Sharp (Clustal, PCGene) of phosphoglucomutase or phosphomannomutase of *Pseudomonas aeruginosa* (PSALGC) (Zielinski *et al.*, 1991), *Neisseria meningitidis* (NMEPGM) (Zhou *et al.*, 1994), *Neisseria gonorrhoeae* (NGOPGM) (Zhou *et al.*, 1994), *Azospirillum brasilense exoC* (ABEXOC), *Escherichia coli cpsG* (ECCPSG), *Escherichia coli rfbk* (ECRFBK) (Marolda and Valvano, 1993), *Salmonella enterica* (SECPSG) (Stevenson *et al.*, 1991), *Salmonella enterica rfbK* (SERFBK) (Lee et al., 1992), *Escherichia coli rfbk2* (ECRFB2) (Jayaratne *et al.*, 1994), *Xanthomonas campestris* (XCXANA) (Köplin *et al.*, 1992). Stars indicate conservative changes (S-T-A, L-V-I-M, K-R, D-E, Q-N, F-Y-W). The phosphoglucomutase and phosphomannomutase signature ([GA]-[LIVM]-X-[LIVM]-[ST]-[PGA]-S-H-X-P-X-X-X-X-[GN]) is indicated in bold (Dai *et al.*, 1992), the Mg binding site is boxed and the sugar binding site is underlined (Ye *et al.*, 1994)

Table 2 Phosphoglucomutase and phosphomannomutase activity of *A. brasilense* Sp7 and *exoC*.

Strain or mutant	PMM (mU)	PGM (mU)
Sp7	78.5	891
exoC	7	889

PMM: phosphomannomutase activity, one unit corresponds to the conversion of 1 mmol mannose-1-phosphate per minute per mg protein.

PGM: phosphoglucomutase activity, one unit corresponds to the conversion of 1 mmol glucose-1-phosphate per minute per mg protein.

activity (Table 2). Indirect evidence suggests that the *A. brasilense exoC* mutant has a less glycosylated polar flagellum, compared to the wild type strain.

The *A. brasilense exoD* gene encodes a GDP-mannose-dehydrogenase

The *A. brasilense exoD* gene is localised upstream from the *exoC* gene. An open reading frame of 1152 bp was identified. The deduced gene product shows homology to AlgD (GDP-mannose-dehydrogenase; EC 1.1.1.132) of *P. aeruginosa* within the NH_2-terminal two thirds of the protein (results not shown) (Deretic *et al.*, 1987). GDP-mannose-dehydrogenase converts GDP-mannose into GDP-mannuronic acid.

Acknowledgements

This work was supported by the K.U.Leuven Research Fund (Postdoctoral fellowship to DJP), the FGWO (No. 3.0095.93), and Geconcerteerde Onderzoeksacties (GOA 93 Vanderleyden).

References

Adams CW, Fornwald JA, Schmidt FJ, Rosenberg M, Brawner ME (1988) Gene organization and structure of the *Streptococcus lividans gal* operon. J. Bacteriol. 170: 203-212.

Boyd J, Oza MN, Murphy JR (1990) Molecular cloning and DNA sequence analysis of the diphtheria tox iron-dependent regulatory element (*dtxR*) from *Corynebacterium diphtheriae*. Proc. Natl. Acad. Sci. USA 87: 5968-5972.

Buendia AM, Enenkel B, Köplin R, Niehaus K, Arnold W, Pühler A (1991) The *Rhizobium meliloti exoZ/exoB* fragment of megaplasmid 2: ExoB functions as a UDP-glucose 4'-epimerase and ExoZ shows homology to NodX of *Rhizobium leguminosarum* bv. strain TOM. Mol. Microbiol. 5: 1519-1530.

Coyne MJ, Russell KS, Coyle CL, Goldberg JB (1994) The *Pseudomonas aeruginosa algC* gene encodes phosphoglucomutase, required for the synthesis of a complete lipopolysaccharide core. J. Bacteriol. 176: 3500-3507.

Croes C, Van Bastelaere E, Declercq E, Eyers M, Vanderleyden J, Michiels K (1991) Identification and mapping of loci involved in motility, adsorption to wheat roots, colony morphology, and growth in minimal medium on the *Azospirillum brasilense* Sp7 90-MDa plasmid. Plasmid 26: 83-93.

Dai JB, Liu Y, Ray WJ, Konno M (1992) The crystal structure of muscle phosphoglucomutase refined at 2.7-angstrom resolution. J. Biol. Chem. 267: 6322-6337.

Deretic V, Gill JF, Chakrabarty AM (1987) *Pseudomonas aeruginosa* infection in cystic fibrosis: nucleotide sequence and transcriptional regulation of the *algD* gene. N ucleic Acids Res. 15: 4567-4581.

De Troch P, Keijers V, Vanderleyden J (1994) Sequence analysis of the *Azospirillum brasilense exoB* gene, encoding UDP-glucose 4'-epimerase. Gene 144, 143-144.

Houng HSH, Kopecko DJ, Baron, LS (1990) Molecular cloning and physical and functional characterization of tha *Salmonella typhimurium* and *Salmonella typhi* galactose utilization operons. J. Bacteriol. 172: 4392-4398.

Jayaratne P, Bronner D, MacLachlan R, Dodgson C, Kido N, Whitfield C (1994) Cloning and analysis of duplicated *rfbM* and *rfbK* genes involved in the formation of GDP-mannose in *Escherichia coli* O9:K30 and participation of *rfb* genes in the synthesis of the group I K30 capsular polysaccharide. J. Bacteriol. 176: 3126-3139.

Jennings MP, van der Ley P, Wilks KE, Maskell DJ, Poolman JT, Moxon ER (1993) Cloning and molecular analysis of tha *galE* genes of *Neisseria meningitidis* and its role in lipopolysaccharide biosynthesis. Mol. Microbiol. 10: 361-369.

Köplin R, Arnold W, Hötte B, Simon R, Wang G, Pühler A. (1992) Genetics of xanthan production in *Xanthomonas campestris*: the *xanA* and *xanB* genes are involved in UDP-glucose and GDP-mannose biosynthesis. J. Bacteriol. 174: 191-199.

Lee SJ, Romana LK, Reeves PR (1992) Sequence and structural analysis of the *rfb* (O antigen) gene cluster from a group C1 *Salmonella enterica* strain. J. Gen. Microbiol 138: 1843-1855.

Lemaire H-G, Müller-Hill B. (1986) Nucleotide sequences of the *galE* gene and the *galT* gene of *E. coli*. Nucl. Acids. Res. 14: 7705-7711.

Marolda CL, Valvano, MA (1993) Identification, expression, and DNA sequence of the GDP-mannose biosynthesis genes encoded by the O7 *rfb* gene cluster of strain VW187 (*Escherichia coli* O7:K1). J. Bacteriol. 175: 148-158

Maskell DJ, Szabo MJ, Butler PD, Williams AE, Moxon, ER (1991) Molecular analysis of a complex locus from *Haemophilus influenzae* involved in phase-variable lipopolysaccharide biosynthesis. Mol. Microbiol. 5: 1013-1022.

Metzger M, Bellemann P, Bugert P, Geider K. (1994) Genetics of galactose metabolism of *Erwinia amylovora* and its influence on polysaccharide synthesis and virulence of the fire blight pathogen. J. Bacteriol. 176: 450-459.

Michiels K, Vanderleyden J, Van Gool A, Signer ER (1988) Isolation and characterization of *Azospirillum brasilense* loci that correct *Rhizobium meliloti exoB* and *exoC* mutants. J. Bacteriol. 170: 5401-5404.

Michiels K, De Troch P, Onyeocha I, Van Gool A, Elmerich C, Vanderleyden J (1989) Plasmid localization and mapping of two *Azospirillum brasilense* loci that affect exopolysaccharide synthesis. Plasmid 21: 142-146.

Michiels KW, Croes CL, Vanderleyden J (1991) Two different modes of attachment of *Azospirillum brasilense* Sp7 to wheat roots. J. Gen. Microbiol. 137: 2241-2246.

Peng H-L, Fu T-F, Liu S-F, Chang H-Y (1992) Cloning and expression of the *Klebsiella pneumoniae* galactose operon. J. Biochem. 112: 604-608.

Poolman B, Royer TJ, Mainzer SE, Schmidt BF (1990) Carbohydrate utilization in *Streptococcus thermophilus*: characterization of the genes for aldose 1-epimerase (mutarotase) and UDP-glucose 4'-epimerase. J. Bacteriol. 172: 4037-4047.

Robertson BD, Frosch M, Van Putten JPM (1993) The role of *galE* in the biosynthesis and function of gonococcal lipopolysaccharide. Mol. Microbiol. 8: 891-901.

Stevenson G, Lee S, Romana LK, Reeves PR (1991) The cps gene cluster of *Salmonella* strain LT2 includes a second mannose pathway: sequence of two genes and relationship to genes in the *rfb* gene cluster. Mol. Gen. Genet. 227: 173-180.

Uttaro AD, Cangelosi GA, Geremia RA, Nester EW, Ugalde RA (1990) Biochemical characterization of avirulent *exoC* mutants of *Agrobacterium tumefaciens*. J. Bacteriol. 172: 1640-1646.

Wieringa RK, Terpstra P, Hoi WGJ (1986) Prediction of the occurrence of the ADP-binding ßaß-fold in proteins using an amino acid sequence fingerprint. J. Mol. Biol. 187: 101-107.

Ye RW, Zielinski NA, Chakrabarty AM (1994) Purification and characterization of phosphomannomutase/phosphoglucomutase from *Pseudomonas aeruginosa* involved in biosynthesis of both alginate and lipopolysaccharide. J. Bacteriol. 176: 4851-4857.

Zhou D, Stephens DS, Gibson BW, Angstrom JJ, McAllister CF, Lee FK, Apicella MA (1994) Lipooligosaccharide biosynthesis in pathogenic *Neisseria*: cloning, identification, and characterization of the phosphoglucomutase gene. J. Biol. Chem. 269: 11162-11169.

Zielinski NA, Chakrabarty AM, Berry A (1991) Characterization and regulation of the *Pseudomonas aeruginosa algC* gene encoding phosphomannomutase. J. Biol. Chem. 266: 9754-9763.

Biodegradation of polysaccharides by *Azospirillum*

Veerle Keijers, Sofie Dobbelaere, My Ali Bekri and Jos Vanderleyden

F.A. Janssens Laboratory of Genetics,
K.U.Leuven
W. De Croylaan 42
3001 Heverlee
Belgium

Abstract

Breakdown of plant polysaccharides by root colonising bacteria can provide them with new carbon sources. Within the genus *Azospirillum*, breakdown of pectin (and polygalacturonic acid) is best documented for *A. irakense*. From a gene library of *A. irakense* KBC1 we have isolated several cosmid clones that express polygalacturonase and pectate lyase activity in *E. coli* and *A. brasilense*. Genes encoding these enzymes appear to be closely linked in the *A. irakense* genome.

Furthermore we introduced the alpha-amylase gene of *Streptomyces venezuelae* into *A. brasilense* to study heterologous gene expression. The gene was found to be expressed from its own promoter in *A. brasilense*, and the transconjugants were able to hydrolyse soluble starch. As a side effect we observed that these transconjugants were also able to use additional carbohydrates as a carbon source, compared to the wild type strain. Currently we are testing several possibilities explaining this phenomenon.

Introduction

Associative type of beneficial bacteria plant interactions, exemplified by *Azospirillum* plant root interactions, are difficult to study because of the absence of an easily detectable plant phenotype. Recently we designed a strategy to visualise the *Azospirillum*-plant interaction by using *Azospirillum* strains expressing the *Escherichia coli gusA* gene (Vande Broek et al 1993). This allowed us to test the effect of various bacterial mutants and/or environmental parameters on *Azospirillum* wheat root interactions. Moreover, the methodology of reporter genes, such as *gusA*, makes it also possible to monitor expression of specific *A. brasilense* genes during plant root association, even in a heterologous background. This was demonstrated with the *A. brasilense nifH::gusA* fusion in *A. brasilense* and *Rhizobium meliloti* (Vande Broek et al 1992, 1993). From these studies it was concluded that *A. brasilense* strain Sp245 rarely infects plant roots and that the amount of carbon sources in the root exudates constitutes a major factor limiting associative nitrogen fixation. In this paper we discuss some possibilities on how to improve *Azospirillum* plant root association, based on biodegradation of plant polysaccharides.

NATO ASI Series, Vol. G 37
Azospirillum VI and Related Microorganisms
Edited by I. Fendrik et al.

Biodegradation of polysaccharides by natural isolates of *Azospirillum*

Azospirillum strains are unable to degrade starch and cellulose. Some strains of *A. brasilense* and *A. lipoferum* were shown to be able to grow on xylan (Halsall and Gibson 1985). Degradation of polygalacturonic acid and pectin has been reported for *A. brasilense* and *A. lipoferum* (Tien et al 1981; Plazinski and Rolfe 1985), and some characteristics of a polygalacturonic acid transeliminase of *A. brasilense* have been studied (Tien et al 1981). However, these species cannot grow on pectin or polygalacturonic acid, and the reported activities are extremely low when compared with other pectinolytic bacteria. *A. irakense*, the species most recently described (Khammas et al 1989), can grow and fix nitrogen when pectin is the sole carbon source (Khammas and Kaiser 1991). Interestingly, some *A. irakense* isolates were obtained from surface sterilised rice roots.

Pectinolytic activity of *Azospirillum irakense*

Khammas and Kaiser (1991) analysed the pectinolytic activity of seven *A. irakense* isolates, and gave evidence for the presence of two types of pectinolytic enzymes. All strains tested have inducible Ca^{2+} dependent pectate lyase activity. Six strains showed also pectin methylesterase activity. So far, none of the corresponding enzymes have been purified. Most likely *A. irakense* produces pectate lyases of both endo- and exo- activities, since it can grow on pectin as sole carbon source.

Cloning of *Azospirillum irakense* genes encoding pectinolytic activity

A cosmid library of the *A. irakense* LMG10653 genome was constructed in vector pLAFR1 and used for transfection of *Escherichia coli* DH5α cells. Approximately 3000 clones were plated on LB-agar plates containing 1% (w/v) pectin (Sigma). Plates were overlaid with a 2% (w/v) solution of hexadecyltrimethylammonium bromide. Sixteen clones showed a clear halo, indicating degradation of pectin. DNA restriction analysis of the corresponding cosmids revealed two common *Eco*RI fragments, 9.6 and 9.2 kb, for all positive clones. A restriction map of the region encoding pectinolytic enzymes could be established. The culture supernatants of some of the *E. coli* clones were tested for specific pectinolytic activities. For all clones tested, both pectate lyase and polygalacturonase activity were demonstrated spectrophotometrically, using the thiobarbituric assay (Sherwood 1966). An analysis is shown in Fig. 1.

Two cosmid clones were transferred to *A. brasilense* Sp245, and conferred, as expected, pectin degrading activity to the transconjugants. It is not known yet how many different pectinolytic enzymes and how many different genes are involved in the overall pectinolytic activity of *A. irakense*. In pathogenic bacteria, like *Erwinia* species each class of pectinolytic enzymes is represented by different isoenzymes (Salmond 1994). It is not very likely that a similar complexity will hold for *A. irakense*, since all the clones that encode pectinolytic activity map in one region of the *A. irakense* genome.

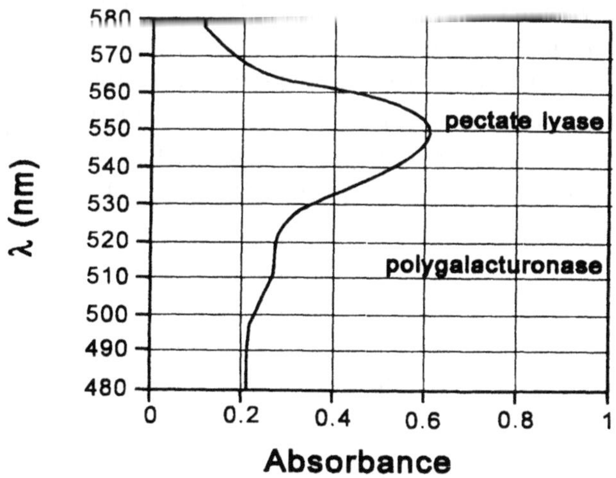

Fig. 1 Pectinolytic activity of *E. coli* transformants containing *A. irakense* DNA.

Cloning of an α-amylase gene in *Azospirillum brasilense*

In an attempt to widen the carbohydrate metabolism of *Azospirillum*, we have initiated a study on heterologous gene expression in *A. brasilense*, using α-amylase as a case study. Many bacteria are able to degrade starch. The biochemistry and genetics of starch degradation is best described for Gram-positive bacteria. A gene of *Streptomyces venezuelae* was chosen, given the similar high GC content of the genomes of *Azospirillum* and *Streptomyces*. A 4.8-kb insert of construct pIJ5050 (Virolle et al 1988), encoding the α-amylase of *S. venezuelae*, was recloned in pLAFR3. The resulting construct, pFAJ501, was mobilised in *A. brasilense* strains

Sp7 and Sp245. Although similar results were obtained for both transconjugants, specific data given below hold for Amy6, which is Sp245(pFAJ501).

Extracellular α-amylase activity of Amy6

Fig. 2 gives a schematic representation of the α-amylase activity of Amy6 tested with the Lugol assay. Amy6 was grown for 24 hrs in different minimal media, as specified in Fig. 2, and 10 microliter samples of the culture supernatants were tested on agar plates containing 0.5% (w/v) soluble starch (Janssen Chimica). Plates were incubated for 48 hrs at 28°C and subsequently stained with Lugol solution. The appearance of a clearing zone indicates the degradation of starch. From these data it is clear that Amy6 has α-amylase activity. No significant effect is seen from the addition of maltose or glucose to the growth medium. In *S. venezuelae* maltose induces and glucose represses the expression of the α-amylase. A slightly higher α-amylase activity can be noticed from culture supernatants in which starch was added during growth of the cultures. For comparison, 10 microliters of a solution containing 1 en 10 μg α-amylase per ml are used as a control. Wild type strain Sp245 did not show α-amylase activity under any of the conditions tested.

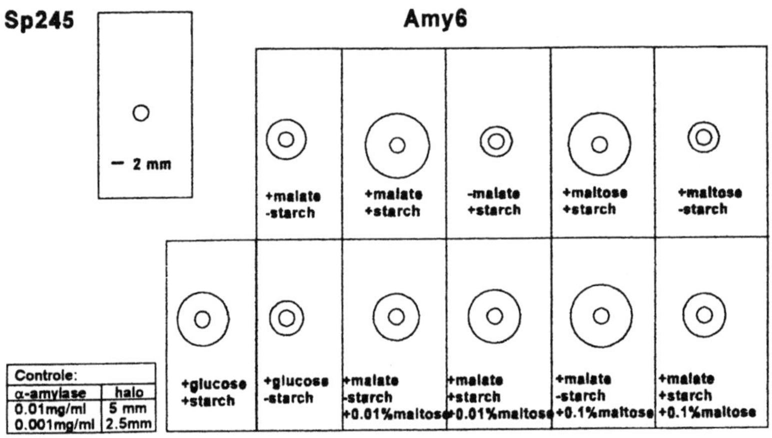

Fig. 2 α-amylase activity, of Sp245 and Amy6, grown in different minimal media, analysed by lugol-staining.

Nitrogen fixation activity of Amy6

In order to test whether α-amylase expression has an impact on the nitrogen fixation capacity of Amy6, the acetylene reduction activity (ARA) was measured. The strain was therefore inoculated in different nitrogen free minimal media, as specified in Fig. 3. It can be seen that Sp245 and Amy6 have comparable ARA on minimal medium containing malate. Substantial ARA is also present in cultures grown with glucose, maltose or starch as sole carbon source. In neither of these conditions, wild type strain showed any ARA. It was somewhat surprising to find that Amy6 can fix nitrogen on these carbohydrates, since none of them can be used as carbon source by *A. brasilense* (Khammas et al, 1989). In the case of glucose this is due to the lack of a glucose uptake system (Martinez-Drets et al, 1984). Indeed, also in our experiments Sp245 does not fix nitrogen in the presence of glucose. Here too, an effect can be seen of the presence of starch in the culture medium.

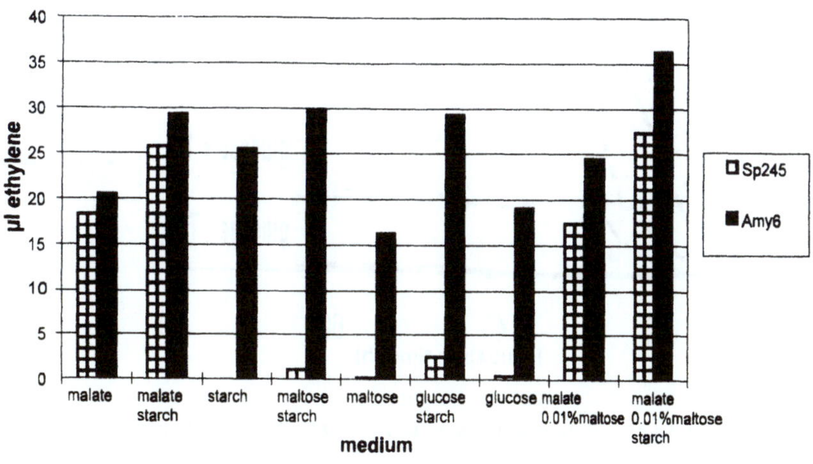

Fig. 3 Acetylene reduction activity of Sp245 and Amy6 in different minimal media

Growth of Amy6 on various carbon sources

In view of the results obtained above, growth characteristics of wild type Sp245, strain Sp245(pLAFR3) and strain Sp245(pFAJ501) were compared, as shown in Fig.4. Strain Sp245(pLAFR3) was included as a control since expression of tetracycline resistance gene might alter the membrane characteristics of *Azospirillum* and therefore explain growth of Amy6 on glucose. When Sp245(pLAFR3) is inoculated in minimal medium containing glucose and tetracycline, a slight growth can be observed. No growth is observed in the absence of tetracycline, indicating that expression of the tetracycline gene indeed allows some uptake of glucose by *A. brasilense*. Investigation of the ARA of these strains revealed the same effect of tetracycline (data not shown). Analogous results obtained for mannose and maltose suggest that in fact we are dealing with a general phenomenon.

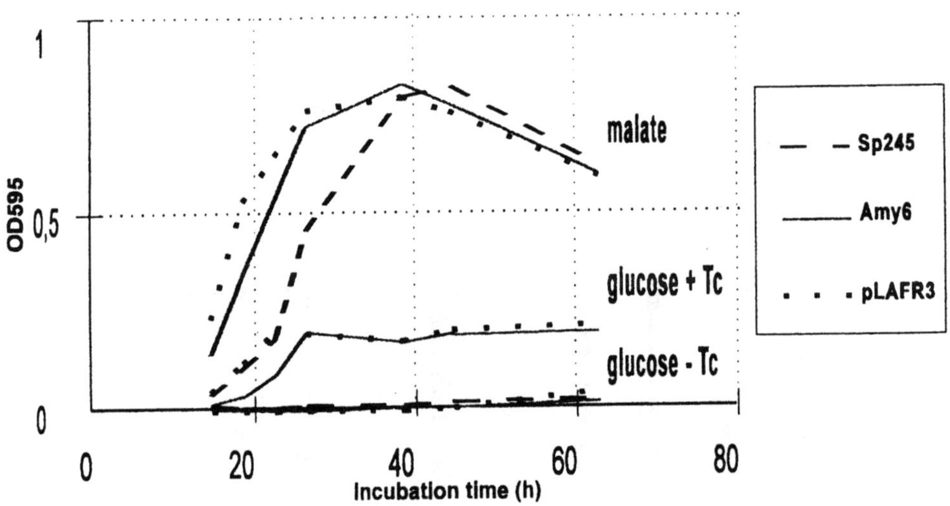

Fig. 4 Influence of Tc and carbon sources on growth of Sp245, Amy6 and pLAFR3.

Conclusions

We have reported on the cloning and expression of genes that should allow *A. brasilense* strains to use a broader range of carbon sources, more particularly polysaccharides. With respect to pectinolytic activity, we are in the process of identifying all the genes that encode pectinolytic activity in *A. irakense*. It remains to be determined what the effect will be on plants when expressing these genes in *A. brasilense*. *A. irakense* and *A. brasilense* differ with respect to auxin biosynthesis, which is absent in *A. irakense*.

Expressing α-amylase activity in *A. brasilense* gave a surprising result, in that the engineered strain was able to grow and fix nitrogen on glucose and other carbohydrates. It remains to be determined to what extent expression of the tetracycline resistance can cause the uptake of these carbohydrates in *A. brasilense* cells.

References

Halsall DM, Gibson AH (1985) Cellulose decomposition and associated nitrogen fixation by mixed cultures of *Cellulomonas gelida* and *Azospirillum* species or *Bacillus macerans*. Appl Environm Microbiol 50:1021-1026

Khammas KM, Ageron E, Grimont PAD, Kaiser P (1989) *Azospirillum irakense* sp. nov., a nitrogen fixing bacterium associated with rice roots and rhizosphere soil. Res Microbiol 140:679-693

Khammas KM, Kaiser P (1991) Characterisation of a pectinolytic activity with *Azospirillum irakense*. Plant and Soil 137:75-79

Martinez-Drets G, Del Gallo M, Burpee C, Burris RH (1984) Catabolism of carbohydrates and organic acids and expression of nitrogenase by *Azospirilla*. J Bacteriol 159:80-85

Plazinski J, Rolfe BG (1985) Analysis of the pectinolytic activity of *Rhizobium* and *Azospirillum* strains isolated from *Trifolium repens*. J Plant Physiol 120:181-187

Salmond GPC (1994) Factors affecting the virulence of soft rot *Erwinia* species: the molecular biology of an opportunistic phytopathogen. InKado CI, Crosa JH (eds) Molecular Mechanisms of Bacterial Virulence. Kluwer Academic Publishers pp193-206

Sherwood RT (1966) Pectin lyase and polygalacturonase production by *Rhizoctonia solani* and other fungi. Phytopathology 56:279-286

Tien TM, Diem HG, Gaskins MH, Hubbell DH (1981) Polygalacturonic acid transeliminase production by *Azospirillum* species. Can J Microbiol 27:426-431

Vande Broek A, Michiels J, de Faria SM, Milcamps A, Vanderleyden J (1992) Transcription of the *Azospirillum brasilense nifH* gene is positively regulated by NifA and NtrA and is negatively controlled by the cellular nitrogen status. Mol Gen Genet 232:279-283

Vande Broek A, Michiels J, Van Gool A, Vanderleyden J (1993) Spatial-temporal colonisation patters of *Azospirillum brasilense* on wheat root surface and expression of the bacterial *nifH* during the association. Molec Plant-Microb Inter 6:592-600

Virolle M.J., Long C.M., Chang S. and Bibb M.J. (1988) Cloning, characterisation and regulation of an α-amylase gene from *Streptomyces venezuelae*. Gene, 74 (1988) 321-334.

Genetical aspects of indole acetate production in *Azospirillum brasilense* Sp245.

Elena Katzy, Lilia Petrova, Igor Borisov and Valery Panasenko.
Laboratory of Microbial Genetics, Institute of Biochemistry and Physiology of Plants and Microorganisms, Russian Academy of Sciences
prospekt Entuziastov 13, 410015 Saratov
Russia

Abstract

An *Azospirillum brasilense* Sp245 spontaneous mutant, Sp245.5, lacking indigenous plasmids of 85 (p85), 120 and 130 Mda produced about two times less indole-3-acetic acid (IAA) in the presence of tryptophan (Trp) than the wild type strain. Positive influence of p85-pSUP5011 cointegrate molecule transfer into *A. brasilense* Sp245.5 on IAA production in the transconjugants obtained was distinguishable only in suboptimal for IAA biosynthesis conditions. On the contrary, IAA production in *Agrobacterium tumefaciens* GMI9023 (p85-pSUP5011) increased several times in comparison with *A. tumefaciens* GMI9023 and was even greater than in the donor *A. brasilense* Sp245.160 mutant or in the wild type Sp245 strains. Transfer of p85-pSUP5011 into *Pseudomonas putida* 53a resulted in increase of IAA excretion and decrease of anthranilic acid (Ant) release. A suggestion about the dependence of an hypothetical IAA pathway probably encoded by p85 on the presence of unidentified products of Trp breakdown through Ant or of Ant itself is given.

Key words : *Azospirillum brasilense* Sp245, *Agrobacterium tumefaciens, Pseudomonas putida,* anthranilic acid, indole-3-acetic acid, plasmid DNA.

Introduction

Bacteria of the genus *Azospirillum* are diazotrophs able to associate with roots of different grasses. Being in an intermediate position between symbiotic and free-living nitrogen-fixing bacteria, azospirilla are of substantial interest for basic research. Genetic studies on *Azospirillum* and their plasmids have been initiated in the early eighties. Only a 90 MDa plasmid of *A. brasilense* Sp7 has been studied in detail (Croes *et al.*, 1991). It is still

NATO ASI Series, Vol. G 37
Azospirillum VI and Related Microorganisms
Edited by I. Fendrik et al.

necessary to evaluate significance of other *Azospirillum* plasmids for the plant-bacterial interaction. Recently, it was shown that an 85 MDa plasmid of *A. brasilense* Sp245 is likely to be involved in the regulation of aromatic amino acid metabolism and IAA production (Katzy *et al.*, 1990a). In the present study, some more data in favour of p85 involvement in IAA production are obtained.

Materials and Methods

Bacterial strains and plasmids.

Azospirillum brasilense Sp245 was obtained from Dr. J. Döbereiner. Its mutant derivative Sp245.160 harbouring cointegrate p85-pSUP5011 was described previously (Katzy *et al.*, 1990b). An Sp245 spontaneous mutant Sp245.5 which has lost p85, p120 and p130 after long-term storage of Sp245 in a rich medium was characterized earlier (Katzy *et al.*, 1994). Transconjugants of *Agrobacterium tumefaciens* GMI9023 and *Pseudomonas putida* 53a harbouring p85-pSUP5011 mobilized from Sp245.160 (GMI9023-160 and 53a-160, respectively) were obtained earlier (Katzy, 1992).

Testing IAA production *in vivo*.

Testing IAA production *in vivo* was done as described previously (Katzy *et al.*, 1990a). In a number of experiments on *P. putida*, King's B medium (King *et al.*, 1954) supplemented with Trp, 0.02 % was used. Salkowski's reagent was used for colorimetric estimation of IAA (Tang and Bonner, 1947).

Determination of anthranilic acid production.

Ant was examined as described (Crawford and Gunsalus, 1966) at 330 nm excitation and 396 nm emission wavelengths with a Hitachi spectrofluorometer.

Cell-free extract preparation and IAA production *in vitro*.

Bacteria were grown on malate salt agar medium (MSM) (Döbereiner and Day, 1976) supplemented with trypton, 0.3 %. After 48 hrs of incubation at 32°C cells were collected and washed twice with 0.1 M potassium phosphate buffer (pH 7.4) containing EDTA, 1 mM and dithiothreitol, 1 mM. The bacteria were resuspended in the same buffer (9 g wet weight per 9 ml), frozen and disrupted in a French pressure cell. The extract was cleared up by centrifugation at 6,000 g for 30 min at 4°C. The supernatant was collected and diluted with the buffer to the final volume of 20 ml. One ml of the extract was added to 3 ml of each of 9 assay mixtures to give in the final volume 1 mM of each of the compounds listed below, except pyridoxal phosphate (Pph), the concentration of which was 0.1 mM: Trp (I); Trp, α -ketoglutaric acid (KGA) (II); Trp, L-glutamic acid (Glu) (III); Trp, KGA, Glu (IV); Trp, KGA, Pph (V); Ant (VI); Ant, KGA, Glu (VII); Ant, Trp, KGA, Glu (VIII) and only the buffer (IX). Concentration of protein in all assay mixtures was 0.6 mg/ml as determined by the method of Lowry et al. (1951). The reactions were carried out at 32°C for 90 min and stopped by the addition of 2N HCl to pH 3.0. Reaction volumes were divided into two equal parts, one of which was used for thin layer chromatography (TLC) and the other one for estimation of IAA produced. All the assays were done in triplicate.

Results and Discussion

In a previous paper (Katzy *et al.*, 1990a) we have described a Tn5-Mob mutant of *A. brasilense* Sp245 (Sp245.63) in which decreased IAA production was coupled with enhanced Ant excretion in the presence of Trp. It was shown that kanamycin resistance (KmR) of Tn5-Mob comigrated with p85 of the mutant. As a result of p85-Tn5-Mob transfer from *A. brasilense* Sp245.63, *A. tumefaciens* GMI9023 gained the ability to produce Ant on the media with Trp or trypton. To facilitate analysis of plasmid properties, we constructed *in vivo* other mobilizable derivatives of p85 (Katzy, 1992) and used one of them in the present study.

<u>IAA production in *A. brasilense* Sp245 mutants with altered plasmid pattern</u>.

Data on IAA production in A. brasilense Sp245 and its mutant derivatives Sp245.160, Sp245.5 and Sp245.5-160 (with cointegrate p85-pSUP5011 mobilized from Sp245.160) (Fig.1) are given in Table 1.

Table 1. IAA production in *A. brasilense* Sp245 and its mutants with altered plasmid pattern

	IAA (µg/ml)[a] produced in	
Strain	MSM + Trp, 0.02 %	MSM + trypton, 0.3 %
Sp245	28.5 ± 2.2	8.7 ± 1.4
Sp245.160	35.3 ± 2.5	8.8 ± 3.1
SP245.5	14.9 ± 3.7	6.5 ± 1.0
Sp245.5-160	18.2 ± 2.1	14.0 ± 2.5

[a]Values are given as the mean of 5 to 11 independent determinations ± standard deviation

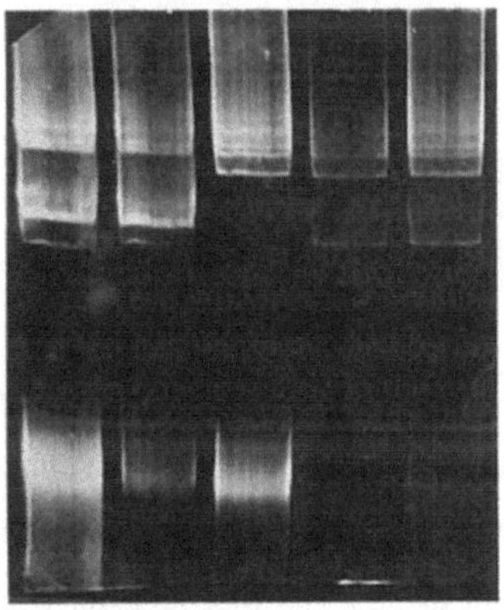

Fig.1. Plasmid pattern of *A. brasilense* Sp245 (1), Sp245.160 (2), Sp145.5 (3), and Sp245.5-160 (4,5)

The wild type strain and its mutant derivatives were inoculated at the same density, and grew at similar rates. We were unable to detect IAA production when bacteria were grown in MSM without supplements. The highest IAA concentrations were found in supernatants of the cultures grown in MSM with Trp. Strain Sp245.5 produced about two times less IAA in MSM+Trp than the wild type strain and the difference was significant. When p85-pSUP5011 was transferred into Sp245.5 from Sp245.160 with slightly increased IAA excretion, the level of IAA produced in the transconjugants in MSM+Trp was almost unaffected. IAA excretion by Sp245, Sp245.5 and Sp245.160 grown in MSM with trypton was decreased and the difference between the strains became not significant. Surprisingly, IAA production in strain Sp245.5460 grown in MSM with trypton was almost equal to that in MSM with Trp.

The conclusion derived from the assays in MSM with trypton that transfer of p85 into Sp245.5 results in two-fold increase in IAA excretion by the transconjugants is not consistent with the data obtained in the assays of bacteria grown in MSM+Trp. One of the possible explanations to this discrepancy is that a normal control mechanism of IAA production was disrupted in Sp245.5. Transfer of p85 alone was not enough for its restoration to the wild type.

It is worth-while to mention here that Sp245.5 has profound changes in a number of properties as compared with the wild type strain: it is not able to reduce nitrite and does not respond chemotactically to sugars and amino acids except glutamine and alanine; its lipopolysaccharides (LPS) do not react antigenically with antiserum against LPS of the parent strain. The mutant and Sp245 have the same colony morphology, acetylene reduction activity and pattern of outer membrane proteins except, an additional 44 kDa polypeptide in Sp245.5 (Katzy *et al.*, 1994). So far, our attempts to obtain stable complementation of LPSd and Che- mutations in Sp245.5 using recombinant plasmids from a gene bank of Sp245 in a broad host range vector plasmid pVZ361 (repRSF1010) were unsuccessful: initially restored wild type LPS and Che phenotypes were lost again after the third passage of transconjugants through selective agar.

Because of the contradiction in data obtained on *A. brasilense* Sp245.5 and Sp245.5-160, we had to search other approaches towards understanding the role of p85 in IAA production.

Influence of p85 on IAA excretion by *A. tumefaciens* and *P. putida* strains.

It has been shown earlier (Katzy, 1992) that Km^R mobilizable derivatives of p85 are stably inherited by *A. tumefaciens* and *P. putida* cells in the selective conditions. Therefore, we used these strains and their transconjugants harbouring p85 in the present study.

Changes in Ant and IAA excretion by *P. putida* cells as a result of p85 transfer were as follows: when bacteria were grown in MSM+Trp, there was decrease in Ant release from 0.32 to 0.11 μg/ml and increase in IAA production from 0.28 to 0.54 μg/ml in *P. putida* 53a-160 in comparison with *P. putida* 53a. When bacteria were grown in King's B medium, decrease in Ant excretion was from 3.16 μg/ml in *P. putida* 53a to 0.83 μg/ml in *P. putida* 53a-160. In the latter case, quantitative determination of IAA produced was not done, but TLC analysis revealed the absence of IAA spots in *P. putida* 53a and clearly seen spots with Rf and colour reactions characteristic of IAA in *P. putida* 53a-160 (data not shown). There are at least three possible explanations to the changes in IAA excretion in *P. putida* 53a-160. One is that p85 carries genes for IAA biosynthesis via a pathway independent of Ant. The second possibility is that p85 encodes an IAA biosynthetic pathway dependent on Ant or (a) product(s) of Ant conversion. Another possibility is that there was repression of *P. putida* 53a own IAA biosynthesis by Ant and decrease in Ant concentration caused by p85 promoted derepression.

Data in favor of p85 involvement in IAA production were also obtained on *A. tumefaciens* GMI9023 and its transconjugants harbouring p85 (GMI9023-160). There was an increase in IAA excretion from 27.3 μg/ml in *A. tumefaciens* GMI9023 to 73.6 μg/ml in GMI9023-160. Surprisingly, the level of IAA produced in the new host was even greater than in the donor *A. brasilense* Sp245.160 mutant strain or in the wild type Sp245 strain. This phenomenon could be explained by relaxed control of copy number and/or p85 expression in the new host.

IAA production by cell-free extracts of *A. brasilense* Sp245.

Results obtained earlier (Katzy *et al.*, 1990a; 1992) and in the present study led us to the suggestion that an hypothetical IAA pathway probably encoded by p85 might be dependent on unidentified products of Trp breakdown through Ant or on Ant itself. To test this suggestion further, experiments on the estimation of IAA production from Ant and Trp were done *in vitro* using *A. brasilense* Sp245 cell-free extracts. It was impossible to detect IAA biosynthesis from Trp without simultaneous addition of KGA or Glu, whereas in the presence of Ant alone 0.864 nM of IAA per mg of protein per min was produced. The quantity of IAA synthesised from Trp in the presence of KGA or KGA+Pph (1.992 nM $mg^{-1}min^{-1}$) was about twice as much as in the presence of Glu (0.912 nM $mg^{-1}min^{-1}$) or KGA+Glu (0.929 nM $mg^{-1}min^{-1}$). The data suggested that IAA was produced from Trp via transamination reaction. When assay mixtures contained only Ant or Ant+KGA+Glu, the quantity of IAA produced (0.864 and 0.888 nM $mg^{-1}min^{-1}$, respectively) was almost the same as in the presence of Trp+KGA+Glu. When Trp was added together with Ant+KGA+Glu, IAA production was only slightly reduced (to 0.759 nM $mg^{-1}min^{-1}$). TLC analysis revealed IAA in all the mixtures, except that with Trp alone and without supplements (data not shown). Therefore, Ant did not suppress IAA production by cell-free extracts of *A. brasilense* Sp245. Moreover, Ant seemed to be used in an IAA biosynthetic pathway independent of Trp transamination reaction. However, addition of Ant (0.02 and 0.004 %) to MSM did not stimulate IAA production in *A. brasilense* Sp245 *in vivo*. This could be explained by the existence of permeability barriers for exogenously added Ant.

Acknowledgement.

The research described in this publication was made possible in part by Grant N°RNQ000 from the International Science Foundation.

References

Crawford, I.P., Gunsalus, I.C. (1966) Inducibility of tryptophan synthetase in Pseudomonas putida. Proc. Natl. Acad. Sci. USA. 56:717-724.

Croes, C., Van Bastelaere, E., DeClercq, E., Eyers, M., Vanderleyden, J., Michiels, K. (1991) Identification and mapping of loci involved in motility, adsorption to wheat roots, colony morphology, and growth in minimal medium on the Azospirillum brasilense Sp7 90-MDa plasmid. Plasmid, 26:83-93.

Döbereiner, J., Day, J.M. (1976) In: Newton W.E. and Nymans, C.J. (ed.) Symposium on nitrogen fixation. Washington State University Pullman, p. 518-538.

Katzy, E.I. (1992) Plasmid p85 from Azospirillum brasilense Sp245: Studies on the possible host range and incompatibility with plasmids from Azospirillum brasilense Sp7. Molekul Genetika 9:8-10.

Katzy, E.I., Borisov, I.V., Mashkina, A.B., Panasenko, V.I. (1994) Plasmid content effect on the behavioural reactions in Azospirillum brasilense Sp245 associated with Gramineae. Molekul Genetika 2:29-32.

Katzy, E.I., Brodnikova, N., Egorenkov, D., Panasenko, V., Ignatov, V. (1992) Anthranilate and indole-3-acetate production in Azospirillum brasilense Sp245 mutant derivatives. Symbiosis 13:147-150.

Katzy, E.I., Iosipenko, A.D., Egorenkov, D.A., Zhuravleva E.A., Panasenko, V.I., Ignatov, V.V. (1990a) Involvement of Azospirillum brasilense plasmid DNA in the production of indole acetic acid. FEMS Microbiol. Lett., 72:1-4.

Katzy, E.I., Zhuravleva, E.A., Panasenko, V.I. (1990b) Transposon mutagenesis, elimination and mobilization of plasmids from the nitrogen-fixing bacteria Azospirillum brasilense Sp245. Molekul Genetika 2:29-32.

King, E.O., Ward, M.K., Raney, D.E. (1954) Two simple media for the demonstration of pyocianin and fluorescin. J. Lab. Clin. Med., 44:301-307.

Lowry, O.H., Rosebrough, N.J., Farr, A.L., Randall, R.J. (1951) Protein measurement with the folin phenol reagent. J. Biol. Chem., 193:265-275.

Tang, Y.W., Bonner, J. (1947) The enzymatic inactivation of indole-acetic acid. I. Some characteristics of the enzyme contained in pea seedlings. Arch. Biochem., 13:11-25.

Auxin biosynthesis and denitrification in plant growth promoting bacteria

W. Zimmer, K. Kloos, B. Hundeshagen, E. Niederau and H. Bothe
Botanisches Institut, Universität zu Köln
Gyrhofstr. 15
D- 50931 Köln
Germany

Abstract

An indole-3-pyruvate decarboxylase (=IPDC) is known to be involved in the production of indole-3-acetic acid (=IAA) in the root associated *Enterobacter cloacae* strain FERM BP-1529 and in *Azospirillum brasilense*. Segments of the IPDC gene could be amplified by PCR from several different enterobacterial species which convert tryptophan to IAA. Five of the cloned segments were sequenced. All sequences showed strong homology to the IPDC-gene but were significantly different from the published sequence from *A. brasilense* Sp 245. Conserved region, detected in the alignment of the IPDC-gene segments were used as primers for PCR to identify this gene in *Zea mays*. An amplification product and a positive hybridization signal were the first genetical hints for the existence of this gene in higher plants. As the ability to produce the plant hormone was found not to be limited to nitrogen fixing plant growth promoting bacteria, but occurred also in the non-nitrogen fixing, free living *Klebsiella aerogenes*, the bacterial production of IAA might have other functions than promoting plant growth.
To study the distribution of the denitrification ability in the genus *Azospirillum* by DNA-hybridization, oligonucleotides were synthesized for the genes for the dissimilatory cytochrome cd_1 nitrite reductase (*nirS*), the N_2O reductase (*nosZ*) and the dissimilatory Cu-nitrite reductase (*nirK*). These primers were used for PCR with the different *Azospirillum* species. In case of *A. irakense* KA3 an amplificate was obtained with the *nirK* oligonucleotides. Cloning, sequencing and hybridization of the fragment established the identity of the segment as part of the Cu-nitrite reductase gene, the occurrence of which was unknown for *Azospirillum*. In case of *A. brasilense* Sp7, a segment of the cytochrome cd_1 nitrite reductase gene, *nirS*, and a segment of the N_2O reductase gene, *nosZ*, could be amplified by PCR with strong sequence homology to the corresponding genes from *P. stutzeri*. Whereas the *nosZ* segment cross-hybridized also with DNA from all other *A. brasilense* and *A. lipoferum* strains investigated and with *A. halopraeferens*, the *nirS* segment gave only positive hybridization signals for *A. brasilense* Sp7 and strain Cd. *nirS* could be localized on a cosmid of a genomic gene bank of strain Sp7.

Production of the auxin indole-3-acetic acid by species of the Enterobacteriaceae.

Among free living, root associated and symbiotic bacteria, the production of the plant hormone indole-3-acetic acid (=IAA) is catalyzed either by (i) a tryptophan (=Trp) monoxygenase and an indole-3-acetamide hydrolase as in the phytopathogenic *Pseudomonas syringae* (Yamada et al. 1985), (ii) a nitrilase as in *Alcaligenes faecalis* (Kobayashi et al. 1993), (iii) an unspecific Trp side chain oxidase as in *Pseudomonas fluorescens* (Oberhänsli et al. 1991), or by (iv) an indole-3-pyruvate decarboxylase (=IPDC) as in *Enterobacter cloacae* strain FERM BP-1529 isolated from cucumber (Koga et al. 1991) and in *Azospirillum* (Bothe et al. 1994, Costacurta et al. 1994) where anthranilate is probably involved in the regulation of the IAA biosynthesis (Zimmer et al. 1991). Other strains of *Enterobacter* were described to be incapable in IAA formation (Koga et al. 1991). However, studying the ability of IAA formation among different species of the *Enterobacteriaceae*, also other *Enterobacter cloacae* strains (NCIMB11461 and NCIMB11463), *E. agglomerans* strains (strain 333 and 339, from Prof. W. Kling-

NATO ASI Series, Vol. G 37
Azospirillum VI and Related Microorganisms
Edited by I. Fendrik et al.

müller), *Pantoea agglomerans* IMET11328 (from Dr. S. Ruppel/Dr. C. Scholz), *Klebsiella aerogenes* DSM681 and *K. oxytoca* DSM3539 were found to convert Trp to IAA in amounts comparable to *Azospirillum* (Zimmer et al. 1994). In contrast, *E. coli* K12 was incapable in this conversion (Zimmer et al. 1994). As it was intended to identify the conserved regions of the IPDC known to be involved in the IAA production in *E. cloacae* FERM BP-1529, oligonucleotide primers were synthesized for different regions of the corresponding gene. One pair of these oligonucleotides allowed to amplify a segment of the predicted size by PCR with DNA of each of the seven IAA producing enterobacterial species but not with *E. coli* K12 indicating that the IPDC gene is spread among several species of the *Enterobacteriaceae* (Zimmer et al. 1994).

Figure 1: Identification of conserved regions at the alignment of enterobacterial IPDC amino acid sequences

```
                                    ------>
E.agg.333   NGIAGSYAEYVPVVHIVGAPALTSQRKGELLHHTLGDGEFSHFMRMSAPVSVAQASLTPENALAEIDRVIEEVMYHSRPGYLLLPSDVA
E.agg.339   NGIAGSYAEYVPVVHIVGAPRLTSQRKGELLHHTLGDGEFCHFMRMSAPVSVAQASLTPENALAEIDRVIEEVMYSSRPGYLLLPSDVP
E.clo.1529  NGIAGSYAEHVPVLHIVGAPGTAAQQRGELLHHTLGDGEFRHFYHMSEPITVAQAVLTEQNACYEIDRVLTTMLRERRPGYLMLPADVA
E.clo.11461 NGVAGSYAEYVPLLHIVGAPCSGVQQRGELLHHTLGDGDFHHFYRMSEPVTAARAILTAQNACYEIDRVLEVMLMQSRPGYLMLPADVA
E.clo.11463 NGIAGSYAEYVPLLHIVGAPCSGVQQRGELLHHTLGDGDFHPFYRMSESVTAARAILTAQNACYEIDRVLEVMLTQSRPGYLMLPADVA
K.aer.681   NGIAGSYAEHIPVLHIVGAPSTGAQQPGELLHHTLGDGDFPSFARMTEQITCSQALLTAGNAANEIDRVLRDMLTHHRPGYLIVPADVA
            **.****** .*..******   *. ***********.*  * .*.   .. ..* **  ** *****.  ..   *****..*.**.

E.agg.333   ALPVSTRAHALPARQPPFSPSSLEA
E.agg.339   ALPVSTRAHALPARQPPFSPSSLEAFTAAAQTQLRGANRVSLLADFLADRFKVKKALEQWMEEVPLGTSTLLMGKGLFNEQQAHFAGTY
E.clo.1529  KKAATPPVNALTHKQAHADSACLKAFRDAAENKLAMSKRTALLADFLVLRHGLKHALQKWVKEVPMAHATMLMGKGIFDERQAGFYGTY
E.clo.11461 KKPATPPVNALTTPPFPVNEACLNAFRQHAQAMLSVGSSVALLADFLAA-FGFIQMLQCWMTDQLPTHTTVILGKGIVKRKHGGFSYPY
E.clo.11463 KKPATPPVNALTIPPFPVNEACLNA
K.aer.681   R-AGTLPQPALRVEPPAVKPAC-RV
              . .  **     .   *  *.  *  .  .******       *. *. .    .*...***.   ... *  *
                    <-----
E.agg.339   SGAASAPSTKEAIEGADVVITVGVKFTDTITAGFYQQRPSGNASTSALFAARAIRSSITGRSKNGQRDAPPPPLNLAFRPGRLILLNTT
E.clo.1529  SGSASTGAVKEAIEGADTVLCVGTRFTDTLTAGFTHQLTPAQTIEVQPHAARVGDVWFTGIPMNQAIETLVELCKQHVHAGLMSSSSGA
E.clo.11461 HGAGGQRGPVPAIEGADTVICIGTRFTDTVTAGFTQQFARERTIEIQPFAARVGDVWFSGIPMREAINALIPLVRARAHAAANHRPPQA
E.clo.11463                                                             VGDVWFSGIPMREAINALIPLVRARAYSAANHRPLQV
K.aer.681                                                               VGDHWFSGVPMDNALAALMTLCPPWRRSAYA-SGRGA
             *...  .   ****** *. .* .****.**** .*     .     ***    ..* .     .        ..

E.agg.339   HCRWSRDNSLGQYAFWQQIQDSLQPGDVLVAEQGTACFAAAALNLPQDLQFVVQPLWGSIGYTLAA
E.clo.1529  IPFPQPDGSLTQENFWRTLQTFIRPGDIILADQGTSAFGAIDLRLPADVNFIVQPLWGSIGYTLAA
E.clo.11461 ARAPVHQGTLNQETLWQTVQQFIRPDDILLADQGTAAFGAASLRLPAGAELLVQPLWGSIGYTLAA
E.clo.11463 RHAPVHQGTLNQETLWQTVQAFIRPDDILLADQGTAAFGAASLRLPAGAELLVQPLWGSIGYTLAA
K.aer.681   EAEEGPEGELTQKNFWATVQGPLRPGDIILADQGTAAFGIAAVKLPSEASLIVQPLWGSIGYTLAA
                 .  * *  .*  .*  ..*.*...*.***. *.   . **   ..**************
```

Sequences are from *Enterobacter agglomerans* 333 (*E.agg*.333, this work), *E. agglomerans 339* (*E.agg*.339, this work), *E. cloacae* FERM BP-1529 (*E.clo.*1529, translated from EMBL-No. D90214), *E. cloacae* NCIMB11461 (*E.clo.*11461, this work), *E. cloacae* NCIMB11463 (E.clo.11463, this work) and *Klebsiella aerogenes* (*K.aer*.681, this work). An asterisk indicates identical residues in all five sequences and a point indicates conservative changes.

Identification of conserved regions in the IPDC-sequences. After cloning of the IPDC segments from *E. agglomerans* 333 and 339, *E. cloacae* NCIMB11461 and NCIMB11463 and from *K. aerogenes* DSM681 the origin of the segments from these strains was successfully verified by DNA-hybridization (Zimmer et al. 1994). In order to identify conserved regions in the IPDC-gene, the segments were sequenced completely or partially (EMBL No. X80713, X80714, X80715, X80716 and X80717). As it was intended using such regions to detect the gene also in other bacterial families or even in plants, the deduced amino acid sequences were aligned and homologous and conservative amino acid exchanges were calculated by the program CLUSTAL V (Fig. 1). All sequences showed significant homology to the published IPDC-gene of *E. cloacae* FERM BP-1529, establishing that the amplified segments code for the same enzyme (Fig. 1). New oligonucleotide primers were synthesized against the identified conserved regions in the IPDC-sequence (Fig. 1). Using these primers it was possible to amplify a segment of the predicted size from *Zea mays* DNA by PCR, which positively cross-hybridized with the *Enterobacter* IPDC-segments resulting in a weak hybridization signal (data not shown). This is the first genetical indication of the presence of the IPDC gene in a higher plant.

Figure 2: Tree and distant matrix of the IPDC amino acid sequences of *Enterobacteriaceae* and *Azospirillum* compared to PDC sequences.

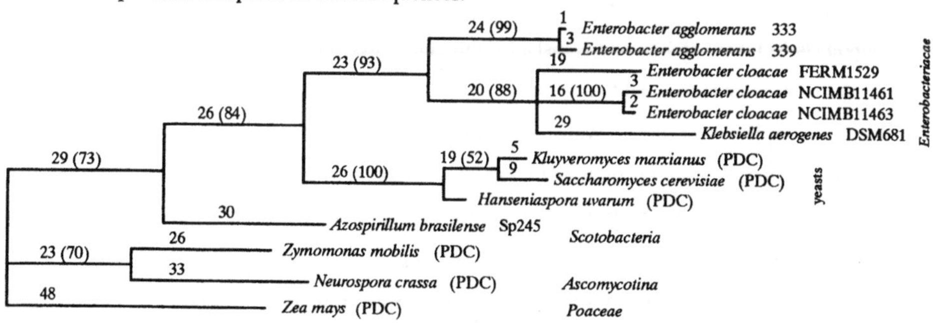

The tree was calculated by PAUP (Vers. 3.0) using a CLUSTAL V alignment of the part of the protein sequence where all 13 sequences were available (see Fig. 1). The values above the branches are the branch lengths and the probability of the grouping in brackets (in %). The amino acid sequence of the pyruvate decarboxylase (PDC) from *Zea mays* (EMBL No. P28516) was used as an outgroup for the calculation of the tree. The abbreviations of the species are as given in Fig. 1. The sequence of *Azospirillum brasilense* Sp245 (A.bras.245, translated from EMBL No. L26240) and the fungal pyruvate decarboxylase sequence of *Kluyveromyces marxianus* (*K.mar.*(PDC), EMBL No. L09727) are included in the distance matrix.

Relation of enterobacterial IPDC- and fungal PDC-sequences to the corresponding A. brasilense *gene.* The tree calculation made for the enterobacterial IPDC-sequences is in good correlation with the systematical relationship of the strains (Fig. 2) with the exception that the *Klebsiella aerogenes* sequence showed a closer distance in the matrix to the *E. cloacae* sequences than the *Enterobacter agglomerans* segments. Unexpectedly, the published sequence of *A. brasilense* Sp245 (Costacurta et al. 1994) is significantly different from the enterobacterial sequences. It is even more distant from the enterobacterial IPDC sequences than the pyruvate decarboxylase (PDC) of yeasts, described to catalyze the related but different decarboxylation of pyruvate to acetaldehyde (Fig. 2). Therefore the enzyme encoded by the published IPDC-sequence of *Azospirillum* can have a distant specificity spectrum than the enzymes from the other organisms and possibly has some additional physiological functions.

Role of IAA production in soil bacteria. In contrast to the *Enterobacter* strains, *Klebsiella oxytoca* and *K. aerogenes* are described as free living non-associative organisms. The ability of IAA-formation and the presence of sequences homologous to the IPDC-gene in these soil bacteria causes some doubt that IAA production has only a role in the bacteria-plant interaction. An alternative function might be the use of Trp as a carbon source or the decreasement of growth inhibition caused by high concentrations of Trp, an idea forwarded for *Azospirillum* (Bar and Okon, 1992).

Distribution of genes for denitrification among *Azospirillum* species.

In addition to nitrogen fixation and the production of phytohormones some associative bacteria might affect plant growth by denitrification. In this reaction, nitrate is reduced via nitrite to the gaseous nitrogen compounds NO, N_2O and N_2 and is lost as a nitrogen source for plants. Therefore, the non denitrifying strains of the root associated bacterium *Azospirillum* are believed to be more beneficial for plant growth promotion. However, nitrite, which is the first intermediate in this pathway, has been shown to mimic effects of IAA in several plant tests for auxins (Zimmer et al. 1988). There is a controversal discussion about the ability of denitrification in the different strains and species of *Azospirillum*. Whereas most *A. lipoferum* and many *A. brasilense* strains were reported to be denitrifiers (Tarrand et al. 1978, Neyra et al. 1977), *A. irakense* and *A. amazonense* strains were originally described as non denitrifying species (Khammas et al. 1989, Magalhaes et al. 1983). Later on, denitrification was found also in *A. amazonense* QR242 and Y1, but the activity was lower than in the *A. lipoferum* and *A. brasilense* strains (Neuer et al. 1985). The physiological conditions for the dissimilatory reduction of nitrite to NO were intensively studied for *A. brasilense* Sp7, where it is catalyzed by a cytochrome cd_1 containing enzyme (for a review see Hartmann and Zimmer 1993). Whereas the reduction of nitrate, nitrite and N_2O is coupled to an energy formation at the respiratory chain and supports growth of *Azospirillum* under anaerobic conditions, *Azospirillum* does not grow reducing NO to N_2O (Voßwinkel et al. 1991).

Isolation of nitrite reductase and N_2O reductase gene segments. A genetical approach served to solve the question, which *Azospirillum* species and strains are denitrifiers. As it was aimed to identify the involved genes in the different species by DNA hybridization, suitable gene probes had to be isolated. Therefore oligonucleotides were synthesized against conserved regions in the published DNA-sequences of (i) the cytochrome cd_1 nitrite reductase gene of *Pseudomonas aeruginosa* NCTC6750, (ii) the Cu-nitrite reductase of *Alcaligenes faecalis* S-6, *Pseudomonas aureofaciens* ATCC13985 and *Pseudomonas aeruginosa* G-179, (iii) the Cu-containing N_2O reductase from *Pseudomonas stutzeri* ZoBell, *Pseudomonas aeruginosa* NCTC6750 and *Alcaligenes eutrophus* H16.

Figure 3: Alignment of the deduced amino acid sequence translated from the PCR segment of *Azospirillum irakense* KA3 and *Alcaligenes xylosoxidans* with the *nirK* oligonucleotide primers

```
A.fa. GMVPWHVVSGMNGAIMVLPREGLHDGKGKALTYDKIYYVGEQDFYVPRDENGKYKKYEAPGDAYEDTVKVMRTLTPTHVVFNGAVGALTG
P.au. GMVPWHVVSGMNGALMVLPRDGLRDPQGKLLHYDRVYTIGESDLYIPKDKDGHYKDYPDLASSYQDTRAVMRTLTPSHVVFNGRVGALTG
P.ae. GMVPWHVTSGMNGAIMVLPRDGLKDHKGHELVYDKVYYVGEQDFYVPKDENGKFKKYESAGEAYPDVLEAMKTLTPTHVVFNGAVGALTG
A.xy. GMVPWHVTSGMSGTVMVLPRDGLKDPQGKPLHYDRAYTIGEFDLYIPKGPDGKYKDYATLAESYGDTVQVMRTLDAVHIVFNGKVGALTG
A.ir. GMVPWHVTSGMLRHIMVLPRDGLKDEKGKPIHYDRAYTIGEFDLYVPKDANGKYKDYESAGDSYADTGGGMKGLSH-HVVFNGAVGALTV
      ******* ***    +*******+** *   *  + ***+ * +** * *+*+    * +* *      +* *     *+ *   *+***** *****
```

Sequences are from *Alcaligenes faecalis* S-6 (*A. fa.*, translated from EMBL-No. D13155), *Pseudomonas aureofaciens* ATCC13985 (*P.au.*, translated from EMBL-No. Z21945), *Pseudomonas aeruginosa* G-179 (*P.ae.*, translated from EMBL-No. M97294), *Alcaligenes xylosoxidans* NCIMB11915 (*A. xy.*, this work) and *Azospirillum irakense* KA3 (*A. ir.*, this work). Asterisk and plus indicate identical and similar amino acids, respectively, in all five sequences (amino acids regarded as similar are: S-T-A; L-V-I-M; K-R; D-E; Q-N; F-Y-W).

In case of the Cu-nitrite reductase oligonucleotides, a segment of the predicted size could be amplified by PCR from *Alcaligenes xylosoxidans* NCIMB11015 and unexpectedly also from *Azospirillum irakense* KA3. After cloning and sequencing of the segment, the sequence comparison established the identity as segments of the Cu-nitrite reductase gene (Fig. 3). DNA-hybridization studies among the different *Azospirillum* species revealed positive signals only in case of *A. irakense* (see Table 1) and therefore established the origin of the PCR product from this strain. Whereas *A. xylosoxidans* is known to have this enzyme (Smith and Tiedje 1992), the identification of the Cu-nitrite reductase in *Azospirillum irakense* KA3 is quite surprising, as it is described as a non denitrifying *Azospirillum* species (Khammas et al. 1989). Moreover, these are the first strong hints for the presence of the copper type enzyme in the genus *Azospirillum*, where up till-now only the cytochrome cd_1 containing nitrite reductase was found (Danneberg et al. 1986 and 1989, Lalande et al. 1987, Zimmer et al. 1984).

Figure 4: Alignment of the deduced amino acid sequence translated from the PCR segment of *Azospirillum brasilense* Sp7 amplified with the *nirS* oligonucleotide primers

```
P.st. IYFERCAGCHGVLRKGATGKNLEPHWEKTEDGKKIEGGTLKLGTKRLENIIAFGTEGGMVNY--DDILTAEEINLMARYIQHTPDIPPEFSLQDMKDSWNLI
P.ae. IYFQRCAGCHGVLRKGATGKPLTPDITQ------------QRGQQYLEALITYGTPLGMPNWGSSGELSKEQITLMAKYIQHTPPQPPEWGMPEMRESWKVL
P.de. IYFERCAGCHGVLRKGATGKALTPDLTR------------DLGFDYLQSFITYGSPAGMPNWGTSGELTAEQVDLMANYLLLDPAAPPEFGMKEMRESWQVH
A.br. IYFQRCAGCHGVLRKGATAKNLEPDATK------------KLGQARLEKILAFGTE-GMPNF--EDILSKDEIKSLATYIQMTPDSPPEWGMTEMKNSWKVA
      *** ************** * * *                *   *  ++++*  ** *+     *+ + +  +* *+   *  ***+ + +*+ ** +

VPVERR--RQMNKVNLENVFAITLRDA---QLWDGDTHEIWKILDTGYAVHISRLSASGR-MSTPSAGWLTTIIDMWYPEPTTVATVRLGP-IRSVDVSKFKGYEDKY
VKPEDRPKKQLNDLDLPNLFSVTLRDAGQIALVDGDSKKIVKVIDTGYAVHISRMSASGRYLLVIGRDARIDMIDLWAKEPTKVAEIKIGIEARSVESSKFKGYEDRY
VAPEDRPTQQENDWDLENLFSVTLRDAGQIALIDGTTYEIKSVLDTGYAVHISRMSASGRYLFVIGRDGKVNMIDLWMKEPATVAEIKIGSEARSIETSKMEGWEDKY
IPPEKRPTKKMNNVNLDNVFSVTLRDAGKVALIDGDSKKIWNIVKSAYAVHISRWSASGRYVYVIGRDGKLDMIDLWMEKPEVVATIKTGLDARSVDTSKFKGYEDAY
+  * *    *   * *+*+++++++    * ** +  *  ++ + ******* ***** +         +***+   *  ** ++ *   ***+ **  *+** *
```

Sequences are from *Pseudomonas stutzeri* ZoBell (*P.st.*, translated from EMBL-No. M80653), *Pseudomonas aeruginosa* NCTC6750 (*P.ae.*, translated from EMBL-No. X16452), *Paracoccus denitrificans* (*P.de.*, translated from EMBL-No. U05002), and *Azospirillum brasilense* KA3 (*A. br.*, this work). Asterisk and plus indicate identical and similar amino acids, respectively, in all four sequences as described in the legend to Fig. 3.

The oligonucleotides for the cytochrome cd_1 nitrite reductase and the N_2O reductase allowed to amplify a segment of the predicted size by PCR in case of *Azospirillum brasilense* Sp7 and from *Pseudomonas stutzeri* which was used as a control strain. The amplification products were cloned and sequenced. Strong homologies were found to the published sequences in case of the nitrite reductase segment of *A. brasilense* and *P. stutzeri* (Fig. 4) and for the N_2O-reductase segments (39% identical and 43% similar residues, data not shown) establishing that the PCR products corresponded to the desired gene regions.

Table 1: Distribution of the nitrite- and nitrous oxide- reductase genes within the genus *Azospirillum*, monitored by DNA-hybridization.

strain	NO_2^--reductase		N_2O-reductase
	nirS	*nirK*	*nosZ*
A. brasilense Sp7	~20	-	>20
A. brasilense 7030	-	-	>20
A. brasilense Cd	>20	-	>20
A. lipoferum Sp59b	-	-	~18
A. amazonense Y1	-	-	-
A. halopraeferens AU4	-	-	~19
A. irakense KA3	-	5, 2.5	-

Genomic DNA, completely digested by *Eco*RI, was blotted onto nylon membrane. Hybridization was performed in 5 x SSC, 0.02% SDS at 68°C for 16h with digoxigenin labeled *nir*S-segments (cytochrome cd_1 nitrite reductase gene) from *Pseudomonas aeruginosa* DSM6195 and *A. brasilense* Sp7, a *nir*K-segment (Cu-nitrite reductase gene) from *Alcaligenes xylosoxidans* NCIMB11015 and *nosZ*-segments (nitrous oxide reductase gene) from *Pseudomonas stutzeri* ZoBell (Bothe et al. 1994) and *A. brasilense* Sp7. The sizes of the positive hybridizing signals on the Southern blot fragments are given in kb. (-) indicates no hybridization signal.

Study of the distribution of the denitrification genes in the genus Azospirillum. DNA-hybridization with the N_2O-reductase segment from *P. stutzeri* and *A. brasilense* revealed positive signals for all *A. brasilense* and *A. lipoferum* strains and for *A. halopraeferens* but not for *A. amazonense* and *A. irakense* (Table 1). In contrast, the cytochrome cd_1 nitrite reductase gene segment from *P. aeruginosa* and *A. brasilense* Sp7 hybridized only against digested DNA from *A. brasilense* Sp7 and the closely related strain *A. brasilense* Cd (Table 1). The fact, that *A. lipoferum* Sp59 and *A. amazonense* Y1 definitively known as denitrifiers from physiological investigations (Neuer et al. 1985) did not show any hybridization signal either with the two different *nirS* or the two different *nirK* probes, is unexpected and surprising. The segment of the nitrite reductase gene of *A. brasilense* Sp7 which was used as a probe could possibly represent a non conservative part of the gene which allowed to detect the gene only in the closely related species. Therefore this experiment will be repeated with a longer segment of the gene. The diverse nitrite reductases of *Azospirillum* possibly have different phylogenetic origins and may have been introduced into *Azospirillum* via conjugational transfer from different non related denitrifying soil bacteria. Interestingly, *A. brasilense* 7030, which is a mutant from *A. brasilense* Sp7 without the 115 MDa megaplasmid, did not show a hybridization signal with the *nirS*-probes (Table 1). This is a strong hint for a plasmid localization of *nirS* at least in *A. brasilense* Sp7 which could explain the differences in the distribution of the denitrification genes in the genus *Azospirillum*. However, it has to be studied in more detail, whether nitrite reductase genes are always localized on plasmids, which would explain the existence of non-denitrifying strains and closely related denitrifying strains.

Isolation of a cosmid clone carrying the nirS *gene from* Azospirillum. Aiming to obtain a longer segment of *nirS* for further hybridization studies, the PCR segment of the cytochrome cd_1 nitrite reductase gene from *A. brasilense* Sp7 was used for a screening of a genomic gene bank of *A. brasilense* Sp7, which was constructed cloning the genomic fragments after a partial *Sal*I digest into the broad host range vector pVK100. Among 500 clones one positive signal could be detected which allowed the localization of the *nirS* gene on the map of the corresponding cosmid pVI2 (Fig. 5).

Figure 5: Localization of the nitrite reductase gene *nirS* on the restriction map of pVI2, a cosmid of a genomic bank of *Azospirillum brasilense* Sp7

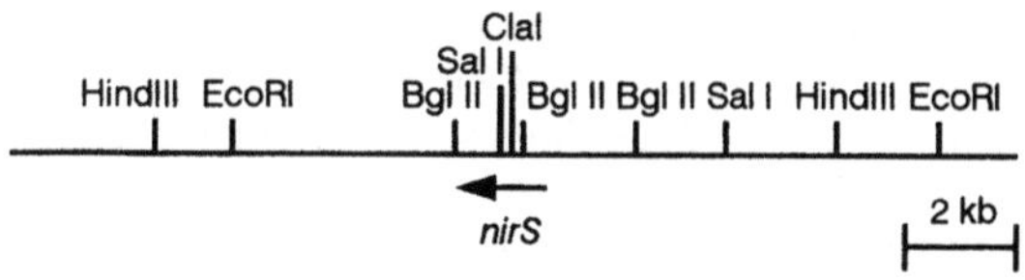

The cosmid pVI2 is a derivative of the broad host range vector pVK100, harbouring a fragment of the *A. brasilense* Sp7 genome partially digested by *Sal*I in the *Xho*I site of the kanamycin-resistance gene.

As the gene is located in the middle of the cloned segment, the cosmid probably contains the entire nitrite reductase cluster of *A. brasilense* Sp7. Therefore, the organisation of the genes and their regulation can be studied now. A subcloned *Bgl*II fragment of this cosmid, intragenic in the *nirS* gene, will be used to repeat the screening of the *Azospirillum* strains for the presence of the cytochrome cd_1 nitrite reductase gene. Moreover, transfer of this cosmid into weak denitrifying strains such as *A. brasilense* Sp245 and a study of the influence on the growth of the host plant will allow to elucidate the role of denitrification in the association.

Acknowledgement

This work was kindly supported by grants of the Deutsche Forschungsgemeinschaft to H. Bothe (No. Bo 441/18-1) and to W. Zimmer (No. Zi 337/3-2).

References

Bar, T.; Okon, Y. (1992) Induction of indole-3-acetic acid synthesis and possible toxicity of tryptophan in *Azospirillum brasilense* Sp7. Symbiosis 13: 191-198

Bothe, H; Zimmer, W.; Kloos, K; Kaldorf, M. (1994) *Azospirillum* and related organisms: Ecological, physiological, biochemical and genetical aspects, Plant Soil, submitted

Costacurta, A.; Keijers, V.; Vanderleyden, J. (1994) Molecular cloning and sequence analysis of an *Azospirillum brasilense* indole-3-pyruvate decarboxylase gene. Mol. Gen. Genet. 243: 463-472

Danneberg, G.; Kronenberg, A.; Neuer, G.; Bothe, H. (1986) Aspects of nitrogen fixation and denitrification by *Azospirillum*. Plant Soil 90: 193-202

Danneberg, G.; Zimmer, W.; Bothe, H. (1989) Energy transduction efficiencies in nitrogenous oxide respirations of *Azospirillum brasilense* Sp7. Arch. Microbiol. 151: 445-453

Hartmann, A.; Zimmer, W. (1993) Physiology. In: *Azospirillum*-Grass Association. Okon, Y. (ed.) CRC Press Inc. pp. 15-39

Khammas, K.M.; Ageron, E.; Grimont, P.A.D.; Kaiser, P. (1989) *Azospirillum irakense* sp. nov., a nitrogen-fixing bacterium associated with rice roots and rhizosphere soil. Res. Microbiol. 140: 679-693

Kobayashi, M.; Izui, H.; Nagasawa, T.; Yamada, H. (1993) Nitrilase in biosynthesis of the plant hormone indole-3-acetic acid from indole-3-acetonitrile: cloning of the *Alcaligenes* gene and site directed mutagenesis of cysteine residues. Proc. Natl. Acad. Sci. USA 90: 247-251

Koga, J.; Adachi, T.; Hidaka, H. (1991) Molecular cloning of the gene for indolepyruvate decarboxylase from *Enterobacter cloacae*. Mol. Gen. Genet. 226: 10-16

Lalande, R.; Knowles, R. (1987) Cytochrome content of *Azospirillum brasilense* Sp7 grown under aerobic and denitrifying conditions. Can. J. Microbiol. 33: 151-156

Magalhaes, F.M.; Baldani, J.I.; Souto, S.M.; Kuykendall, J.R.; Döbereiner, J. (1983) A new acid-tolerant *Azospirillum* species. Ann. Acad. Bras. Cien. 55: 417-430

Neuer, G.; Kronenberg, A.; Bothe, H. (1985) Denitrification and nitrogen fixation by *Azospirillum*. III. Properties of a wheat-*Azospirillum* association. Arch. Microbiol. 141: 364-370

Neyra, C.A.; Döbereiner, J.; Lalande, R.; Knowles, R. (1977) Denitrification by N_2-fixing *Spirillum lipoferum*. Can. J. Microbiol. 23: 300-305

Oberhänsli, T.; Defago, G.; Haas, D. (1991) Indole-3-acetic acid (IAA) synthesis in the biocontrol strain CHA0 of *Pseudomonas fluorescens*: role of trytophan side chain oxidase. J. Gen. Microbiol. 137: 2273-2379

Smith, G.B.; Tiedje, J.M. (1992) Isolation and characterization of a nitrite reductase gene and its use as a probe for denitrifying bacteria. Appl. Environ. Microbiol. 58: 376-384

Tarrand, J.J.; Krieg, N.R.; Döbereiner, J. (1978) A taxonomic study of the *Spirillum lipoferum* group, with descriptions of a new genus, *Azospirillum* gen. nov. and two species, *Azospirillum lipoferum* (Beijerinck) comb. nov. and *Azospirillum brasilense* sp nov.. Can. J. Microbiol. 24: 967-980

Voßwinkel, R.; Neidt, I.; Bothe, H. (1991) The production and utilization of nitric oxide by a new, denitrifying strain of *Pseudomononas aeruginosa*. Arch. Microbiol. 156: 62-69

Yamada, T.; Palm, C.J.; Brooks, B.; Kosuge, T. (1985) Nucleotide sequences of *Pseudomonas savastanoi* indole acetic acid genes show homology with *Agrobacterium tumefaciens* T-DNA. Proc. Natl. Acad. Sci. USA 82: 6522-6526

Zimmer, W.; Aparicio, C.; Elmerich, C. (1991) Relationship between tryptophan biosynthesis and indole-3-acetic acid production in *Azospirillum*: identification and sequencing of a *trpGDC* cluster. Mol. Gen. Genet. 229: 41-51

Zimmer, W.; Hundeshagen, B.; Niederau, E. (1994) Demonstration of the indolepyruvate decarboxylase gene in different auxin producing species of the *Enterobacteriaceae*, Can. J. Microbiol. (in press)

Zimmer, W.; Roeben, K.; Bothe, H. (1988) An alternative explanation for the plant growth promotion by bacteria of the genus *Azospirillum*. Planta 176: 333-342

Zimmer, W.; Penteado Stephan, M.; Bothe, H. (1984) Denitrification by *Azospirillum brasilense* Sp7. I. Growth with nitrite as respiratory electron acceptor. Arch. Microbiol. 138: 206-211

Azospirillum Phylogeny Based on *rrs* (16S rRNA Gene) Sequences

René Bally, Jacqueline Haurat and Philippe Normand

Laboratoire d'Ecologie Microbienne du Sol,
URA C.N.R.S. 1450, Université Claude Bernard Lyon -I-
43 Bd. du 11 novembre 1918
69622 Villeurbanne cedex
France

Abstract:

Partial 16S rDNA sequence was determined for the five bacterial species belonging to the genus *Azospirillum* and were compared to ribosomal sequences from other bacteria belonging to the alpha subdivision of the Proteobacteria. A tree obtained by using the Neighbor-joining method was presented. The five *Azospirillum* species, A. *lipoferum, A. brasilense, A. irakense, A. amazonense* and *A. halopræferens,* were grouped into a significant cluster which has been found to be closely related to *Magnetospirillum magnetotacticum* and *Rhodospirillum rubrum.*

Keywords: *Azospirillum, Magnetospirillum, Rhodospirillum,* Proteobacteria, PCR, ribosomal, phylogeny

Abbreviations: rRNA, ribosomal RNA; *rrs,* ribosomal RNA small-encoding gene; PCR, polymerase chain reaction; bp, base pair(s); rDNA, ribosomal DNA;

Introduction

Nitrogen fixing bacteria of the genus *Azospirillum* are associated with roots of grasses from various geographical origins. The genus *Azospirillum* is generally described as composed of five species on the basis of physiological, morphological and molecular properties (DNA-DNA hybridizations): *A. lipoferum, A. brasilense* (Tarrand et al. 1978), *A. amazonense* (Magalhaes et al. 1983), *A. halopræferens* (Reinhold et al. 1987) and *A. irakense* (Khammas et al. 1989).

rRNA sequences are used to measure phylogenetic and evolutionary relationships among bacteria. The "GenBank release of February 1994 listed 2857 complete or partial 16S sequences and 540 for the 23S gene, making the former the reference of choice. The 16S rRNA gene is a patchwork of conserved and non-conserved regions making it a suitable molecule for employing new approaches for bacterial characterization such as PCR amplification and DNA sequencing.

In this study we have used the same approach to measure the phylogenetic relationships between of the genus *Azospirillum*, the closely related bacteria *Rhodospirillum* and *Magnetospirillum* and some bacteria belonging to the alpha subdivision of the Proteobacteria (Woese, 1987).

NATO ASI Series, Vol. G 37
Azospirillum VI and Related Microorganisms
Edited by I. Fendrik et al.

Materials and Methods:

Bacteria:
Azospirillum spp. listed in Table 1 were grown in nutrient broth (Difco), pH 6.8 and maintained in Nfb malate medium except for *A. halopraeferens* which was grown on SMI medium (Reinhold et al., 1987).

Table 1. List of bacterial strains used.

species	strain	origin	source or reference
Azospirillum lipoferum	4B	rice, France	Bally *et al.*, 1983
	Sp.BR.17	Maize, Brazil	LMG1264
	USA5a	soil isolate	LMG4384
	$59b^T$	wheat, brazil	Tarrand *et al.*, 1978
	CRT1	maize, France	LIPHA (France)
Azospirillum brasilense	$Sp7^T$	*Digitaria*,Brazil	LMG1263
	Sp245	wheat, Brazil	Dobereiner, 1984
Azospirillum amazonense	$Y1^T$	*Digitaria*,Brazil	Baldani (Brazil)
Azospirillum irakense	KBC1	rice, Irak	Khammas *et al.*, 1989
Azospirillum halopraeferens	$Au4^T$	*Leptochloa fusca* Pakistan	Rheinhold *et al.*, 1987

DNA manipulations:
The total DNA of mid-log phase bacteria was isolated according to Simonet et al. (1990). Total DNA was amplified using primers FGPS6-63 (5'-GGAGAGTTAGATCTTGGCTC-3') and FGPS659'-64 (5'-CACCGCTACACCAGGAATTC-3'), which contain sites for *Bgl* II and *Eco* RI respectively.

P.C.R. amplification:
Double- stranded amplification was performed using a standard protocol (Nazaret et al. 1991) yielding intense DNA bands of the expected size (i.e. 650 bp, not shown). Template DNA was mixed with 20mM of dNTP, 2mM of primers and 2.5 units of Taq DNA polymerase in the reaction buffer for a final volume of 50 ml overlayed by parrafin oil. The reaction was carried out for 30 cycles: (denaturation 93°C, 1 mn; annealing 55°C, 1 mn; extension 72°C, 2 mn).

Sequencing:
The dsDNA fragments were sequenced in both directions according to Sanger et al. (1977) as modified by Winship (1989) using primers FGPS6-63 (described above) and FGPS 305'-78 (5'-CCAGTGTGGCCGGTCGCCCTCTC-3') and a T7 sequencing kit (Pharmacia, Sweden). In the case of *A. halopraeferens* and *A. lipoferum USA5a* strains, the sequences were unreadable due to ladders and artifactual bands, the amplified fragments were therefore digested

by *Bgl* II and *Eco* RI which have sites in the primers (underlined), cloned into *Bam* HI-*Eco* RI cut pBluescript II SK- vector (Stratagene, La Jolla, CA) and transformed into *E. coli* DH5αF' (BRL) as described by Maniatis et al. (1982). These primers frame a structural domain (region "A") of the 16S rDNA gene known to be highly variable (Pernodet et al., 1989) which contains helices 6, 9 and 10 (hypervariable regions V1, V2 and V3 of Raué et al., 1988)

Sequencing analysis:
The 230 bp DNA fragments were aligned using the Clustal (Higgins and Sharp, 1988) program. Pairwise evolutionary distances between DNA fragments were computed according to Jukes and Cantor (1969) using perfect matches only and the length of the shorter sequence without gaps as denominator with the Poisson correction for multiple substitutions (Nei, 1987). The sequences considered for distance calculation exclude positions 25 to 41, and positions 131 to 154 (Fig. 1). The Neighbour-Joining method (Saitou and Nei, 1987) was used to obtain unrooted gene trees. Rooted trees were obtained using the KITSCH algorithm of the Phylogeny Inference Package (Felsenstein, 1985) and the UPGMA (Sneath and Sokal, 1973) algorithm.

Results

The alignment of partial DNA sequences of the *Azospirillum* spp. strains 16S rRNA is given in Fig. 1. The five *Azospirillum lipoferum* strains have between 3 and 17 substitutions, resulting in a difference of between 1.5 and 8.6 and in genetic distances of between 0.015 and 0.091 substitutions/site. The two closest strains are BR17 and CRT1. In contrast, the two *A. brasilense* strains would be in the lower range with 3.5 % of differences. The tree obtained with the above-given alignment using the Neighbour-joining method is presented in Fig. 2. The five *Azospirillum* species *A. lipoferum, A. brasilense, A. amazonense, A. irakense* and *A. halopræferens* strains are grouped into a coherent cluster which is close to *Magnetospirillum magnetotacticum* (formerly *Aquaspirillum;* Eden et al., 1991).

```
                         1          11         21         31         41         51
 1 A-lipoferum-4B        ACACATGCAA GTCGAACGGT GCCT-TC--- ------GGGC ACAGTGGCGC ACGGGTGAGT
 2 A-lipoferum-BR17      ACACATGCAA GTCGAACGAA GGCT-TC--- ------GGCC TTAGTGGCGC ACGGGTGAGT
 3 A-lipoferum-USA5a     ACACATGCAA GTCGAACGAA GGCT-TC--- ------GGCC TTAGTGGCGC ACGGGTGAGT
 4 A-lipoferum-59b       ACACATGCAA GTCGAACGAG GGTT-TCCTT CGGGG-GGCC TTAGTGGCGC ACGGGTGAGT
 5 A-lipoferum-CRT1      ACACATCCAA GTCGTTCGAA GGCT-TC--- ------GGCC TTAGTGGCGC TCGGGTGAGT
 6 A-brasilense-SP7      ACACATGCAA GTCGAACGAA GGCT-TC--- ------GGCC TTAGTGGCGC ACGGGTGAGT
 7 A-brasilense-SP245    ACACATGGGA GTCGAACGAG GGCT-TC--- ------GGCC TTAGTGGCGC ACGGGTGAGT
 8 A-amazonense-Y1       ACACATGCAA GTCGAACGAA GGCT-TC--- ------GGCC TTAGTCGCGC ACGGGTGAGT
 9 A-irakense-KBC1       ACACATGCAA GTCGAACGCA ---T-CGC-- -AACA----- TGAGTGGCGC ACGGGTGAGT
10 A-halopraeferensAu4   ACACATGCAA GTCGAACGAA GGCT-TC--- ------GGCC TTAGTGGCGC ACGGGTGAGT
11 Magnetospirillum-ma   ACACATGCAA GTCGAACGAA ---G-TC-TT CGGAC----- TTAGTGGCGC ACGGGTGAGT
12 Azorhizobium          ACACATGCAA GTCGAACGGG CCC-TTCG-- --------GG TCAGTGGCAG ACGGGTGAGT
13 Rhizobium-meliloti    ACACATGCAA GTCGAGCGCC --CGCAAG-- ---------G GGAGCGGCAG ACGGGTGAGT
14 Rhodospirillum-rubr   ACACATGCAA GTCGAACGCA ---T-CCCTT CGGGA----- TGAGTGGCGC ACGGGTGAGT
15 Methylosinus-tricho   ACACATGCAA GTCGAACGGG -CGCAGC--- GCATGCGTAG ACAGTGGCAG ACGGGTGAGT
16 Bradyrhizobium        ACACATGCAA GTCGAGCGGG -CGTAGC--- ----AATACG TCAGCGGCAG ACGGGTGAGT
17 Rhodopseudomonas      ACACATGCAA GTCGAACGGG -CGTAGC--- ----AATACG TCAGTGGCAG ACGGGTGAGT
18 Erythrobacter-longu   ACACATGCAA GTCGAACGAA CCCT-TC--- -------GGG TTAGTGGCGC ACGGGTGGCT
```

```
                           61         71         81         91         101        111
 1 A-lipoferum-4B          AACACGTG-G GAACCTGCCT TTCGGTTCGG AATAACGTTT GGAAACGA-A CGCTAACACC
 2 A-lipoferum-BR17        AACACGTG-G GAACCTGCCT TTCGGTTCGG AATAACGTCT GGAAACGG-A CGCTAACACC
 3 A-lipoferum-USA5a       AACACGTG-G GAACCTGCCT TATGGTTCGG AATAACGTCT GGAAACGG-A CGCTAACACC
 4 A-lipoferum-59b         AACACGTG-G GAACCTGCCT TTCGGTTCGG AATAACGTCT GGAAACGG-A CGCTAACACC
 5 A-lipoferum-CRT1        AACACGTG-G GAACCTGCCT TTCGGTTCGG AATAACGTCT GGAAACGG-A CGCTAACACC
 6 A-brasilense-SP7        AACACGTG-G GAACCTGCCT TTCGGTTCGG GATAACGTCT GGAAACGG-A CGCTAACACC
 7 A-brasilense-SP245      AACACGTG-G GAACCTGCCT TTCGGTTCGG GATAACGTCT GGAAACGG-A CGCTAACACC
 8 A-amazonense-Y1         CACGCGTG-G GAACCTGCCC TTTGGTTCGG AATAACTCCG GGAAACTG-G AGCTAATACC
 9 A-irakense-KBC1         AACACGTG-G GAATGTGCCC TTTGGTTCGG AATAACGTCT GGAAACGG-A CGCTAATACC
10 A-halopraeferensAu4     AACACGTG-G GAACCTGCCT TTCGGTTCGG AATAACTCCG GGAAACCTGG AGCTAACACC
11 Magnetospirillum-ma     AACACGTG-G GAATATACCT CTTGGTGGGG AATAACGTCG GGAAACTG-A CGCTAATACC
12 Azorhizobium            AACGCGTG-G GAACGTGCCC TTCAGTTCGG AATAACCCAG GGAAACTT-G GGCTAATACC
13 Rhizobium-meliloti      AACGCGTG-G GAATCTACCC TTTTCTACGG AATAACGCAG GGAAACTT-G TGCTAATACC
14 Rhodospirillum-rubr     AACACGTG-G GAACGTACCT TGGAGTGCGG AATAATCTTT GGAAACGA-G GACTAATACC
15 Methylosinus-tricho     AACGCGTG-G GAACGTAC-T TTCGGTTCGG AATAACTCAG GGAAACTW-A AGCTAATACC
16 Bradyrhizobium          AACGCGTG-G GAACGTACCT TTTGGTTCGG AACAACACAG GGAAACTT-G TGCTAATACC
17 Rhodopseudomonas        AACGCGTG-G GAACGTACCT TTTGGTTCGG AACAACACAG GGAAACTT-G TGCTAATACC
18 Erythrobacter-longu     AACGCGTG-G GAACCTGCCT TTAGGTTCGG AATAACTCAG AGAAATTT-G AGCTAATACC

                           121        131        141        151        161        171
 1 A-lipoferum-4B          GGATACGTCC ---------- ----TTCGGG ---AGAAAGT TTACGCCGAG AGAGGGGCCC
 2 A-lipoferum-BR17        GGATACGCCC ---------- ----TTTGGG ---GGAAAGT TTACGCCGAG AGAGGGGCCC
 3 A-lipoferum-USA5a       GGATGTGCCC CCCAAG---- -CGATT-GGG TGGGGAAAGT TTACGCCATG AGAGGGGCCC
 4 A-lipoferum-59b         GGATACGCCC ---------- ----TATGGG ---GGAAAGT TTACGCCGAG AGAGGGGCCC
 5 A-lipoferum-CRT1        GGATACGNNC ---------- ----TTTGGG ---GGAAAGT TTACGGCGAG AGAGGGGCCC
 6 A-brasilense-SP7        GGATACGTCC CCCAG----A GAGATTTGGG CGGAGAGAGT TTACGCCGAG AGCGGGGCCC
 7 A-brasilense-SP245      GGATACGTCC ---------- ----TTCGGG ---AGAAAGT TTACGCCGAG AGAGGGGCCC
 8 A-amazonense-Y1         GGATGAGCCT GATGGTTGTG GAGACTGTCA GGGAAAGATT TATCGCCGAA GGAGGGGCCC
 9 A-irakense-KBC1         GGATGTCCCC ---------- ----TGCGGG -GGAAAGATT TATCGCCAAA GGAGCAGCCC
10 A-halopraeferensAu4     GGATACGCCC TTCGGGGGAA AG--TTTACG CCGAGAGAT- ---------- ---GGGCCCG
11 Magnetospirillum-ma     GCATACGCCC ---------- ----TTCGGG -GGAAAGATT TATCGCCGAG AGATTAGCCC
12 Azorhizobium            GGATACGTCC ---------- -----GAAAG GAGAAAGATT GATCGCTGAA GGATCGGCCC
13 Rhizobium-meliloti      GTATGAGCCC ---------- ----TTCGGG -GGAAAGATT TATCGGGAAA GGATGAGCCC
14 Rhodospirillum-rubr     GCATACGCCC ---------- ----TTAGGG -GGAAAGATT TATCGCTCCA AGATCGGCCC
15 Methylosinus-tricho     GGATACGCC- ---------- ----TTAAGG GGGAAAGATT TATTCGCGAA AGATCGGCCC
16 Bradyrhizobium          GGATAAGCCC ---------- ----TTACGG -GGAAAGATT TATCGCCGAA AGATCGGCCC
17 Rhodopseudomonas        GGATAAGCCC ---------- ----TTACGG -GGAAAGATT TATCGCCGAA AGATCGGCCC
18 Erythrobacter-longu     GGATAATGTC ---------- -----TTCGG ACCAAAGATT TATCGCCTTT AGATGGGCCC

                           181        191        201        211        221        231
 1 A-lipoferum-4B          GCGTCGGATT AGGTAG---T TGGTGTGGTA ACGGCGCACC AAGCCGACGA TCCGTAGCTG
 2 A-lipoferum-BR17        GCGTCGGATT AGGTAG---T TGGTGAGGTA AAGGCTCACC AAGCCTTCGA TCCGTAGCTG
 3 A-lipoferum-USA5a       GCGTCGGATT AGGTAG---T TGGTGTGGTA ACGGCGCACC AAGCCGACGA TCCGTAGCTG
 4 A-lipoferum-59b         GCGTCGGATT AGGTAG---T TGGTGAGGTA ATGGCTCACC AAGCCTGCGA TCCGTAGCTG
 5 A-lipoferum-CRT1        GCGTCGGATT AGGTAG---T TGGTGAGGTA ATGGCTCACC AAGCCTTCGA TCCGTAGCTG
 6 A-brasilense-SP7        GCGTCCGATT AGGTAG---T TGGCGGGGTG ATGGCCCACC AAGCCGACGA TCGGT-GCTG
 7 A-brasilense-SP245      GCGTCCGATT AGGTAG---T TGGTGGGGTA ATGGCCCACC AAGCCGACGA TCGGTAGCTG
 8 A-amazonense-Y1         GCGTCCGATT AGGTAG---T TGGTGAGGTA ACGGCTCACC AAGCCGACGA TCGGTAGCTG
 9 A-irakense-KBC1         GCGTCTGATT AGGTAG---T TGGTGGGGTA AAGGCCTACC AAGCCGACGA TCAGTAGCTG
10 A-halopraeferensAu4     AGCTCGGATT AGGTAG---T TGGTGGGGTA ATGGCT-ACC AAGCCGACGA CCGAGC----
11 Magnetospirillum-ma     GCGTCCGATT AGCTAG---T TGGTGAGGTA ATGGCTCACC AAGGCGACGA TCGGTAGCTG
12 Azorhizobium            GCGTCTGATT AGCTAG---T TGGTGAGGTA ATGGCTCACC AAGGCGACGA TCAGTAGCTG
13 Rhizobium-meliloti      GCGTTGGATT AGCTAG---T TGGTGGGGTA AAGGCCTACC AAGGCGACGA TCCATAGCTG
14 Rhodospirillum-rubr     GCGTCCGATT AGCTAG---T TGGCGGGGTA ATGGCCCACC AAGGCGACGA TCGGTAGCTG
15 Methylosinus-tricho     GCGTCCGATT AGCTAGCTGT TGGTGAGGTN AAGGCTNACC AAGGCGACGA TCGGNAGCTN
16 Bradyrhizobium          GCGTCTGATT AGCTAG---T TGGTGAGGTA ATGGCTCACC AAGGCGACGA TCAGTAGCTG
17 Rhodopseudomonas        GCGTCTGATT AGCTAG---T TGGTGAGGTA ATGGCTCACC AAGGCGACGA TCAGTAGCTG
18 Erythrobacter-longu     GCGTTAGATT AGATAG---T TGGTGGGGTA ATGGCCTACC AAGTCGACGA TCTATAGCTG
```

		241	251
1	A-lipoferum-4B	GTCTGAGAGG	GCG
2	A-lipoferum-BR17	GTCTGAGAGG	ATG
3	A-lipoferum-USA5a	GTCTGAGAGG	ATG
4	A-lipoferum-59b	GTCTGAGAGG	GCG
5	A-lipoferum-CRT1	GTCTGAGAGG	ATG
6	A-brasilense-SP7	GTCTGAGAGG	GCG
7	A-brasilense-SP245	GTCTGAGAGG	GCG
8	A-amazonense-Y1	GTCTGAGAGG	GCG
9	A-irakense-KBC1	GTCTGAGAGG	ATG
10	A-halopraeferensAu4	----------	---
11	Magnetospirillum-ma	GTCTGAGAGG	ATG
12	Azorhizobium	GTCTGAGAGG	ATG
13	Rhizobium-meliloti	GTCTGAGAGG	ATG
14	Rhodospirillum-rubr	GTCTGAGAGG	ATG
15	Methylosinus-tricho	GTCTGAGAGG	ATG
16	Bradyrhizobium	GTCTGAGAGG	ATG
17	Rhodopseudomonas	GTCTGAGAGG	ATG
18	Erythrobacter-longu	GTCTGAGAGG	ATG

Fig. 1. Lineup of the 16S partial DNA sequences of the five *Azospirillum* species(*A. lipoferum* strains: 4B, BR17, USA 5a, 59b, and CRT1, *Azospirillum brasilense strains:* sp7 and sp245, *Azospirillum irakense* strain KBC1, *Azospirillum amazonense* strain Y1 and *Azospirilum halopræferens* Au4) with *Magnetospirillum magnetotacticum* (GenBank no. M58171) (Eden et al. 1991) *Erythrobacter longus* ATCC 33941 (GenBank no. M55493), *Rhizobium meliloti* NZP 4017 (GenBank no. M55495), *Bradyrhizobium japonicum* USDA 59 (GenBank no. M55489), *Azorhizobium caulinodans* ORS 571 (GenBank no. M55491), *Rhodopseudomonas palustris* 2.1.6 (GenBank no. M55496), *Rhodospirillum rubrum* ATCC 11170 (GenBank no. M55497) (Young et al. 1991), and *'Methylosinus trichosporium'* (GenBank no. M29024) (Tsuji et al. 1989). Numbering starts at position 51 relative to the *Escherichia coli* sequence given in Raué et al. (1988). Underlined bases at coordinates 94, 103, 105 and 203 are particular to *Azospirillum* and to *Magnetospirillum* species.

Discussion

This sequencing work was undertaken to determine the structure of genus *Azospirillum* and its relation to other genera of the alpha subdivision of the Proteobacteria, and to see if generic nucleotides positions could be exploited to identify these bacteria in complex substrates such as the soil or the rhizosphere where they are normally found.

The genus *Azospirillum* appears to be coherent since the five strains tested are closer to one another than they are to the rest of the strains examined. A more comprehensive analysis of more *Azospirillum* species and of different genera in the same alpha subdivision of the Proteobacteria should help pinpoint the relation of *Azospirillum* to other taxonomical units. The % similarity detected between the five *Azospirillum* species is rather low (67%, distance = 0.43 substitution/site), compared with the *Bradyrhizobium japonicum* -"*Photorhizobium*" genus (97%, 0.033 s/s; Young et al. 1991), with *Frankia* genus (92%, most important difference among 40 strains tested 0.097 s/s, Nazaret et al. 1991).

The resulting tree is entirely consistent with the one given by Young et al. (1991) which used a slighthly different part of the 16S rDNA gene. Young et al. (1991) avoided using positions 19-38 because of uncertainties in the alignment. We found that the potential error is not very important since the trees obtained using the whole fragment or avoiding the gap-region were not different. Young did not have the *Magnetospirillum magnetotacticum* sequence available, which proved to lie closest to *Azospirillum. Magnetospirillum* and *Azospirillum* share certain characteristics in common such as morphology and nitrogen fixation, which would appear to derive from a shared ancestor. Nitrogen fixation may not be the most striking feature to highlight a common ancestry since it is widespread among bacteria and has a very ancient phylogeny (Normand and Bousquet, 1990). The particular S-shape morphology may be more indicative of a recent ancestor.

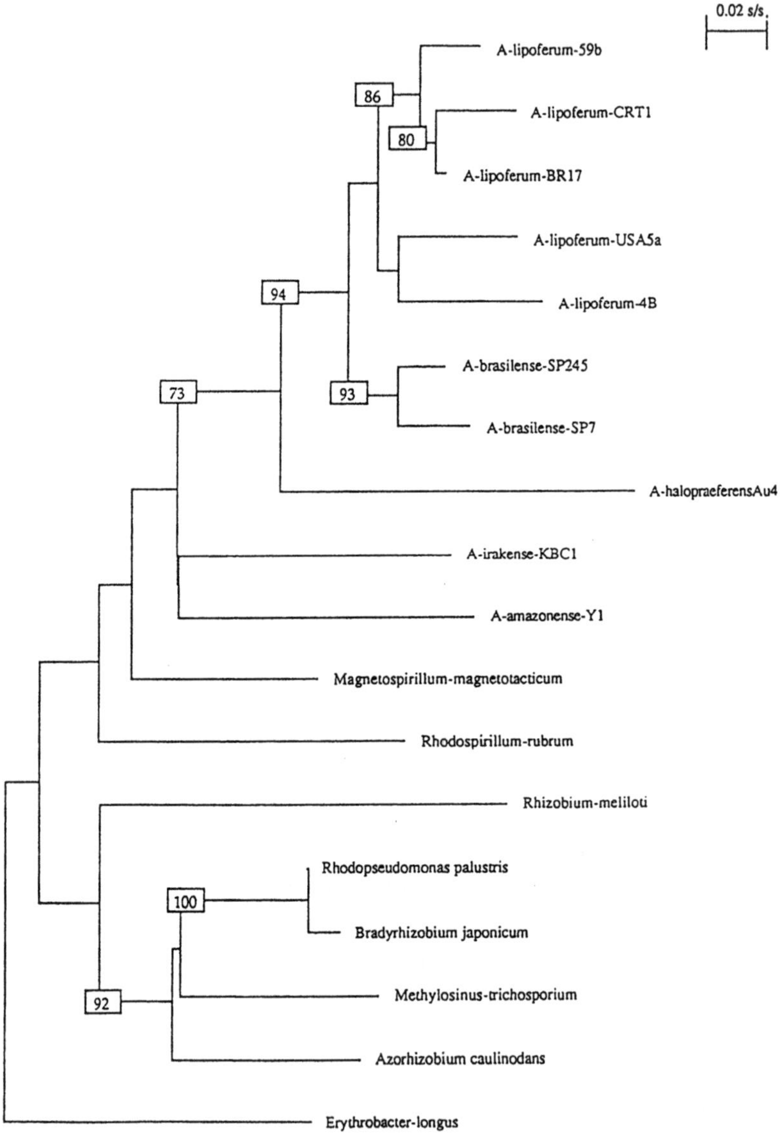

Fig. 2. Neighbour-Joining tree:
The Neighbour-Joining algorithm was used on the lined-up sequences of Fig. 1, considering segments 1-21, 41-130 and 154-253, and avoiding positions with gaps. The box represents the only group with a bootstrap result above 70%.

References:

Bally. R., Thomas-Bauzon. D., Heulin. T., Balandreau. J., Richard. C. and De Ley. J. 1983. Determination of the most frequent N2-fixing bacteria in a rice rhizosphere. Can. J. Microbiol. 29: 881-887.

Eden, P.A., Schmidt, T.M., Blakemore, R.P., and Pace, N.R. 1991. Phylogenetic analysis of *Aquaspirillum magnetotacticum* using polymerase chain reaction-amplified 16S rRNA specific DNA. Int. J. Syst. Bacteriol. 41: 324-325

Felsenstein, J. 1985. Confidence limits on phylogenies: an approach using the bootstrap. Evolution 39: 783-791

Higgins, D.G., and Sharp, P.M. 1988. Clustal: a package for performing multiple alignment on a microcomputer. Gene 73: 237-244.Jukes, T.H., and Cantor, C. R. 1969. Evolution of protein molecules. In: Mammalian protein metabolism. H.N. Munro, ed. Academic Press, New York, pp. 21-132

Khammas, K. M., Ageron, E., Grimont, P. A. D., and Kaiser, P. 1989. *Azospirillum irakense* sp. nov. , a nitrogen-fixing bacterium associated with rice roots and rhizosphere soil. Res. Microbiol. 140: 679-693.

Li, W.T. 1981. Simple method for constructing phylogenetic trees from distance matrices. Proc. Natl. Acad. Sci. USA 78: 1085-1089

Magalhaes, F.M.M., Baldani, J. I., Souto, S. M., Kuykendall, J. R., and Döberreiner, J 1983. A new acid tolerant *Azospirillum* species. An. Acad. Bras. Cienc. 55: 417-430

Maniatis. T., Fritsch. E.F., Sambrook. J. 1982. Molecular cloning, a laboratory manual. Cold Spring Harbor Laboratory, Cold Spring Harbor, NY.

Nazaret, S., Cournoyer, B., Normand, P., and Simonet, P. 1991. Phylogenetic relationships among Frankia genomic species determined by use of amplified 16S rDNA Sequences. J. Bacteriol. 173: 4072-4078

Nei, M. 1987. Molecular Evolutionary Genetics. Columbia University Press, New York

Normand, P., and Bousquet, J. 1990. Phylogeny of nifH sequences in Frankia and in other nitrogen-fixing microorganisms. J. Mol. Evol. 29: 436-447

Raué, H.A., Klootwijk, J., and Musters, W. 1988. Evolutionary conservation of structure and function of high molecular weight ribosomal RNA. Prog. Biophys. molec. Biol. 51: 77-129

Reinhold, B., Hurek, T., Fendrik, I., Pot, B., Gillis, M., Kersters, K., Thielemans, S., and De Ley, J. 1987. *Azospirillum halopraeferens* sp. nov., a nitrogen-fixing organism associated with roots of Kallar grass (Leptochloa fusca Kunth). Int. J. Syst. Bacteriol. 37: 43-51

Saitou, N., and Nei, M. 1987. The neighbor-joining method: a new method for reconstructing phylogenetic trees. Mol. Biol. Evol. 4: 406-425

Sanger, F., Nicklen S., and Coulson, A.R. 1977. DNA sequencing with chain-terminating inhibitor. Proc. Natl. Acad. Sci. USA 74: 5463-5467

Simonet, P., Normand P., Moiroud. A., and Bardin. R. 1990. Identification of *Frankia* strains in nodules by hybridization of polymerase chain reaction products with strain-specific oligonucleotide probes. Arch. Microbiol. 153: 235-240

Sneath, P.H.A., and Sokal, R.R. 1973. Numerical taxonomy. W. H. Freeman, San Francisco

Tarrand, J.J.,Krieg, N.R., and Döberreiner, J. 1978. A taxonomic study of the *Spirillum lipoferum* group, with descriptions of a new genus, *Azospirillum* gen.nov. and two species , *Azospirillum lipoferum* (Beijerinck) comb.nov.and *Azospirillum brasilense* sp.nov. Can.J Microbiol. 24: 967-980.

Winship, P.R. 1989. An improved method for directly sequencing PCR amplified material using dimethyl sulfoxide. Nucl. Acids Res. 17: 1266

Woese, C.R. 1987. Bacterial evolution. Microbiol. Rev. 51: 221-271

Young, J.P.W., Downer, H.L., and Eardly, B.D. 1991. Phylogeny of the phototrophic *Rhizobium* strain BTAi1 by polymerase chain reaction-based sequencing of a 16S rRNA gene segment. J. Bacteriol. 173: 2271-2277.

Involvement of *lon* Protease of *Azospirillum brasilense* in iron uptake

Elena Mori, Renato Fani, Monica Fulchieri,
Enzo Gallori, Marco Bazzicalupo.
Dipartimento di Biologia Animale e Genetica,
Università di Firenze, via Romana 17,
50125 Firenze, Italy.

Abstract

A Tn5 derived mutant of *A.brasilense* impaired in the ability to take up iron from the medium was characterized. The mutant showed an altered pattern of membrane proteins, in particular two of the proteins, that in the parental strain are induced by iron starvation, were missing in the mutant. A 3.4 kb DNA region of *A. brasilense*, containing the gene disrupted by Tn5 in the mutant strain, was cloned and sequenced. The analysis of the nucleotide sequence shoved that this DNA region encodes for a polypeptide with very high degree of homology with the product of the gene *lon* of *Escherichia coli*. The possible involvement of this gene in the response to heat-shock was suggested by the higher resistance to heat treatment of the mutant compared with the parental strain.

Introduction

Siderophores production is a particularly interesting trait for the Gram-negative, nitrogen-fixing bacterium *Azospirillum brasilense*. Under conditions of iron limitation this bacterium is able to produce siderophores that are an important survival factor in a crowded and highly competitive environment such as rizosphere (Bachhawat and Gosh 1987a, Hartmann 1989, Mori et al. 1992).

By Tn5 transposon mutagenesis we have previously isolated five mutants of *A. brasilense* strain SPF94 altered in siderophore production. The mutants were selected by the lack of halo around colonies growing on Chrome Azurol S plates. (Schwyn and Neilands 1987). Mutants were characterized by their ability to grow in the presence or absence of iron. When cells were grown in iron-free medium, these strains showed reduced growth rate and final growth yield compared to parental strain SPF94 (Mori et al.1992). We report here the further characterization of one of these mutants and the cloning and sequencing of a 3.4 kb DNA region of *A. brasilense* containing the gene, presumably involved in iron uptake, that was disrupted by Tn5 in this particular strain.

NATO ASI Series, Vol. G 37
Azospirillum VI and Related Microorganisms
Edited by I. Fendrik et al.

Results

Membrane proteins involved in response to iron starvation. Iron starvation induced in *A. brasilense* the biosynthesis of five membrane proteins with molecular weight ranging from 78 to 90 kDa (Bachhawat and Gosh 1987b). The membrane protein pattern of one of the five mutants, SPFS5, was positively different in respect with that of wild type strain (Fig.1). In particular one of the major proteins induced by iron limitation was absent in the mutant.

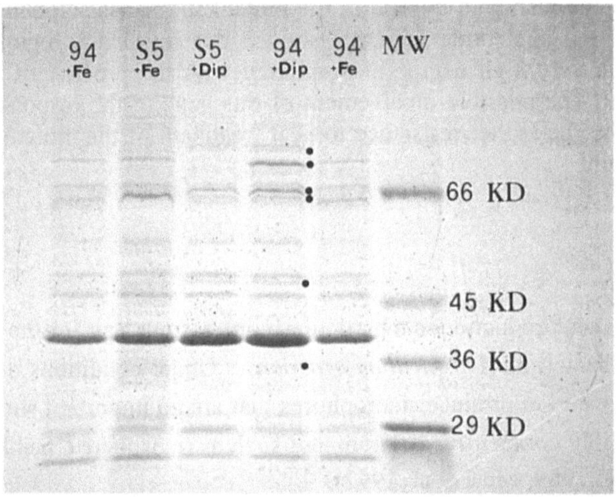

Figure 1. SDS-PAGE of the outer membrane proteins of *A.brasilense* strains SPF94 (94) and SPFS5 (S5). grown in iron supplemented (+Fe) and iron deprived (+Dip) medium. Iron deprivation was obtained with the addition of 70 μM dipyridyl. Proteins were purified and electrophresed as described (Bachhawat and Gosh 1987b). Proteins loaded in the 5 wells were from left to right: 20, 50, 25, 25, 20 μg respectively.

Molecular cloning of Azospirillum lon *gene.* Restriction mapping of chromosomal DNA of strain SPFS5 showed that Tn5 was inserted in a *Sal*I-*Sal*I fragment of about 3 kb. A sequence adjacent to the transposon was cloned and used as probe to identify the *Sal*I fragment from a genomic library of the parental strain. The nucleotide sequence of the entire fragment was determined by the Sanger method (Sanger et al.1977). In order to determine the complete sequence of the open reading frame (ORF) found in the *Sal*I fragment (see below), a larger *Pst*I fragment of parental strain DNA was cloned and partially sequenced.

The analysis of the sequence, whose total length is 3480 nt, revealed the presence of a long ORF (2430 nt) and of an upstream interrupted ORF (558 nt). The two open reading frames are separated by an intergenic region of 266 bp (Figure 2).

Sequence analysis. The analysis of the longest ORF (from nt 832 to 3261) showed that it encodes a putative polypeptide of 810 aminoacids with a calculated molecular weight of 90135 D. The comparison of the sequence of the putative polypeptide with the EMBL GenBank revealed a very high degree of sequence homology with the *lon* gene products from different sources.

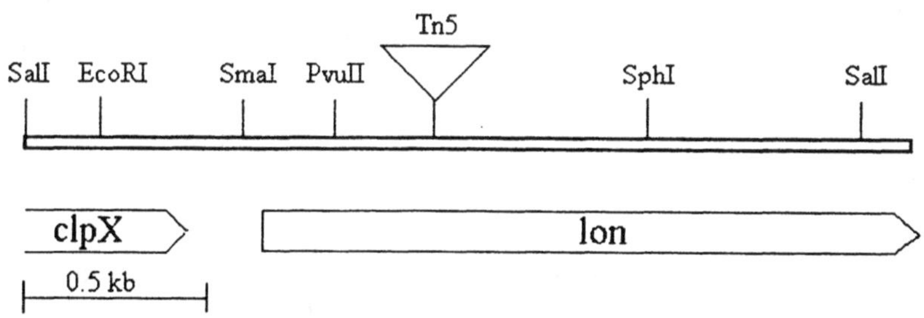

Figure 2. Restriction map of the DNA fragment containing part of *clpX* and the entire *lon* gene. Bars with arrowhaeds indicate the translated sequence and the direction of transcription.

Aminoacid sequence of *A. brasilense lon* gene product shows 61%, 72% and 52% identity and 79%, 76% and 79% similarity with those of *Escherichia coli* (Chin et al 1988), *Myxococcus xanthus* (Tojo et al. 1993) and *Bacillus brevis* (Ito et al. 1992) respectively.

The *lon* gene in *E. coli* encodes a protease that is one of the heath-shock protein identified in *E. coli*. Lon contains a DNA binding domain, an ATPase domain, a proteolytic domain and a regulatory domain that binds unfolded polypeptides (Chin et al. 1988). The specific role of the Lon protease in the cellular physiology is not yet really understood.

DNA sequence near the transposon showed that the *lon* gene of *A. brasilense* strain SPFS5 was inactivated by Tn5 insertion at nt 705 corresponding to codon 235 which is immediatly after the putative DNA binding domain at the border between the highly and less conserved regions (Fig.2 and 3).

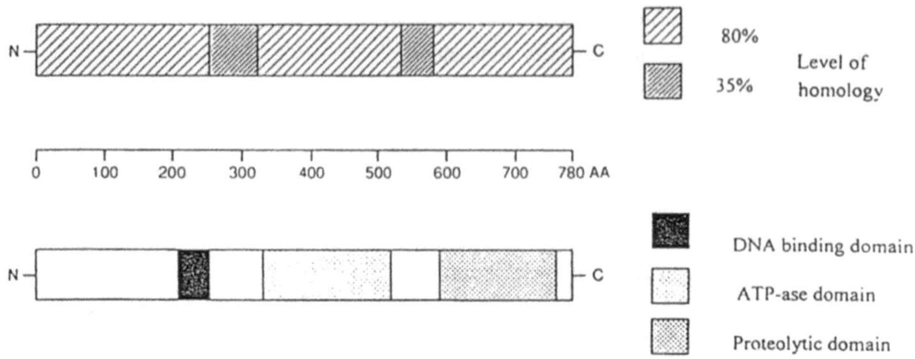

Figure 3. Regions of homology of *lon* gene products of *E. coli*, *B. brevis*, *M. xanthus*, and *A. brasilense* (up). Localization of different domains of the protein (down).

The shortest ORF encodes a putative polypeptide of 186 aminoacids showing very high degree of sequence homology with the carboxy-terminal region of the *clpX* gene product of *E. coli* (Gottesman 1993). ClpX protein is the ATPase subunit of the energy-dependent serine-protease ClpXP that, like Lon, belongs to the family of heat-shock proteins (Parsell and Lindquist 1993).

Accurate alignment of the *lon* aminoacidic sequences of *E. coli*, *M. xanthus*, *B. brevis* and *A. brasilense*, revealed three highly conserved regions and two other regions with very low degree of sequence homology (Fig.3). The DNA binding, the ATPase, and the proteolytic domains are located inside the most homologous regions in the four *lon* product (Fig.3).

Is the lon gene product of A. brasilense involved in response to heat-shock? In order to test whether or not the *lon* gene of *A. brasilense* is involved in the heat-shock response as its *E. coli* homologous is, we compared resistance of parental and *lon* mutant strains to high temperature. Results of the experiment are reported in Table 1 and clearly indicated that strain SPFS5 is much more resistant than parental strain to 5 minutes of heath-shock at 51°C.

This result indicate a kind of involvment of the *lon* gene product of *A. brasilense* in the cellular response to heat. However, the higher resistance of the mutant suggests that the product of *lon* gene could be responsible, at least in part, of the cell death after heat-shock, rater than in the recovery of cellular life as the heat-shock proteins are supposed to be.

Table 1. Survival of *A.brasilense* strains to heat-shock treatment.[a]

	Survival of strains [b]	
Treatment	SPF94	SPFS5
Control, 33°C	100 [c]	100 [c]
51°C for 5 minutes	11	105
42°C for 30 minutes plus 51°C for 5 minutes	55	85

[a] The experiment was carried out by rapidly shifting 0.5 ml of log-phase cultures to water bath at different temperatures for the times indicated, after the treatments cultures were kept at 33°C for 15 minutes, diluted and plated for counting.

[b] Survival is expressed as % of the CFU of the control kept at 33°C. Data are means of three replicates.

[c] 100 % corresponded to 4×10^7 CFU/ml

References

Bachhawat AK and Gosh S (1987a) Iron transport in *Azospirillum brasilense* : Role of the Siderophore Spirilobactin. J Gen Microbiol **133**: 1759-1765

Bachhawat A and Ghosh S (1987b) Isolation and Characterization of the Outer Membrane Proteins of *Azospirillum brasilense*. J Gen Microbiol **133**:1751-1758.

Chin DT, Goff SA, Webster T, Smith T and Goldberg A (1988) Sequence of the *lon* gene in *Escherichia coli*. J Biol Chem **263**: 11718-11728.

Gottesman S, Clark WP, de Crécy-Lagard V and Maurizi MR (1993) *ClpX* an alternative subunit for ATP-dependent *Clp* protease of *E. coli*. J Biol Chem **268**: 22618-22626.

Hartmann A (1989) Ecophysiological aspects of growth and nitrogen fixation in *Azospirillum* spp. In: *Nitrogen fixation with non legumes*. F.A. Skinner et al.,eds. Kluwer Academic Publishers. pp.123-136.

Ito K, Udaka S and Yamagata H (1992) Cloning, characterization, and inactivation of the *Bacillus brevis lon* gene. J Bacteriol **174** : 2281-2287.

Mori E, Fani R, Gallori E, Fantappiè O and Bazzicalupo M (1992) Mutants of *Azospirillum brasilense* altered in the uptake of iron. Symbiosis **13** : 115-122

Parsell DA and Lindiquist S (1993) The function of the heat-shock proteins in stress tolerance: Degradation and reactivation of damaged proteins. Annu Rev Genet **27**: 437-496.

Sanger F, Nicklen S and Coulson AR (1977) DNA sequencing with chain terminating inhibitors. Proc Nat Acad Sci U.S.A. **74:** 5463-5466.

Schwyn B and Neilands JB (1987) Universal chemical assay for the detection and determination of siderophores. Anal Biochem **160**: 47-56.

Tojo N, Inouye S and Komano T (1993) Cloning and nucleotide sequence of the *Myxococcus xanthus lon* gene: Indispensability of *lon* for vegetative growth. J Bacteriol **175**: 2271-2277.

Cloning and Analysis of an *Azospirillum brasilense* Iteron and *hslUV* Operon Containing Region

Pietro Liò, Marco Bazzicalupo, Annamaria Grifoni,
Elena Mori and Renato Fani

Department of Animal Biology and Genetics "Leo Pardi"
University of Florence
Via Romana 17
I-50125 Firenze
Italy

Abstract

A 3 kb DNA region from *Azospirillum brasilense* strain SPF94, located upstream of the histidine biosynthetic cluster, was cloned and analysed. The 5' region of this sequence contained two putative ORF sharing a very high degree of sequence homology with the *Escherichia coli hslUV* heat-shock operon.

The region between the *A. brasilense hslUV* genes and the histidine operon did not contain any ORF but the analysis performed by mutual informational algorithm permitted to detected 8 tandemly repeated sequences of 21 bp. Since iterons are known to act as protein binding sites, we suppose that this sequence could play a role in regulating the histidine or the*hslUV* operons. Moreover the computational analysis of the DNA topology has revealed that this region is remarkably curved, a feature generally associated to protein binding sites.

Introduction

DNA sequencing of chromosomal regions with unknown function did often result in the recognition of new interesting genes either in prokaryotes and

NATO ASI Series, Vol. G 37
Azospirillum VI and Related Microorganisms
Edited by I. Fendrik et al.

eukaryotes. On the other hand the great power that DNA sequence analysis has reached nowadays, regarding both comparison of sequences in huge databanks and studies linking structure and function of nucleic acids, allowed in many cases researchers to predict the function of the DNA region they have sequenced. In the search for additional histidine biosynthetic genes in a DNA region upstream of the *his Azospirillum brasilense* operon (Fani et al. 1993), we have sequenced about 3 kb that, although do not contain any gene apparently involved in histidine biosynthesis, revealed the presence of two interesting genetic elements: a heat-shock operon and a protein binding regulatory sequence.

Results and Discussion

Identification of A. brasilense hslUV *operon*

The nucleotide sequence of the 3 Kb DNA region upstream the *his* operon was determined and its analysis showed the presence of several open reading frames (ORF) on both strands.

None of the putative proteins encoded by these ORF showed a significant degree of homology with any of the known *his* gene products, suggesting that if additional *his* genes with known function exist in the *A. brasilense* genome they are located in other chromosomal regions. Neverthless two of these ORF (ORF 546 and ORF 978, the latter resulting truncated), located about 2 Kb far from the first gene of the *his* operon (*hisBd*) (see Figure 3 of Fani et al., this book), showed a high degree of nucleotide sequence homology (58.4%) with the *Escherichia coli hslUV* heat-shock bicistronic operon which encodes the two proteins HslU (19.1 kDa) and HslV (49.6 kDa) (Chuang et al. 1993; Plunkett et al. 1993). The putative proteins encoded by ORF546 and ORF978 compared with the corresponding *E. coli* proteins showed a percentage of identical (58% and 55% for HslV and HslU, respectively) and similar amino acids (68% and 62% for HslV and HslU, respectively) (Figure 1), which is very high if

HslV

M	T	S	H	D	P	I	Q	W	H	G	T	T	I	L	S
V	R	K	N	G	Q	V	V	I	A	G	D	G	Q	V	S
V	G	Q	T	V	M	K	S	N	A	R	K	V	R	W	L
A	G	G	T	V	M	G	G	F	A	G	A	T	A	D	A
L	T	L	F	E	R	L	E	T	K	L	E	K	H	P	G
Q	L	L	R	A	C	V	E	M	A	K	D	W	R	T	D
R	Y	L	R	R	L	E	A	M	M	A	V	A	D	R	N
V	S	L	I	L	T	G	N	G	D	V	L	E	P	E	D
G	L	I	G	I	G	S	G	G	A	Y	A	L	S	A	A
R	A	L	V	D	V	D	G	L	D	A	E	A	V	A	R
K	A	M	K	I	A	A	G	I	C	V	Y	T	N	E	N
V	T	L	E	K	L	*									

HslU

M	S	A	P	S	I	A	A	A	T	A	A	F	S	P	R
E	I	V	S	E	L	D	R	Y	I	V	G	Q	H	E	A
K	R	A	V	A	I	A	L	R	N	R	W	R	R	Q	Q
L	P	E	G	L	R	E	E	V	L	P	K	N	I	L	M
I	G	P	T	G	V	G	K	T	E	I	A	R	R	L	A
K	L	A	Q	A	P	F	L	K	V	E	A	T	K	F	T
E	V	G	Y	V	G	G	D	V	V	E	Q	I	V	R	D
L	V	E	A	A	I	G	L	T	K	E	R	L	R	K	E
V	A	A	K	A	E	L	R	A	E	E	R	V	V	D	A
L	C	G	E	N	A	S	P	E	T	R	Q	K	F	R	K
M	L	R	E	G	T	L	N	D	R	E	I	E	I	Q	V
A	D	T	A	G	G	L	P	T	F	D	I	P	G	M	P
A	A	Q	M	G	M	L	N	L	N	D	I	F	G	K	A
M	G	G	R	T	K	S	R	R	M	S	V	A	E	S	Y
T	V	L	M	A	E	E	S	D	K	L	L	D	Q	E	K
V	V	S	E	A	I	Q	A	V	E	Q	N	G	I	V	F
L	D	E	I	D	K	I	S	A	R	S	D	Y	K	G	G
A	D	V	S	R	E	G	V	Q	R	D	L	L	P	L	I
E	G	T	T	V	S	T	K	H	G	A	V	K	T	D	H
I	L	F	I	A	S	G	A	F	H	L	S	K	P	S	D
L	L	P	E	L	Q										

Figure 1. Amino acid sequence of the putative proteins encoded by *A. brasilense* ORF546 and ORF978 homologous to the *E. coli* HslV and HslU proteins, respectively. Identical residues to that of the homologous *E. coli* proteins in the two sequences are bold and underlined; conserved residues are bold.

compared to other homologous gene products in these bacteria. Moreover we found a perfect conservation of an *ATP/GTP* binding motif (GPTGVGKT) in the HslU protein, suggesting that it plays analogous function in *A. brasilense* and *E. coli.* Although the *hslUV* operon clearly belong to the family of heat-shock genes in *E. coli*, its function has not been elucidated up to now, neither homologous genes have been found in other species.Interestingly, the organization of the two genes in the operon is strongly conserved in *E. coli* and in *A. brasilense,* athough they lie in different chromosomal regions. Therefore we hypothesize a concerted evolution of the *hslUV* operon that could have acted as a single functional and/or structural block in the evolution of these genomes.

Identification of an iteron

The region between the *hslUV* and *his* operons did not show a significant degree of similarity with any sequence contained in the EMBL GenBank. Neverthless by analyzing the sequence with mutual information (Gutell et al. 1992), we found that this region contains eight 21-bp repeated DNA sequences (*iteron*). The analysis revealed that the consensus sequence of the eight *A. brasilense* repeats, AGNGT*CAAGTG*, matches very well the consensus sequence (*CANNTG*) for the binding of many *b-HLH* (basic region, helix-loop-helix region) protein domains (Cahill et al. 1992).

Besides, the theoretical analysis of DNA curvature, using the algorithm developed by Bolshoy et al. (1991) has revealed that the DNA region containing the repeats is remarkably curved and that the curvature increases at each CAANTG repeat (Figure 2), a property also associated to protein binding sites. Because the 21-bp repeats exactly correspond to twice the DNA helix turn, it can be suggested that the consensus sequence (*CAANTG*) faces a putative binding protein at the same position, with respect to the major or minor groove structure

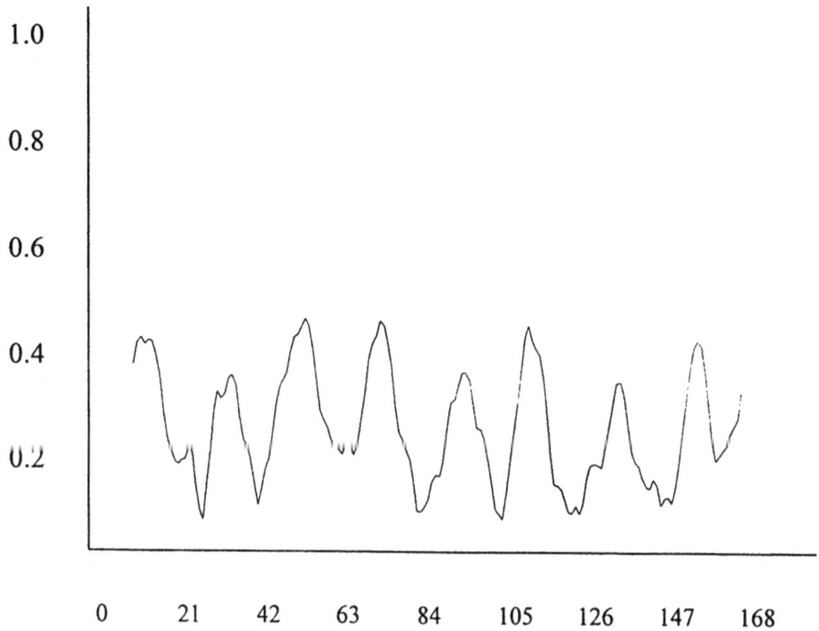

Figure 2. Theoretical DNA curvature map of the region containing the repeats: in abscissa the base position; in ordinate the measure of the curvature.

every two DNA helical turns. The requirement of DNA bending in transcription and in site specific recombination has been extensively described (Lavigne et al. 1991; Martin et al. 1993).

The working hypothesis is that the *iteron* located between the *hslUV* and the histidine biosynthetic operons could play a role in the regulation in one of these operons. Experiments carried out in order to check this hypothesis are in progress.

References

Bolshoy A, McNamara P, Harrington RE and Trifonov EN (1991) Curved DNA without A-A: experimental extimation of all 16 DNA wedge angles. Proc Natl Acad Sci USA 88: 2312-2316.

Cahill SJA, Benfield PA, Fairman R, Wasserman ZR, Brenner S L, Stafford WF, Altenbach C, Hubbel WL and DeGrado WF (1992) Molecular characterization of helix-loop-helix peptides. Science 255: 979-981.

Chuang SE, Burland V, Plunkett G, Donna LD and Blattner RF (1993) Sequence analysis of four new heat-shock genes constituting the*hslTS*/*ibpAB* and *hslUV* operons in *Escherichia coli*. Gene 134: 1-6.

Fani R, Alifano P, Allotta G, Bazzicalupo M, Carlomagno MS, Gallori E, Rivellini F and Polsinelli M (1993) The histidine operon of *Azospirillum brasilense*: organization, nucleotide sequence and functional analysis. Res Microbiol 144: 187-200.

Gutell RR, Power A, Hertz GZ, Putz JE and Stormo GD (1992) Identifying constraints on the higher-order structure of RNA: continued development and application of comparative sequence analysis methods. Nucl Acids Res 21: 5785-5795

Lavigne M, Herbert M, Kolb A and Buc H (1992) Upstream curved sequences influence the initiation of transcription at the *Escherichia coli* galactose operon. J Mol Biol 224: 293-306.

Martin JP and Espinosa M (1993) Protein-induced bending as a transcriptional switch. Science 260: 805-808.

Plunkett G, Burland V, Donna LD and Blattner RF (1993) Analysis of the *Escherichia coli* genome. DNA sequence of the region from 87.2 to 89.2 minutes. Nucl Acids Res 21: 3391-3398.

Cloning of a *Streptomyces* Endoglucanase Gene in *Azospirillum brasilense*

Brunella Perito, Renato Fani and Giorgio Mastromei

Department of Animal Biology and Genetics "Leo Pardi"
University of Florence
Via Romana 17
I-50125 Firenze
Italy

Abstract

The *eglS* gene from *Streptomyces rochei* A2, encoding an endoglucanase, was cloned in the conjugative vector pAF700 obtaining the 22.5 kb recombinant plasmid pCSF121. This plasmid was transferred to *Escherichia coli* DH1 and *Azospirillum brasilense* Sp6. *E. coli* [pCSF121] showed no endoglucanase activity, whereas *A. brasilense* [pCSF121] showed a low endoglucanase activity, which was cell associated. *A. brasilense* [pCSF121] was not able to use carboxymethyl-cellulose as carbon source. No significative difference in nitrogen fixing activity was found between *A. brasilense* [pCSF121] and *A. brasilense* [pAF700]. The different behavior of *E. coli* [pCSF121] and *A. brasilense* [pCSF121] in the expression of the *eglS* gene suggests that *A. brasilense* might recognize some *Streptomyces* regulative signal, i.e. for transcription, internal to the cloned fragment.

Introduction

Cellulose is the most abundant natural polymer, mainly present as the major constituent of plant biomass. Microorganisms capable of hydrolyzing cellulose comprise numerous bacteria, actinomycetes and fungi. The best studied cellulolytic system of fungi involves three major classes of enzymes (Coughlan, 1990) which act cooperatively: cellobiohydrolase, endo-1,4-β-glucanase and cellobiase or β-glucosidase.

Streptomyces rochei A2 is a cellulolytic actinomycete from which we have isolated and characterized the *eglS* endoglucanase gene (Perito et al., 1994). In *E. coli* the *eglS* gene is expressed only when its 5' end is correctly oriented respect to the vector *lac* promoter, suggesting the absence of a *Streptomyces* promoter recognized by *E. coli*.

NATO ASI Series, Vol. G 37
Azospirillum VI and Related Microorganisms
Edited by I. Fendrik et al.

The EglS protein is organized into two domains: an N-terminal catalytic domain and a C-terminal cellulose binding domain. In *E. coli*, the endoglucanase activity is also expressed when the *eglS* gene is cloned in a truncated form of 977 bp, encoding only the catalytic domain (plasmid pCSF13).

Azospirillum brasilense, is a Gram-negative bacterium able to fix nitrogen in association with the roots of agronomically important grasses. It does not degrade cellulose, but some strains, as *A. brasilense* Sp6, show β-glucosidase activity.

The purpose of this work was to verify if the expression of a foreign endoglucanase activity, together with the endogenous β-glucosidase, could confer to *A. brasilense* cellulase activity, which in turn could improve the capacity of the *Azospirillum* cells to interact and colonize plant roots.

Materials and Methods

Strains and media used have been already described (Fani et al., 1988; Perito et al., 1994). All methods for DNA manipulation were based on Sambrook et al. (1989). The Congo red test on agar plates was described by Teather and Wood (1982). Endoglucanase activity was determined using carboxymethyl-cellulose (CMC) as substrate and the amount of reducing sugars produced was determined by the dinitro-salicylic acid (DNS) method (Miller et al., 1960). The activity was expressed as μmoles of glucose released in 1 min (international unit = I.U.). Chloramphenicol resistance was determined on agar plates after 48 h of growth at 37°C. Nitrogenase activity of *A. brasilense* cultures was assayed by adding acetylene to the flasks and measuring ethylene production, as described by Mugnai et al. (1994).

Results and Discussion

The *S. rochei* A2 977 bp *eglS* fragment, encoding the endoglucanase catalytic domain, was extracted from plasmid pCSF13 (Perito et al., 1994) and flanked with two *Pst* I restriction sites. This fragment, of about 1000 bp, was cloned in the unique *Pst* I site of the broad host range conjugative vector pAF700 (Fani et al., 1988), to obtain the 22.5 kb plasmid pCSF121 (Fig. 1). In this construction the fragment has 165 nucleotides of *Streptomyces* DNA before the ATG start codon and lies between the *tet* vector promoter (recognized both by *E. coli* and *A. brasilense*) and the vector promoterless *cam* gene.

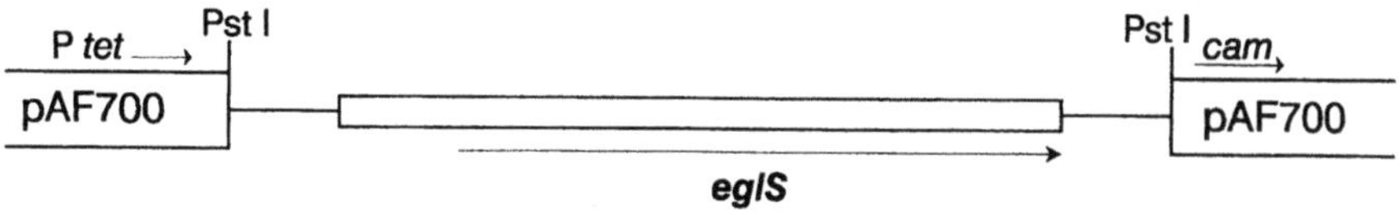

Fig. 1. Structure of plasmid pCSF121.

Plasmid pCSF121 was transferred to *E. coli* DH1 by transformation and then to *A. brasilense* Sp6 by conjugation, which still represents the most efficient way for genetic transfer in *Azospirillum*. *E. coli* [pCSF121] was negative to the Congo red test, indicating the lack of endoglucanase activity, whereas *A. brasilense* Sp6 [pCSF121] was positive. The lack of detectable endoglucanase activity in *E. coli* might be due to the plasmid low copy number (Fani et al., 1988). On the other hand, the expression of *eglS* by *A. brasilense* could be due to the use of *Streptomyces* regulative signals, present in the cloned fragment, which might increase transcription in *Azospirillum*.
The endoglucanase activity was measured by the DNS method in cell extracts and in supernatants of *A. brasilense* [pCSF121] cultures (Tab. I). It was lower than that determined in *E. coli* [pCSF13] and was exclusively cell associated, whereas in *E. coli* [pCSF13] was mostly released into the medium.
The ability of *A. brasilense* Sp6 [pCSF121] to grow with CMC as carbon source was tested in MSP medium. The results showed that the microorganism could not use CMC as carbon source and they are in agreement with the intracellular localization of the endoglucanase which, therefore, cannot degrade the extracellular CMC. The nitrogen fixing capability was determined by measuring the nitrogenase activity of *A. brasilense* [pCSF121] (Fig. 2). No significative variation in the fixation rate was observed respect to the control *A. brasilense* [pAF700].

Tab. I. Endoglucanase activity in *E. coli* and *A. brasilense* clones

Strain	I.U./ml	
	Medium	**Cell Extract**
E. coli [pCSF13] -IPTG	0.046	0.014
E. coli [pCSF13] +IPTG	0.082	0.012
A brasilense [pCSF121]	0.000	0.006

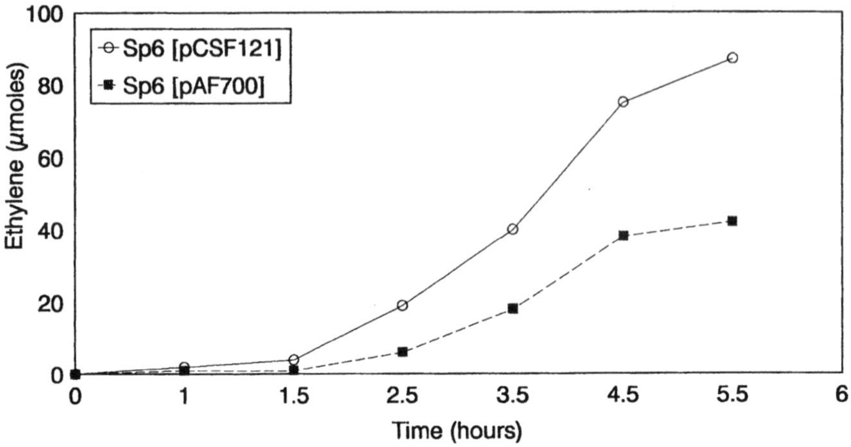

Fig. 2. Nitrogen fixing activity of *A. brasilense* Sp6 [pCSF121] and [pAF700]. Nitrogenase activity is expressed as μmoles of ethylene produced by 80 ml cultures, grown as described by Mugnai et al. (1994).

The chloramphenicol resistance of *E. coli* [pCSF121] and *A. brasilense* [pCSF121] was tested on agar plates. The results reported in Tab. II showed that *E. coli* [pCSF121] was more sensitive to the antibiotic than *E. coli* [pAF700]. Probably, the cloned *Streptomyces* fragment decreases *cam* transcription. On the contrary, *A. brasilense* [pCSF121] was more resistant than *A. brasilense* [pAF700]. These data again suggest a possible recognition by *A. brasilense* of *Streptomyces* transcription signals, which could improve transcription both of *eglS* and *cam* genes.

In conclusion, we have shown that *A. brasilense* Sp6 [pCSF121] expresses the *S. rochei* A2 *eglS* gene. The endoglucanase activity is low and cell associated and does not allow the use of extracellular cellulosic substrates. Its presence does not affect the nitrogen fixing activity of the host. The different behavior of *E. coli* [pCSF121] and *A. brasilense* [pCSF121] in the expression of *eglS* and *cam* genes suggests that *A. brasilense* might recognize some *Streptomyces*

Tab. II. Chloramphenicol minimal inhibitory concentration for *E. coli* and *A. brasilense* clones

Strain	Chloramphenicol (μg/ml)
E. coli [pAF700]	50
E. coli [pCSF121]	15
A. brasilense [pAF700]	10
A. brasilense [pCSF121]	30

regulative signals, i.e. for transcription, internal to the cloned fragment. This is not surprising, since both *Azospirillum* and *Streptomyces* genomes have a G+C content of about 70% (Tarrand et al., 1978; Hopwood and Kieser, 1990) and may share similar regulative sequences.

Acknowledgments

This work was supported by grants from the Italian Ministry of University and Scientific and Technological Research.

References

Coughlan, M. P. (1990). Cellulose degradation by fungi. In: Fogarty, W. M. and Kelly, C. T. (Eds), Microbial enzymes and biotechnology. Elsevier Applied Science, London and New York, pp. 1-36.

Fani R., Bazzicalupo M., Ricci F., Schipani C. and Polsinelli M. (1988). A plasmid vector for the selection and study of transcription promoters in *Azospirillum brasilense*. FEMS Microbiol. Lett. 50: 271-276.

Hopwood, D. A. and Kieser T. (1990). The *Streptomyces* genome. In: Drlica, K. and Riley, M. (Eds), The bacterial chromosome. American Society for Microbiology, Washington, D. C., pp. 147-162.

Miller, G. L., Blum, R., Glennon, W. E. and Burton, A. (1960). Measurement of carboxymethylcellulase activity. Anal. Biochem. 2: 127-132.

Mugnai, M., Bazzicalupo, M., Fani, R., Gallori, E., Paffetti, D. and Pastorelli, R. (1994). Factors affecting nitrogen fixation and *nif* gene transcription in *Azospirillum brasilense*. FEMS Microbiol. Lett. 120: 133-136.

Perito B., Hanhart E., Irdani T., Iqbal M., McCarthy A.J. and Mastromei G. (1994). Characterization and sequence analysis of a *Streptomyces rochei* A2 endoglucanase gene. Gene 148: 119-124.

Sambrook, J., Fritsch, F. E. and Maniatis, T. (1989). Molecular cloning: a laboratory manual. Cold Spring Harbor Laboratory, Cold Spring Harbor, N.Y.

Tarrand J.J., Krieg N.R., Dobereiner J. (1978). A taxonomic study of the *Spirillum lipoferum* group, with description of a new genus, *Azospirillum* gen. nov. and two species, *Azospirillum lipoferum* (Beijerinck) comb. nov. and *Azospirillum brasilense* sp. nov. Can. J. Microbiol. 24: 967-980.

Teather, R. M. and Wood, P. J. (1982). Use of Congo red polysaccharide interactions in enumeration and characterization of cellulolytic bacteria from the bovine rumen. Appl. Environm. Microbiol. 43: 777-780.

Initiation of Monocot Plant Development *In vitro* For Studying Plant-Microbe Interaction

Gábor Gyulai, Lilja Murenyetz[1], János Janovszky[2], Márta Tárczi[3], Ilona Rácz[4]
Department of Genetics and Plant Breeding
University of Agricultural Sciences
H-2103 Gödöllő
Hungary

Abstract

In vitro plant regeneration cycles from salt tolerant forage grasses of *Puccinellia limosa, P.distand*, the hybrid of *A. repens* x *B. inermis* and the *Festuca arundinacea* were initiated *in vitro*. Plant regeneration pathways were found to be somatic embryogenesis. All the crucial steps in plant development were characterized from the scutellar embryo stage to the coleoptilar stage. The somaclonal variation of the hybrid *A. repens* x *B. inermis* was characterized by cytological analysis. Although chromosomes paired essentially as 21 bivalents, the frequency of univalents was found to be higher in the hybrid than in the hybrid-somaclones. The polyamine content (spermine, spermidine, putrescine) was measured to compare embryogeneic and non embryogenic cell line capacities.

Introduction

For the *in vitro* study of the interaction of monocot plants and microbes aseptic plant regeneration cycles are needed. In the present work plant regeneration for forage grasses of the salt tolerant *Puccinellia distans, P.limosa*, the hybrid *A. repens* x *B. inermis* and the *Festuca arundinacea* are presented to supply experimental tools for the study of monocot plant-microbe interaction.

1, Institute of Bioorganic Chemistry, Puschinovo, Russia, R-142292; 2, Irrigation Research Centre, Szarvas, Hungary, H-5540; 3, F.R. Research Institute of Gödöllő University, Kompolt, Hungary, H-3356; 4, Department of Plant Physiology, Eötvös Loránd University, Budapest, Hungary, H-1145

NATO ASI Series, Vol. G 37
Azospirillum VI and Related Microorganisms
Edited by I. Fendrik et al.

Materials and Methods

Plant regeneration cycles: Callus initiation and plant regeneration of *P. limosa* (Schur.) Holmbg. and *P. distans* (L.) Parl. were carried out on modified MS-media as previously described (Heszky et al. 1989; Binh et al. 1989). For the hybrid *A. repens* (L) Beauv. x *B. inermis* Leyss. cv. *nanus (Agromus)* medium-F was used (Gyulai et al. 1992). In the spikelet culture of *F. arundinacea* Schreb. a modified medium-F was used with the application of kinetin (0.1 mg/l) and 2.4-dichlorophenoxy acetic acid (2.5 mg/l). **Cytological analysis:** Chromosomes were counted in root tip cells pretreated for 24 h at 4 °C in 2% sucrose then fixed in 95% ethanol:glacial acetic acid = 3:1 for 3 h. Specimens were stained in 4% acetocarmine for 3 days. For softening, the samples were hydrolised in boiling acetic acid (45%) then squashed in 45% acetic acid before microscopic examination. **Polyamine (PA) analysis:** PA analysis was carried out by TL chromotography according to Nielsen (1990). Standard deviations were calculated from four independent experiments.

Results and Discussion

Spikelet culture of *Festuca arundinacea* Schreb.: The technique of spikelet-originated plant regeneration (Rédei & Rédei 1955; Kasperbauer et al. 1980;) can be resulted in new genotypes of androgenic or gynogenic haploids and doubled haploids. The regeneration of new *Festuca* genotypes in our experiments was successful on the medium-F2 (Table 1). The regenerants were potted in the greenhouse for genetical analysis. Among the regenerants numerous somatic-albinos were obseved.

Table 1. Spikelet-originated plant regeneration of *F. arundinacea* cv. Sz.1-3.

	F.a. cv.Sz.1.		*F.a.* cv.Sz.2.		*F.a.* cv.Sz3.	
Media	explants	regenerants	explants	regenerants	explants	regenerants
E1B5	480	-	800	-	560	-
F1	170	-	620	-	80	-
F2	380	-	320	70	400	110

E1B5 = modified MS (Binh et.al 1989) medium; F1 = medium-F (Gyulai et al. 1992) containig 2.4-D (0.25 mg/l) and kinetin (0.1 mg/l); F2 = the same basal medium containing 2.4-D (2.5 mg/l) and kinetin (0.1 mg/l)

Cytolological analysis of the hybrid Agromus: The normal and abnormal chromosome numbers in the hybrid plant of *Agromus* and in its hybrid-somaclones showed significant differences (Table 2). In the progenises of the hybrid somaclones SC 1-6 more mixopbid plants were found than in the *Agromus.* Moreover, the SC2 somaclone had a progeny with the chromosome number of 2n=28. These results suggest that somatic embryogenesis influences the chromosome numbers towards instability which can be resulted in new types of somaclones (Fehér et al. 1989).

Table 2. Distribution of mitotic chromosome numbers in the vegetative progenies of the hybrid *Agromus* (AB) and its somaclones (SC1-6)

Chr. No.	*Agromus*		*Agromus* Somaclones											
	AB		SC1		SC2		SC3		SC4		SC5		SC6	
	No	%	No	%	No	%	No	%	No	%	No	%	No	%
2n = 42	17	85	7	70	3	23	8	73	5	71	10	77	6	75
mixoploid	3	15	3	30	9	69	3	27	2	29	3	23	2	25
2n = 28	-	-	-	-	1	8	-	-	-	-	-	-	-	-

Meiotic chromosome pairing of the hybrid *Aromus:* The meiotic chromosome pairing of *Agromus* and its somaclones showed a significant pattern (Table 3). In the *Agromus,* chromosomes paired as 21 bivalents and the univalents frequency was higher than in the somaclones. A low frequency of multivalents (trivalents to hexavalents) was also observed. Although the somaclones (SC2, SC9, SC10, SC11) had fewer bivalents, the frequency of multivalents was greater in the *Agromus.* The somaclone SC8 showed a significant difference in the meiotic chromosome pairing, supplying a promising new somaclone (Gyulai et al.1992).

Table 3. Meiotic chromosome pairing in the hybrid *Agromus* (AB) (2n=42) and its somaclones SC7-11 (2n = 42). No of counted cells = 7-49.

plants	I.	II.	III.	IV.	VI.	VII.	VIII.	X.	XII.	XIV
AB	1.29	16.29	0.57	0.43	0.71	-	-	-	-	-
SC7	0.06	6.2	0.06	0.67	0.20	0.02	0.08	0.10	0.06	0.02
SC8	0.71	14.28	0.14	1.14	0.71	-	-	-	-	-
SC9	-	8.42	-	0.58	0.19	0.08	0.15	0.04	0.04	0.08
SC10	0.03	7.77	0.03	0.83	0.20	-	0.13	0.07	0.03	-
SC11	0.13	12.25	0.5	1.5	0.38	-	0.63	-	0.25	-

Cytological analysis of the *Puccinellias:* For the initiation of plant regeneration from *P.distans* and *P.limosa,* two *P.distans* and one *P.limosa* genotype was used (Binh & al. 1989; Heszky & al. 1984). Because of the genotypic diversity of *Puccinellias*, first the chromosome numbers of the basic materials were analysed (Table 4). The first genotypes of *P.distans* were found as 2n=21 and 28, the second genotypes of *P.distans* as 2n=21, and the *P.limosa* as 2n=28. The analysis of regenerants is under development.

Table 4. Distribution of metaphase chromosome numbers in two genotypes of P*uccinellia distans* (L.) Parl. and the *P.limosa* (Shur.)Hombg.

	P.distans 1 (2n=21,28)		*P.distans* 2 (2n=21)		*P.limosa* (2n=28)	
Chr.No	No	%	No	%	No	%
21	65	40	172	46	6	4
28	66	40	33	21	130	87
35	-	-	-	-	3	2
aneuploids	33	20	51	33	10	7

Polyamine (PA) analysis*:* For cell line selection, the polyamine content was found to be a good marker to separate embryogenic lines from non-embryogenic lines. When the polyamine concentrations were monitored through *in vitro* embryogenesis of *B. inermis,* the highest PA content was found in the early stages of embryo development which declined during the later embryo phases. The embryogenic cell line showed extremely high PA content compared to the non-embryogenic line (Table 5). This phenomenon went together with chlorophyll pigmentation on the surfaces of calli.

Table 5. Polyamine (PA) content of the embryogenic and non-embryogenic calli of *B. inermis* Leyss. cv.*nanus*

	nmolPA/g fresh weight		
Developmental phases	Spermine	Spermidine	Putrescine
non-embriogenic calli	44.2	59.6	1970.1
embryogenic calli	253.4	260.4	2573.7

Standard deviations were less than 5% of the mean value in each case.

When the polyamine content of the regenerated *Agromus* somaclones was analyzed (Table 6), the SC1 and the SC3 somaclones showed lower putrescine content. In contrast, the polyamine patterns of hybrid somaclones were similar to the parental plant of *B. inermis* however the somaclones phenologically show the characters of *A. repens.*

Table 6. Polyamine (PA) content (μmol/g fresh weight) of the parental plants of *Agropyron repens* (A), *Bromus inermis* (B), the hybrid of *A. repens* x *B. inermis* (AB) and its hybrid somaclones (SC1-SC9)

PA	A	B	AB	SC1	SC2	SC3	SC4	SC5	SC7	SC9
Spm	0.19	0.14	0.15	0.14	0.13	0.17	0.16	0.13	0.14	0.13
Spd	0.35	0.29	0.46	0.37	0.46	0.45	0.35	0.45	0.51	0.33
Put	2.46	1.28	1.15	0.77	1.10	0.65	1.05	1.02	1.27	0.93

Spm = spermine, Spd = spermidine, Put = putrescine. Standard deviations were less than 5% of the mean value in each case.

To conclude, *in vitro* plant regeneration for four grass species was developed to supply experimental tools for the study of monocot plant-microbe interaction. The chromosome number of the regenerants were counted and polyamine content was measured to characterize the embryogenic capacity.

Acknowledgments: The authors would like to thank Mark Davies for his comments on the manuscript. This work was supported in part by the Hungarian Fund of OMFB No. 94-97-44-0657 an the Fund of AMFK No.520/94.

References

Binh DQ, Heszky LE, Gyulai G, Kiss E, Csillag A (1989) Plant regeneration from callus of Puccinellia distans (L) Parl. Plant Cell Tissue Organ Cult 18 :195-200.

Fehér A, M Hangyel-Tárczi, I Bócsa, D Dudits (1989) Somaclonal chromosome variation in tetraploid alfalfa. Plant Sci 60 : 91-99.

Gyulai G, Janovszky J, Kiss E, Lelik L, Csillag A, Heszky L (1992) Callus initiation and plant regeneration from inflorescence primordia of the intergeneric hybrid Agropyron repens (L) Beauv. x Bromus inermis Leyss. cv. nanus on a modified nutritive mediumn. Plant Cell Rep 11 : 266-269.

Rédei G, Rédei G (1955) Rearing wheats from ovaries cultured in vitro. Acta Bot Hung 2 : 183-186.

Heszky LE, Binh DQ, Kiss E, Gyulai G (1989) Increase of green plant regeneration efficiency by callus selection in Puccinellia limosa (Schur.) Holmbg. Plant Cell Rep 8 : 174-177.

Kasperbauer MJ, Buckner RC, Springer WD (1980) Haploid plants by anther-panicle culture of tall fescue. Crop Sci 20 : 103-107.

Nielsen KA (1990) Polyamine content in relation to embryo growth and dedifferentation in barley (Hordeum vulgare L.). J Exp Bot 41: 849-854.

Chitinase Expression in Strawberry Root Colonized by *Azospirillum Brasilense* and V. A. Mycorrhiza

Carlos H Bellone, Silvia C de Bellone, Raul O Pedraza

Departament of Ecology University of Tucuman Avda Roca 1900 4000 5 N de Tucuman Argentina

Abstract

Spores of *Glomus intraradix* and *Azospirillum brasilense* were inoculated on 5 varieties of Strawberry. In all case, both microorganisms develop chitinase expression on plants, finding that the highest values of this enzyme are produced when the inoculation is altogether with both microorganisms. Associated to this process it was found increments in the number of bacteria and nitrogenase activity with the dual inoculation.

Introduction

The obligate symbionts V. A. mycorrhizal fungi are exceptionally common among terrestrial flowering plants (Genderman 1968, Newman et al 1987). On the other hand, several studies have pointed out that *Azospirillum* colonizes many species of root plants (Döbereiner et al 1980, Okon 1985, Okon et al 1994), and that the dual inoculation produces a beneficial effect on crops.

The fungi involved in this symbiosis have chitin in their wall cell (Bomfante-Fasolo et al 1982), and, the bacterias have N-acetyl glucosamine, a monomeric component of chitin which has been implicated as an inductor in the interactions of the chitinase, and phytoalexine production as a response of defence against fungi, bacterias and virus which commonly invade the plants (Bowles 1990, Roby et al 1987, Metraux et al 1986, Schulumbaum et al 1986, Boller et al 1983).

The mycorrhizal fungus increases the chitinase activity (Spanu et al 1989) besides some other enzymes such as cellulase and pectinase (Bellone 1983, Bellone et al 1993). Equally, *Azospirillum* increases the pectinolitic and cellulolitic activity in root tissues by itself or combined with mycorrhizal fungi (Bellone 1983, Bellone et al 1993) but there is

NATO ASI Series, Vol. G 37
Azospirillum VI and Related Microorganisms
Edited by I. Fendrik et al.

not information about the chitinase production during the interaction of both microorganisms. The pectic fragments obtained by enzymatic action are considerated to be responsible for the elicit chitinase accumulation (Broekart et all 1988) and defensive response regulation (Benhamou et al 1990).

In this experiment we examine the combined effect of VAM *G. intraradix* and *A. brasilense* in strawberry root plants and the chitinase production.

Methods

Five different varieties of strawberry: Chandler, Pájaro, Oso grande, Selva and Fern, were cultivated "in vitro" using the Murashige and Skoog media (1962). We used plants from meristems just in order to eliminate pathogens of the tissues, therefore to avoid a probable interference in the chitinase expression. Seedlings from the 5 varieties (5 month old) were inoculated with sterile spores of *G. intraradix* and with *A. brasilense*, both isolated from the rhizosphere of commercial crops.

Each seedling has got 100 spores and 1 x 10^5 bacterias cultured in NFb media (Döbereiner et al 1980). They have been in greenhouse under a day/night grade of 14/10 hs, with temperature of 27° C during the day and 20° C during the night. They were watered with sterile tap water and with Hoagland type nutrient solution once a week. The plants inoculated with both microorganisms and the controls were tested 4 week after the inoculation.

The infection produced by *Glomus* was estimated by the Giovaneti and Mosse method (1980) and *Azospirillum* by its presence in NFb media (Döbereiner et al 1980). The results are given as a percentage of root infected by the fungus and by the presence of *Azospirillum* in roots cultivated in NFb media.

For the chitinase assay, it was extracted from the roots with a 50 mM Tris buffer, adjusted to pH 8,0 with HCl, 500 mM NaCl and 0,2% of Triton X 100. The chitinase was assayed as described by Boller et al (1983), and the proteins were determined according to Bradford (1976), using bovine serum albumin as standard. All experiments were repeated 5 times, using for each treatment 10 replication. Nitrogenase activity was measured using the acetylene reduction assay (ARA) 15 day after of the inoculation. The plants were sealed into 200 ml bottles stoppered with suba seals and 10% acetylene was injected into the bottles. After 24 h the ethylene concentration in the bottles was measured on a Carle gas chromatograph fitted with a 50 cm porapak column and a hydrogen flame ionization detector.

Results

Table 1. Chitinase and nitrogenase activity. Presence of *Azospirillum* and % of infection by *Glomus* in strawberry roots.

Treatment		Chandler	Pajaro	Oso grande	Selva	Fern
Mycorrhiza	(1)	32±12	42±10	38±14	28±11	36±14
(M)	(3)	325±53	230±32	279±48	317±40	265±58
Azospirillum	(2)	+	+	+	+	+
(A)	(3)	98±18	112±28	69±12	76±19	96±14
	(4)	22±4	18±6	20±8	30±12	16±16
	(1)	69±30	78±28	86±31	75±25	92±32
M + A	(2)	+	+	+	+	+
	(3)	462±69	342±78	378±61	441±69	392±65
Control		-	-	-	-	-

1:% of mycorrhiza - 2: presence (+) or absence (-) of *Azospirillum*.

3: Chitinase activity (nkat/mgr protein ± SE).

4: Nitrogenase activity (nM the etylene/mgr protein ± SE).

The plants express chitinase activity only by the induction of both microorganisms, as it is shown in table I.

Glomus colonized about 32 to 42% of the roots during the period of cultivation in greenhouse.

In plants inoculated with *Glomus* and *Azospirillum* altogether around 69 to 92% of the root length was colonized by *Glomus*, at the same time there was also an increment of the Nitrogenase activity.

Discussion

The roots of strawberry plants were extensively colonized by *Glomus* and *Azospirillum*, and that ability does not differ to much between the varieties used, and it was not affected by the chitinase expression studied here. The microscopical analysis did not reveal any conspicuous difference in the structure of hyphae, vesicles and arbuscules in any of the varieties tested. The induction of chitinase in strawberry plants, as well as some other enzymes such as proteinases and glucanases inducted by the same microorganisms in other plants, suggest that the signalling system, which induce and regulate these enzymes, would belong to several genes associated to the same process.

The different amount of chitinase induced by both microorganisms is due to the fungi structures, as they make possible to find more target regions to induce chitinase activity than the bacteria.

We still do not know the role of the inducer in the root system, whether it is involved directly in the chitinase production just because of their presence or whether by the substances they produce during the infection into the roots. According to this supposition we suggest that both microorganisms associated to strawberry plants can be able to use host defensive compounds as recognition cues. Thus, it seems possible to use the antifungal properties of chitinase production induced by *Glomus* and *Azospirillum*, without reducing their symbiotic potential.

Inductive expression of chitinase, in combination with other enzymes may provide a better protection against a wide range of pathogens.

References

Benhamou N Chamberland H Panze E J (1990) Implication of pectic components in cell surface interaction between tomato root cells and *Fusarium oxysporum spradicis lycopersisi*. Plant physiol 92: 995-1003.

Bellone C H (1983) Incrementos de la actividad celulolitica en tejidos no diferenciados de cana de azucar (*Saccharum* spp L) cultivados "in vitro" e inoculados con dos especies de *Azospirillum*. Revista Agron6mica del Noroeste Argentino vol XX (1 - 4): 5 - 15.

Bellone C H Bellone S C de (1993) Nitrogenase cellulase and pectinase activity in sugar cane roots inoculated together with mycorrhiza and nitrogen fixing bacteria. VI international symposium of the nitrogen fixing microorganisms in leguminous plant. Egypt.

Bonfante-Fasolo P Grippioli R (1982) Cytochemical and biochemical observation on the cell wall of the spore of *Glomus epigaeous*. Protoplasma 123:140-151.

Boller T Gehri A Manch F Vogelli V (1983) Chitinase in bean leaves induction by ethylene purification properties and possible functions. Planta 157:22-31.

Bradford M M (1976) A rapid and sensitive method for the quantification of microgram quantities of protein utilizing the principle of protein-dye-binding. Anal Biochem 72:248-254.

Döbereiner J De Polli H (1980) Nitrogen fixation in rhizocoenosses. International symposium on roots soil system. Londrina. Brasil.

Bowles D J (1990) Defense-related proteins in higher plants. Annu Rev Biochem 59:873-907.

Broekaert W F Peumans W J (1988) Pectic polysaccharides elicit chitinase accumulations in tobacco. Physiol Plant 74:740-44.

Genderman J W (1968) Vesicular-Arbuscular mycorrhiza and plant growth. Anual Review of Phytopathology 6:397-418

Giovanneti M Mosse B (1980) An evaluation of techniques for measuring vesicular-arbuscular mycorrhizal infection in roots. J New phytol 84:489-500.

Metraux J P Boller T (1986) Local and systemic induction of chitinase in cucumber plants in response to viral bacterial and fungal infections. Physiol Plant Phathol 28:161-69.

Murashige T Skoog F (1962) A revised medium for rapid growth and bioassays with tobacco tissue cultures. Plant physiol 15: 473- 497.

Newman E I Reddell P (1987) The distribution of mycorrhizas among families of vascular plants. New phytol 106:745-751.

Okon Y (1985) *Azospirillum* as a potential inoculated for agriculture. Trends Biotechnol 3:223-228

Okon Y, Labandera-Gonzáles C A (1994) Agronomic application of *Azospirillum*: an evaluation of 20 years world wide field inoculation. Soil Biol. Biochem (in press).

Roby D Gadelle A Toppan A (1987) Chitin oligosaccharides as elictiors of chitinase activity in melon plants. Biochem. Biophys Rev Commun 143:885-92.

Schulumbaum A Manch F Vogeli V Boller T (1986) Plant chitinases are potent inhibitors of fungal growth. Nature 324: 365-367.

Spanu P Boller T Ludwig A Wiemken A Faccio A (1989) Chitinase in roots of mycorrhizal *Allium porrum* regulation and localization. Planta 177:447-55.

SECTION III:

INTERACTIONS WITH OTHER MICROORGANISMS

Endophytic bacteria: their possible role in the host plant

S. Di Fiore[1] and M. Del Gallo
Dept. of Food Technology, Environmental and Microbiological Sciences,
University of Molise,
I-86100 Campobasso, Italy

Abstract

The presence of bacteria in the internal tissues of healthy plants has largely been ignored in the past, although in the last fifty years more and more isolations of microorganisms from plant roots, leaves, stems and even spikes and seeds, have been reported.

Little is known about their biological role inside the host plants. Up to now endophytes have given rise to a new biological studies, as they can be employed as plant protecting or plant growth-promoting agents and as nitrogen fixers to supply the associated plant with nitrogen.

The aim of this paper is to compare different approaches to the study of endophytes and to formulate a hypothesis about their role. We maintain that they are mostly a natural associative component of the plant micro-ecosystem, able to balance and to enhance their host fitness to the surrounding environment.

First results about their applications in agriculture are discussed.

Introduction

During the last fifty years many bacteria have been isolated from the interior of crop plants. Since it has always been maintained that healthy plants are normally bacteria-free, it is not an easy matter to find adequate explanations - with the exception of some well known diazotrophic bacteria - for this large number of isolates from the host tissues so far. However, these findings have recently stimulated new interest among researchers.

Because of the lack of knowledge about their biological role inside the plant, several terms have been employed to define these microorganisms: endogenous bacteria (Sturdy and Cole, 1974), xylem-residing bacteria (Gardner *et al.*, 1982;

---

1 Biotechnology and Agriculture Sector, ENEA-Casaccia, Via Anguillarese 301, I-00060 Rome.

NATO ASI Series, Vol. G 37
Azospirillum VI and Related Microorganisms
Edited by I. Fendrik et al.

Gagnéet *al.*, 1987), latent bacteria (Hayward, 1974). In accordance with Kloepper *et al.* (1992), we believe that it is not necessary to use specific expressions such as "endorhizosphere" to define the environment inside the root in which rhizobacteria are frequently isolated: they are too restrictive or can easily generate misunderstandings. However, it is difficult to create new exhaustive terms since many microorganisms have been isolated from the whole plant.

Up to now the only term generally in agreement among the scientific community seems to be "endophytes".

Endophyte is a composite Greek word that means "inside the plant". In the literature this word is utilised to define those microorganisms - mostly bacteria and fungi - able to enter plant tissues, and to establish themselves inter- and intracellularly. Endophytes have the property to establish active relationship with their hosts and, as suggested in two recent reviews (Kloepper and Beauchamp, 1992; Kloepper *et al.*, 1992), they could be defined as "colonists". This term is useful since the colonisation is an active process involving steps both outside and inside the host (Sprent and De Faria, 1988), survival and multiplication being two of the most important. But, in a broad ecological sense, colonisation means also the aptitude of an organism to spread over (in this case inside) a habitat. So, this term could be selective in regard to those organisms which, though inside the plant, are not allowed to spread over different tissues or organs of their host.

How should these microorganisms be considered? Is their isolation a sufficient requisite to call them endophytes? And why do plants contain such a mixed microflora inside their tissues?

Through critical analysis of the literature we will try to answer these questions and to put forward some hypotheses.

Biocenoses or latent infections? Two approaches to endophytes

There are at least two different approaches to the study of endophytes: the "phytopathological" and the "ecological".

Both have some common steps:

i) isolation of these microorganisms from plants and identification,

ii) estimation of their numerical consistence,

iii) their localisation.

However, there are some important differences.

The first approach, the phytopatological, has the epidemiological purpose of verifying the pathogenicity of these microorganisms and preventing their spreading into the surrounding environment.
After the microorganisms' isolation from symptomless or from generally retained non-host healthy plants, the so-called "sensitivity tests" are carried out, which consist in inoculating all isolates singularly by injection or by artificial wounds on plants retained to be susceptible. At the same time "hypersensitivity tests" on tobacco leaves are carried out. After the appearance of the first symptoms the microorganisms are reisolated from the damaged tissues and their degree of pathogenicity is estimated.
Alternatively, seedlings of the original hosts are inoculated with a strong inoculum (10^6 to 10^8 cells per plant) and plants are grown in abiotical stress conditions (water stress, high temperatures or low pCO_2) to test the appearance of pathological symptoms. In other words, artificial plant-growth conditions are recreated without taking into account that defensive barriers of stressed plants are generally lowered (Schroth and Weinhold, 1986; Sturdy and Cole, 1974). However, even though more sophisticated phytopathological approaches has been described - Ercolani (1984), for example, stressed the importance of considering both host resistance and dose/response relationship - the above mentioned methods are the most commonly used in endophytes investigations. Thus, most of the microorganisms isolated so far inside the plant are recognized as latent infective bacteria and their symptomless hosts constitute a dangerous infection source for susceptible crops (Bugbee *et al.*, 1987; Hayward, 1974).
The "ecological" approach, instead, has the aim of exploring the biological significance of these microorganisms inside the plant and of understanding which role they play in association with the plant. This approach could be invaluable for a potential exploitation of these microorganisms in agriculture.
The most important differences between the two approaches consist in the fact that the first is focused on the single microorganism and its pathogenicity; on the contrary, the second approach is mostly focused on the association between host plant and related endophytes. This second approach considers both plant and microorganisms as a whole ecosystem tending toward equilibrium.

The plant as a new complex micro-ecosystem

Plants can be considered as a complex micro-ecosystem where "different habitats can be exploited by a wide variety of bacteria" (McInroy and Kloepper,

1994). These habitats are not only represented by plant external surfaces - where epiphytic bacteria predominate - but even by internal tissues where many

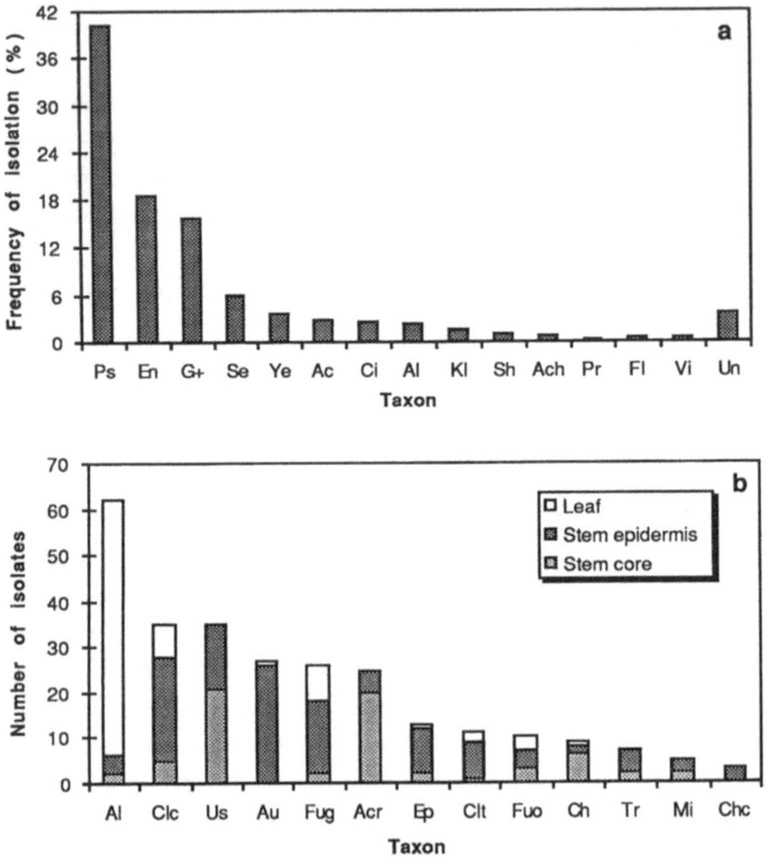

Figure 1. **a)**: Isolation frequency of bacteria derived from xylem sap of *Citrus jambiri* . Ps: *Pseudomonas*; En: *Enterobacter*; G+: Gram-positive species (*Bacillus, Corynebacterium, Arthrobacter, Actinomycetes*); Se: *Serratia*; Ye: *Yersinia*; Ac: *Acinetobacter*; Ci: *Citrobacter*; Al: *Alcaligenes-Moraxella*; Kl: *Klebsiella*; Sh: *Shigella*; Ach: *Achromobacter*; Pr: *Providencia*; Fl: *Flavobacterium*; Vi: *Vibrio*; Un: Unknown (elaborated from: Gardner *et al.* 1982). **b)**: number of fungal isolates in *Zea mays* inside different plant tissues. Al: *Alternaria alternata*; Clc: *Cladosporium cladosporioides*; Us: *Ustilago* sp.; Au: *Aurebasidium pullulans*; Fug: *Fusarium graminearum*; Acr: *Acremonium strictum*; Ep: *Epicoccum purpurascens*; Clt *Cladosporium tenuissimum*; Fuo: *Fusarium oxysporum*; Ch: *Chaetomium* sp.; Tr: *Trichoderma harzianum*; Mi: *Microdochium bolleyi*; Chc: *Chaetomium cochliodes* (elaborated from Fisher *et al.* 1992).

microorganisms penetrate and survive. We can compare each organ and each plant-tissue to a habitat where the different bacteria fill a different ecological niche, and represent a nutritional and associative component within a biological community.

Inside the plant micro-ecosystem different microbial species, both bacteria and fungi (Fisher *et al.*, 1992), are able to interact and to establish an equilibrium. Some of them can be considered as dominant species (Van Peer *et al.* 1990) and are represented by those microorganisms most frequently isolated, and in large numbers, from the host. As well as the dominants, there is a large variety of species that cannot easily be isolated because of their low numerical consistence. These are considered as rare species. In all reports dealing with endophytes it is possible to find these two groups of microorganisms: in the figure 1 isolation frequency of endophytic bacteria (a) and fungi (b) is shown. These data confirm the presence and the distribution of dominant and rare species supporting the micro-ecosystem hypothesis. Furthermore, applying a microbial ecology's concept, it is possible to roughly consider the two groups as r- and K-strategists respectively (Andrews and Harris, 1986).

At the base of each ecosystem there must be biodiversity, as vast as possible: many rare species with few dominant ones is a basic definition of the biological component of an ecosystem. Biodiversity is recognized as the most important condition to reach its climax.

According to Odum (1963), to understand an ecosystem it is not only important to know both abiotics and biotic components, but particularly their spatial-temporal variations and interrelationship and - we can also add - the processes leading to its establishment. This could be helpful to understand in which way climax conditions are reached. Furthermore, it is essential to study these variations in a natural ecosystem, where all the factors are interacting simultaneously.

Applying these concepts to the plant micro-ecosystem we can hypothesize that each plant species has its own associated microorganisms: neutral symbiont, beneficial and pathogenic (Döbereiner, 1988b; Fisher *et al.*, 1992; Schroth and Weinhold, 1986). Each host-associated microbial community is the result of a co-evolutionary process (Misaghi and Donndelinger, 1990) developed during the course of the phylogenesis. As proposed for the selection of new fungal pathogenic *formae speciales* and bacterial pathovars for the evolution of biotypes and races, the gene-for-gene theory (Tosa, 1992) could be adapted to beneficial endophytic microorganisms.

With the exception of seed-borne microorganisms, most of the endophytes start colonization when the seed germinates. A wide variety of molecules recognized by a pool of microorganisms are produced by the seedling. This pool is not only made up of specific host-associated microorganisms but even by a generic saprophytic and/or parasitic microflora. As soon as seed germinates competition starts, until species with the greatest host "affinity" - both dominant and rare - will establish in its rhizosphere displacing or preventing less specific colonists to increase in number. Such a situation is brought about with plant growth-promoting rhizobacteria (PGPR) inocula on crop plants when biological control of pathogenic microorganisms and increases in yield are obtained (Kloepper and Schroth 1981a, 1981b; Suslow and Schroth, 1982).

After endophytes have colonised the root exterior, they enter their host: this property can be considered a way to avoid external competition and to have a higher substrata availability; thus, they have been considered as "fortunate winners of the game" (Hartmann, 1988). However, competitive interaction between microorganisms could happen inside internal tissues too, until equilibrium among the species is obtained.

Microorganism interactions could be not only competitive. Because of their normal intercellular localisation and since this microhabitat is rich in plant metabolites, a kind of nutritive chain might be established: on this basis, the presence of dominant and rare species, r- and K-strategists, could be explained as the result of a nutritional biological succession, even inside the plant.

The plant, however, is a living organism and the associate endophytic community colonisation is the result of a dynamic process involving both host and microorganisms, all along the plant's life cycle.

Endophytes and their hypothetical role

Once inside the plant not all microorganisms spread through all tissues. As previously outlined, this could be due either to interactions with other microorganisms, or to the different nutritional requirements of each microorganism that let them inhabit different niches - represented by a tissue and, more specifically, by intercellular spaces inside each tissue. There are few reports, instead, on the intracellular presence of endophytes. This seems to be an occasional finding, except for particular cases such as *Rhizobium* or *Alcaligenes faecalis* (You *et al.*, 1991) where both bacteria are enveloped by the plant in specialised structures. Other microorganisms can easily reach xylematic vessels passing through the cortical layer and root endodermis. Following the plant growth cycle these bacteria can be translocated into the upper organs and can reach even the reproductive structures (Fogher and Civardi, 1989; Misaghi and Donndelinger, 1990; Reeser and Kostka, 1988; our laboratory, unpublished data). Some establish in the seeds giving origin to the so-called "seed-born" microorganisms. Furthermore, it is possible to presume that some microorganisms are transported epiphytically during seedling emergence or can reach the upper part of the plant from the air. In this case, penetration could be through natural openings on the leaves (e.g. stomata, hydathodes, nectarthodes) or through stem lenticels (Kluepfel, 1993). A completely different way of penetration is described by Ashbolt and Inkerman (1990) on sugar cane via mealybug, and by Kluepfel (1993), via different insects.

Mycorrhizas, are the most likely common vector of endophytes. Described for the first time by Paula *et al.* (1991), this very effective technique is becoming the usual way of inoculating endophytes inside crop plants (Isopi *et al*, 1995).

Independent of the way they penetrate, all microorganisms which establish itself inside a specific host should be considered "endophytes". This aspect is of great importance because only microorganisms isolated more than once from internal tissues of their host plant can be considered real endophytes. We can extend a commonly accepted concept of rhizospheric microorganisms (Kloepper *et al.* 1992) to the generic endophytic microflora: microorganisms living in the plant micro-ecosystem are not "a passive chance encounter" with a host plant, but are the result of an active process (colonisation).

One of the most important aspects related to the study of endophytes is their isolation. There are two rules for the commonly used isolation techniques: avoid contamination and reproduce biodiversity. For example it is generally recognised that some microorganisms are embedded within a mucigel layer on the root surface, so even the most effective sterilizing agents or washing techniques cannot prevent the isolation of bacteria which are not endophytic (Hayward, 1974). Thus, isolation is not always a sufficient requisite to ensure the endophytic nature of a microorganism. Whenever possible it would be better to mark and follow them after reinoculation in field or in greenhouse experiments.

Nevertheless, commonly used sterilising agents can easily penetrate leaves, stem and root tissues, thus reducing quantitative and qualitative estimation of microorganisms. A correct isolation technique should take into account the differences in the colonisation pattern during the whole plant-growth cycle. It

would be better to repeat isolation several times at different growth stages for a clear picture of the endophytes' population dynamic.
However, many microorganisms cannot be isolated because of: unsuitable growth media, different growth property (slow growing species hidden by fast growing ones, that is K-strategists hidden by r-strategists), very low numerical consistence, different metabolism (i.e. aerobic or anaerobic). Concerning this aspect, to our knowledge, there are only few reports on isolation of anaerobic endophytes (Ercolani and Casolari, 1966; Zeikus and Ward, 1974).
In general, these aspects should be taken into account before drawing a wrong conclusion about the endophytic nature of certain isolates. Interesting suggestions are reported in a recent review by Kloepper and Beauchamp (1992).

At present it is not easy to understand the role of endophytic microflora. Our micro-ecosystem hypothesis is mostly based on biodiversity within the microbial community, outside and inside the host plant. We believe that biodiversity of plant-associated microbial species is a basic condition to establish plant fitness in each environment. Plant breeding, linked to an unbalanced agricultural environment, has enormously increased yields, but has probably altered this internal equilibrium and has led to the appearance of pathogenic responses, selecting a few species. Healthy plants, in fact, naturally have well-regulated associative endophytic microflora. The loss of this equilibrium could favour disease development. The presence of so-called "latent infective bacteria" could finally find an explanation: bacteria which are normally in a low number, are continuously selected and suddenly proliferate not being antagonised by naturally occurring microflora.

In order to explain the role of endophytic bacteria we can make some considerations. In table 1 a large number of endophytes isolated from healthy plants is presented. Noted are:
i) more then two-thirds of the isolates are gram negative bacteria;
ii) most have aerobic or microaerobic metabolism;
iii) most have flagella;
iv) none has been isolated from phloematic tissues;
v) even though some of these microorganisms are considered to be plant pathogens, they were isolated from symptomless or from non-susceptible host plants.

Since it is not possible to draw out common features among the different plant inhabitants that could explain the presence of different endophytes, the only way is to consider them generically as "associative microflora". This term has previously been suggested (Kloepper and Beauchamp, 1992) to describe a generic relationship between the host plant and the related microorganism. According to Lynch *et al.* (1979) associations can be described in terms of dependency (obligate or facultative), durability (persistent or transient), specificity (specific or non-specific) and location (ecto or endo-organisms).

In terms of dependency it is possible to distinguish between obligate and facultative endophytes. Most of the microorganisms quoted in table 1 can be considered as facultative endophytic-associated microorganisms because of their capability to live outside plant tissues as rhizospheric bacteria. There are also some well documented examples of obligate endophytes: *Corynebacterium sepedonicum* has been recognised as a normal obligate endophyte of sugar beet (*Beta vulgaris*). This bacterium is able to follow the whole biological cycle of its host, to reach the seeds and to be transmitted as a seed-born microorganism (Bugbee *et al.* 1987). *Herbaspirillum seropedicae* and/or *Pseudomonas* (*Herbaspirillum*) *rubrisubalbicans* is another example known to live only in the internal tissues of their host (sugar cane, sorghum). It has been shown that this/these specie/s are unable to survive outside the plants, only inside some crop residues (Gillis *et al.*, 1991; Baldani *et al.*, 1992).

In terms of durability the endophytes can be considered as permanent since, once inside the plant, they survive as long as possible until the host can support their growth. Thus, it is not important to distinguish between microorganisms able to survive during the complete plant life cycle and those that can be isolated only for a short period of time. Both are permanent: the only difference being their fitness to the host's environment. The former have gained an evolutive advantage which is mostly due to the niche they hold inside the plant (e.g. those xylem-residing bacteria easily translocated to the stem). The latter occupy a less durable environment.

It is not easy to speculate on endophytes specificity. Except for some particular and well documented case such as *Rhizobium*-legumes symbioses, actynorrhizal symbioses and few mycorrhizal symbioses, most of the endophytes seem to colonize more than a host. Some taxa listed in table 1 have been isolated from several hosts, but this is not enough to assert that they are aspecific. Specificity cannot be observed at species level, but it is mostly a strain level-related property. Instead of specificity in many cases it seems more correct to use compatibility and

Table 1. Endophytic bacteria isolated from living trees, healthy crops and weeds

taxon		host plant	references
Acetobacter	*diazotrophicus*	sugar cane (*Saccharum officinarum*)	Gillis *et al.* (1989); Döbereiner *et al* (1988a)
		sweet potato (*Ipomoea batatas*)	Döbereiner *et al.* (1988a)
		cameroon grass (*Pennisetum purpureum*)	Döbereiner *et al.* (1988a)
Achromobacter	spp.	lemon (*Citrus jambhiri*)	Gardner *et al.* (1982)
Acinetobacter	*baumanii*	cotton (*Gossypium hirsutum*)	McInroy and Kloepper (1994)
	lwoffii	lemon	Gardner *et al.* (1982)
Agrobacterium	*radiobacter*	cotton	McInroy and Kloepper (1994)
	spp.	potato (*Solanum tuberosum*)	De Boer and Copeman (1974)
Alcaligenes	spp.	sweet corn (*Zea mays*); cotton	McInroy and Kloepper (1991)
Arthrobacter	*globiformis*	*Azolla* spp.	Forni *et al.* (1992)
	spp.	sweet corn	McInroy and Kloepper (1991)
		cotton	McInroy and Kloepper (1994)
		lemon	Gardner *et al.* (1982)
Aurobacterium	*barkeri*	sweet corn; cotton	McInroy and Kloepper (1994)
	saperdae	sweet corn; cotton	McInroy and Kloepper (1994)
Azoarcus	spp.	kallar grass (*Leptochloa fusca*)	Hurek *et al.* (1991);
		rice (*Oryza sativa*)	Hurek *et al.* (1991)
Azospirillum	spp.	many crops	many reports
Bacillus	*brevis*	cotton	Misaghi and Donndelinger (1990)
	cereus	cotton; corn (*Zea mays*)	Mahaffee *et al.* (1994b)
	megaterium	sweet corn; cotton	McInroy and Kloepper (1994)
		potato	Hollis (1951)
	pumilus	cotton	McInroy and Kloepper (1994) Misaghi and Donndelinger (1990)
		potato	Sturdy and Cole (1974)
	sphaericus	potato	Sturdy and Cole (1974)
	subtilis	sugar beet (*Beta vulgaris*)	Jacobs *et al.* (1985)
	spp.	potato	De Boer and Copeman (1974)
		cotton	Misaghi and Donndelinger (1990)
		lemon	Gardner *et al.* (1982)
		spanish oaks (*Quercus texana*)	Brooks *et al.* (1988)
Brevibacterium	spp.	sweet corn; cotton	McInroy and Kloepper (1991)
Burkholderia	*cepacia*	lemon	Gardner *et al.* (1982)
	pickettii	sweet corn	McInroy and Kloepper (1994)
	solanacearum	cotton	McInroy and Kloepper (1994)
Cellulomonas	spp.	cotton	McInroy and Kloepper (1991)
Citrobacter	*freundii*	potato	Sturdy and Cole (1974);
		lemon	Gardner *et al.* (1982)
	testaceum	sweet corn; cotton	McInroy and Kloepper (1994)
	spp.	cucumber (*Cucumis sativus*)	Meneley and Stanghellini (1974)
Clavibacter	*xyli* sub. *cynodontis*	bermuda grass (*Cynodon dactylon*)	Kostka *et al.* (1988)
	spp.	cotton	Misaghi and Donndelinger (1990) McInroy and Kloepper (1991)
Corynebacterium	*sepedonicum*	sugar beet	Bugbee *et al.* (1987)
	spp.	lemon	Gardner *et al.* (1982)
		sugar beet	Jacobs *et al.* (1985)
Curtobacterium	spp.	sweet corn; cotton	McInroy and Kloepper (1991)
Enterobacter	*aerogenes*	lemon	Gardner *et al* (1982)
(Pantoea)	*agglomerans*	corn	Fogher and Civardi (1989); Fisher *et al.* (1992)
	asburiae	bean (*Phaseolus vulgaris*)	Mahaffee *et al.* (1994a)
	cloace	corn	Fogher and Civardi (1989); Fisher *et al.* (1992)
		cucumber	Meneley and Stanghellini (1974)
		lemon	Gardner *et al.* (1982)
		potato	Hollis (1951)
	sakazakii	lemon	Gardner *et al.* (1982)
Erwinia	*carotovora*	cucumber	Meneley and Stanghellini (1975)
	herbicola	sugar beet	Jacobs *et al.* (1985)
	spp.	corn	Fogher and Civardi (1989)
		cucumber	Meneley and Stanghellini (1974)
		cotton	Misaghi and Donndelinger (1990)
Flavimonas	*oryzihabitans*	sweet corn	McInroy and Kloepper (1994)

(*continued*)

taxon		host plant	reference
Flavobacterium	spp.	potato	De Boer and Copeman (1974)
		lemon	Gardner *et al.* (1982)
		sweet corn; cotton	McInroy and Kloepper (1991)
Herbaspirillum	*seropedicae*	many gramineous plants	many reports
	rubrisubalbicans	(*)	(*)
Hydrogenophaga	*flava*	sweet corn; cotton	McInroy and Kloepper (1994)
	pseudoflava	sweet corn; cotton	McInroy and Kloepper (1994)
Klebsiella	*aerogenes*	potato	Sturdy and Cole (1974)
	ozaene	potato	Sturdy and Cole (1974)
	terrigena	corn	Fisher *et al.* (1992)
Kluyvera	spp.	swett corn; cotton	McInroy and Kloepper (1991)
Lactobacillus	spp.	sugar beet	Jacobs *et al.* (1985)
		sweet corn; cotton	McInroy and Kloepper (1991)
	spp.	lemon	Gardner *et al.* (1982)
Listeria	*innocua*	cotton	McInroy and Kloepper (1994)
Methanobacterium	spp.	american elm (*Ulmus americana*)	Zeikus and Ward (1974)
		black willow (*Salix nigra*)	Zeikus and Ward (1974)
		eastern cottonwood (*Populus deltoides*)	Zeikus and Ward (1974)
Methylobacterium	spp.	sweet corn; cotton	McInroy and Kloepper (1991)
Microbacterium	*imperiale*	sweet corn	McInroy and Kloepper (1994)
Micrococcus	spp.	potato;	De Boer and Copeman (1974)
		sweet corn; cotton	McInroy and Kloepper (1991)
Moraxella	spp.	sweet corn	McInroy and Kloepper (1991)
Morganella	spp.	sweet corn; cotton	McInroy and Kloepper (1991)
Ochrobactrum	*anthropi*	sweet corn	McInroy and Kloepper (1994)
Phyllobacterium	spp.	sweet corn; cotton	McInroy and Kloepper (1991)
Providencia	*alcalifaciens*	cotton	McInroy and Kloepper (1994)
	spp.	lemon	Gardner *et al.* (1982)
Proteus	*mirabilis*	cucumber	Meneley and Stanghellini (1974)
Pseudomonas	*aeruginosa*	lemon;	Gardner *et al.* (1982)
		sugar beet	Jacobs *et al.* (1985)
	corrugata	alfalfa (*Medicago sativa*)	Lukezic (1979)
		corn	Fisher *et al.* (1992)
	fluorescens	bean	Mahaffee *et al.* (1994a)
		wheat (*Triticum* spp.)	Peng *et al.* (1994)
		tomato (*Lycopersicon esculentum*)	Van Peer *et al.* (1990)
		lemon	Gardner *et al.* (1982)
		sugar beet	Jacobs *et al.* (1985)
		corn	Fisher *et al.* (1992)
	marginalis	corn	Fisher *et al.* (1992)
	putida	tomato	Van Peer *et al.* (1990)
		lemon	Gardner *et al.* (1982)
		spruce (*Picea glauca*)	O'Neill *et al.* (1992)
	rubrisubalbicans		see *Herbaspirillum*
	viridoflava	corn	Fogher and Civardi (1989)
	spp.	corn	Fogher and Civardi (1989); Fisher *et al.* (1992)
		potato	De Boer and Copeman (1974)
		lemon	Gardner *et al.* (1985)
		tomato	Van Peer *et al.* (1990)
		live oaks (*Quercus fusiformis*)	Brooks *et al.* (1988)
Rothia	spp.	sweet corn; cotton	McInroy and Kloepper (1991)
Salmonella	spp.	sweet corn; cotton	McInroy and Kloepper (1991)
Serratia	*liquefaciens*	lemon	Gardner *et al.* (1982)
	marcescens	corn	Fogher and Civardi (1989)
	spp.	cotton	McInroy and Kloepper (1994)
		orange (*Citrus sinensis*)	Gardner *et al.* (1985)
Shigella	spp.	lemon	Gardner *et al.* (1982)
		sweet corn; cotton	McInroy and Kloepper (1991)
Staphylococcus	*hominis*	spruce	O'Neill *et al.* (1992)
	spp.	sweet corn; cotton	McInroy and Kloepper (1991)
Streptococcus	spp.	cotton	McInroy and Kloepper (1991)
Vibrio	spp.	corn	Fisher *et al.* (1992)
Xanthobacter	spp.	sweet corn	McInroy and Kloepper (1991)
Xanthomonas	spp.	potato	De Boer and Copeman (1974)
		cotton	Misaghi and Donndelinger (1990)
		sugar beet	Jacobs *et al.* (1985)
Yersinia	spp.	lemon	Gardner *et al.* (1982)
		cotton	McInroy and Kloepper (1991)

to consider the endophytic community the result of a compatible association with the host plant. However specificity is one of the most important features to explore because it is essential for an applicative exploitation of endophytes.
On the basis of these observations a general definition can be put forward: endophytes can be considered as permanent associative microorganisms, whether obligate or facultative, able to inhabit healthy plant interior as members of a diversified dynamic community.

Practical applications

Several endophytes have already been exploited for practical application in laboratory, greenhouse or in field experiments, mostly as biocontrol or as plant growth-promoting agents.
Clavibacter xyli subsp. *cynodontis* is one of the most exploited: this bacterium, isolated as natural host of bermuda grass (*Cynodon dactylon*), can colonise many botanical species (Kostka *et al.* 1988); following inoculation it can penetrate, in a short time, the whole plant. These characteristics and the opportunity to express the genetic information transferred from *Bacillus thuringiensis* subsp *kurstaki* (for the delivering of the CryIIA δ-endotoxin) make it possible to utilise this bacterium for the european corn borer (*Ostrinia nubilalis*) control, eliminating chemical treatment (Millner and Kitt, 1993; Mahaffee *et al.*, 1994b).
Similar results were obtained in *in vitro* experiments with *Bacillus cereus* - where the same *B. thuringiensis* genes have been transferred successfully - leading to the control of two lepidoptera species (Mahaffee *et al* 1994b).
Chen *et al* (1994) - in greenhouse conditions - successfully reduced wilt disease (*Fusarium oxysporum* f. sp. *vasinfectum*) symptom's in artificially infected cotton plants by inoculation of naturally occurring endophytes isolated from the same host plant.
In greenhouse experiments it has been demonstrated that rhizospheric and endophytic bacteria associated with some crops with antagonistic properties to phytopathogenic nematodes, are involved in controlling these pests (Kloepper *et al.* 1991).
Endophytic fungi have been also exploited as biocontrol agents: *Acremonium*/grass association has been identified as a symbiotic mutualism with both the plant host and the endophyte benefitting from the relationship.

This endophyte is able to improve fitness (Clay, 1993) and to increase host's associated resistence to biotic (Latch, 1993) and abiotic stresses (Bacon, 1993).

As plant growth-promoters endophytes exert beneficial effects on their host in different ways:

- directly, by producing stimulatory substances, such as phytohormones (Tabacchioni *et al*, 1993) or as nitrogen fixers, thus supplying the host with combined nitrogen (Döbereiner *et al.*, this volume);
- indirectly, exerting antagonistic interactions against major or minor root pathogens; these interactions - mostly due to spatial and nutritional competition or to antibiotic production - have been shown for endophytic PGPR (Brooks *et al.*, 1988; Van Buren *et al.*, 1993) (concerning endophytic PGPR, it would be more appropriate to call them PGPEI). In fact, as previously remarked, many endophytic microorganisms were isolated and originally described as rhizospheric. To our knowledge, except for *Azospirillum* sp., no systematic studies have been performed to verify whether PGPR microorganisms, following inoculation, are able to enter inside host tissues.

Endophytes which have aroused greatest interest so far are undoubtedly the nitrogen fixers since they can be utilised in the field as a natural source of combined and instantly available nitrogen. It has been shown that these microorganisms can substitute chemical fertilizers, or can aid in reducing chemical input in agriculture. Not only *Azospirillum* (Okon and Labandera-Gonzales, 1994) but also *Acetobacter diazotrophicus* (Boddey *et al.* 1991), *Herbaspirillum seropedicae* and *Azoarcus* (Hurek *et al.*, 1993) have been tested for their property to fix nitrogen once inside the plant in greenhouse or in field trials.

Most of the studies carried out to evaluate endophytes for their ability in colonising internal tissues or to exert beneficial interactions, have been based on inoculation techniques unsuitable to agricultural applications (e.g. injection, wound or vacuum inoculation).

The only techniques compatible with the most commonly utilised farm machinery are seed and foliar inoculation. Since few reports have been focused on these kinds of inoculation, to fully exploit them further experiments need to be undertaken.

Of certain interest could also be direct inoculation techniques aimed at transferring microorganisms in micropropagated plants or in seeds: the former technique has been successfully experimented in field trials on micropropagated plantlets of sugar cane, in vitro inoculated with *Acetobacter diazotrophicus*

(James *et al.*, 1994). In vitro inoculation represents probably the best way to study endophytic microorganisms in gnotobiotic conditions. The latter, though very advantageous, could be applied after a complete growth cycle to those plant-bacteria associations where the endophytic species are able to colonise reproductive structures too.
Because of their property to establish symbiotic relationship with most crops and since it has been shown that some spore-associated bacteria are directly transferred into the host tissues, mycorrhiza represent a further possibility to inoculate endophytes in a completely natural way. Such a result has been obtained with co-inoculation of diazotrophic bacteria and VA mycorhiza in sweet potato, sugar cane, sweet sorghum (Paula *et al.*, 1991), sugar beet (Del Gallo *et al.*, 1991) and corn plants (Di Fiore *et al.* 1994).

Conclusion

Possible hypotheses about endophytes' role has been put forward. Further conclusive considerations can be made, however.
During the last fifty years agricultural techniques have been greatly changed to increase crop yield. This has led to the disappearance of old cropping systems which were able to maintain agro-ecosystem's equilibrium. Since monoculture and application of high levels of agri-chemicals became a common way to manage agro-ecosystems, biotic diversity has continually decreased, the balance among the different component has been disrupted, reducing and/or eliminating competitiveness against so-called "pests". The soil microbial pool has decreased, in number and in species, because of the lack of both quantity and quality of soil humus and organic matter. These changes can also be due to the reduced amount of root exudates released by breeded plant, where a higher productivity is reached by diverting the photosynthate to the higher part of the plant in detriment to the root apparatus and to its sphere of influence.
Endophytes can be helpful to partially re-establish biotic diversity inside the plant associated biocommunity, to protect the plant from pests and environmental stress, and to supply the plant combined nitrogen and plant-growth promoting substances. More information, however, are needed about the microbial community of crop plant. Only when a complete picture of naturally occurring plant-colonists and their specific interactions is obtained, it will be possible to reconstruct an equilibrium in the agro-ecosystems.

It is not by mere chance that the most consistent results obtained from inoculation were not with single microorganisms but with multiple inoculation of selected PGPR/PGPEndophytes and mycorrhizal fungi.

Aknowledgements

We wish to thank Franco Zucconi and Florita Botts for the helpfull discussions and critical revision of the manuscript.

References

Andrews JH, Harris RF (1986) r- and K-Selection and microbial ecology. Advances in Microbial Ecology 9:99-147

Ashbolt NJ, Inkerman PA (1990) Acetic acid bacterial biota of the pink sugar cane maelybug *Saccharococcus sacchari*, and its environs. Appl Env Microbiol 56: 707-712

Bacon CW (1993) Abiotic stress tolerances (moisture, nutrients) and photosynthesis in endophyte-infected tall fescue. *In* "*Acremonium*/grass interactions" (Special Issue). Joost R, Quisenberry S (*eds*) Agriculture Ecosyst Environ 44: 123-141

Baldani VLD, Baldani JI, Olivares F, Döbereiner J (1992) Identification and ecology of *Herbaspirillum seropedicae* and the closely related *Pseudomonas rubrisubalbicans*. Symbiosis 13: 65-73

Boddey RM, Urquiaga S, Reis V, Döbereiner J (1991) Biological nitrogen fixation associated with sugar cane. Plant Soil 137: 111-117

Brooks DS, Gonzalez CF, Appel DN (1988) Isolation and evaluation of endophytic bacteria from live oaks for *in vitro* inhibition to *Ceratocystis fagacearum* and colonization of oak trees. (*abstract*) Phytopathol 78: 626

Bugbee WM, Gudmestad NC, Secor GA, Nolte P (1987) Sugar beet as a symptomless host for *Corynebacterium sepedonicum*. Phytopathol 77: 765-770.

Chen C, Banske EM, Musson G, Kloepper JW (1994) Biological control potential and population dynamics of endophytic bacteria in a cotton/*Fusarium* wilt system. *In* "Improving plant productivity with rhizosphere bacteria". Ryder M H, Stephens P M, Bowen G D (*eds*). CSIRO, Australia, *pp* 191-193

Clay K (1993) The ecology and evolution of endophytes. *In* "*Acremonium*/grass interactions" (Special Issue). Joost R, Quisenberry S (*eds*) Agriculture Ecosyst Environ 44: 39-64

De Boer SH, Copeman RJ (1974) Endophytic bacteria flora in *Solanum tuberosum* and its significance in bacterial ring rot diagnosis. Can J Plant Sci 54: 115-122

Del Gallo M, Fabbri P, Marini F, Döbereiner J (1991) First results of inoculation of sugar beet (*Beta vulgaris*) with *Acetobacter diazotrophicus* via V.A.

mycorrhiza.(*abstract*) *In* "6th Internation Symposium on Microbial Ecology", Barcellona , 6-11 September 1992 *pp* 168

Di Fiore S, Fabbri P, Marini F, del Gallo M, Bozzini A (1994) Tripartite interaction between *Zea mays*, *Glomus* spp and Nitrogen-fixing bacteria introduced in plant rhizosphere via mycorrhiza. Abstract book of the NATO Advanced research Workshop on *Azospirillum* and Related Microorganisms. Sárvár, September 4-7, 1994

Döbereiner J, Reis VM, Lazarini AC (1988a) New N_2-fixing bacteria in association with cereals and sugar cane. *In* "Nitrogen Fixation: hundred years after. Bothe H, DE Bruijn FJ, Newton WE (*eds*) Stuttgard: Gustav Fischer *pp*: 717-722

Döbereiner J (1988b) Isolation and identification of root associated diazotrophs. Plant Soil 110: 207-212

Ercolani GL (1984) Infectivity titration with bacterial plant pathogens. Ann Rev Phytopathol 22: 35-52

Ercolani GL, Casolari A (1966) Ricerche di microflora in pomodori sani. Ind Conserve Parma 41: 15-22

Fisher PJ, Petrini O, Lappin Scott HM (1992) The distribution of some fungal and bacterial endophytes in maize (*Zea mays*). New Phytol 122: 299-305

Fogher C, Civardi L (1989) Isolamento e caratterizzazione di endofiti del mais. L'Informatore Agrario 2: 93-95

Forni C, Haegi A, Del Gallo M, Grilli Caiola M (1992) Production of polysaccharides by *Arthrobacter globiformis* associated with *Anabaena azollae* in *Azolla* leaf cavity. FEMS Microb Lett 93: 269-274

Gagné S, Richard C, Rousseau H, Antoun H (1987) Xylem-residing bacteria in alfalfa roots. Can J Microbiol 33: 996-1000

Gardner JM, Chandler JA, Feldman AW (1985) Growth response and vascular plugging of citrus inoculated with rhizobacteria and xylem-resident bacteria. Plant Soil 86: 333-345

Gardner JM, Feldman AW, Zablotowicz M (1982) Identity and behavior of xylem-residing bacteria in rough lemon roots of Florida citrus trees. Appl Env Microbiol 43: 1335-1342

Gillis M, Kersters K, Hoste B, Janssens D, Kroppenstedt RM, Stephan MP, Teixeira KRS, Döbereiner J, De Ley J (1989) *Acetobacter diazotrophicus* sp. nov. a nitrogen-fixing bacteria associated with sugar cane. Int J System Bacteriol 39: 361-364

Gillis M, Döbereiner J, Pot B, Goor M, Falsen E, Hoste B, Reinhold B, Kersters K (1991) Taxonomic relationship between (*Pseudomonas*) *rubrisubalbicans*, some clinical isolates (ef group 1), *Herbaspirillum seropedicae* and (*Aquaspirillum*) *autotrophicum*. *In* "Nitrogen Fixation" Polsinelli M, Materassi R, Vincenzini M (*eds*) Kluwer Academic Publishers, Dordrecht Boston and London *pp* 293-294

Hartmann A (1988) Ecophysiological aspects of growth and nitrogen fixation in *Azospirillum* spp. Plant Soil 110: 225-238

Hayward AC (1974) Latent infection by bacteria. Ann Rev Phytopathol 12: 87-97

Hollis JP (1951) Bacteria in healthy potato tissue. Phytopathol 41: 350-366

Hurek T, Reinhold-Hurek B, Van Montagu M, Kellenberger E (1991) Infection of intact roots of kallar grass and rice seedlings by *Azoarcus*. *In* "Nitrogen Fixation" Polsinelli M, Materassi R, Vincenzini M (*eds*) Kluwer Academic Publishers, Dordrecht Boston and London *pp* 235-242

Hurek T, Burggraf S, Woese CR, Reinhold-Hurek B (1993) 16s rRNA-targeted polymerase chain reaction and oligonucleotide hybridization to screen for *Azoarcus* spp., grass-associated diazotrophs. Appl Env Microbiol 59: 3816-3824

Isopi R, Fabbri P, del Gallo M, Puppi G (1995) Dual inoculation of *Sorghum bicolor* (L.) Moench. ssp *bicolor* with Vesicular Arbuscular Mycorrhizas and *Acetobacter diazotrophicus*. Symbiosis (in the press)

Jacobs MJ, Bugbee WM, Gabrielson DA (1985) Enumeration, location, and characterization of endophytic bacteria within sugar beet roots. Can J Bot 63: 1262-1265

James EK, Reis VM, Olivares FL, Baldani JI, Döbereiner J (1994) Infection of sugar cane by the nitrogen-fixing bacterium *Acetobacter diazotrophicus*. J Exp Bot 45: 757-766

Kloepper JW, Beauchamp CJ (1992) A review of issues related to measuring colonization of plant roots by bacteria. Can J Microbiol 38: 1219-1232

Kloepper JW, Rodriguez-Kàbana R, Mc Inroy JA, Collins DJ (1991) Analysis of populations and physiological characterization of microorganisms in rhizospheres of plants with antagonistic properties to phytopathogenic nematodes. Plant Soil 136: 95-102

Kloepper JW, Schippers B, Bakker PAHM (1992) Proposed elimination of the term *endorhizosphere*. Phytopathol 82: 726-727

Kloepper JW, Schroth MN (1981a) Relationship of in vitro antibiosis of plant growth-promoting rhizobacteria to plant growth and the displacement of root microflora. Phytopathol 71: 1020-1024

Kloepper JW, Schroth MN (1981b) Plant growth-promoting rhizobacteria and plant growth under gnotobiotic conditions. Phytopathol 71: 642-644

Kluepfel DA (1993) The behavior and tracking of bacteria in the rhizosphere. Ann Rev Phytopathol 31: 441-472

Kostka SJ, Reeser PW, Miller DP (1988) Experimental host range of *Clavibacter xyli* subsp. *cynodontis* (CXC) and CXC/*Bacillus thuringiensis* recombinant (CXC/BT). (*abstract*) Phytopathol 78: 1540

Latch GCM (1993) Physiological interactions of endophytic fungi and their hosts. Biotic stress tolerance imparted to grasses by endophytes.*In* "*Acremonium*/grass interactions" (Special Issue). Joost R, Quisenberry S (*eds*) Agriculture Ecosyst Environ 44: 143-156

Lukezic FL (1979) *Pseudomonas corrugata*, a pathogen of tomato, isolated from symptomless alfalfa roots. Phytopathol 69: 27-31

Lynch JM, Fletcher M, Latham MJ (1979) Biological interactions. *In* "Mycrobial Ecology A Conceptual approach". Lynch JM, Poole NJ (*eds*) Blackwell Scientific Publications *pp* 171-189

Mahaffee WF, Kloepper JW, Van Vuurde JWL, Van der Wolf JM, Van den Brink M (1994a) Endophytic colonization of Phaseolus vulgaris by *Pseudomonas fluorescens* strain 89B-27 and *Enterobacter asburiae* strain JM22. *In*

"Improving plant productivity with rhizosphere bacteria". Ryder M H, Stephens P M, Bowen G D (*eds*). CSIRO, Australia, *pp* 180

Mahaffee WF, Moar WJ, Kloepper JW (1994b) Bacterial endophytes genetically engineered to express the CryllA δ-endotoxin from *Bacillus thuringiensis* subsp. *kurstaki*. *In* "Improving plant productivity with rhizosphere bacteria". Ryder MH, Stephens PM, Bowen GD (*eds*). CSIRO, Australia, *pp* 245-246

McInroy JA, Kloepper JW (1991) Analysis of population densities and identification of endophytic bacteria of maize and cotton in the field. *In* Plant Growth-Promoting Rhizobacteria Progress and Prospects Keel C, Koller B, Défago G (*eds*) *pp* 328-331

McInroy JA, Kloepper JW (1994) Novel bacterial taxa inhabiting internal tissue of sweet corn and cotton. *In* "Improving plant productivity with rhizosphere bacteria" Ryder MH, Stephens PM, Bowen GD (*eds*). CSIRO, Australia, *pp* 190

Meneley JC, Stanghellini ME (1974) Detection of enteric bacteria within locular tissue of healty cucumbers. J Food Sci 39: 1267-1268

Meneley JC, Stanghellini ME (1975) Establishment of an inactive population of *Erwinia carotovora* in healty cucumber fruit. Phytopathol 65: 670-673

Millner PD , Kitt DG (1993) Mycorrhizal colonization of corn co-inhabited by biopesticidal-recombinant *Clavibacter xyli* subsp *cynodontis*. Microb Releases 2: 81-84

Misaghi IJ, Donndelinger CR (1990) Endophytic bacteria in symptom-free cotton plants. Phytopathol 80: 808-811

O'Neill GA, Chanaway CP, Axelrood PE, Radley RA, Holl FB (1992) An assessment of spruce growth response specificity after inoculation with coexistent rhizosphere bacteria. Can J Bot 70: 2347-2353

Odum EP (1963) Ecology. Holt Rinehart and Winston Inc.

Okon Y, Labandera-Gonzales CA (1994) Agronomic applications of *Azospirillum*: an evaluation of 20 years worldwide field inoculation. Soil Biol Biochem 26: 1591-1601

Paula MA, Reis VM, Döbereiner J (1991) Interactions of *Glomus clarum* with *Acetobacter diazotrophicus* in infection of sweet potato (*Ipomoea batatas*), sugar cane (*Saccharum* spp.), and sweet sorghum (*Sorghum vulgare*). Biol Fertil Soils 11: 111-115

Peng Y, Zhang Z, Zhang Y, Huang D (1994) Internal root colonization of wheat by introduced genetically modified *Pseudomonas fluorescens*. *In* "Improving plant productivity with rhizosphere bacteria". Ryder MH, Stephens PM, Bowen GD (*eds*). CSIRO, Australia, *pp* 194

Reeser PW, Kostka SJ (1988) Population dynamics of *Clavibacter xyli* subsp. *cynodontis* (CXC) and a CXC/*Bacillus thuringiensis* subsp. *kurstaki* (BT) recombinant in corn (*Zea mays*). (*abstract*) Phytopathol 78: 1540

Schroth MN, Weinhold AR (1986) Root-colonizing bacteria and plant health. HortScience 21: 1295-1298

Sprent JI, De Faria SM (1988) Mechanisms of infection of plants by nitrogen fixing organisms. Plant Soil 110: 157-165

Sturdy ML, Cole LJ (1974) Studies on endogenous bacteria in potato tubers infected by *Phytophthora erythroseptica* Pethybr. Ann Bot 8: 121-127

Suslow TV, Schroth MN (1982) Rhizobacteria of sugar beets: effects of seed application and root colonization on yield. Phytopathol 72: 199-206

Tabacchioni S, Bevivino AM, Chiarini L, Visca P, Del Gallo M (1993) Characteristics of two rhizosphere isolates of *Pseudomonas cepacia* and their plant-growth promoting activity. Microb Releases 2: 161-168

Tosa Y (1992) A model for the evolution of *formae speciales* and *races*. Phytopathol 82: 728-730

Van Buren AM, Andre C, Ishimaru CA, (1993) Biological control of the bacterial ring rot pathogen by endophytic bacteria isolated from potato. (*abstract*) Phytopathol 83: 1406

Van Peer R, Punte HLM, De Wega LA, Schippers B (1990) Characterization or root surface and endorhizosphere Pseudomonads in relation to their colonization of roots. Appl Env Microbiol 56: 2462-2470

You CB, Song W, Wang HX, Li JP, Lin M, Hai WL (1991) Association of *Alcaligenes faecalis* with wetland rice. Plant Soil 137: 81-85

Zeikus JG, Ward JC (1974) Methane formation in living trees: a microbial origin. Science 184: 1181-1183

VAM-Fungi and their Interactions with other Microorganisms

Fritz Schönbeck
Institut für Pflanzenkrankheiten und Pflanzenschutz,
Universität Hannover
30419 Herrenhäuser Str. 2
Germany

Abstract

With relatively few exceptions roots of agricultural plants are infected by vesicular-arbuscular mycorrhizal fungi (VAMF). The mycorrhiza affects other soil microbes, and other microbes affect VAM. This symbiosis may also influence the general resistance of plants to pathogens. The mycorrhizal endophytes as well as root pathogens in the soil environment are part of a complex interacting system. Host plants inoculated with VAMF have been shown to exibit increased resistance to fungal root diseases and wilts. On the other hand diseases on green parts of the plants are usually more severe in mycorrhizal plants indicating that VAM changes the physiology of the whole plant. In the light of the present knowledge VAM is discussed with regard to the use in plant production.

Key words

Rhizosphere and N-fixing bacteria/ VA mycorrhiza/ plant pathogens/ plant health

Introduction

In discussing VAMF with other organisms in the soil and in the plant one should realise that the mycorrhizal fungi are an integral part of the biocenosis of nearly every terrestrial habitat and that they as obligat biotrophic organisms grow only in intimate connection with plants. In contrast to other obligat but pathogenic biotrophics the VAMF spread with extramatrical hyphae in the soil. The extension of hyphae from VAMF goes far beyond the geometrically depletion zone of the plant roots. VAMF hyphae extend up to

NATO ASI Series, Vol. G 37
Azospirillum VI and Related Microorganisms
Edited by I. Fendrik et al.

several cm from an infected root system, that means they extend the region of root hairs. The mycorrhizal system actually bridges across the rhizosphere and provides an organic link between the soil and the roots.

There is a widespread agreement that mycorrhizae are beneficial to the growth and health of plants. Some plant scientists emphasize the keyrole mycorrhizae play in the quest to have healthy plants on earth and sustainable systems of agricultural and forest production. Their description and assessment often sound like an enthusiastic song of praise to VAM. But it is one thing to discover impressive reactions under special conditions and another thing is to make use of it in the practical crop production.

VAMF act as independent organism only to a restrict degree. Their function depends largely from the host plant and the retroaction of their activity touches the plant. Therefore I prefered to put the plant into the centre of this system. The plant is the true target of our efforts and its metabolisms and reactions to biotic or non biotic environmental factors are inevitably altered due to VAM. But at first I will present shortly interactions between VAM and different kinds of soil bacteria.

Results and Discussion

General Rhizosphere Microflora

Mycorrhizal infection has been shown to affect rhizosphere populations. The increase in nutrient absorbing surface area due to VAMF should increase potential surfaces for bacterial colonization, thus supporting a larger rhizosphere population. From many publications it can be concluded that the microbial equilibrium is altered by the formation of VAM. But it is difficult to draw any general conclusions from the studies other than the fact that microbial populations in the rhizosphere are influenced by VAMF. The effect varies with time, the species of VAMF, the host and other environmental parameters and is hardly predictable with certainty

VAM Stimulation / Inhibition

Spreading on the root surface against or in spite of other organisms is an early stage of all mycorrhiza infections. Some of the bacteria described as intimately associated with VAMF either stimulate or inhibit some aspects of mycorrhizal formation. Some have a more neutralistic relationship. Some species enhanced the germination of spores, increased root infection, the development of the extramatrical hyphae. It is speculated that volatile substances produced by the microbes are involved in such effects.

There are some indications that soil microbes might be involved in the inhibition of spore germination and root colonization in non sterile soil. In steamed soil the spores could germinate. For example the development of *Glomus mosseae* decreased in the presence of *Aspergillus niger*. The inhibition may also be caused by volatile substances. But there are other data which put the theory of microbial suppression of VAM into some doubt. There is also evidence that bacteria and bacteria-like-organisms (Spiroplasma) are associated with walls and membranes of VAM spores. This association also may affect the growth of VAMF (Paulitz, Linderman 1991).

Rhizobium

The tripartite relationship of host-*Rhizobium*-VAM has been extensively studied to determine the nature of the relationship. Taking data from the literature, nearly all leguminous plants with nodules formed by *Rhizobium* grew better if they were also colonized by VAMF. Generally, the growth increase is explained by increased photosynthesis and increased carbon flow to the nodule and VAM sinks. The *Rhizobium* system also benefits from the presence of VAM by the correction of that deficiency which is limiting the system, for example the P uptake. But all the components of the tripartite system must function optimally for maximum plant growth to occur. If this is not the case experiments lead to results which are not in accordance with the above given statement, and the symbiotic effectiveness will suffer. To quote our own experiments with broad beans: simultaneous inoculation with *Rhizobium* and VAMF resulted in an inhibition of the nodulation as well as in the mycorrhization (Paulitz, Linderman 1991, Barea et al 1994).

Azotobacter, Azospirillum, Acetobacter

The population of *Azotobacter* in the rhizosphere were enhanced mostly by inoculation with VAMF. Inoculation with *Azotobacter* increased the mycorrhization. The combination of both of them increased root dry weight over treatments with either used alone. Inoculation of plants with both *Azospirillum* and VAMF has been shown to promote plant growth. The increases in growth were followed by increases in N and P contents of the crops. It should be mentioned that not in all cases an enhanced biomass accumulation was found. It is an open question whether *Azospirillum* interacts specifically with members of the *Glomus* species. *Azospirillum* has been isolated from surface sterilized spores of various *Glomus* species. The implication of N-fixing Azospirilla on the spores is not known (Paulitz, Linderman 1991, Barea et al 1994).

Spores of *Glomus clarum* obtained from sugar cane or sweet potatoes contained *Acetobacter diazotrophicus*. This bacterium infected aerial parts of sweet potatoes only when inoculated together with VAM spores. The possibility of infecting seedlings with endophytic diazotrophs via VAM represents an unique and very interesting way of introducing selected bacteria into plants (Okon 1994). (See also Doebereiner, Baldani and DelGallo in this volume).

Plant-Growth-Promoting Bacteria

Plant-growth-promoting-bacteria are rhizosphere-inhabiting bacteria that have a benificial effect on the plant growth. It has been suggested that those rhizobacteria may interact with VAMF to produce the growth-enhancing effects. But considering the published results critically there are not enough substantial data to draw general conclusions with regard to interactions between the three groups of organisms (Paulitz, Linderman 1991).

Phosphate-Solubilizing-Bacteria

VAMF do not seem to take up much P from insoluble sources.Many bacteria produce organic acids and solubilize inorganic and organic forms of phosphorus that are unavailable to plants. In various publications it has been suggested the possibility of an interaction between VAMF and phosphate-solubilizing bacteria. A synergistic increase was observed in dry weight and P-uptake in plants grown in a low P soil and dually inoculated with VAMF and P-solubilizing bacteria. Considerable controversy exists to what extent improved P nutrition by VAMF is due to solubilization of soil P by these bacteria. The interaction observed might be due to the production of plant hormones or vitamins by the bacteria, rather than to P solubilization (Paulitz, Linderman 1991).

Plant Diseases

Even when VAMF have a direct effect on microorganisms it can hardly be valuated without the host plant. The plant is an inevitable part of this interaction. A mycorrhizal plant differs in various aspects from a non mycorrhizal.

Diseases on mycorrhizal and non-mycorrhizal plants have been investigated by several authors. Most of this research was done with root-infecting fungi or nematodes. However, aerial plant parts should also be considered. Although only the roots are infected by the mycorrhizal fungus, the entire plant may be influenced. In studies on mycorrhizal effects the mycorrhizal fungus was inoculated a certain time before the pathogen to be studied. The reactions of mycorrhizal and non-mycorrhizal plants to the pathogen were compared.

Enhanced resistance of mycorrhizal plants to root diseases (caused by fungi or nematodes) was observed in most cases. This was expressed mostly by less host damage and, in some case, decreased infection (in particular with wilt diseases). In these cases alterations seem to take place in the host roots which increase the general resistance to the pathogens.

After inoculating green parts of mycorrhizal and non-mycorrhizal plants with pathogenic fungi, we found in some ways surprising results. With both obligate and facultative parasites we found higher disease severity in the mycorrhizal plants. The numbers of necroses, spots or pustules were increased, the incubation time of bean rust was shortened and the number of spores per pustule were increased.

There have been only a few reports on virus infection in mycorrhizal and non-mycorrhizal plants. In all cases the resistance of the plants to viruses appeared to be reduced by the mycorrhiza. Dividing the root system into two parts, one mycorrhized and the other without mycorrhiza, we found an increased virus content only in the mycorrhizal parts. Individual root cells invaded by the mycorrhizal fungus could still act as host cells for TMV. The virus multiplication increased remarkably in cells containing the fungus in its arbuscular stage (Schönbeck 1979, Dehne 1982).

An important question is the localisation of the mycorrhizal effects inside the plant. Are they restricted to the very mycorrhizal parts or does the plant as a whole react? There is not a single answer. It depends on the kind of the mycorrhizal effect. Considering the alterations in the shoots - resulting for example in an increased transpiration rate or lower resistance - a systemic effect must be assumed. A nutritional effect can also be systemic: the mycorrhizal half of the split root system provides greater phosphorous nutrition to the non mycorrhizal half.

Regarding root diseases there are records in the literature that resistance associated with VAM appears to be localized in some cases but systemic in others. We inoculated mycorrhizal and non mycorrhizal tobacco and lettuce plants with the obligate parasitic fungus *Olpidium brassicae*. After a certain incubation period, both the mycorrhizal and the parasitic fungi were seen to grow in the same individual roots, adjacent cells being infected with either of them, but in several hundreds of root sections examined not a single cell was occupied by both. This is in contrast to the finding that VAMF and viruses can occupy the same individual root cell.

Like single cells, mycorrhizal cell complexes also responded differently from non-mycorrhizal ones to *O. brassicae*: with an increased proportion of mycorrhizal cells the infection decreased. Roots of mycorrhizal plants which were not colonized by the mycorrhizal fungus were found to be infected by *O. brassicae* to the same extent as roots of non-mycorrhizal plants. Thus effects of the mycorrhiza on root diseases caused by fungi seem to be restricted to the mycorrhizal areas. The effective principle seems to move only over a very short distances in the roots. There is no evidence that non-mycorrhizal roots of a mycorrhizal root system behave as mycorrhizal roots with regard to enhanced disease resistance. This conclusion is supported by another finding: it is well known that gene activation is correlated with hypomethylation of the DNA. A reduced level of genomic methylation always occured in mycorrhizal roots but not in the uninfected regions of the roots of mycorrhizal plants.

In the light of these results it can be stated that the protective effects against pathogens are localized essentially in the mycorrhizal sites of the roots. On the other side effects to the green parts of the plant must be considered as systemic.

Conclusions

The benefits arising from VAM to various soil microbes and to plants are well documented in numerous experiments. Soil organisms can improve plant growth mainly through phosphorus nutrition, hormone production, suppression of root pathogens, biological nitrogen fixation, greater ability to withstand environmental stress. But these effects don`t occur at all events and conditions. When we try to evaluate the experimental data we are faced with the general problem that scientists don`t like to publish investigations with negative outcomes. From my own experiences I know that beside the many excellent results indicating benefits to the plants due to mycorrhizal actions frequently an entire inefficiency of the mycorrhiza was found. And mostly we were not able to find out why.

The endeavour to utilize VAM to increase crop productivity is impleted above all by the inadequate reliability of the efficiancy. We are not able to predict most of the beneficial mycorrhizal effects with adequate certaincy, because we don`t know enough about the dependence of the mycorrhizal efficiancy on the fungus, host, environmental conditions and the interactions between all the substantial factors. Recently data have been established that the genetical purity of the fungal genotyp is a prerequisite to obtain reproducible results. That means it is necessary to work with inocula coming from single spore inoculations. To reach optimal effects one has to use special inocula not only for each host species but also for each variety growing under various conditions. And the various environmental conditions must be considered as further factors affecting the effects of VAM.

I wonder whether these problems can ever be solved satisfyingly. For the time beeing I see a successfull use of VAM in less complex systems which can be regulated to some extent like nurseries or the production of various greenhouse crops with transplanted plants.

References

Barea J M, Azcon R, Azcon-Aguilar C (1994) Vesicular-arbuscular mycorrhizal fungi in nitrogen-fixing systems in: Norris J R, Read D, Varma A K (eds) Techniques for mycorrhizal research. Academic Press London.

Dehne H-W (1982) Interaction between vesicular-arbuscular mycorrhizal fungi and plant pathogens. Phytopathology 72: 1115-1119.

Okon Y (ed) (1994) Azospirillum/Plant Associations. CRC Press Boca Raton.

Paulitz T C, Linderman R G (1991) Mycorrhizal interactions with soil organisms in: Arora D K, Rai B, Mukerji K G, Knudsen G R (eds) Handbook of Applied Mycology Vol 1. M. Dekker, Inc. New York.

Schönbeck F (1979) Endomycorrhiza in relation to plant diseases in: Schippers B, Gams W (eds) Soil-Borne Plant Pathogens. Academic Press London.

Interaction of Rhizobia with Non-Legume Crops for Symbiotic Nitrogen Fixation Nodulation.

Edward C Cocking, Shanker L Kothari, Caroline A Batchelor, Sunita Jain, Gordon Webster, June Jones, Joan Jotham and Michael R Davey.
Plant Genetic Manipulation Group
Department of Life Science
University of Nottingham
Nottingham NG7 2RD
U K

Abstract

We have previously demonstrated that naturally occurring rhizobia, such as those isolated from the root nodules of non legume *Parasponia* species, and from stem nodules of tropical legume species, are able to enter the root systems of maize, rice and wheat by 'crack entry'. This 'crack entry' of rhizobia occurs where lateral roots emerge through the root cortex, resulting in the penetration of rhizobia both between and into cells of the cortex of young emerging lateral roots which, as a result, become thicker and shortened. Recently, we have interacted oxygen tolerant *Azorhizobium caulinodans* ORS571, isolated from stem nodules of the tropical legume *Sesbania rostrata*, and also oxygen tolerant *Rhizobium* ORS310 isolated from stem nodules of the tropical legume *Aeschynomene indica*, with the root systems of rice and wheat. This interaction resulted in the development of plants with short, thickened lateral roots, containing intercellular and intracellular rhizobia in cells of the cortex, interspersed amongst normally developing lateral roots. These plants possessed significant nitrogen fixation activity as determined using ethylene production in the acetylene reduction assay. Following the interaction of rice and wheat with Nif^- mutants of ORS571 there was no ethylene production from acetylene, confirming the acetylene reduction assay as a measure of nitrogenase activity. Currently, we are assessing the extent to which this symbiotic nitrogen fixing interaction of *Azorhizobium caulinodans* ORS571 and *Rhizobium* ORS310 with rice and wheat will be responsive to the lipo-oligosaccharide signals controlling specificity of nodulation interactions of these rhizobia with *Sesbania rostrata* and *Aeschynomene indica* respectively. Also we are assessing the extent to which this symbiotic nitrogen fixation will be beneficial to the growth and development of rice and wheat.

Introduction

The symbiotic nodular fixation of atmospheric nitrogen by legumes was discovered more than a century ago. Following this discovery, trials were made attempting the nodulation of non-legume crops with rhizobia isolated from legume nodules, but no conclusive results were obtained (Burrill and Hansen, 1917). In recent years, the increasing interest in sustainable agriculture and the perceived environmental

NATO ASI Series, Vol. G 37
Azospirillum VI and Related Microorganisms
Edited by I. Fendrik et al.

disadvantage of the over-use of nitrate fertilisers, have renewed interest in achieving new nodular symbioses between rhizobia and non-legume crops, especially the cereals. This renewed interest was stimulated by the finding that the cell wall at the tip of root hairs of a wide range of legumes and non-legumes could be rapidly removed by a mixture of cellulase and pectinase (Cocking, 1985). This enzymatic removal of the apical cell wall at the tip of root hairs exposed the plasma membrane, enabling experiments to determine if the specificity of rhizobia for the nodulation of legumes could be altered by removing this cell wall barrier. Using rhizobia that did not normally nodulate *Trifolium repens* (white clover), it was shown that this enzymatic treatment enabled *Rhizobium loti* to induce the formation of nitrogen fixing nodules on clover, and indicated the cell wall at the root epidermis as a target site for the genetic manipulation of *Rhizobium* symbiotic associations (Al-Mallah *et al.*, 1987).

This led to experiments to determine whether previous failures to nodulate non-legume crops with rhizobia were related to the inability of rhizobia to penetrate into their root systems. Rice, wheat and oilseed rape seedlings were treated with cellulase and pectinase and incubated with rhizobia on a medium lacking fixed nitrogen. Nodular structures were formed on their roots; some of the cells of these nodular structures contained rhizobia (Cocking *et al.*, 1990).

Parasponia is the sole non-leguminous plant genus that possesses root nodules and a *Bradyrhizobium-Rhizobium* symbiosis (Becking, 1992). It was also found that rhizobia, isolated from nodules of these non-legume *Parasponia* spp., formed elongate nodules on oilseed rape seedlings without the need for any enzyme treatment. Such nodules resembled swollen lateral roots and contained both intercellular and intracellular rhizobia. This observation indicated that certain rhizobia, such as those from nodules of non-legume *Parasponia* spp., are able to enzymatically erode cell walls at the root epidermis, probably because these rhizobia possess suitably active cellulase and pectinase enzymes. This led to an investigation of the interaction of rhizobia isolated from *Parasponia* nodules with maize, rice and wheat to determine whether a similar nodular interaction would occur without the need for any enzyme treatment.

Interaction Studies Using *Parasponia* Rhizobia

Seeds of the widely cultivated Indica rice variety IR36 were treated with *Parasponia* rhizobia and grown for 4 weeks on a medium lacking fixed nitrogen. Lateral roots were observed that were shorter and thicker than lateral roots formed in rice seedlings in the absence of rhizobia. Examination of sections of suitably fixed and embedded material showed rhizobia between cells. Electron microscopic studies also demonstrated rhizobia within membrane-bounded vesicles in the cytoplasm of cells of the cortex (Cocking *et al.,* 1992). Axenic seedlings of maize and wheat that had been inoculated with *Parasponia* rhizobia (CP283) and grown for several weeks on a medium lacking fixed nitrogen, developed an extensive root system with numerous lateral roots. As in the case of rice, lateral roots were observed on the maize and wheat seedlings which were shorter and thicker than lateral roots on seedlings grown in the absence of rhizobia. These short, thickened lateral roots (STLRs) were readily identified by their staining with the vital redox dye, tetrazolium red. When sections of maize and wheat STLRs were examined using light microscopy, after 3 weeks of interaction with *Parasponia* rhizobia, toluidine blue staining showed that rhizobia were invading at the base of emerging lateral roots. After 8 weeks, cells of the cortex of the STLRs had been extensively invaded by rhizobia, with extensive accumulation of rhizobia within cortical cells (Cocking *et al.,* 1994b). These studies clearly indicated that the invasion pathway of the rhizobia was by 'crack entry' at the point of emergence of lateral root primordia. They also demonstrated that the cortex can be invaded by intercellular rhizobia without infection thread formation, as in the nodulation of peanut (Chandler, 1978).

Interaction Studies Using *Aeschynomene* and *Sesbania* Rhizobia

Azorhizobium caulinodans (ORS571) is known to form stem nodules on the tropical legume *Sesbania rostrata* that are of adventitious (lateral) root origin; the invasion pathway is by 'crack entry' (Tsien *et al.,* 1983). *Rhizobium* ORS310 also forms stem nodules on another tropical legume, *Aeschynomene indica,* that are also of lateral root origin. Again, the invasion pathway is by 'crack entry' (Alazard and Duhoux, 1987). Such rhizobia, isolated from stem nodules, when interacted with the roots of seedlings of maize, rice and wheat, are also able to enter these non-legume crops by 'crack entry' where the lateral roots emerge through the root epidermis. This early stage in the formation of lateral root nodules is readily detectable by light microscopic examination of sections of embedded material (Cocking *et al.*, 1994c). It has been suggested that these particular rhizobia have invasive properties, probably associated with the secretion of bacterial cellulase and pectinase, which enable them to penetrate between the cells of the primordia of emerging lateral roots and, subsequently, into cells of the cortex. More than a century ago, Schneider undertook experiments to force an interaction between rhizobia and the roots of maize. He observed extensive infection by rhizobia in parenchymatous cells near the vicinity of emergence of lateral roots, but further invasion did not occur. However, he did not have available, at that time, these stem nodulating rhizobia (Schneider, 1893).

Assessments of Nodular Nitrogen Fixation

Studies using maize, rice and wheat interacting with *Parasponia, Aeschynomene* and *Sesbania* rhizobia had indicated the likelihood that the 'crack entry' initiation of lateral root nodule formation in these cereals would establish a niche for symbiotic nitrogen fixation. The herbicide 2,4-dichlorophenoxyacetic acid (2,4-D) has been shown to stimulate the formation of nodular structures in cereals (Kennedy and Tchan, 1992), but this para-nodulation appears to lack interaction specificity. The resultant restricted intracellular invasion has limited the use of such para-nodulation, utilising plant growth substances such as 2,4-D, for the development of an effective interaction with rhizobia leading to symbiotic nitrogen fixation.

It is known that the nitrogenase of *Azorhizobium caulinodans* (ORS571), unlike that of most other rhizobia, is tolerant of 3% of oxygen in the gas phase when cells are grown in nitrogen-free liquid medium (O'Gara and Shanmugam, 1976). *Rhizobium* ORS310 is also known to have a nitrogenase that is tolerant to 0.5% oxygen under similar conditions (Alazard, 1990). Because it was anticipated that nitrogen fixation in the nodular interactions involved in the 'crack entry' of rhizobia into emerging lateral roots of non-legume crops might be inhibited by the oxygen levels present, nitrogen fixation studies were undertaken on rice and wheat interacting with *Azorhizobium caulinodans* ORS571 and *Rhizobium ORS310.*

Studies Using Wheat

Seeds of several varieties of wheat, including the variety Wembley (spring wheat) were surface sterilised and germinated aseptically. Seedlings were transferred to Pyrex tubes containing sterile nitrogen-free medium, inoculated with either ORS571 (kindly provided by Dr J K Ladha, IRRI) or ORS310, and grown under controlled conditions. After 2, 4 or 8 weeks, the Pyrex culture tube closures were replaced by Subaseals and acetylene introduced into the culture tubes to give a final concentration of 10% acetylene. Air samples from the culture tubes were assayed for ethylene production using gas chromatography. Ethylene production was not detected in the controls in which uninoculated wheat seedlings were cultured for up to 8 weeks. The ethylene production of wheat seedlings (variety Wembley) interacting with ORS571, are shown in Table 1, (Cocking *et al.*, 1994a). Recently, Nif^- mutants of ORS571 (Denèfle *et al.*, 1987), kindly provided by Dr C Elmerich, Institut Pasteur, have been shown not to produce ethylene under these test conditions when used to inoculate wheat seedlings, confirming ethylene production as a measure of nitrogenase activity. Ethylene production after 2 and 4 weeks by wheat seedlings interacting with ORS571 and ORS310 was comparable to the rates of ethylene production by ORS571 and ORS310 following interaction with their natural hosts (*Sesbania rostrata* and *Aeschynomene indica* respectively) under comparable growth conditions. Evidence has also been obtained for a correlation between wheat STLR formation and ethylene production, with increased rates of ethylene production involving greater *Rhizobium*-nodular interaction. Light microscopy of wheat STLRs has shown the presence of rhizobia both between and within cells of the host plant. Currently, nitrogen fixation activity is being determined using ^{15}N dilution assays (Kothari *et al.*, 1994).

Table 1. Wheat variety Wembley : Inoculation with ORS571

Age of plants, (weeks)	μg N*/l	Number of plants	Range (n moles ethylene/plant/24h)	Mean** (n moles ethylene/plant/24h)
2	0	35	0.4-112	13.7
	20	34	0.3-129	13.3
	400	33	0.2-122	20.2
4	0	33	0.7-9.5	3.1
	20	33	0.4-7.5	2.6
	400	33	1.0-52.0	14.2
8	0	30	0.7-4.6	3.6
	20	30	3.0-3.8	3.4
	400	30	1.3-15.7	11.4

* Supplied as NH_4NO_3

** The mean amount of ethylene produced per plant is the mean for plants that showed acetylene reduction activity to produce ethylene.

Studies Using Rice

Seeds of rice (varieties Lemont and IR42) were surface sterilised after dehusking, germinated aseptically, similarly inoculated with either ORS571 or ORS310 and grown under controlled conditions. Assays for nitrogen fixation activity were carried out as for wheat, using the acetylene reduction procedure. The ethylene production of rice seedlings interacting with ORS571 is shown in Table 2 (Cocking *et al.*, 1994a). Recently, increased rates of ethylene production using ORS310 have been obtained, together with evidence for a correlation between rice STLR development and ethylene production. Currently, nitrogen fixation assays using ^{15}N dilution procedures are also being carried out (Batchelor *et al.*, 1994).

Table 2 Rice variety IR42 : Inoculation with ORS571

Age of plants, (weeks)	µg N*/l	Number of plants	Range (n moles ethylene/plant/24h)	Mean**(n moles ethylene/plant/24h)
2	0	173	0.9-117	35
	20	45	0	0
	400	20	0	0
4	0	42	4.4	4.4
	20	25	0	0
	400	20	0	0
8	0	20	0	0
	20	20	0	0
	400	20	0	0

Rice variety Lemont : Inoculation with ORS571

2	0	171	0.4-265	28.3
	20	45	0.4-23.5	10.9
	400	20	2.6	2.6
4	0	42	4.3	4.3
	20	25	0	0
	400	18	0.5-0.6	0.6
8	0	20	6.6	6.6
	20	20	2.4-5.0	3.6
	400	20	2.5	2.5

* Supplied as NH_4NO_3

** The mean amount of ethylene produced per plant is the mean for plants that showed acetylene reduction activity to produce ethylene.

Future Outlook

Experiments are now being undertaken to grow a range of rice and wheat varieties inoculated with ORS571 and ORS310 to maturity in pots, and to analyse such plants for nodular nitrogen fixation activity. These experiments will also be used to assess the extent to which this nodular interaction will be beneficial to the growth and development of rice and wheat. Nodulation factors (lipo-oligosaccharides) that are secreted by ORS571 and ORS310 in response to plant signals, and known to stimulate nodule fixation in their natural hosts (Spaink, 1992), are being evaluated for their effects on the interaction of *Parasponia, Aeschynomene* and *Sesbania* rhizobia with rice and wheat. It is hoped that these studies will provide a baseline of knowledge for the development of a commercial inoculation system for nitrogen fixation in non-legume crops.

Acknowledgement

This work was supported by the Rockefeller Foundation (International Rice Nodulation Group) and the Overseas Development Administration.

References

Alazard D (1990) Nitrogen fixation in pure culture by rhizobia isolated from stem nodules of tropical *Aeschynomene* species. FEMS Microbiol Lett 68: 177-182.

Alazard D, Duhoux E (1987) Nitrogen-fixing stem nodules of *Aeschynomene afraspera.* Biol Fertil Soils 4: 61-66.

Al-Mallah MK, Davey MR, Cocking EC (1987) Enzymatic treatment of clover root hairs removes a barrier to *Rhizobium*-host specificity. Bio/Technol 5: 1319-1322.

Batchelor CA, Kothari SL, Jain S, Jones J, Jotham JP, Poulton P, Webster G, Davey MR, Cocking EC (1994) Interaction of rhizobia with rice for nodular nitrogen fixation (in prep).

Becking J-H (1992) The *Rhizobium* symbiosis of the non-legume *Parasponia.* In: Biological Nitrogen Fixation (Stacey G, Burris RH, Evans HJ, eds) Chapman and Hall, London, New York, pp 479-559.

Burrill TJ, Hansen R (1917) Is symbiosis possible between legume bacteria and non-legume plants? Univ of Illinois, Agric Exptl Station Bull 202: 113-167.

Chandler MR (1978) Some observations on infection of *Arachis hypogaea* by *Rhizobium.* J Exp Bot 29: 749-755.

Cocking EC (1985) Protoplasts from root hairs of crop plants. Bio/Technol 3: 1104-1106.

Cocking EC, Al-Mallah MK, Benson E, Davey MR (1990) Nodulation of non-legumes by rhizobia. In: Nitrogen Fixation : Achievements and Objectives (Gresshoff PM, Roth LE, Stacey G, Newton WE, eds) Chapman and Hall, London, New York, pp 813-823.

Cocking EC, Batchelor CA, Kothari SL, Jain S, Webster G, Jones J, Jotham J, Davey MR (1994a) The interaction of rhizobia with cereals for symbiotic nitrogen fixation. In: Proceedings Second Asia Pacific Agricultural Biotechnology Symposium, Madras, India, March 1994 (in press).

Cocking EC, Davey MR, Kothari SL, Srivastava JS, Jing Y, Ridge RW, Rolfe BG (1992) Altering the specificity control of the interaction between rhizobia and plants. Symbiosis 14: 123-130.

Cocking EC, Srivastava JS, Cook JM, Kothari SL, Davey MR (1994b) Studies on the nodulation of maize, wheat, rice and oilseed rape : interactions of rhizobia with emerging lateral roots. In: Biological Nitrogen Fixation Novel Associations with Non-Leguminous Crops. Proceedings International Symposium on Nitrogen Fixation with Non-Leguminous Crops (Yanfu N, Kennedy IR, Tingwei C, eds) Qingdao Ocean University Press, Qingdao, China, pp 59-65.

Cocking EC, Webster G, Batchelor CA, Davey MR (1994c) Nodulation of non-legume crops. A new look. Agro-Food-Indust Hi-Tech, Jan/Feb 1994: 21-24.

Denèfle P, Kush A, Norel F, Paquelin A, Elmerich C (1987) Biochemical and genetic analysis of the *nifHDKE* region of *Rhizobium* ORS571. Mol Gen Genet 207: 280-287.

Kennedy IR, Tchan YT (1992) Biological nitrogen fixation in non-leguminous field crops : recent advances. Plant and Soil 141: 93-118.

Kothari SL, Jain S, Batchelor CA, Jones J, Jotham JP, Poulton P, Webster G, Davey MR, Cocking EC (1994) Interaction of rhizobia with wheat for nodular nitrogen fixation (in prep).

O'Gara F, Shanmugam KT (1976) Regulation of nitrogen fixation by rhizobia : Regulation of NH_4^+ assimilation. Biochim Biophys Acta 451:342-352.

Schneider A (1893) A new factor in economic agriculture. Univ of Illinois, Agric Exptl Station Bull 29: 301-319.

Spaink HP (1992) Rhizobial lipo-oligosaccharides : answers and questions. Plant Mol Biol 20: 977-986.

Tsien HC, Dreyfus B, Schmitt EL (1983) Initial stages in the morphogenesis of nitrogen-fixing stem nodules of *Sesbania rostrata*. J Bacteriol 156: 888-897.

Diazotrophic Mixed Cultures of *Azospirillum Brasilense* and *Enterobacter Cloacae*

Paul Kaiser
Institut National Agronomique,
16 rue Claude Bernard,
75231 Paris cedex 15
FRANCE

Abstract
Coculture of *A. brasilense* and *E. cloacae* on semi-solid medium showed two compartments, one for each bacterium, and led to efficient nitrogen fixation. *A. brasilense* in the culture grew and fixed nitrogen using the fermentation products produced by *E. cloacae*, including acetic and succinic acids, acetoin and 2,3-butanediol.

Introduction

Cocultures of diazotrophs and polysaccharide-degrading bacteria or fungi were studied as early as the beginning of the century (Pringsheim, 1908). However, the first important research on cocultures of *Azospirillum* species and cellulolytic bacteria was the much more recent work of Halsall and Gibson (1985). *A. brasilense* fixed 12 to 19 mg of nitrogen when associated with a normal strain of *Cellulomonas*, and 72 mg of nitrogen per g of straw decomposed when cultured with a mutant strain of *Cellulomonas* (Halsall and Gibson, 1986). These mixed cultures allow the use of a wide variety of crop residues (Halsall and Gibson, 1989) which contain various polysaccharides including hemicellulose and pectins. Khammas and Kaiser (1992) showed that pectin can be used efficiently as a carbon source for nitrogen fixation by cocultures of *Azospirillum* species with pectin-degrading bacteria (*Bacillus polymyxa* or *B. subtilis*). The nitrogenase activity was expressed at the expense of the production of pectin degradation and fermentation products, including acetic and succinic acids. Nitrogenase activity (ARA) of various pure cultures and cocultures of a variety of *Azospirillum* species and *B. polymyxa* isolates in association with rice were also measured using the spermosphere model (Khammas et al. 1993). Mean ethylene production was significantly higher in cocultures than in pure cultures. Cocultures can be considered as metabolic associations where the *Bacillus* provides degradation and fermentation products which are used by *Azospirillum* for nitrogen fixation.

We studied *Enterobacter cloacae - A. brasilense* cocultures on glucose. Both these species are frequently found associated with rice roots (Roger and Watanabe, 1986), which exude

NATO ASI Series, Vol. G 37
Azospirillum VI and Related Microorganisms
Edited by I. Fendrik et al.

simple sugars including glucose (Sadhu and Dhas, 1971). However *A. brasilense* uses glucose inefficiently. The nitrogenase activity of cocultures was measured and we investigated the use of glucose fermentation products by *A. brasilense*.

Materials and Methods

Bacterial strains

The strains used were: *Azospirillum brasilense* Sp7 (ATCC 29145) kindly supplied by C. Elmerich; *Enterobacter cloacae* 3081/VI; and *Azospirillum lipoferum* LMG 4384, both kindly supplied by the Unité des Enterobactéries, Pasteur Institute, Paris.

Culture media

Complete nutrient broth (Gauthier and Elmerich, 1977) contained (in grams per litre): Difco nutrient broth (8), $MgSO_4.H_2O$ (0.25), KCl (1) and $MnCl_2$ (0.001) adjusted to pH 7 with 1M KOH. Glucose medium was prepared from two solutions as follows. Solution 1: 6 g K_2HPO_4 and 4 g KH_2PO_4 in 200 ml H_20 adjusted to pH 6.8 with KOH; Solution 2: 0.2 g $MgSO_4.7H_2O$, 0.1 g NaCl, 0.02 g $CaCl_2.2H_2O$, 0.05 g yeast extract, 0.05 g bactopeptone (Difco), 5 g glucose, 0.0001 g biotin, and 10 ml trace elements solution (see below) in 800 ml H_20. For semi-solid medium, 1.8 g of agar was added to 800 ml of solution 2. Sixteen ml aliquots of solution 2 were distributed into 40 ml serum bottles and solutions 1 and 2 sterilised separately. After cooling, 4 ml of solution 1 was added to each aliquot of solution 2. The trace element solution contained (grams per litre): Na_2EDTA (4.1), $FeSO_4.7H_2O$ (2.4), $MnSO_4.2H_2O$ (0.1) , H_3BO_3 (0.1), $NaMoO_4$ (0.1), $ZnSO_4$ (0.01), $CuSO_4$ (0.005).

Inocula

Strains were precultured in nutrient broth for 24 hours. One aliquot (0.05 ml) of these cultures was used to inoculate one bottle of glucose medium. The bottles were incubated without shaking at 30°C for 4 days.

Nitrogenase activity assay

Nitrogenase activity of 20 ml culture samples was determined by the acetylene reduction assay (ARA) as described by Turner and Gibson (1980). Results are expressed as nmoles ethylene produced per hour per 20 ml culture.

Determinations of organic acids and neutral fermentation products

Succinic acid was extracted with chloroform. Acetic acid, acetoin and 2,3-butanediol were extracted with isopropanol and assayed by gas liquid chromatography (Carlier, 1985, Gallagher et al., 1991).

Counting cells

A Thomas chamber under a phase contrast microscope was used to count cells.

Glucose content determinations

Glucose concentrations were determined using PG0 enzymes (Sigma Corporation).

Results and Discussion

Growth and nitrogen fixation by cocultures on glucose

Cocultures on semisolid medium displayed two compartments: a lower layer containing a pure culture of E. cloacae (about 2 cm thick, 10^8 cells/ml), and an upper layer which consisted of a pink film about 1 mm thick which was a dense culture of *A. brasilense* containing few *E. cloacae cells*. Thus it appeared that each of the two strains cannot grow in the presence of the other. This was investigated in liquid cultures. In cultures inoculated simultaneously with both *E. cloacae* and *A. brasilense*, only *E. cloacae* multiplied. The growth of cultures of *A. brasilense* subsequently inoculated with *E. cloacae* was very slow. Thus, *A. brasilense* is unable to grow normally in the presence of *E. cloacae*, and coculture is only possible in semi-solid medium where the two strains can grow separately. This contrasts with *Azospirillum* and *Bacillus* strains (Khammas and Kaiser, 1992), which can easily be cocultured in liquid, pectin-based media, and *Azospirillum* and *Cytophaga* on sand medium (Halsall and Gibson, 1985).

Cocultures of *A. brasilense* and *E. cloacae* on semisolid glucose medium fixed nitrogen. However, the nitrogenase activity of these cocultures was lower than that of *A. lipoferum* cultivated alone on glucose (Table 1). *E. cloacae* consumed more than half the glucose without fixing nitrogen, explaining the lower efficiency of these cocultures than that of *A. lipoferum* cultures. In these cocultures, the nitrogen fixing activity was expressed by *A. brasilense*. However, *A. brasilense* cannot use glucose as a carbon source, in contrast to other *Azospirillum* strains (Khammas et al. 1989). *A. brasilense* cultured on *E. cloacae* culture filtrates similarly expressed strong nitrogen fixing activity. Thus, in the cocultures glucose was used by *E. cloacae* and the resulting fermentation products were used by *A. brasilense* for nitrogen fixation.

Table 1 : Nitrogenase activities of cocultures and pure cultures of *Azospirillum* species and *E. cloacae* on glucose medium (values are the means of two to five experiments).

Strains	ARA after 4 days of culture nmol C_2H_4/h/culture	Glucose consumed %
A. lipoferum	4 090	35
A. brasilense	88	5
E. cloacae	0	62
A. brasilense - *E. cloacae* coculture	3 180	62
A. brasilense on *E. cloacae* culture filtrate	2 380	-

Utilisation of *E. cloacae* fermentation products of glucose by *A. brasilense.*

Enterobacter cloacae produced four main fermentation products from glucose: acetic and succinic acid; acetoin and 2,3-butanediol (table 2). In nutrient-poor, nitrogen-containing medium more acid and less neutral products were produced and the pattern was the inverse in rich, nitrogen-containing media (data not shown). *E. cloacae* was cultivated in liquid glucose medium (see materials and methods) with 0.1 g/l NH_4Cl for 7 days without shaking. The culture was then centrifuged and filter sterilized. Agar was added, and the medium inoculated with *A. brasilense* Sp7. After 4 days the nitrogenase activity and the residual fermentation products were measured.

Strain Sp7 grew well on culture filtrate of *E. cloacae* resulting in a thick and pink-brownish film just below the surface of medium. (On the first two days the film was white). The bacterial cells were motile and contained pHB. *A. brasilense* Sp7 fixed nitrogen (Table 1) and used the fermentation products of *E. cloacae* (table 2). Succinic acid was used most efficiently, then acetic acid and acetoin; less 2,3- butanediol was utilized and only at the end of the culture. Strain Sp7 also utilized a small amount of the residual glucose (data not shown). Thus growth and nitrogen fixation depended mainly on the fermentation products of *E. cloacae*. In cocultures of *Azospirillum* and *Bacillus* species (Khammas and Kaiser 1992) the nitrogenase activity of *Azospirillum* uses the fermentation products of pectin including acetic and succinic acids. Halsall and Gibson (1989) have shown that the degradation products of cellulose (glucose and cellobiose) are poor substrates for the nitrogenase of most *Azospirillum* strains. However, in cocultures of *Azospirillum* and *Cytophaga* on cellulose, the *Azospirillum* utilizes fermentation products of cellulose provided by the *Cytophaga*. It is well established that *Azospirillum* uses organic acids as carbon sources (Krieg and Döbereiner 1984). We confirm this observation and show that other substrates including acetoin and 2,3- butanediol can be used. Dense, pink-brownish cultures could be obtained with these substrates as sole carbon sources.

In conclusion *Azospirillum* can be associated with a wide variety of sugar- or polysaccharide-degrading bacteria. The cocultures can be considered as metabolic associations where the sugar-degrading bacteria produce degradation and fermentation products which can be used by *Azospirillum* species.

Table 2 : Quantities of fermentation products of glucose in 7 day cultures of *E. cloacae* and in 4 day culture filtrates of *E. cloacae* inoculated with *Azospirillum brasilense* Sp7 (values are means of two experiments).

Strains	acetic acid (mM)	succinic acid (mM)	acetoin (mM)	2,3-butanediol (mM)
E. cloacae	1.1	8.5	2	4.8
A. Brasilense Sp_7 on culture filtrates from *E. cloacae*	0.2	1.3	0.4	2
% fermentation products consumed by *A. brasilense*	82	85	80	58

Acknowledgements
Thanks are due to P. Piquerel for technical assistance. This work was supported by research funds from Institut Pasteur, Paris.

References

Carlier J.P. (1985) La chromatographie en phase gazeuse des produits de fermentation, son application dans le diagnostic des bactéries anérobies. Bull. Instit. Pasteur (Paris) **83**, 57-69.

Gallagher I.H.C., Hancock E.M., and Sissons C.M. (1991) Isopropanol as an alternative to ethers in the extraction of C_2-C_6 fermentation acids for gas liquid chromatography. Letters in Appl. Microbiol. **13**, 230-232.

Gauthier D. and Elmerich C. (1977) Relationship between glutamine synthetase and nitrogenase in *Spirillum lipoferum*. FEMS Microbiol. Letters. **2**, 101-104.

Halsall D.M. and Gibson A.H. (1985) Cellulose decomposition and associated nitrogen fixation by mixed cultures of *Cellulomonas gelida* and *Azospirillum* species or *Bacillus macerans*. Appl. Env. Microbiol. **50**, 1021-1026.

Halsall D.M. and Gibson A.H. (1986) Comparison of two *Cellulomonas* strains and their interaction with *Azospirillum brasilense* in degradation of wheat straw and associated nitrogen fixation. Appl. Env. Microbiol. **51**, 855-861.

Halsall D.M. and Gibson A.H. (1989) Nitrogenase activity of a range of diazotrophic bacteria on straw, straw breakdown products and related compounds. Soil Biol. Biochem. **21**, 291-298.

Khammas K.M., Ageron E., Grimont P.A.D. and Kaiser P. (1989) *Azospirillum irakense sp. nov.*, a nitrogen-fixing bacteria associated with rice roots and rhizosphere soil. Res. Microbiol. **140**, 679-693.

Khammas K.M. and Kaiser P. (1992) Pectin decomposition and associated nitrogen fixation by mixed cultures of *Azospirillum* and *Bacillus* species. Can. J. Microbiol. **38**, 794-797.

Khammas K.M., Tran V.V. and Kaiser P. (1993) Nitrogen fixation in rice plantlets inoculated with mixed cultures of *Azospirillum* and *Bacillus* species. Sixth Int. Symp. on BNF with non legumes, Ismailia, Egypt, In press.

Krieg N.R. and Döbereiner J. (1984) Genus *Azospirillum,* Tarrand, Krieg and Döbereiner 1979. In Bergey's manual of systematic bacteriology. Vol. 1 . Edited by N.R. Krieg and J.G. Holt. Williams and Wilkings Co., Baltimore, Md. pp 94-104.

Pringsheim H. (1908) Zentralbl. J. Bakt. 20, 248.

Roger P.A. and Watanabe I. (1986) Technologies for utilizing biological nitrogen fixation in wetland rice: Potentialities, current usage, and limiting factors. Fert. Res. **9**, 39-77.

Sadhu M.K. and Dhas T.M. (1971) Root exudates of rice seedlings. Influence of one variety on another. Plant and Soil. **34**, 817-820.

Turner G.L. and Gibson A.H. (1980) Measurement of nitrogen fixation by indirect means. In: Methods for evaluating biological nitrogen fixation. pp 111-138. Bergersen F.I. (ed.). John Willey and Sons Ltd, Chichester, New-York.

Molecular Study of the Interaction of *Azospirillum lipoferum* with Wheat Germ Agglutinin

E Karpati[1], P Kiss[2], M Afsharian[1], F Marini[3], S Buglioni[3], I Fendrik[4] and M Del Gallo[5]
[1]Department of Biotechnology
Gödöllö University of Agricultural Sciences
2103 Gödöllö, Hungary

Abstract

Capsular receptors of *Azospirillum lipoferum* SpBr17 involved in interaction with wheat germ agglutinin have been studied. In the capsular fraction of bacteria two putative receptors were detected, a 25 kDa protein/glycoprotein and a complex of about 100 kDa. No attachment of wheat germ agglutinin to the corresponding capsular components was found in the wheat germ agglutinin-nonbinding mutant of *A. lipoferum*, as expected. The Tn2706 transposon insertion of the mutant has been localized.

Keywords

Azospirillum, wheat germ agglutinin, lectin, wheat germ agglutinin receptor, glycoprotein, polysaccharides, cell capsule

Introduction

Wheat germ agglutinin (WGA), the best known among the closely related lectins of gramineous plants, is found on wheat embryos and root tips of adult wheat plants (Mishkind

[2]Department of Chemistry and Biochemistry, Gödöllö University of Agricultural Sciences, 2103 Gödöllö, Hungary
[3]ENEA Casaccia, Biotechnology and Agriculture Sector, Via Anguillarese 301, 00060 Roma, Italy
[4]Institute of Biophysics, University of Hannover, Herrenhauser Str 2, 30419 Hannover, Germany
[5]DISTAAM, Faculty of Agriculture, University of Molise, 86100 Campobasso, Italy

NATO ASI Series, Vol. G 37
Azospirillum VI and Related Microorganisms
Edited by I. Fendrik et al.

et al. 1980). It specifically binds terminal N-acetyl-ß,D-glucosamine oligomer and N-acetyl-D-neuraminic acid residues (Wright 1980).

In the colonization of wheat roots by *Azospirillum*, WGA may represent a particular binding site for bacterial receptors. In this attachment some bacterial glycocalyx components and organella, such as capsular proteins/glycoproteins, polysaccharides, fimbriae and flagella, are supposed to be involved (Del Gallo and Fendrik 1994). Additional biological activities are also triggered by the interaction of WGA with carbohydrate determinants of cell surfaces (Chochola et al. 1993).

Recognition and binding of WGA by receptors of *A. lipoferum* SpBr17 has been investigated in this paper. Moreover, localization of a WGA-receptor encoding DNA region has also been described.

Materials and methods

A. lipoferum SpBr17R, a rifampicin resistant derivative of SpBr17 (ATCC 29709) (our laboratory), *E. coli* HB101(pAW1142::Tn2706) (Sm^R, Tc^R, Cm^R) (Olasz et al. 1993) and *E. coli* JF1754(pRK2013) (Km^R) (Figurski and Helinski 1979) strains were used.

A. lipoferum was grown on solid minimal medium containing 37 mmol malate and 19 mmol NH_4Cl (Okon et al. 1976). For Tn2706 mutagenesis, *A. lipoferum* and *E. coli* strains were grown on nutrient agar or broth (Difco). Streptomycin (Sm), chloramphenicol (Cm), tetracycline (Tc), rifampicin (Rif) and kanamycin (Km) were used in 50, 10, 20, 20 and 50 μg/ml concentrations, respectively.

Binding of WGA to *A. lipoferum* was tested by epifluorescence microscopy (Del Gallo et al. 1989) using fluorescein isothiocyanate (FITC) labeled WGA (Sigma).

Tn2706 (Cm^R) carried by suicide pAW1142 (Tc^R) (Olasz et al. 1993) and pRK2013 helper plasmid were conjugated into *A. lipoferum* by triparental plate mating (Ruvkun and Ausubel 1981). WGA-nonbinding (WGA^-) mutant was selected among the Cm^RRif^R transconjugants in the absence of fluorescence after FITC-WGA treatment.

Capsule was solubilized by shaking of 4-day old cells in 20 mmol Hepes buffer, 1 % of Triton X-100 pH 7.0, for 18 hours at +4°C. Protease inhibitors (Protease Inhibitors Set,

Boehringer) and 10 μmol 3,3'-dithiobis-(sulfosuccinimidylpropionate) (DTSSP) crosslinker (Pierce) were also added. Capsule-free cells were removed by centrifugation at 8000 rpm for 20 minutes at +4°C. Supernatant was concentrated with Centricon-10 centrifugal concentrator (MW cut-off 10,000). Protein content was measured according to Lowry et al. (1951). Electrophoresis of capsular proteins was performed by the Laemmli method (1970), except for the omission of ß-mercaptoethanol in the sample buffer. After the electrophoretic migration, the 7.0 % polyacrylamide gels were stained with Coomassie blue R-250. Lectin staining of proteins on Western blots was done with DIG-labeled WGA (DIG Glycan Differentiation Kit, Boehringer).

Purification of plasmid and genomic DNAs, restriction endonuclease digestion, agarose electrophoresis and Southern blot were according to Sambrook et al. (1989). Random primed labeling of DNA probes with DIG-dUTP, hybridization and detection by enzyme-immunoassay were performed with DIG-DNA labeling and detection kit (Boehringer).

Results

Testing of WGA-binding with wild type and WGA-nonbinding (WGA$^-$) mutant cells of *A. lipoferum*. FITC-WGA stained preparations of both the wild type and WGA$^-$ mutant were observed for fluorescence with epifluorescence microscope. In the case of the wild type, binding of WGA produced yellow-green glistening contours around the cells and this fluorescence was mostly coupled with the capsules (Figure 1A). WGA-binding was obtained with up to 96 % of stained cell population.In contrast, WGA$^-$ mutant showed almost completely dark phenotype (Figure 1B). Less than 1 % of population was with dimmed fluorescence.

WGA-binding components in the cell capsule. Results presented in Figure 1, in agreement with earlier observations (Del Gallo et al. 1989), revealed that WGA was attached by the capsular layer of the cells. Thus, the capsular WGA-binding components, especially proteins and glycoproteins, have been further investigated. Prior to capsule solubilization, the capsular structures, including also the putative WGA-binding sites, have been stabilized by covalent crosslinking among the terminal amino groups of capsular proteins. The solubilized capsular

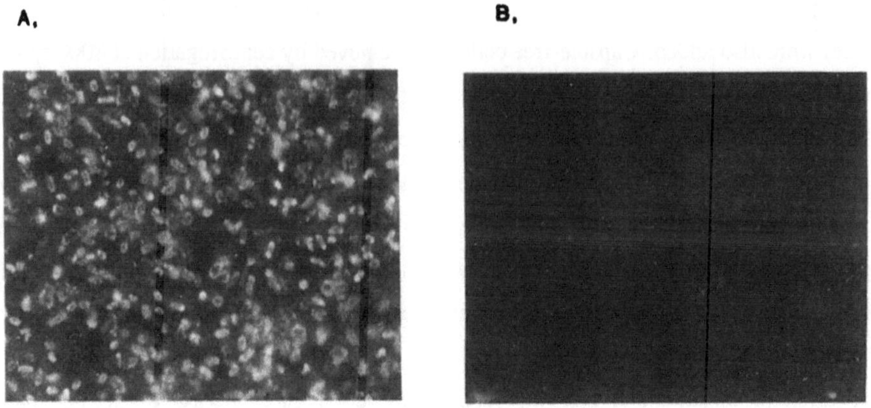

Figure 1. Binding of WGA to *A. lipoferum* cells. FITC-WGA stained wild type (A,) and WGA-nonbinding (WGA$^-$) mutant (B,) cells were observed with epifluorescence microscope. Both pictures present $4x10^2$ cells, approximately.

material was separated by SDS-PAGE. The migration patterns of capsule preparations from both the wild type and WGA$^-$ mutant cells are shown in Figure 2A. Western blots of the same patterns were probed with digoxigenin-conjugated WGA. WGA-binding components are presented in Figure 2B.

Six distinct protein bands and a complex of about 100 kDa, probably produced by crosslinking, were revealed. Protein bands gave poorer staining with Coomassie blue than expected considering the protein amount loaded, probably due to their high sugar contents. Protein profiles from the wild type and WGA$^-$ mutant cells were comparable.

WGA-binding signals were only detected on the wild type pattern. Two WGA-binding components of the capsule have been ascertained. The first corresponded to an approximately 25 kDa protein/ glycoprotein and the second to the crosslinked complex of about 100 kDa. No binding of WGA to the corresponding capsular components in the WGA$^-$ mutant was found, as expected.

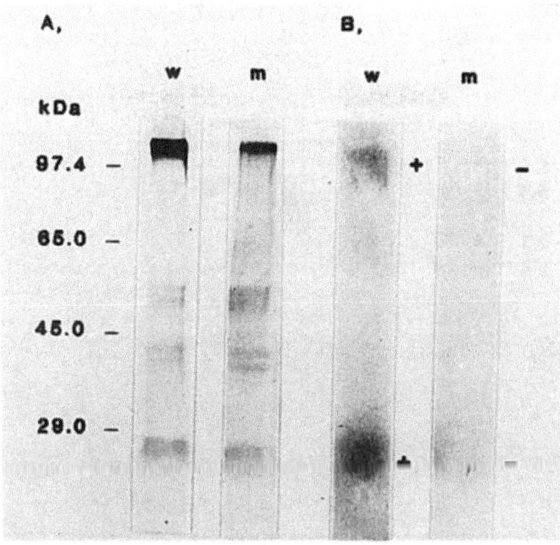

Figure 2. Migration pattern of capsular proteins and the WGA-receptors. Proteins were separated by SDS-PAGE and stained with Coomassie blue. (A,) protein profiles; (B,) Western blots probed with DIG-WGA; (w) wild type (50 μg of protein); (m) WGA$^-$ mutant (42 μg of protein); (+) WGA-binding component, (-) lack of WGA-binding.

WGA$^-$ mutant and its impaired gene site. WGA$^-$ mutant strain (Figure 1B) was isolated from Tn2706 mutagenized population of *A. lipoferum*. It was obtained after screening about 7×10^3 Tn2706 inserted transconjugants for lack of WGA-binding. This mutant strain was also tested for the presence of calcofluor binding polysaccharides, flocculation, nitrogenase activity and root colonization ability. Results obtained were comparable to those of wild type (data not shown).

The impaired gene of the mutant strain was detected by localization of Tn2706 insertion. Total DNA of the mutant was digested with *Eco*RI and *Bam*HI. After gel electrophoresis the DNA fragments were blotted to nitrocellulose membrane and hybridized both with Tn2706 and pAW1142 DNA probes (Figure 3a). Four hybridizing fragments were revealed on *Eco*RI profile with Tn2706 probe. Two fragments of identical size were detected on *Bam*HI profiles with both Tn2706 and pAW1142 probe. Moreover, pAW1142 possesses single *Eco*RI and single *Bam*HI sites, while Tn2706 has single *Eco*RI but no *Bam*HI sites.

Location of the insertion is shown in Figure 3b.

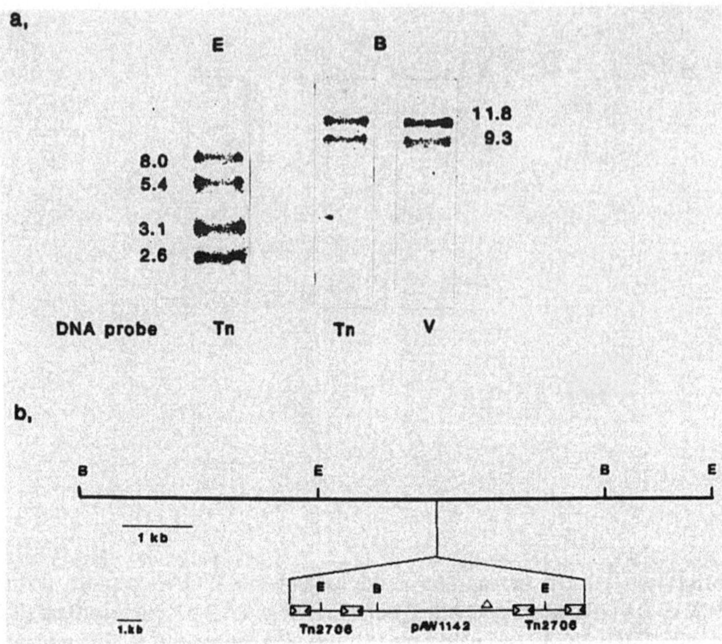

Figure 3. Localization of Tn2706 insertion in the WGA⁻ mutant strain. (a,) Detection of genomic DNA fragments carrying Tn2706 insertions. Total DNA digested with *Eco*RI and *Bam*HI and separated by gel electrophoresis was hybridized with DIG-labeled Tn2706 (Tn) and pAW1142 (V) DNA probes. (b,) Location of Tn2706 insertion. The upper line shows the genomic DNA and location of the insertion. The bottom part represents the inserted cointegrate including two Tn2706 copies linked by pAW1142 with 0.8 kb undefined deletion (Δ).

The mutant carries a 14.0 kb single insertion. It has been localized on a 5.1 kb *Eco*RI and a 7.1 kb *Bam*HI fragment of genomic DNA. This insertion includes a cointegrate of two Tn2706 copies linked by vector plasmid pAW1142 with 0.8 kb undefined deletion.

Discussion

Establishment of association between *Azospirillum* and gramineous host plants includes a range of interactions between the associative partners. One of the most important steps in this process is the interaction between the root lectins and carbohydrate surface components of bacterium glycocalyx. According to lectin hypothesis in *Rhizobium*-legume symbiosis (Bohlool and Schmidt 1974, Dazzo and Hubbel 1975) lectins serve as plant determinants of host range specificity.

The role of lectins in the association between *Azospirillum* and gramineous plants is almost unknown.

Wheat germ agglutinin (Mishkind et al. 1980) with well-defined carbohydrate binding specificity is the best studied among the closely related gramineous lectins. Besides its role in cellular recognition, many physiological and metabolic activities were also shown to be triggered by binding of WGA. They concern the glucose transport and lipolysis, cell growth inhibition, control of morphological states and activation of transcriptional factors (Wright 1992, Wright and Jaeger 1993).

Increased induction of glutamine synthetase, nitrogenase activities and ammonia excretion by binding of WGA was found in *A. brasilense* Sp245 (Antonyuk et al. 1993). WGA receptors were suggested both in *A. brasilense* and *A. lipoferum* by epifluorescence microscopy. Binding of WGA varied both with culture age and carbon source of growth medium (Del Gallo and Haegi 1990). A certain degree of specificity in lectin recognition by *Azospirillum* was also indicated (Del Gallo et al. 1989). Lipopolysaccharide-protein and polysaccharide-lipid complexes from capsule and culture liquid of *A. brasilense* were shown to interact with WGA (Konnova et al. 1994).

In this paper the wheat germ agglutinin and *A. lipoferum* have been chosen as a working system to investigate the lectin recognition and binding.

Particularly, capsular proteins/glycosylated proteins of *A. lipoferum* were studied as putative WGA receptors. They have been ascertained by direct detection of lectin binding to the separated capsular material.

Two WGA-binding receptors have been revealed in the capsule, a 25 kDa protein probably with glycosylated side chains and a complex of about 100 kDa. This complex may include multiples of 25 kDa protein/glycoprotein and also other proteins of glycocalyx in a crosslinked structure.

The binding of WGA to both capsule receptors could only detected when capsular structures were stabilized by covalent crosslinking prior to capsule preparation. Thus, the spherical location of receptor components and/or of glycosylated side chains within the binding site for WGA is also apparently required for the biological activity of the receptor.

There was no binding of WGA to the capsular components of WGA⁻mutant corresponding to the two wild type WGA receptors, although both of them were present in the capsule. Therefore, it can be concluded that the 25 kDa receptor protein/glycoprotein has been impaired by the Tn2706 insertion in the mutant.

Since the molecular weight of this mutant receptor protein/ glycoprotein is identical to that of the wild type, it is not a truncated form of the molecule, due to polar effect of transposon insertion. However, the mutant receptor might also carry defects in its amino acid sequence and therefore conformation of sites to be glycosylated on the molecule. Glycosylation of the receptor is a prerequisite for WGA-binding. Thus, these impaired sites cannot be recognized by the glycosylating enzyme. Moreover, the absence or defect of the glycosylating enzyme itself cannot be excluded either.

The impaired gene encoding for the WGA-receptor or involved in its posttranslational modification has been localized on a 5.1 kb *Eco*RI and a 7.1 kb *Bam*HI DNA fragment.

The 14 kb inserted cointegrate in the WGA^- mutant probably was produced by transpositional fusion of target sequences (Stalder and Arber 1989) on recipient DNA and previously dimerized Tn2706 donor plasmid DNA.

Further molecular analysis of the wild type and mutant receptors and the gene localized will provide more information about the role of WGA recognition and binding in root colonization by *A. lipoferum.*

Complex surface carbohydrate structures of *Azospirillum* probably play a key role in cellular recognition and adhesion between the bacterium and plant root. They can serve as special surface patterns for attachment of host plant lectins such as wheat germ agglutinin. Selective binding of such molecules as putative plant signals can transmit information from the plant into the bacterium, through complicated and almost unknown signalling pathways. Thus, lectin binding studies by *Azospirillum* give a model system for more precise understanding of this plant-bacterium association.

Acknowledgments

The authors wish to thank F. Altieri for advices in SDS-PAGE and helpful discussions and M. de Zamaroczy for critical review of the paper.

This work was partially supported by a PHARE-ACCORD fellowship to E.K. from National Committee for Technological Development, Hungary.

References

Antonyuk LP, Fomina OR, Galkin MA, Ignatov VV (1993) The effect of wheat germ agglutinin on dinitrogen fixation, glutamine synthetase activity and ammonia excretion in *Azospirillum brasilense* Sp245. FEMS Microbiol Lett 110:285-290

Bohlool BB, Schmidt EL (1974) Lectins: a possible basis for specificity in the *Rhizobium*-legume root nodule symbiosis. Science 185:269-271

Chochola J, Fabre C, Bellan C, Luis J, Bourgerie S, Abadie B, Champion S, Marvaldi J, El Battari A (1993) Structural and functional analysis of the human vasoactive intestinal peptide receptor glycosylation. J Biol Chem 268:2312-2318

Dazzo FB, Hubbell DH (1975) Cross reactive antigens and lectin as determinants of symbiotic specificity in the *Rhizobium*-clover association. Appl Microbiol 30:1017-1033

Del Gallo M, Negi M, Neyra CA (1989) Calcofluor- and lectin binding exocellular polysaccharides of *Azospirillum brasilense* and *Azospirillum lipoferum*. J Bacteriol 171:3504-3510

Del Gallo M, Haegi A (1990) Characterization and quantification of exocellular polysaccharides in *Azospirillum brasilense* and *Azospirillum lipoferum*. Symbiosis 9:155-161

Del Gallo M, Fendrik I (1994) The Rhizosphere and *Azospirillum*. In: Okon Y (ed) *Azospirillum*/plant associations. CRC Press, Boca Raton London Tokyo, pp 57-75

Figurski DH, Helinski DR (1979) Replication of an origen-containing derivative of plasmid RK2 dependent on a plasmid function provided in trans. Proc Natl Acad Sci USA 76:1648-1652

Konnova SA, Makarov OE, Skvortsov IM, Ignatov VV (1994) Isolation, fractionation and some properties of polysaccharides produced in a bound form by *Azospirillum brasilense* and their possible involvement in *Azospirillum*-wheat root interactions. FEMS Microbiol Lett 118:93-100

Laemmli UK (1970) Cleavage of structural proteins during the assembly of the head of bacteriophage T4. Nature 227:680-685

Lowry OH, Rosebrough MJ, Farr AL, Randall RJ (1951) Protein measurement with folin phenol reagent. J Biol Chem 193:265-275

Miskind M, Keegstra K, Palevitz BA (1980) Distribution of wheat germ agglutinin in young plants. Plant Physiol 66:950-955

Okon Y, Albrecht SL, Burris RH (1976) Factors affecting growth and nitrogen fixation of *Spirillum lipoferum*. J Bacteriol 127:1248-1254

Olasz F, Stalder R, Arber W (1993) Formation of the tandem repeat $(IS30)_2$ and its role in IS30-mediated transpositional DNA rearrangements. Mol Gen Genet 239:177-187

Ruvkun GB, Ausubel FM (1981) A general method for site-directed mutagenesis in prokaryotes. Nature 289:85-88

Sambrook J, Fritsch E, Maniatis T (1989) Molecular cloning: a laboratory manual. Cold Spring Harbor Laboratory, Cold Spring Harbor USA

Stalder R, Arber W (1989) Characterization of *in vitro* constructed IS30-flanked transposons. Gene 76:187-193

Wright CS (1980) Location of the N-acetyl-D-neuraminic acid binding site in wheat germ agglutinin. J Mol Biol 139:53-60

Wright CS (1992) Crystal structure of a wheat germ agglutinin/ glycophorin-sialoglycopeptide receptor complex. J Biol Chem 267:14345-14352

Short-Term Attachment of Bacteria from the *Rhizobiaceae* Family to the Roots of Cereals

Galina K. Solovova, Vladimir A. Velikov, Yurii V. Krivopalov, and Mikhail I. Chumakov
Laboratory of Microbial Genetics
Institute of Biochemistry and Physiology of Plant and Microorganisms
Russian Academy of Sciences
Pr. Entuziastov 13, Saratov 410015, Russia

Abstract

The attachment of *Agrobacterium* cells to wheat roots is an active process and it was possible to inhibit *A. radiobacter* attachment to wheat roots by different substances. The *chvB* Tn*5* *A. tumefaciens* and Calcofluor-dark Tn*5* *A. radiobacter* mutants poorly attached to wheat roots. The ability of *A. rhizogenes* LMG 150 to attach to the roots of various rice cultivars was different. The competitiveness deficient mutant of *R. meliloti* strain SHM-105 had the ability to attach to rice roots 6 times less as compared with the wild strain. Presumably, agrobacterial rhicadhesin do not participate directly in the attachment process to the monocotyledonous wheat root surface. The flagella of *A. radiobacter* are involved in attachment to the monocotyledonous cell surface, since non-motile, Fla minus T5 mutants of *A. radiobacter* 5D-1, have low attachment ability to wheat roots.

Keywords

attachment, *Agrobacterium tumefaciens*, *A. radiobacter*, *A. rhizogenes*, *Rhizobium meliloti*, wheat, rice roots

Introduction

It is known that bacteria of the *Rhizobiaceae* family are characterized by their ability to infect plants. *Rhizobium* forms nitrogen-fixing nodules on the roots of legumes, and *Agrobacterium* forms tumors on the stems and leaves of dicotyledonous plants.

Bacterial adsorption and attachment to the plant cell surface is the first real contact in the interactions. The initial steps of agrobacterial and rhizobial interactions with the dicotyledonous plant surface have been amply covered.

NATO ASI Series, Vol. G 37
Azospirillum VI and Related Microorganisms
Edited by I. Fendrik et al.

It seems evident that *Agrobacterium* and *Rhizobium* attachment-deficient mutants are uncapable of forming either tumors or nodules on dicotyledonous plants. For example, *R. meliloti ndvA* and *ndvB* attachment-defective mutants form non-effective nodules on alfalfa. *A. tumefaciens chvA* and *chvB* mutants in attachment are all avirulent (Cangelosi *et al.*, 1987). Adsorption and attachment of agrobacteria and rhizobia to the monocotyledonous plant surface are still imperfectly understood. It is considered that agrobacteria and rhizobia fail to adsorb specifically to the monocotyledonous cell surface since they lack a receptor for binding to the monocotyledonous plant cell surface (Lippincott *et al.*, 1977; Pueppke, 1984). We used *A. radiobacter* strain 5D-1 as a model organism in experiments on attachment.

In 1986, we isolated an *A. radiobacter* 5D-1 strain from surface sterilized wheat roots (Chumakov *et al.*, 1992). A similar strain was reisolated from surface sterilized wheat roots in 1994. It means that *A. radiobacter* possibly belongs to the root-associated bacteria.

The genes required for attachment of the pathogenic *Agrobacterium* are located on a chromosome. *A. radiobacter* differs from *A. tumefaciens* only in that it lacks a Ti-plasmid. It has previously been shown that *A. tumefaciens* and *R. meliloti* have similar chromosome maps (Hooykaas *et al.*, 1982). Consequently, *A. radiobacter* is genetically close to the genera *Rhizobium* and *Agrobacterium*.

Materials and methods

Bacteria. We used *Agrobacterium tumefaciens* LMG 187, *A. rhizogenes* LMG 150 (provided by Dr. De Ley (Belgium); *A. radiobacter* 5D-1 isolated from surface-sterilized wheat roots (Chumakov *et al.*, 1992). Plants. Seeds of wheat (*Triticum aestivum* cv. *Saratovskaya* 29) and rice (the cultivars used are listed in Results and Discussion) were sterilized with 4 % $Ca(OCl)_2$ for 20 min, then washed with sterile water and placed on 1.5 % agar on Petri dishes. Two-day-old seedlings were placed in Petri dishes on filters moistened with sterile water. Two-week-old wheat plants were used for inoculation.

Determination of bacterial adsorption to wheat roots. The *Agrobacterium* cultures were grown for 18 h at 30°C on glucose medium 16 containing (g/l): glucose - 10.0; yeast extract - 10.0; $(NH4)_2SO_4$ - 1.0; K_2HPO_4 - 0.25; agar - 15.0; pH 7.0. ^{14}C-labelled glucose (20 ml, 10 megabequarel (Mbk)/ml) was added to the cell suspension (2.0 ml) at an optical density OD_{600}

= 2.0, and incubated for 2 h at 30°C. Then the suspension was centrifuged at 3000 g for 15 min. Cells were resuspended in phosphate buffer (PB) to optical density OD_{600} = 0.1.

The excised wheat roots (100-200 mg) were incubated in Petri dishes with 3-5 ml of ^{14}C-labelled bacterial suspension. Next, the roots were placed on a filter paper in a funnel, washed with 20-30 ml PB with the following composition (g/l): K_2HPO_4 - 0.4; KH_2PO_4 - 0.4; NaCl - 9.0; pH 7.0; and dried at 100°C for 2 h. The radiation activity of the roots with the cells adsorbed was measured on a liquid scintillation counter Rack-Beta 1211/1212 LKB, Wallac, Finland). The roots unincubated with ^{14}C-labelled glucose and the supernatant of ^{14}C-labelled bacterial suspension were used as the control.

Treatment of bacterial cells. The following substances were used for the treatment of bacterial cells (10^7-10^8 cells/ml): trypsin (Reanal) - 1 mg/ml; cellulase (from *Trichoderma viridae*, Boehringer Mannheim) - 1.7 mg/ml; glutaraldehyde (Serva) - 1 (v/v). Bacterial cells at an optical density OD_{600} = 0.5 were incubated for 15 min, then centrifuged at 3000 g for 15 min, and resuspended in PB.

Results and discussion

The attachment of *A. tumefaciens* and *A. radiobacter* to wheat roots proceeded most actively during the first 90-120 min, beginning within a few seconds after inoculation. Then the process slowed down and became stable. The earlier findings (Matthysse *et al.*, 1981) suggest that the attachment dynamics of *A. tumefaciens* to the dicotyledonous (carrot) cells has a similar dependence. This type of adsorption is a theoretical monolayer type. Scanning electron microscopy (SEM) revealed that the attachment process proceeds in another way (Solovova *et al.*, 1994). The surface of wheat roots was not covered by one layer of the bacteria, but bacteria were concentrated in two and more layers in the attachment zone making the "mosaic type" of attachment.

It seems that the number of the sites of attachment of *A. radiobacter* 5D-1 to wheat roots is restricted, since increasing the concentration of *A. radiobacter* cells in the suspension more than 8-17 x 10^6/ml did not lead to increased numbers of the attached cells (data not shown).

It was possible to inhibit *A. radiobacter* attachment to wheat roots by different substances. Treating *A. radiobacter* cells with respiratory inhibitors resulted in decreased numbers of the bacterial cells attached to wheat roots. The number of heat-killed agrobacteria attached to wheat roots decreased as well. We obtained Tn*5* mutants of *A. radiobacter* 5D-1 defective in attachment to wheat roots. The attachment-deficient Tn*5* *A. tumefaciens* mutants poorly attached to wheat roots. The numbers of pathogenic and non-pathogenic agrobacteria attached to the wheat root surface were similar.

The bacteria grown on different media had different abilities to attach to wheat roots. Possibly, different bacterial surface structures were expressed on different media. The age of bacterial cells was also important for the manifestation of attachment. *A. radiobacter* 5D-1 cells in the stationary phase of growth (20-22 h) had a higher attachment ability (data not shown). From these results we can assume that the attachment of *Agrobacterium* cells to wheat roots is an active process.

Attachment of *R. meliloti, Agrobacterium rhizogenes, A. radiobacter*, and *A. tumefaciens* to rice roots

A. radiobacter strain 5D-1 attached to rice roots more actively than did *A. tumefaciens* strain LMG187, and *A. rhizogenes* strain LMG140. We tested the ability of *A. rhizogenes* LMG150 to attach the roots of various rice cultivars. The cultivars Lemont, IR54 and Taipei 309 had a higher number of attached cells to the rice root surface, as distinct from Nippobare. The competitiveness-deficient mutant of *R. meliloti* strain SHM-105 had the ability to attach to rice roots 6 times less as compared with the wild strain.

We speculate that the attachment of bacteria of the *Rhizobiaceae* family to the cereal root surface is an active process.

Bacterial surface polysaccharides which are involved in attachment of *A. radiobacter* 5D-1 to the root surface of cereals

Succinoglucan (expolysaccharide 1, EPS I, SG) is a high molecular weight polymer composed of polymerized octasaccharide subunits. *A. tumefaciens, A. rhizogenes, A. radiobacter* and *R. meliloti* can produce succinoglucan (SG). The SG-deficient mutant of *R. meliloti* Rm1021 induced nodule formation on alfalfa. *R. meliloti* Rm2011 wild type strain can produce fibrils while the SG-deficient mutant cannot. SEM revealed fibril-like structures (FLS) on the cell surface of *A. radiobacter* and *A. tumefaciens* during their attachment to wheat roots.

Within 5-15 min after incubation of *A. radiobacter* with wheat roots, most bacterial cells had not FLS, but within 90-120 min after incubation most cells had FLS (data not shown). It is apparent that FLS can be elaborated by the bacterial cell without the plant on a glucose-containing medium. This was shown in a model system with bacterial cells separated from glucose by the nylon membrane. We demonstrated that under these conditions non-pathogenic bacteria can abundantly produce FLS connecting the cells to the nylon membrane and to each other. Calcofluor can bind to polysaccharides with 1.2; 1.3; 1.6 types of bonds and change the adsorption maxima of polysaccharides in the ultraviolet area. Previous data (Gonsales *et al.*, 1993) revealed that the *R. meliloti* Tn*5* mutants which have the dark colony on the medium with Calcofluor cannot produce the EPS I. The Calcofluor-dark Tn*5* mutants of *A. radiobacter* strain 5D-1 had a decreased ability to attach to wheat roots.

SEM showed that most mutant cells did not have FLS (data not shown). Therefore, we propose that FLS are composed of these polysaccharides and involved in bacterial attachment to monocotyledonous plants. Studies by Matthysse *et al.* (1981) indicated that pathogenic agrobacteria can produce cellulose fibril-like structures during attachment to dicotyledonous plants. The treatment of wild strain 5D-1 with cellulase decreased the attachment ability 1.5-fold, but treating the cells of strain 5D-1 by cellulase-colloidal gold complex did not give any positive reaction in dot-blot analysis. SEM showed that the attached bacterial cells retained FLS after cellulase treatment. Possibly the FLS of *A. radiobacter* consisted of polysaccharides whose structure differed from that of cellulose. It seems that the wheat root lectin cannot serve as a receptor for *A. radiobacter* EPS, since dot-blot analysis with WGA (wheat germ agglutinin) labelled with colloidal gold did not give any positive reaction with *A. radiobacter* cells.

<u>*A. radiobacter* surface proteins are involved in the attachment process</u>

Rhicadhesin in a common Ca^{2+}-dependent protein for bacteria of the family *Rhizobiaceae* (Swart *et al.*, 1994). Previous workers (Smit *et al.*, 1987) showed that Ca^{2+}-dependent surface proteins are involved in bacterial attachment to the legume root hairs. It was shown that *A. tumefaciens chvB* attachment-deficient mutant with disturbed capacity to synthesize cyclic β-1,2 glucan from UDP-glucose failed to synthesize active rhicadhesins. We speculate rhicadhesins to be directly involved in the attachment process, since rhicadhesin addition restored the attachment ability. The attachment of non-pathogenic agrobacteria to wheat roots was studied on media with varying concentrations of Ca^{2+} ions and no differences were observed. Presumably, rhicadhesins do not participate directly in the attachment process to the monocotyledonous wheat root surface. It is necessary to emphasize that the method of evaluating attachment ability described by Smit *et al.* (1987) and the one used in our experiments are different. Smit *et al.* (1987) counted the percent of different types of root hair colonization, while we counted the number of the cells attached to wheat roots.

Table 1. The effect of Ca^{++} ions in growth medium on the attachment *A. radiobacter* 5D-1 to wheat roots

Ca^{++} concentration (mM)	No. of adsorbed bacteria (10^6 cells/g root)
0	177 ± 22
1,4	233 ± 31
7,0	175 ± 39
14,0	190 ± 31

Treatment of *A. radiobacter* 5D-1 cells with trypsin and glutaraldehyde, which can disturb protein structure, decreased the attachment ability of strain 5D-1 to wheat roots. The role of flagella in the attachment of agrobacterial cells was not described before. By Tn*5* mutagenesis we obtained a non-motile mutant (362) of *A. radiobacter* 5D-1. Transmission electron microscopy revealed that this mutant lacked flagella (data not shown). The attachment of these mutants to wheat roots decreased dramatically. We can assume that flagella of *A. radiobacter* are involved in attachment to the monocotyledonous cell surface.

Table 2. Attachment of Tn*5* mutants of *A. radiobacter* 5D-1 to wheat roots

Strain	No. of attached cells(10^6 cells/g root)
5D-1 (wild strain)	128 ± 20
207 che (-)	27 ± 13
362 mot (-)	0

Acknowledgements: We acknowledge Professor B. Simarov (Russia) for *R. meliloti* attachment-deficient mutants; Thanks are due to Professor De Ley (Belgium) for kindly providing *A. tumefaciens* LMG 187, and *A. rhizogenes* LMG 150 and Dr. Khush (Philippines) for the generous gift of rice seeds. This research was supported in a part by Russian Ministry of Science and Technology Policy and Russian Fond of Basic Research (grant 94-04-13373).

References

Cangelosi GA, Hung L, Puvanesarajah V, Stacey G, Ozga DA, Leigh JA, Nester EW (1987) Common loci for *Agrobacterium tumefaciens* and *Rhizobium meliloti* exopolysaccharide synthesis and their roles in plant interactions. J Bacteriol 169: 2086-2091

Gonsalez JE, Glucksmann A, Reuber TL, Walker GC (1993) Exopolysaccharides and *Rhizobium meliloti-Alfalfa* interactions. J Bacteriol 176: 1047-1051

Breedveld MW, Miller KJ (1994) Cyclic B-glucans of the family *Rhizobiaceae*. Microbiol Rev 58: 145-161

Hooykaas PJJ, Peerbolte R, Regensburg-Tuink AJG, de Vries P., Schilperoort RA (1992) A chromosomal linkage map of *Agrobacterium tumefaciens* and a comparison with the maps of *Rhizobium spp.* Mol Gen Genet 188: 12-17

Lippincott BB, Whatley MH, Lippincott JA (1979) Tumor induction of *Agrobacterium* involves attachment of bacterium to a site on the host plant cell wall. Plant Physiol 59: 388-390

Matthysse AG, Holmes KV, Gurlitz (1981) Elaboration of cellulose fibrils by *Agrobacterium tumefaciens* during attachment to carrot cells. J Bacteriol 145: 583-595

Pueppke SG (1984) Absorption of slow- and fast-growing rhizobia to soybean and cowpea root. Plant Physiol 75: 924-928

Solovova GK, Krivopalov YV, Velikov VA, Chumakov MI (1994) Attachment of *Agrobacterium radiobacter* to wheat roots. Microbiologiya (submitted for publication)

Smit G, Kijne JW, Lugtenberg BJJ (1987) Both cellulose fibrils and a Ca^{2+}-dependent adhesin are involved in the attachment of *Rhizobium leguminosarum* to pea root hair tips. J Bacteriol 169: 4294-4301

Swart S, Lugtenberg BJJ, Smit G, Kijne JW (1994) Rhicadhesin-mediated attachment and virulence of an *Agrobacterium tumefaciens* chvB mutant can be restored by growth in highly osmotic medium. J Bacteriol: 3816-3819 providing *A. tumefaciens* MO 137, and [illegible] strains LMG [illegible], [illegible] for the generous gift of tree seeds. This research was supported [illegible] [illegible] [illegible]

References

[illegible]

Interactions Between *Azospirillum* and *Phialophora radicicola*

F. Flouri, K. Sini[1] and C. Balis
Laboratory of General and Agricultural Microbiology,
Agricultural University of Athens,
Iera Odos 75, 118 55 Athens,
Greece.

Abstract

Phialophora radicicola can restrict colonization of cereal and grass roots by *Gaeumannomyces graminis*. *A. brasilense* and *A. lipoferum* were grown in *Phialophora* culture filtrate medium. The parameters of growth showed that *A. lipoferum* is able to grow on *Phialophora* culture filtrate medium. In the case of *A. brasilense* the growth was poor in the medium. Wheat plants were inoculated with *Azospirillum* and *Phialophora*, either individually or in combination.Fresh and dry weight of the wheat plants were determined. The results showed that the dual inoculation of wheat by *A. lipoferum* and *P. radicicola* produced significantly higher dry matter production than their corresponding control.

Introduction

N_2 fixing bacteria of the genus *Azospirillum* are able to associate closely with roots of plant species of the ecologically and agriculturally very important family Gramineae. Growth and establishment of *Azospirillum* is to a large degree dependent on interactions with rhizosphere population. Interactions with common, indigenous and predominant microbial species have been scarcely studied (Bashan and Levanony, 1990). The most studied microbial interactions of *Azospirillum* refer to *Rhizobium* and vesicular-arbuscular mycorrhizal (VAM) fungi. Inoculation of crop plants with (VAM) fungi and *Azospirillum* was found to have a synergistic beneficial effect on plant growth (Barea *et al*, 1983; Subba Rao *et al* 1985a,b; Pacovsky *et al*, 1985). The overall impression given by the literature is that the research is almost exclusively restricted to VAM fungi and relatively little attention has been given into the study of the effect of other fungi naturally occuring in the rhizosphere of cereals.

A fungus that is commonly found in grassland is *Phialophora radicicola* Cain. It is an avirulent root parasite of grasses and cereals, with an ectotrophic infection habit closely similar to that of *Gaeumannomyces graminis*, an aggressive pathogen (take-all) of cereal and grass roots. *P. radicicola* was firstly isolated from infected maize roots in Canada by McKeen. Scott (1970) repeatedly isolated it from samples taken at random from old grassland, which suggests that it is a widespread fungus on grass roots. Scott (1970),

[1]Present Adress: Innstituti Studimit Tokave, Tirana, Albania

NATO ASI Series, Vol. G 37
Azospirillum VI and Related Microorganisms
Edited by I. Fendrik et al.

Balis (1970) and Deacon (1973) showed that this fungus, now named *P. radicicola* var. *graminicola* (Deacon, 1974) can restrict colonization of cereal and grass roots by *G. graminis*. Despite all the research into the use of microorganisms for the control of soil-borne pathogens, many questions about their interactions with nitrogen fixers need an answer.

The major objective of the investigation presented in this paper was to examine the interactions between *Azospirillum* spp and *Phialophora radicicola* and to determine the effect of such interactions, if any, on plant growth.

Materials and Methods

Biological material

Azospirillum brasilense Sp7 (ATCC 29145), *Azospirillum lipoferum* Sp 59 (ATCC 29707) and *Phialophora radicicola* var *graminicola* (ATCC 28231), were obtained from the American Type Culture Collection, Rockville, Md. The two *Azospirillum* strains were maintained in nitrogen free malic acid medium with biweekly (*A. lipoferum*) or monthly (*A. brasilense*) transfers whereas fungus was maintained in PDA. Seeds of wheat (*Triticum aestivum* L) cultivar Vergina were obtained from the Agronomy Department of Agricultural University of Athens.

Culture Media and growth conditions

A preliminary experiment was carried out in Petri dishes under axenic conditions in order to observe directly any bacterial-fungus interactions. For this experiment, the bacteria strains mentioned above were transfered from maintenance medium to Petri dishes with PDA. Each dish was then inoculated centrally with an 8 mm disc cut from the periphery of a *Phialophora* culture. To examine the effect of the fungus culture filtrate medium on the growth of bacteria, *Phialophora radicicola* was grown in a medium consisted of: $MgSO_4 \cdot 7H_2O$, 0.5g; $FeCl_3$, 0.2mg; $ZnSO_4$, 0.2mg; $MnSO_4 \cdot 4H_2O$ 0.1mg; KH_2PO_4, 1.0g; asparagine, 2.0g; glucose, 10g; thiamine, 10mg; distilled water, 1L. Glucose and thiamine were filter sterilized and added to the autoclaved and cooled medium (Balis, 1970). The medium was dispensed in 100 ml quantities into 250ml erlenmeyer flasks. Each flask was then inoculated with water agar inoculum disks and incubated at 27 °C. After 25 days mycelial biomass were removed and the culture filtrate was tested as growth medium for *Azospirillum*. Azo medium (Krieg and Döbereiner, 1984) consisted of (in grams per litre):

succinic acid, 5.0; K_2HPO_4, 0.5; $FeSO_47H_2O$, 0.01; $Na_2MoO_42H_2O$, 0.002; $MgSO_47H_2O$, 0,2; NaCl, 0.1; $CaCl_22H_2O$, 0.026; $(NH_4)_2SO_4$ 1.0. The pH was adjusted to 7.0 with KOH. Yeast extract and biotin were added in different treatments at concentrations of 500 and 0.1mg/ml. *Azospirillum* cultures were initiated with 1ml liquid inoculum from a 48h old culture in Azo medium and grown at 27 °C on a reciprocal shaker in 100 ml spectrophotometer culture flasks containing 10 ml of medium. Turbidity of the growing cultures was measured in a Spectronic 20 Spectrophotometer at 3h intervals. Growth parameters were calculated from the experimental data by the following equation:

$$y = \frac{A}{1+e^{KA(T-t)}}$$

where: A is the maximal value which the population can reach, T is the time at which the population attains the A/2 level (inflextion point) and KA is the growth rate (=μ).

Host inoculation experiment

Wheat seeds were planted in pots containing 500 g of sand with a moisture content of 30%. An inoculum of bacterial suspension (10^6 cfu/g sand) was applied at planting. Plants that received *Phialophora* were inoculated by sowing seeds over agar inoculum disks (one disk *per* seed). There were six treatments with and without *A. brasilense, A. lipoferum* and *P. radicicola* var *graminicola*. Each treatment was replicated four times and each pot contained four plants. The pots were kept in a growth room with 14h artificial light and day-night temperatures of 22-17 °C and irrigated from above with distilled water every second day to restore pots to their original weights. Three weeks after sowing, plant parts were cut, oven dried at 80 °C for 5 days and weighted.

Results

Effect of *Phialophora* culture filtrate on growth of *Azospirillum*

The preliminary experiment in the Petri dishes showed no mixed-culture effects as antagonism or inhibition. Therefore, in a second experiment *A. brasilense* and *A. lipoferum* were grown in culture filtrate of *P. radicicola* in order to obtain information on possible infuence of fungus metabolites on bacterial growth. Table 1 shows the determined growth parameters. *A. brasilense* could not grow on *Phialophora* medium but, *Phialophora* culture filtrate appeared to support its growth.

Table 1. Growth parameters of *A. brasilense* and *A. lipoferum* on various nutrient media.

Treatment	Nutrient medium	Parameters of the kinetics growth curve			
		KA(=μ)	A	T	r^2
A. brasilense Sp7	Azo medium	0.24	1.27	19.3	0.99
	Azo medium+biotin	0.23	1.42	19.6	0.99
	Azo medium+yeast extract	0.38	1.82	22.3	0.99
	Phialophora medium	-	-	-	-
	Phialophora culture filtrate	0.19	0.30	11.8	0.99
	Azo medium+*Phialophora* culture filtrate (1:1)	0.35	1.59	19.1	0.99
A. lipoferum Sp59	Azo medium	0.28	0.34	17.8	0.99
	Azo medium+biotin	0.28	0.66	18.0	0.99
	Azo medium+yeast extract	0.34	1.65	21.5	0.99
	Phialophora medium	0.31	0.86	19.6	0.99
	Phialophora culture filtrate	0.20	0.18	11.8	0.99
	Azo medium+*Phialophora* culture filtrate (1:1)	0.37	1.44	14.2	0.99

In Azo medium and *Phialophora* culture filtrate, *A. brasilense* had a similar specific growth rate (μ=KA), 0.24 and 0.19 respectively, but a marked effect was found to have taken place on the A value. The observed A value was about 4 times as high for Azo medium as for *Phialophora* culture filtrate. *A. lipoferum* grew equally well on Azo medium+biotin and on *Phialophora* medium. The specific growth rate for *A. lipoferum* on Azo medium was not significant different compared to that on *Phialophora* culture filtrate. It is known (Okon *et al*, 1976) that addition of small amounts of yeast extract enhances the growth rate and shortens the lag phase. Data in table 1 show a similar pattern of response toward yeast extract and clearly indicate that a mixture of Azo medium and *Phialophora* culture filtrate (1:1) increased specific growth rate and A value of both *Azospirillum*.

Plant growth responses to inoculation

Our data (Table 2) demonstrate that within 3 weeks, wheat root dry wheights did not vary among treatments. However, dry weight of inoculated plants, in general, was higher than that of the control plants. *A. brasilense* inoculation did not have significant effect on shoot dry weight but in combination with *P. radicicola* produced significantly higher shoot dry matter. These effects were more prominent in *A. lipoferum* treatments.

Table 2. Yields for wheat as influenced by inoculation with *Phialophora* and *Azospirillum*.

Treatment	Shoot dw (mg)	Root dw (mg)	Total dw (mg)	Shoot/Root ratio
Control (non inoculated)	17.00 a	31.00 d	48.00	0.55
A. brasilense Sp7 (ATCC 29145)	19.50 a	28.75 d	48.25	0.68
A. lipoferum Sp59 (ATCC 29707)	22.75 ab	29.00 d	51.75	0.78
P. radicicola (ATCC 28231)	23.50 ab	33.25 d	56.75	0.71
P. radicicola + Sp7	22.00 ab	36.00 d	58.00	0.61
P. radicicola + Sp59	29.00 abc	33.00 d	62.00	0.88

Data are given on a per plant basis and represent mean of 4 replicates. Values not fol lowed by identical letters are significantly different, P=0.05.

It is worth noticing that single inoculation of wheat with *P. radicicola* resulted in significant increase in shoot dry weight. However, this increase was reduced, not significantly, in plants inoculated with both *P. radicicola* and *A. brasilense*. Plants inoculated with *P. radicicola* and *A. lipoferum* recorded the highest dry weight of shoot and whole plant. Inoculation with one or both endophytes did not significantly alter shoot-to-root ratios.

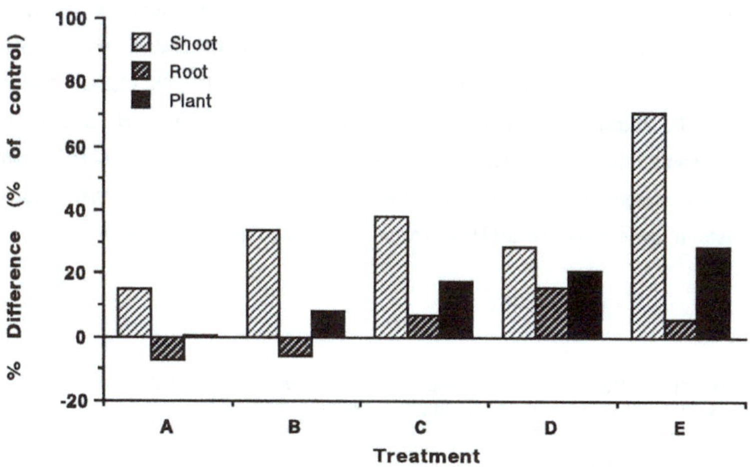

Figure 1. Effects of different treatments on dry weights of plants (A=Sp7; B=Sp59; C=*Phialophora*; D=Sp7+*Phialophora*; E=Sp59+*Phialophora*).Growth of inoculated plants relative to control was calculated as percent difference: [(inoculated plant-control)/control]x100

Figure 1 summarizes the effects of different treatments on the shoot, root and total dry weight of the host plant. The results are expressed as per cent difference of control plants. Inoculation with Sp7, Sp59, *Phialophora*, Sp7+*Phialophora* and Sp59+*Phialophora* resulted in 15, 34, 38, 29, 71 per cent increase in shoot dry weight and 0.5, 8, 18, 21, 29 per cent increase in whole plant dry weight respectively over that of uninoculated control.

Discussion

The results show that the culture filtrate of *Phialophora radicicola* var *graminicola* generally exerts a stimulatory effect on the growth of both *Azospirillum* species. *Phialophora* culture filtrate and addition of yeast extract behave similarly in improving growth parameters. Biotin requirement for growth has been used as a differentiation characteristic between *A. brasilense* and *A. lipoferum*. Presence of asparagine eliminated the need for biotin of *A. lipoferum*. According to Balis (1970), *P. radicicola* can synthesize biotin when growing on asparagine as a nitrogen source. It seems likely that *A. lipoferum* can also synthesize biotin by a process involving asparagine. In conclusion, it appears that the growth of *Azospirillum* on *Phialophora* culture filtrate can be considered as metabolic association, where the *Phialophora* produce degradation products, which can be used by *Azospirillum*. The observed response of wheat plants to *Azospirillum* inoculation agrees with earlier findings. A notable result is the beneficial effect of *Phialophora* on plant growth. Balis (1970), observed that wheat and barley seedlings inoculated with *P. radicicola* grow rather better than the corresponding uninoculated controls. To explain the possible mechanisms of *P. radicicola* on pant growth promotion, further research and study is necessary. Wheat plants inoculated with *P. radicicola* and *A. lipoferum* recorded significant increase in dry weight of shoots over the plants inoculated with either organism alone. This suggests a synergistic or additive effect of the two microorganisms.

References

Balis C (1970) A comparative study of *Phialophora radicicola* an avirulent fungal root parasite of grasses and cereals. Annals of Applied Biology 66:59-73.

Barea JM, Bonis AF and Olivares J (1983) Interactions between *Azospirillum* and VA mycorrhiza and their effects on growth and nutrition of maize and ryegrass. Soil Biology and Biochemistry 15: 705-709.

Bashan Y and Levanony H (1990) Current status of *Azospirillum* inoculation technology: *Azospirillum* as a chalenge for agriculture. Canadian Journal of Microbiology 36: 591-608.

Deacon JW (1973) *Phialophora radicicola* and *Gaeumannomyces graminis* on roots of grasses and cereals. Transactions of the British Mycological Society 61:471-485.

Deacon JW (1974) Further studies on *Phialophora radicicola* and *Gaeumannomyces graminis* on roots and stem bases of grasses and cereals. Transactions of the British Mycological Society 63: 307-327.

Krieg NR and Döbereiner J (1984) Genus *Azospirillum In* Krieg NR and Holt JG (eds) Bergey's Manual of Systematic Bacteriology, Vol. 1, pp. 94-104, Williams and Wilkins, Baltimore.
Okon Y, Albrecht SL and Burris RH (1976) Factors affecting growth and nitrogen fixation of *Spirillum lipoferum*. Journal of Bacteriology 127: 1248-1254.
Pacovsky RS, Fuller G and Paul EA (1985) Influence of soil on the interactions between endomycorrhizae and *Azospirillum* in sorghum. Soil Biology and Biochemistry 17: 525-531.
Scott PR (1970) *Phialophora radicicola,* an avirulent parasite of wheat and grass roots. Transactions of the British Mycological Society 55: 163-167
Subba Rao NS, Tilak KVBR and Singh CS (1985a) Synergistic effect of vesicular-arbuscular mycorrhizas and *Azospirillum brasilense* on the growth of barley in pots. Soil Biology and Biochemistry 17: 119-121.
Subba Rao NS, Tilak KVBR and Singh CS (1985b) Effect of combined inoculation of *Azospirillum brasilense* and vesicular-arbuscular mycorrhiza on pearl miller (*Pennisetum americanum)*. Plant and Soil 84: 283-286.

Establishment of Two Nitrogen-Fixing Bacteria on Roots of Kallar Grass Using Alginate-Coated Seeds in Mixed Inoculation

Tina Stein, Joerg Ueckert and Istvan Fendrik
Institute of Biophysics
University of Hannover
Herrenhäuser Straße 2
D- 30457 Hannover
Germany

Abstract

Nitrogen-fixing bacteria were immobilized in alginate layers on the surface of seeds of Kallar grass (*Leptochloa fusca* (L.) Kunth). It was investigated in *in-vitro* experiments whether these bacteria colonized the germinating plant faster than bacteria that were added with the plant's medium. Two different species of bacteria were compared: *Herbaspirillum seropedicae* and *Azospirillum brasilense*. After monoxenic inoculation both species were recovered from the roots in similar numbers. But in experiments with mixed inoculation, one bacterium immobilized on the seeds, the other species in the medium and vice versa, the endophyte *H. seropedicae* proved to be superior regardless of the method of inoculation.

Introduction

Nitrogen-fixing bacteria are most efficient in supporting plant growth when they colonize roots in very large numbers. Therefore, it has been tried to bring bacteria as close as possible to the developing root of cereal plants to provide them with the best starting conditions for colonization (Döbereiner, 1992).

In previous field investigations bacteria embedded in gel beads or peat were used as inoculum for wheat (Bashan, 1986; Fages, 1990). They were well protected in the matrix and were released slowly to establish in the rhizosphere. The dehydrated gel beads can be stored for a long period with the microorganisms remaining viable.

We investigated whether bacteria immobilized directly on the seeds have an advantage in colonizing the root over bacteria that were added with the plant medium. *Herbaspirillum seropedicae*, *Azospirillum brasilense* Sp7 and Kallar grass (*Leptochloa fusca* (L.) Kunth)

NATO ASI Series, Vol. G 37
Azospirillum VI and Related Microorganisms
Edited by I. Fendrik et al.

were chosen for the experiments. *Herbaspirillum seropedicae* was isolated from surface sterilized roots and stems of maize, rice and sorghum (Baldani *et al.*, 1986), while *Azospirillum brasilense* was first found in the rhizosphere of *Digitaria decumbens* and is associated with several forage grasses. Kallar grass is a highly salt-tolerant grass used as a pioneer plant in Pakistan on salt-affected low-fertility soils.

In these *in-vitro* experiments surface sterilized seeds were covered with an alginate layer that contained nitrogen-fixing bacteria in large numbers. Alginate is a non-toxic polymer from brown algae (*Phaeophyceae*) such as *Laminaria* and *Macrocystis* composed of D-mannuronate and D-guluronate monomeres linked in 1-4′ position. The gelstrength can be varied by the concentration and by using alginate from different species of algae. Addition of Ca^{2+}-ions starts the polymerization and since no heating is necessary bacteria can be embedded without being damaged by high temperatures (Kierstan and Bucke, 1977). For subsequent investigation the gel can be resolved by adding phosphate ions or complexing agents.

Materials and Methods

Seed coating. Seeds of Kallar grass (*Leptochloa fusca* (L.) Kunth) were surface sterilized with 70% ethanol and NaOCl containing 1.5% active chloride and subsequently rinsed in 0.01 M $CaSO_4$ for neutralization. For 2 hours the seeds were soaked in sterile 2.5% $CaCl_2$ solution at room temperature and gentle shaking (100 rpm) for Ca^{2+} uptake. A preculture of *Herbaspirillum seropedicae* was centrifuged and washed with PBS. A 3% sodium alginate solution (Protanal LF 20/60) was autoclaved and 10 ml were added after cooling to 1 ml of the bacterial suspension. Under rapid shaking the suspension with 10^{12} cells/ml (on average) was given to the seeds. For 1 hour the mixture was incubated (100 rpm, room temperature) and Ca^{2+}-ions were released from the seeds. Alginate chains were linked by exchanging sodium and calcium embedding bacteria and attaching them to the seeds.

Medium inoculation. For competition experiments a second species of bacteria (*Azospirillum brasilense* SP7) was inoculated simultaneously with the plant medium. A modified Hoagland nutrient solution was used: 6 g agar, 0.294 g $CaCl_2 \cdot 2\ H_2O$, 0.197 g $MgSO_4 \cdot 7\ H_2O$, 0.149 g KCl, 0.015 g Fe-EDTA, 18.9 mg NH_4Cl (= 5 mg N), 2.86 mg H_3BO_3,

1.88 mg $MnSO_4 \cdot H_2O$, 0.22 mg $ZnSO_4 \cdot 7\ H_2O$, 0.13 mg $Na_2MoO_4 \cdot 2\ H_2O$, 0.08 mg $CuSO_2 \cdot 5\ H_2O$, 2.0 g MOPS-buffer (Morpholinopropane-sulfonic acid), 0.71 g KH_2PO_4, H_2O to 1.0 l, pH 7.1. To simulate salt stress conditions in Kallar grass fields, 2.4 g/l Na_2SO_4 and 1.2 g/l NaCl were added (Reinhold *et al.*, 1986). After sterilization a washed preculture of *Azospirillum brasilense* was mixed with the warm medium to obtain 10^8 cells/ml on average.

Culture conditions. The medium solidified in sterilized glasstubes (20 cm · 3 cm with gas permeable silicon caps) and five alginate-coated seeds were placed on the surface. The soft matrix allowed free development of roots and did not restrict the motion of bacteria. Plants grew for one month in a climate chamber with 10/14 hours day/night cycle at 30/26 °C.

Experiments with *Herbaspirillum* coated seeds and *Azospirillum* in the medium and vice versa were carried out. For single inoculation seeds were coated with either bacterium. Sixty g of sterilized sand irrigated with Hoagland solution as described above was used instead of Hoagland agar.

Reisolation of bacteria. Plants were removed from tubes and the bacteria were reisolated. The roots were homogenized (Bühler homogenizer) in PBS solution. An aliquot of the suspension of homogenized roots and microorganisms was spread on nutrient agar plates containing kongo red. The dye allowed a distinction between the two different bacteria. While *Azospirillum* accumulated the kongo red and grew with dark red colour, *Herbaspirillum* appeared in the same orange shade as the agar with a darker centre. Occasional contaminants were detected as white colonies. The rest of the suspension was dried at 70 °C and the plant's dry weight was determined.

Results and Discussion

Table 1 shows the number of bacteria recovered from Kallar grass roots after one month of cultivation.

After monoxenic inoculation in alginate layers on seeds of Kallar grass both bacteria species, *Herbaspirillum seropedicae* and *Azospirillum brasilense* were recovered from the roots in comparable numbers. In mixed-inoculation experiments *Herbaspirillum* was recovered in numbers exceeding those of *Azospirillum* (see table 1, ratio *Herbaspirillum/Azospirillum*).

Herbaspirillum was dominant even after seed coating with *Azospirillum*. On the other hand, colonization of Kallar grass roots by *Azospirillum* was suppressed in the presence of *Herbaspirillum*. Under these circumstances *Herbaspirillum* may have been forced to penetrate the root. In the plant interior bacteria do not have to compete for the plant's nutrients with other bacteria and multiplication may be enhanced. Experiments with surface sterilized roots have to be carried out to verify this hypothesis. In mixed culture *Herbaspirillum* showed better colonization when embedded in alginate layer. In our experiments two nitrogen-fixing bacteria, both able to form a close association with *gramineae*, were in competition. The endophyte *Herbaspirillum* proved to be superior. This superiority in colonization outweighed the advantage given by seed coating regarding the composition of the bacteria population on and within the root.

Table 1. Bacteria reisolated from Kallar grass roots after one month of cultivation.

	[cfu/g root dry weight]		
	Herbaspirillum seropedicae	*Azospirillum brasilense*	ratio *Herbasp./Azosp.*
monoxenic inoculation:			
(bacteria on the seeds)	$1.4 \cdot 10^8$	$1.9 \cdot 10^8$	
mixed inoculation:			
Herbaspirillum on the seeds + *Azospirillum* in the medium	$4.2 \cdot 10^8$	$3.0 \cdot 10^6$	140
Azospirillum on the seeds + *Herbaspirillum* in the medium	$1.5 \cdot 10^8$	$3.7 \cdot 10^6$	40

The figures are means derived from 6-11 replicates.

Nevertheless, seed coating seems to be a convenient method to improve colonization of plant roots with beneficial bacteria. The influence of seed coating may be even greater in field experiments. In contrast to our experiments in culture vessels the bacteria in the mobile phase are easily washed away or are able to migrate away from the seed. The number of other bacteria in the environment is lower and consequently there is less competition. Furthermore, it may be expected that the effect increases with the size of the coated seeds. It must be kept in mind, that Kallar grass seeds are only 1.5 mm long and that the surrounding alginate layer is rather thin.

References

Baldani, J.I., Baldani, V.L.D., Seldin, L. and Döbereiner, J. (1986) Characterization of *Herbaspirillum seropedicae* gen. nov., sp. nov., a root-associated nitrogen-fixing bacterium, *Int. J. Syst. Bacteriol.* 36: 86-93

Bashan, Y. (1986) Alginate beads as synthetic inoculant carriers for slow release of bacteria that affect plant growth, *Appl. Environ. Microbiol.* 51: 1089-1098

Döbereiner, J. (1992) History and new perspectives of diazotrophs in association with non-leguminous plants, *Symbiosis* 13: 1-13

Fages, J. (1990) An optimized process for manufacturing an *Azospirillum* inoculant for crops, *Appl. Microbiol. Biotechnol.* 32: 473-478

Kierstan, M. and Bucke, C. (1977) The immobilization of microbial cells, subcellular organelles and enzymes in calcium alginate gels, *Biotechnol. Bioeng.* 19: 387-397

Reinhold, B., Hurek, T., Niemann, E.-G. and Fendrik, I. (1986) Close association of *Azospirillum* and diazotrophic rods with different root zones of Kallar grass, *Appl. Environ. Microbiol.* 52: 520-526

First Results on Individual and Dual Inoculation with *Azospirillum* - *Glomus* on Wheat

Alessandro Gori and Franco Favilli
Dipartimento di Scienze e Tecnologie Alimentari e Microbiologiche
Sezione Microbiologia Applicata, University of Florence Piazzale delle Cascine 27
50144 Firenze, Italy
Tel. 39 (55) 352051, Fax 39 (55) 330431

Abstract

Wheat (*Triticum durum* L. cv. Creso) plants were grown in a soil with organic matter and available nitrogen and phosphorous. Plants were inoculated individually either with *Azospirillum brasilense* strain Cdr (rifampicin resistant), and with an arbuscular mycorrhizal (AM) fungus *Glomus* sp. strain A6 or with dual inoculation with both microorganisms. After 12 weeks plants were harvested and data recorded. Fresh and dry weigth of shoots and roots were higher in dual inoculated plants compared with individual inoculated and control.

Keywords:*Triticum durum, Azospirillum brasilense, Glomus* sp., inoculation, transplanting

Introduction

Positive effects of dual inoculation of graminaceous plants by vesicular-arbuscular mycorrhizal fungi and diazotrophic bacteria have been investigated by a number of authors (Pacovsky R.S. *et al.*, 1985; Pacovsky R.S., 1989; Pacovsky R.S., 1989; Subba Rao N.S. *et al.*, 1985; Subba Rao *et al.* 1985).

However, the presence of both mycorrhizae and *Azospirillum* spp. on the same root systems has rarely been reported for wheat plants (Singh C.S. *et al.*; Al- Nahidh S. and Gomah A.H.M., 1991).

In order to evaluate the possibility of using microbial inoculants for field experiments, the effect of single and dual inoculation of *Azospirillum* spp. and *Glomus* spp. on wheat plants was investigated in pot experiments.

NATO ASI Series, Vol. G 37
Azospirillum VI and Related Microorganisms
Edited by I. Fendrik et al.

Materials and Methods

Wheat sterile seedlings: wheat seeds (*Triticum durum* L. cv. Creso) were surface sterilized 15 min with $HgCl_2$(0.1%), then rinsed in sterile water and germinated in Petri dishes on wet filter paper at 22 °C in the dark.

Azospirillum inoculum: *Azospirillum brasilense* Cdr mutant strain rifampicin resistant was used; the strain was grown in Erlenmeyer flasks on Antibiotic medium (Oxoid) supplemented with 20 μg/ml of rifampicin in a rotary shaker (180 rpm, 32 °C, exponential phase of growth). The culture was washed twice with phosphate buffer (0.1 M, pH 7.2) and resuspended in 200 ml of the same buffer to a final concentration of $7.5 \cdot 10^8$ CFU/ml.

Mycorrhizal inoculum: the mycorrhizal fungus used was *Glomus* sp. strain A6; the inoculum consisted of infested soil containing sporocarps, spores, infected root fragments and extramatrical mycelium.

Soil: the soil used was a peaty fertile soil passed through 3 mm sieve and steam sterilized for 90 min to eliminate natural endophytes.

Pots of 16 cm in diameter were filled with this treated soil and used for a pot culture experiment.

Transplanting and inoculation: two plants in each pot were transplanted placing 20g of mycorrhyzal inoculum in the transplantig hole for each pot to inoculate with the fungus while 10 ml of inoculum were distributed for each pot to inoculate with *Azospirillum*. The uninoculated treatment and inoculated plants with mycorrhizae only, received 10 ml of phosphate buffer. The plants were put outdoor and only one plant per pot was allowed to grow. After 12 weeks the plants were collected and fresh and dry weigth of shoots and roots was recorded.

Results

Shoots and roots fresh and dry weigth of the inoculated plants were slightly higher than the uninoculated treatments. Double treatments (*Azospirillum* + *Glomus*) showed the highest rate of both shoot and root dry weigth (Fig. 1,2,3,4).

Data analysed by analysis of variance on five replicates (Table 1) showed, after 12 weeks, that root dry weigth of double inoculated plants was significantly higher ($P < 0.05$) compared with those of the other treatments.

Table. 1 Shoot/Root fresh and dry weigth (values having the same letter within a column are not significantly different P > 0.05).

TREATMENT	SHOOT FRESH WEIGTH	SHOOT DRY WEIGTH	ROOT FRESH WEIGTH	ROOT DRY WEIGTH
CONTROL	**3.37 a**	**0.84 a**	**3.46 a**	**0.45 a**
AZOSPIRILLUM	**5.76 a**	**1.00 a**	**6.10 a**	**0.85 a**
GLOMUS	**3.96 a**	**0.90 a**	**7.48 a**	**0.94 a**
AZOSPIRILLUM + GLOMUS	**6.33 a**	**1.36 a**	**9.29 a**	**1.57 b**

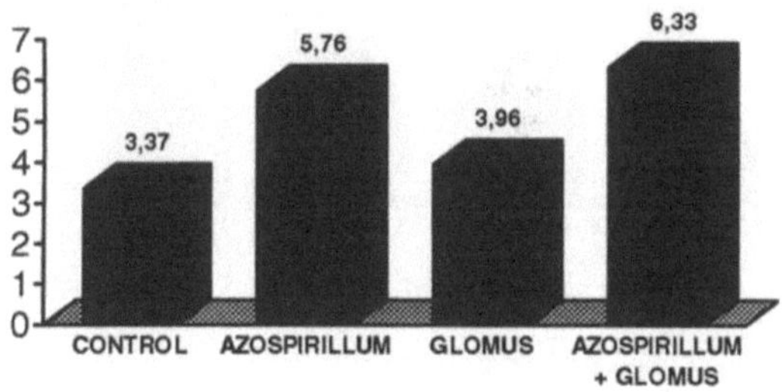

Fig. 1. Shoot fresh weigth means of five replicates (g)

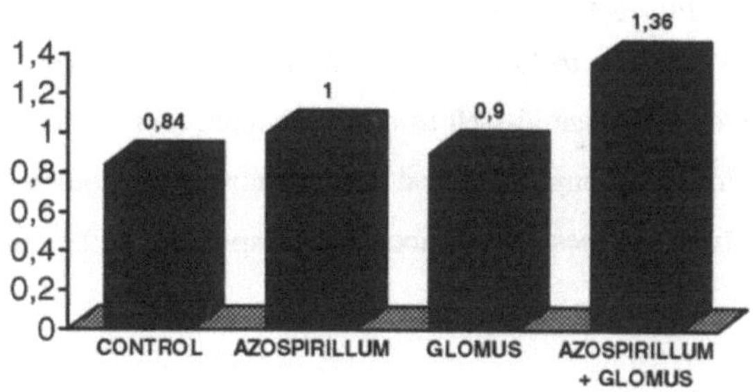

Fig.2. Shoot dry weigth means of five replicates (g)

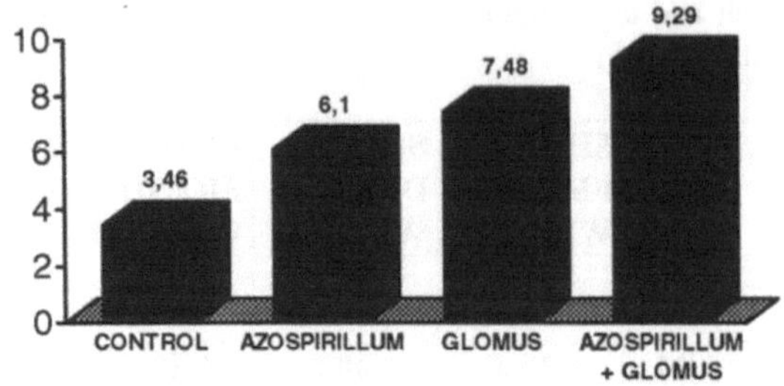

Fig. 3. Root fresh weigth means of five replicates (g)

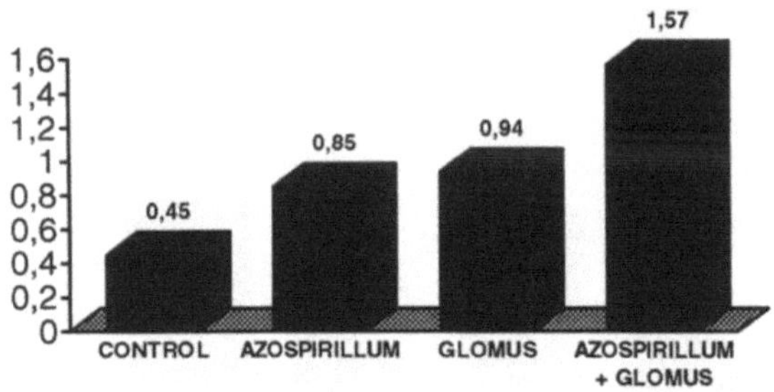

Fig. 4. Root dry weigth means of five replicates (g)

Conclusions

Dual inoculation with *Azospirillum brasilense* Cdr and *Glomus* sp. strain A6 had a positive effect on wheat plants These results showed that dual inoculated wheat plants take great advantage of the inoculation treatment as well as other graminaceous plants. These preliminary results encurage further investigations on several wheat cultivars and other *Azospirillum* and mycorrhizae strains to find the best plant-inoculants responses to be employed in field conditions.

Acknowledgements

The presentation of the work at the NATO Advanced Research Workshop on *Azospirillum* and Related Microorganisms Sârvâr/Hungary, September 4-7,1994 has been possible by a financial support of the Cassa di Risparmio di San Miniato.

The authors thank Prof. Manuela Giovannetti that kindly provided for *Glomus* sp. strain A6.

References

Al-Nahidh S. and Gomah A.H.M.(1991). Response of wheat to dual inoculation with VA-mycorrhiza and *Azospirillum*, fertilized with NPK and irrigated with sewage effluent. Arid Soil Res. Rehabil. 5: 83-96.

Pacovsky R.S.,Fuller G. and Paul E.A.(1985). Influence of soil on the interactions between endomycorrhizae and *Azospirillum* in sorghum. Soil Biol. Biochem. 17: 525-531.

Pacovsky R.S.(1989). Diazotroph establishment and maintenance in the *Sorghum - Glomus - Azospirillum* association. Can.J. Microbiol. 35: 977-981.

Pacovsky R.S. (1989). Influence of inoculation with *Azospirillum brasilense* and *Glomus fasciculatum* on sorghum nutrition. In Nitrogen fixation with non-legumes. Kluwer Academic Publishers pp. 235-239.

Singh C.S., Amawate J.S., Tyagi S.P. and Kapoor A. (1990). Interaction effect of *Glomus fasciculatum* and *Azospirillum brasilense* on yields of various genotypes of wheat (*Triticum aestivum*) in pots. Zentralbl. Mikrobiol. 145: 203-208.

Subba Rao N.S., Tilak K.V.B.R. and Singh C.S. (1985). Effect of combined inoculation of *Azospirillum brasilense* and vesicular-arbuscular mycorrhiza on pearl millet (*Pennisetum americanum*). Plant and Soil 84: 283-286.

Subba Rao N.S., Tilak K.V.B.R. and Singh C.S.(1985). Synergistic effect of vesicular-arbuscular mycorrhizas and *Azospirillum brasilense* on the growth of barley in pots. Soil Biol. Biochem. 17: 119-121.

Morphogenesis of Strawberry Roots Infected by *Azospirillum Brasilense* and V. A. Mycorrhiza.

Carlos H. Bellone, Silvia C. de Bellone
Departament of Ecology, University of Tucuman
Avda. Roca 1900
4000-S. M. de Tucuman
Argentina.

Abstract

Strawberry seedling were inoculated with *Azospirillum brasilense* and *Glomus intraradix*, separately and altogether. The bacteria colonize the roots increasing the number and size of root hairs, producing branched, curled and spatulated development, among other morphologies, as well as infection-thread formation in some of them. VAM fungi normally colonize the roots through the joining of epidermal cells and do not modify the morphology of root hairs. When the inoculation is dual, *A. brasilense* colonizes the root faster than *G. intraradix*, nevertheless, both microorganisms interact positively, increasing the internal and external colonization process of the root.

Introduction

Many kinds of nitrogen-fixing bacterias and *Azospirillum* among them, normally are not able to produce nodulation but they have a potential capacity to colonize the inner part of different species of plants (Baldani et al 1986, Döbereiner et al 1980, Patriquin et al 1983), besides fixing nitrogen, they have the ability to produce and induct phytohormones (Bellone 1983) and enzymes such as glucanases (Tiem et al 1981, Bellone 1983) which could be involved in the development of the root system and to cause a beneficial effect to inoculated plants (Baldani et al 1986) unknowing at the present the mechanisms involved in the penetration of the bacteria into the cortex.

In nodulated plants, during the first stages of the symbiotic development, normally the bacterias infect the roots through the deformed root hairs or by intercell spaces of the

NATO ASI Series, Vol. G 37
Azospirillum VI and Related Microorganisms
Edited by I. Fendrik et al.

epidermis, making one or many infection-threads until reaching the cortex (Callahan et al 1981 and Bhuvaneswari et al 1980).

It has been observed that the same plant species cannot be infected by both pathway, indicating that probably the plants may control the way of entrance (Sprent et al 1990).

Besides, the VAM fungi have a wider host spectrum among all known symbionts, colonizing normally the roots through the joining of epidermal cells (Harley et al 1983 and Newman et al 1987).

Both kind of microorganisms can infect the same plant at the same time through the secretion of factors necessary for growing and attraction of the microsymbionts, showing these last ones a high control on the root morphology (Barea et al 1983).

Object of this work was to analyze the structural changes which occure in Strawberry plant roots after inoculation with *Azospirillum brasilense* and the VAM *Glomus intraradix.*

Methods

Five different varieties of Strawberry plants: Chandler, Pájaro, Oso Grande, Selva and Fern, were cultivated "in vitro" using the Murashige and Skoog media. We used plants from meristems just in order to eliminate the pathogens and other microorganisms of the tissues; therefore to avoid a probable interference.

Seedlings from the 5 varieties (5 month old) were inoculated with sterile spores of *Glomus intraradix* and with *Azospirillum brasilense*, both isolated from the rhizosphere of commercial crops. Each seedling got 100 spores and $1x10^5$ bacterias cultured in NFb media (Baldani et al 1980). They have been in greenhouse under a day/night grade of 14/10 hs, with temperature of 27°C during the day and 20°C during the night. They were watered with sterile tap water and with a Hoagland and Arnon type nutrient solution once a week. Each treatment has been repeated 10 times.

The plants inoculated with both microorganisms and the controls were tested first at 10 and 24 hs from the inoculation and then every 24 hs during one week. Inverted microscopy and Transmission Scanning Electron microscopy have been used to analyse the structure and morphology of the root hairs.

Results

It has been observed that *Azospirillum brasilense* infects the seedling roots first than *Glomus intraradix.*

A careful examination of the complete root system indicates that there are some changes in the root development because of the infection with both microorganisms.

The root hair growth increased more in mycorrhized than in non-mycorrhized roots and the same happened about the root hair number.

Table 1. Effect of the infection by *Azospirillum* and *Glomus* on the root hairs production in Strawberry roots.

	% root hair lenght				
Treatment	Chandler	Pajaro	Oso Grande	Selva	Fern
Mycorrhiza	32,14 ±10	28,35±16	35,70±14	48,40±18	40,20±17
Azospirillum	52,15±14	68,13±30	59,14±22	82,30±34	68,30±33
M+A	28±12	31±13	39±16	36±16	42±14
Control	15±7	17±8	18±9	19±10	21±14

$$\%\text{root hair length} = \frac{\text{root hair length}}{\text{total root length}} \times 100$$

The stimulation on the production and development of the root hairs seems to be an effect of the mycorrhizal infection, even the morphology of them was not modified.

About *Azospirillum*, the increase in length and number of root hairs was higher than with *Glomus*, but in this case, there were some changes in the root hair structure and morphology. After 10 hs of the inoculation, the root hairs show the beginning of a curling, branching and other characteristics. These phenomena increased 24 hs after inoculation, as well as an advanced development in curling and branching; there is also a formation of spatulated root hairs, tip globulated, with double growth, spiral, etc. The branching type is found at any point of growing, but normally it happens in regions close to the nucleous. During these structural and morphological changes there is an active cytoplasmatic movement in all the root hairs, besides this, we could observe a high citoplasmatic density around the nucleous.

In the observation at 48 hs after the inoculation there were some infection-threats in some of the root hairs, one each 100 approximately. The amount of infection-threats was increased in roots colonized by VAM fungus, one each 45 root hair approximately. It is possible to find also in the tips some refringent regions at the beginning of the infection-threats.

The region where the bacterias penetrate the root hairs wall is characterized by a loosening of fibriform, where the deposition of material in the infection-threat formation begins.

Posterior analysis, indicate that *Azospirillum* penetrated the root hair wall by a certain process which involves the degradation of polysaccharides in the contact point with the bacteria. Finally we could find that the infection-threats located in the cortex release the bacterias which colonize other cell distributed around them, coexisting with fungal structures.

Discussion

The main information about the infection process is in the interactions between *Rhizobia* or *Frankia* and root plants. Although some other numerous bacterias can be associated and can penetrate the roots, there was not clear evidence, so far, of how was the infection process, specially about *Azospirillum.*

The results show the increase of epidermal cells differt in number and lenght because of the action of both microorganisms; it is necessary to stress that inspite of the differentiation process is more sensitive to the bacterias than to the fungus, this last one of somehow helps the infection process.

In the same way, the nucleous of the root hairs answer to the bacterial stimulus before the infection happens. However we still do not know if the stimulus can be of external or internal origin even the deformations occur only when the bacterias are presents. In our experiment we observed a high amount of root hairs colonized by bacterias and a low amount of them with infection-threats formation. Even though this happens also on legumes and non legumes because of *Rhizobia* and *Frankia* activity, there could be some important differences due to the anatomical and physiological differences of the plants.

There are some cases where *Rhizobia* do not get out from the infection-threat, but there are some situations where *Rhizobia* is released from the infection-threat and colonize root cells of the host, where they get involved as to what happens with VAM fungi, *Frankia* hyphae and what we found in this work about *Azospirillum.*

This last situation make us think about the position of this bacteria in the evolutive process among the other symbionts.

References

Bhuvaneswari T V and Solheim B 1985. Root hair deformation in the white clover-*Rhizobium trifolii* symbiosis. Physiol plant 63:25-34

Baldani V I D , Alvarez M A de B , Baldani J T and Döbereiner J 1986. Stablishment of inoculated *Azospirillum* spp in the rhizosphere and in roots of field grown wheat and sorghum. Plant and soil 90:35-46

Barea J M Bonnis A F and Olivares J 1983. Interaction betwen *Azospirillum* and v. a. mycorrhiza and their effects on growth and nutrition of maize and ryegrass. Soil biol Biochem 15: 705-709

Bellone C H 1983. Incrementos de la actividad celulolitica en tejidos no diferenciados de cana de azucar (*Saccharum* spp L) cultivados in vitro e inoculados con dos especies de *Azospirillum.* Revista Agrondmica del Noroeste Argentino xX (1-4) 5-15

Bellone C H. Incrementos en la biosintesis de auxinas en raices de maiz (Zea mais) inoculados con *Azospirillum.* Revista Agronómica del Noroeste Argentino XX (1-4) 71-78

Callaham D A and Torrey J G 1981. A structural basis for infection of root hairs of *Trifolium repens* by *Rhizobium.* Can J Bot 59:1647- 1664

Döbereiner J and De Polli H 1980. Nitrogen fixation in rhizocoenoses. International symposium on roots soil system. Londrina Brasil

Harley J L and Smith S E 1983. Mycorrhizal symbiosis. Academic Press. London

Hoagland D R and Armon D I 1950. The water-culture method for growing plants without soil. University of California 32 p.

Murashige T, Skoog F 1962. A revised medium for rapid growth and bioassays with tobacco tissue cultures. Plant physiol 15:473-497

Newman E I and Reddell P 1987. The distribution of mycorrhizas among families of vascular plants. New Phytol 106:745-751

Patriquin D E, Döbereiner J and Jain D K 1983. Sites and process of association betwen diazotrophs and grasses. Can J Microbiol 29; 900-915

Sprent J I and Faria S N 1990.Machanisms of infection of plant by nitrogen fixing organisms. In nitrogen fixation with non legumes. Skinner F A, Boddey R. M. and Fendrik I. Edit Kluwer Academic Publishers

Tiem T M, Diem H G, Gaskins M H and Hubbell D H 1981. Polygalacturonic acid transeliminase production by *Azospirillum* species. Can J Microb 27:426-431

Specific interaction between chickpea (*Cicer arietinum* L.) and three chickpea-*Rhizobium* strains inoculated singularly and in combination with *Azospirillum brasilense* Cd

Paola Fabbri[1] and Maddalena Del Gallo
Dept. Food Technology, Environmental and Microbiological Sciences
Faculty of Agriculture, University of Molise
I-86100 Campobasso, Italy.

Abstract

Three chickpea Italian cultivars were inoculated with three different chickpea-*Rhizobium* strains singularly and in combination with *Azospirillum brasilense* Cd to test both the symbiosis efficiency and the specific interactions occurring among bacteria and plant. Coinoculation of *Azospirillum* and *Rhizobium* enabled to obtain, particularly on chickpea winter cultivars Califfo and Sultano, significant increases in different plant parameters, in comparison with plants inoculated only with *Rhizobium*. Chickpea spring cultivar Calia, instead, responded very poorly to the inoculation of both *Rhizobium* and *Azospirillum*.
Nitrogen content of the plant aerial part showed highly significant differences among plants inoculated with both diazotrophs or with chickpea-*Rhizobium* alone.

Introduction

Recent studies on *Rhizobium*-Legumes symbiosis demonstrated that plant host genotype can greatly influence nodulation and nitrogen-fixation efficiency in *Medicago sativa*, *Phaseolus vulgaris* and other leguminous plants (Miller and Sirois, 1982). The host plant plays an essential role in the establishment of the associative process: a number of host gene-products which support the symbiotic state have been identified. Some of these gene products are present in the infected cells while others are in the nodule, suggesting organ level co-operation in the nodules. Host specific nodule

---

[1]Biotechnology and Agriculture Sector, ENEA-Casaccia, Via Anguillarese 301, I-00060 Rome.

NATO ASI Series, Vol. G 37
Azospirillum VI and Related Microorganisms
Edited by I. Fendrik et al.

proteins such as leghaemoglobin are involved in oxygen, carbon and nitrogen metabolism within the nodules (Appleby, 1984; Verma and Nadler, 1984). The normal nodulation pathway has been outlined as attachment to root hairs, root hair curling, infection thread formation within root hairs, growth of the thread toward the inner root cortex, initiation of a nodule meristem and nodule organogenesis (Bauer, 1981; Vincent, 1980). In this model, nodulation involves a series of suitable plant responses to bacterial invasion: exchanges of signal molecules between host and bacterium are not limited to the initial stages of the interaction. The inability of one of the partners to answer appropriately, at a particular symbiotical stage, may hamper the associative process (Abe *et al.*, 1982; Dazzo *et al.*, 1984).

Variations in the colonization of different plant by rhizosphere populations has been described: in some case the plant can be easily colonized, while certain grasses rarely showed up to 50% of colonization. Influence of plant genotypes on rhizosphere populations was demonstrated already by Neal *et al.* (1970), but there are few reports concerning the variability of interactions between plant growth promoting rhizobacteria (PGPR) and plant genotypes. In fact, differences were found in the establishment of *Azospirillum* strains in wheat roots (Baldani and Döbereiner, 1983) and in the rhizosphere of *Zea mays* (O'Hara *et al.*, 1981).

If the *Rhizobium*-Legumes symbioses have received most of the attention in the past by far, the *Rhizobium-Azospirillum*-legume symbiosis is at present attracting increasing attention also because of the economic and ecological benefits derived from cropping systems using efficient nodulated legumes. There are few studies on this tripartite association, and information about the ecology of this system and the effect on the plant of this co-inoculation are still lacking.

Okon and Kapulnik (1986) have shown that *Azospirillum* inoculation can benefits plant growth and significantly increases yields of several grain, forage an legume crops by 10-35%. Legumes grown in the presence of both *Rhizobium* and *Azospirillum* will normally become infected with specific *Rhizobium* species and can be simultaneously infected with *Azospirillum*. *Azospirillum*, in fact, carries in its genome parts which complement *Rhizobium meliloti* genes coding both for symbiosis and for host specificity (Fogher *et al.*, 1985). It has been thus hypothesized that *Azospirillum* may modify, at early stages, the infection. Most of the researches carried out to analyse the effect of combined inoculation, indicated a trend of growth promotion in *Medicago polimorfa* and in alfalfa (*Medicago sativa*) and an increase in the susceptibility of the plant to rhizobial infections (Yahalom *et al.*, 1991; Itzigsohn *et al.*, 1993)

This increased susceptibility of legumes to *Rhizobium* infection was ascribed to a stimulating effect of auxins and other plant growth regulators excreted by *Azospirillum*. Plant hormones production induce the formation of a large number of epidermal cells that differentiated into roots hairs, favouring *Rhizobium* primary infections (Schmidt *et al.*, 1988; Rai, 1983; Yahalom *et al.*, 1987).

In a previous paper (1991), we found that *Azospirillum* have a more remarkable positive influence on chickpea-*Rhizobium* symbiosis when it was inoculated on chickpea seedlings 66 hours before than *Rhizobium*. Negative effect of co-inoculation, which prevent clover nodulation, occurred when *Azospirillum* cells outnumbered *Rhizobium* cells in the mixture (Plazinski and Rolfe, 1985b).

In the present paper we describe a greenhouse study on the effect of the application of *A. brasilense* Cd on the symbiosis between *Rhizobium* strains and several chickpea Italian winter and spring cultivars.

Chickpea production in the Mediterranean regions has remained static in the past fifteen years. However, there is a prospect of increasing the production by adapting chickpea to winter sowing. We have examined how *Azospirillum brasilense* can affect the interaction between *Rhizobium* and plant genotype, by evaluating its influence on different parameters: i) nodules formation; ii) nitrogenase activity of nodules (performed by acetylene reduction assays, ARA); iii) total plant nitrogen content, and iv) plant biomass production.

Material and method

Organisms and growth conditions

Azospirillum brasilense Cd (ATCC 29729) was grown on malate minimal medium (Okon *et al.*, 1976) supplemented with 0,1 % NH_4Cl and 0,01% yeast extract, pH 6, 8 at 30 C, up to the late logarithmic growth phase.

Chickpea-*Rhizobium* USDA 3377 (Nitragin 27A2), USDA 3379 (Nitragin 27A8), USDA 3380 (Nitragin 27A9) were kindly supplied by Peter Van Berkum (United States Department of Agriculture, Beltsville Experimental Station, Beltsville, MD, USA) Cultures were grown on yeast mannitol agar, (YEM, Vincent, 1970) at pH 6,8 for 72 hrs at 28 C.
Cicer arietinum cultivars utilised were: Calia (spring sowing), Califfo and Sultano (both winter sowing) obtained from Paola Crinò, Plant Breeding Laboratory, ENEA- Casaccia.
Surface sterilised (12% H_2O_2 for 10 min followed by 6 rinsing with sterile water) chickpea seeds were germinated on moist sterile filter paper in Petri dishes at 25 C in the dark.

Greenhouse experiment

4 l plastic pots (20x20x40 cm) were filled with a 2:1 mixture of river sand and perlite, autoclaved by two 1 hr cycles.
Uniform 6-day-old seedlings were transplanted into the pots (four plants per pot) and were inoculated at the time of sowing with either one of the three *Rhizobium* alone or in combination with *A. brasilense.*
Azospirillum was applied as a 1 ml bacterial suspension for each seedling at the concentration of 10^7 cfu x ml^{-1}, while *Rhizobium* strain was applied as 1 ml suspension containing 10^6 bacteria x ml^{-1}.
From the different inoculum/cultivar combinations 18 treatments were obtained: Calia-3380, Calia-3380-Cd, Calia-3379, Calia-3379-Cd, Calia-3377, Calia-3377-Cd; Califfo-3380, Califfo-3380-Cd, Califfo-3379, Califfo-3379-Cd, Califfo-3377, Califfo-3377-Cd; Sultano-3380, Sultano-3380-Cd, Sultano-3379, Sultano-3379-Cd, Sultano-3377, Sultano-3377-Cd.
Pots were arranged in a completely randomised design, 8 replicates per each of the 18 treatments. Plants were kept in a greenhouse under the following conditions: 27 C during the day and 22 C at night with a 13 hrs light period and constant humidity at 70%. During the growth period plants were watered every 48 hrs alternatively with a N-free Jensen mineral solution (Vincent, 1970) and with tap water. Plants were harvested 5 weeks after inoculation.

Plant analyses

At harvesting, shoots were cut at the crown, and roots carefully removed from the soil by gentle shaking and rinsing; nodules were removed and counted.

Nodule dry weight was determined by drying at 60 C for 48 hrs. Fresh and dry weight of roots and shoots were measured separately after drying at 80 C for 72 hrs.

N_2-fixation was measured by evaluating acetylene reduction activity in an atmosphere of 10% acetylene: fresh nodules were placed, immediately after cutting, in sealed bottles (50 ml). The ethylene produced was measured after 30 min of exposure using a Perkin Elmer 8410 gas chromatograph with a flame ionization detector.

Total nitrogen was measured on plant aerial part by macro-Kijeldahl (Vincent, 1970).

Statistical analyses

Statistically significant differences (P=0.05) among data were determined by analysis of variance, and means were analysed using Duncan's Multiple Range Test. In each group of data different letters show significantly different values.

Result and discussion

Chickpea-*Rhizobium* strains 3379, 3380 and 3377 inoculated on the different chickpea genotypes Calia, Califfo and Sultano exhibited differences in their N_2-fixing ability as demonstrated by ARA.

The results (figure 1 and table 1) showed that nodule and A.R.A. of 35-day-old plants differed among the 9 chickpea-*Rhizobium* treatments, but significant differences (P=0,05) were observed only for some cultivar/strain combinations, i.e. Califfo and Sultano. All *Rhizobium* strains, instead, formed - alone as well as in combination with *Azospirillum* - less efficient symbioses with the spring cultivar Calia. Positive influence of *Azospirillum* inoculation was quite pronounced both on Califfo- and Sultano-*Rhizobium* symbioses, and even more pronounced when the cultivar Califfo was coinoculated with the most efficient strain, USDA 3379. *Azospirillum*, when applied in combination with *Rhizobium* (Fig. 1 A), affected chickpea

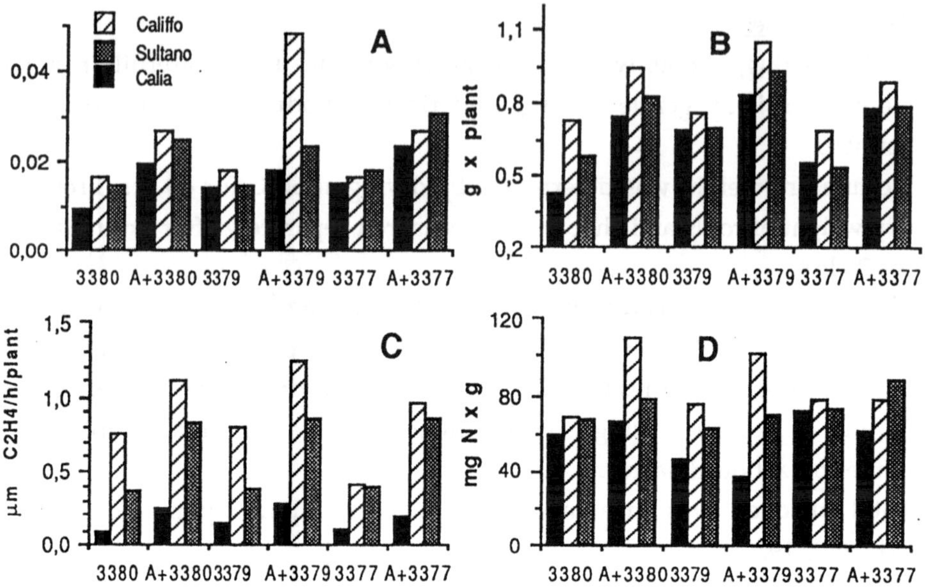

Figure 1. **A**: plant nodule dry weight of three chickpea varieties (Califfo and Sultano=winter cultivars, Calia=spring cv.) inoculated with three different chickpea-*Rhizobium* strains (USDA 3377, 3379, 3380) and *Azospirillum brasilense* Cd (A). **B**: total plant d.w. **C**: acetylene reduction activity. **D**: Nitrogen content in the aerial part (mg x g, dry weight).

nodulation both anticipating nodule formation (data not shown) and increasing nodule number, size and dry weight. Total N content in the aerial part (Tab. 1 and figure 1) of coinoculated Califfo and Sultano plants showed improved nutrient utilization and reached the flowering stage significantly earlier than those not inoculated with *Azospirillum* (data not shown).

The principal mechanism by which *Azospirillum* exerts its effect on the symbiosis has not been clarified, yet. Different modes of action have been proposed: interaction through the host plant, directly or indirectly, not through direct interaction with *Rhizobium*. Stimulation of nodulation may also occurs as result of a direct plant response to *Azospirillum* inoculation. This capability of *Azospirillum* could be ascribed either to modifications induced by the bacterium on young roots, which enter into physiological states susceptible to infection by microsymbiont, or to other unknown

Table 1 . Nodule dry weight, acetylene reduction activity, plant total nitrogen, shoot and root dry weight of three chick pea varieties, Califfo, Sultano and Calia, inoculated with three different chickpea-*Rhizobium* strains (USDA 3377, 3379, 3380) and *Azospirillum brasilense* Cd (A). Different letters indicate statistically different data (P=0.05).

Cicer arietinum, Rhizobium, Azosp.	nodules d.w. (g x plant)		A.R.A. (nM C2H4xhxpl)		total N (mg x plant)		root d.w. (g x plant)		shoot d.w. (g x plant)	
Calia 3380	0.0093	G	0.10	C	15	C	0.1802	G	0.2549	G
Calia 3380 Azospir.	0.0195	CDEF	0.26	C	23	C	0.3865	CDE	0.3525	DE
Calia 3379	0.0141	FG	0.14	C	16	C	0.3475	E	0.3450	DEF
Calia 3379 Azospir.	0.0181	DEF	0.28	C	15	C	0.4308	BCD	0.4095	BCD
Calia 3377	0.0152	EFG	0.11	C	20	C	0.2766	F	0.2797	FG
Calia 3377 Azospir.	0.0233	BCDE	0.19	C	23	C	0.4113	BCDE	0.3725	DE
Califfo 3380	0.0167	DEFG	0.75	B	25	C	0.3647	DE	0.3683	DE
Califfo 3380 Azosp.	0.0269	BC	1.11	A	42	A	0.4766	B	0.4731	AB
Califfo 3379	0.0183	DEF	0.80	B	29	B	0.3850	CDE	0.3810	DE
Califfo 3379 Azosp.	0.0483	A	1.25	A	50	A	0.5624	A	0.4955	A
Califfo 3377	0.0169	DEFG	0.41	C	28	C	0.3400	EF	0.3580	DE
Califfo 3377 Azosp.	0.0272	BC	0.96	B	32	AB	0.4750	B	0.4103	BCD
Sultano 3380	0.0145	FG	0.37	C	21	C	0.2742	F	0.3167	EFG
Sultano 3380 Azosp.	0.0248	BCD	0.82	B	31	AB	0.4290	BCA	0.3977	CD
Sultano 3379	0.0145	FG	0.38	C	22	C	0.3541	E	0.3498	DE
Sultano 3379 Azosp.	0.0234	BCDE	0.85	B	32	AB	0.4792	B	0.4624	ABC
Sultano 3377	0.0181	DEF	0.40	C	20	C	0.2710	F	0.2711	G
Sultano 3377 Azosp.	0.0308	B	0.86	B	30	AB	0.4587	BC	0.3382	DEF

mechanisms. According to Elmerich *et al.*, (1991) *Azospirillum* can produce also '*nod*-genes products' inducing transcriptional activation of nodulation genes in rhizobia, directly enhancing nodulation. Although *Azospirillum* is not a good competitor in rhizosphere environment (Del Gallo and Fendrik, 1994), according to Barak *et al.*, (1983) it could play a role in the symbiosis, also when is coinoculated with *Rhizobium*, because it has a strong microaerophilic attraction to the rhizosphere niche and high mobility and can reach faster than the latter the root.

Combined *Rhizobium* and *Azospirillum* treatments significantly increased, at different extent, plant biomass production in comparison with *Rhizobium* alone. Shoot dry weight (table 1) of cultivar Califfo inoculated both with *Azospirillum* and *Rhizobium*, strain 3379, was significantly greater than any other treatment. Califfo was the cultivar which responded, for all the parameter considered, better to the inoculation, either single or in combination with *Azospirillum*.

According to Itzigsohn et al., (1993) a possible explanation for the strong increase in shoot weight, when the plants are coinoculated with both bacteria, is that *Azospirillum* by promoting the growth, stimulate the plant to overcome the initial stress related to nodule development. This could subsequently lead to vigorous establishment of the plant.

Positive marked effects of inoculation of *Azospirillum brasilense* Cd on the three Italian chickpea cultivars has been found on various root parameters including root length, and particularly on the root elongation zone, the "region of infection". Significant increases in root dry weight were found only when the same cultivars were coinoculated with *Azospirillum* in comparison with single chickpea-*Rhizobium*. No significant differences, instead were found on Calia and Sultano, among the *Rhizobium* strains in combination with *Azospirillum*. This could means that there is not a synergistic effect between the two bacterial species, but there is a direct effect of *Azospirillum* on the plant. The analysis of variance showed, instead, a higly significant interaction effect of the combination *Azospirillum*-3379-Califfo. Shoot/root ratio (figure 2) was mostly affected by the presence of

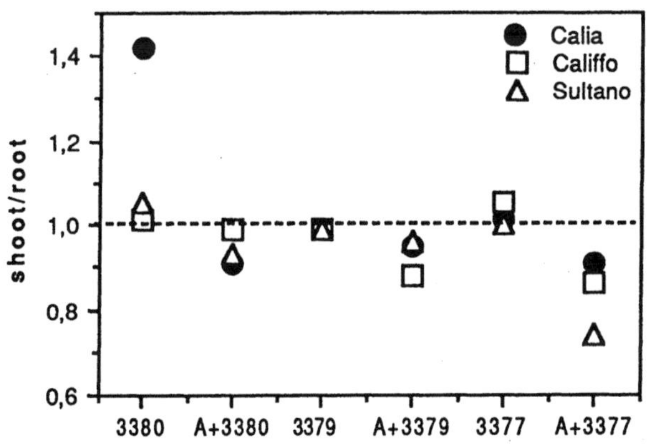

Figure 2. Shoot/root ratio of three chickpea varieties (Calia, Califfo and Sultano) inoculated with three strains of *Rhizobium* (3380, 3379 and 3377) and with *A. brasilense.*

Azospirillum: this likely reflects root stimulation exerted by this bacterium on the inoculated plant. Strain 3380, the less effective on Calia, seems also to influence the less its root. Strain 3377, instead, is the strain which let *Azospirillum* to influence the most root production, with all tested cultivar.

Conclusions

The response to the inoculation of certain inoculum/cultivar combinations suggest a strong plant-genotype bacterial-genotype interaction which could affect nodulation process and plant growth under pots conditions.

In particular:

- Cultivar Calia was less efficient in its symbiotical capability with different *Rhizobium* strains, in all parameter analysed, both in the presence and in the absence of *Azospirillum;*
- Cultivar Califfo was nodulated in a decreasing extent by strains 3379, 3380 and, in a even lower extent, by strain 3377;
- Cultivar Sultano, instead, responded better to 3377 inoculation;
- All chickpea-*Rhizobium* symbioses tested are positively affected, at different extent, by the presence of *Azospirillum*.

Biological processes of rhizosphere colonization could be greatly influenced by plant-root exudates and bacteria utilize exudates from a given genotype differentially. Attractive responses of different *Rhizobium* strains to various organic- and amino-acids, and to different concentrations of mono- and poly-saccharides present in the root exudates suggest that chemotaxis is one of the mechanism responsible for the movement of bacteria towards root surfaces. Experimental evidences (Rai and Prasad, 1983) showed that root exudates from different chickpea varieties may influence the symbiosis by modulating the processes which characterize the first steps of host infection by *Rhizobium*. Different chickpea-*Rhizobium* strains show differences in their chemotactic responses towards the same organic molecules. Reinhold *et al* (1985) investigated strain-specific chemotaxis of some azospirilla and found differences both at species and at strain level. If different strains had a different requirements, this could result in different patterns of association with the roots of even a single plant variety. This could explain the different ability of chickpea cultivars to interact with *Rhizobium* strains and *A. brasilense*.

We have observed that the use of simultaneous combined inoculations by chickpea-*Rhizobium* and *Azospirillum* improve early nodulation, increase nitrogen fixation, plant biomass and reduce water stress. However, the selection of the right plant cultivars and micro-symbionts isolates seems to be a relevant aspect for a practical use of this tripartite symbiosis. To get the maximum from this association it is essential to better understand the contribution and the synergistic actions of each of the component.

References

Abe MA, Amemura A, Hihashi S (1982) Studies on cyclic β-1,2 glucan obtained from periplasmic space of *Rhizobium trifolii* cells. Plant Soil 64:315-324

Appleby CA (1984) Leghemoglobin and *Rhizobium* respiration. Annu Rev Plant Physiol 35:443-478

Baldani VLD, Döbereiner J (1983) Effects of *Azospirillum* inoculation on root infection and nitrogen incorporation in wheat. Can J Microbiol 29: 924-929

Barak R, Nur I, Okon Y (1983) Detection of chemotaxis in *Azospirillum brasilense*. J Appl Bacteriol 53:399

Bauer WD (1981) Infection of Legumes by Rhizobia. Annu Rev Plant Physiol 32:407-449

Dazzo FB, Truchet GL, Hrabak EM (1984) Specific enhancement of clover root hair infection by trifoliin-A-binding lipopolysaccharide from *Rhizobium trifolii*. *In* "Advances in nitrogen fixation research" Veeger, Newton WE (*eds*) Nijhoff/Junk Publishers, The Hague *pp* 413

Del Gallo M, Fabbri P (1991) Effect of soil organic matter on chickpea inoculated with *Azospirillum brasilense* and *Rhizobium leguminosarum* bv *Cicer*. Plant Soil 137:171-175

Del Gallo M, Fendrik I (1994) The rhizosphere and *Azospirillum*. *In* "*Azospirillum*/plant associations" Okon Y (*ed*). CRC Press, Boca Raton *pp* 57-76

Elmerich C, de Zamaroczy M, Vieille C, Delorme F, Onyeocha I, Liang YY, Zimmer W (1991) *Nif* and *nod* genes in *Azospirillum*. *In* "Nitrogen fixation" Polsinelli M, Materassi R, Vincenzini M (*eds*) Kluwer Academic Publishers, Dordrecht Boston and London *pp* 79-87

Fogher C, Dusha I, Barbot P, Elmerich C (1985) Heterologous hybridization of *Azospirillum* DNA to *Rhizobium nod* and *fix* genes. FEMS Microbiol Lett 30:245-249

Itzigsohn R, Kapulnik Y, Okon Y, Dovrat A (1993) Physiological and morphological aspects of interactions between *Rhizobium meliloti* and alfalfa (*Medicaga sativa*) in association with *Azospirillum brasilense*. Can J Microbiol 39: 610-615

Miller HH, Sirois JC (1982) Relative efficacy of different alfalfa cultivar *Rhizobium meliloti* strain combinations for symbiotic Nitrogen-fixation. Appl Environ Microbiol 43: 764-768

Neal JL, Atkinson TG, Larson RI (1970) Changes in the rhizosphere microflora of spring wheat induced by disomic substitution of a chromosome. Can J Microbiol 16:153-158

O'Hara GW, Dovey MR, Lucas JA (1981) Effect of inoculation of *Zea mays* with *Azospirillum brasilense* strains under low temperature conditions. Can J Microbiol 27: 871-877

Okon Y, Albrecht SL, Burris RH (1976) Carbon and ammonia metabolism of *Azospirillum lipoferum*. J Bacteriol 128:592-597

Okon Y, Kapulnik Y (1986) Development and function of *Azospirillum* inoculated roots. Plant Soil 90:3-16

Okon Y, Kapulnik Y, Sarig S (1988). Field inoculation studies with *Azospirillum* in Israel. *In* "Biological Nitrogen Fixation: Recent Development" Subba Rao (*ed)* Oxford and IBH Publ CO, New Delhi *pp* 175-195

Plazinski J, Rolfe BG (1985b) *Azospirillum-Rhizobium* interaction leading to a plant growth stimulation without nodule formation. Can J Microbiol 31:1026-1030

Rai R (1983) Efficacy of associative N_2-fixation by streptomycin-resistant mutants of *Azospirillum brasilense* with genotypes of chickpea *Rhizobium* strains. J Agric Sci 100:75- 80

Rai R, Prasad V (1983) Chemotaxis of *Rhizobium* strains to root exudate of Chickpea (*Cicer arietinum* L.) genotypes and their interaction response on nodulation, nodulins, leghaemoglobin and grain yield in calcareous soil. J Agric Sci Comb 101: 75-81

Reinhold B, Hurek T, Fendrik I (1985) Strain-specific chemotaxis of *Azospirillum* spp. J Bacteriol 162:190-195

Schmidt W, Martin P, Omay SH, Baugerth F (1988) Influence of *Azospirillum brasilense* on nodulation of Legumes. *In* "*Azospirillum* IV: Genetics, Physiology, Ecology" Klingmüller W(*ed*) Springer Verlag, Berlin Heidelberg *pp* 92-100

Verma DPS, Nadler K (1984) Legume-*Rhizobium* symbiosis: host's point of view. *In* "Plant gene research, Genes involved in microbe-plant interactions" Verma DPS, Hohnt (*eds*) Springer-Verlag, New York

Vincent JM (1970) Manual for the Practical Study of Root Nodule Bacteria, IBP Blackwell Scientific Publications *pp* 1-169

Vincent JM (1980) Factors controlling the legume-*Rhizobium* symbiosis. *In* "Nitrogen Fixation" Newton WE and Orme-Johnson WH (*eds*) 2:103

Yahalom E, Okon Y, Dovrat A (1987) *Azospirillum* effects on susceptibility to *Rhizobium* nodulation and on nitrogen fixation of several forage legumes. Can J Microbiol 33:510-514

Yahalom E, Dovrat A, Okon Y, Czosnek H (1991) Effect of inoculation with *Azospirillum brasilense* strain Cd and *Rhizobium* on the root morphology of Burr medic (*Medicago polymorpha* L.). Isr J Bot 40:155-16

SECTION IV:

PHYSIOLOGY AND MOLECULAR BIOLOGY

Interactions between Partners in the Association "Wheat - *Azospirillum brasilense* Sp 245"

Vladimir Ignatov, Galina Stadnik, Olga Iosipenko, Nikolai Selivanov, Alexander Iosipenko and Elena Sergeeva
Institute of Biochemistry and Physiology of Plants and Microorganisms Russian Academy of Sciences
Pr. Entuziastov 13 Saratov 410015
Russia

Abstract

We report here on the effect of association between wheat plants and bacteria of the genus *Azospirillum* on the physiology and biochemistry of the associated partners. Lectin studies revealed strain-specific interaction of a wheat lectin with azospirilla. Inoculation of wheat seedlings with different *Azospirillum* strains and phytohormone (indole-3-acetic acid and abscisic acid) treatments brought about changes in specific lectin activity and in wheat root growth parameters. At concentrations of 0.5 - 2.0 mg/L, wheat lectin was stimulatory to the synthesis of indole-3-acetic acid in *Azospirillum* cells grown on a nitrogen-free medium under microaerophilic conditions. Changes in the synthesis of hydroxyproline-rich proteins and in the activities of exoglycosidases due to *Azospirillum* inoculation were also noted. These findings are in support of the views that the effect of *Azospirillum* on plants (wheat) is largely due to microbial production of indole-3-acetic acid, as well as to nitrogenase activity.

Introduction

First discovered about two decades ago (Döbereiner, 1976), bacteria of the genus *Azospirillum* have become the subject of a systematic research aiming to elucidate their role as associative nitrogen fixers for wild and cultivated cereals. The *Azospirillum* - plant interaction process can be divided as follows: i) chemotaxis; ii) contact recognition which is foremost linked with carbohydrate - protein interaction; iii) exchange of the partners' metabolites, and iv) alterations in the metabolism of the associated partners.

The latter two aspects of plant-microbe interactions have up to now been neglected, and are therefore dealt with in the current work. The plant root system excretes into the environment amino acids, proteins, sugars, organic acids, vitamins, and other bacterium-beneficial substances affecting the growth, development and physiology of a microbial cell population. In turn, cells of the nitrogen fixers excrete ammonium ions and secondary metabolites, among which an important place is taken by phytohormones.

NATO ASI Series, Vol. G 37
Azospirillum VI and Related Microorganisms
Edited by I. Fendrik et al.

This report addresses several aspects of the mutual influence of the associated partners. The specific goals were threefold: i) to explore wheat lectin effects on the vital functions of *Azospirillum*; ii) to study the changes in lectin activity and in the development of wheat seedling roots induced by phytohormones and by *Azospirillum* inoculation; and iii) to investigate the accumulation of the extracellular proteins (glycosidases and hydroxyproline-rich proteins) in wheat seedling roots upon *Azospirillum* inoculation.

Materials and methods

The *Azospirillum* strains employed in this study were *A. brasilense* strains Sp 245, Sp 245.90, Sp 245.63, Sp 75, and Sp 7; and *A. lipoferum* Sp 43 (obtained from the Culture Collection of this Institute). Information on cultivation of organisms, as well as on experiments to elucidate the effect of lectin on the physiological and biochemical properties of *Azospirillum*, has been reported elsewhere (Antonyuk *et al.*, 1993; Iosipenko *et al.*,1994). Seeds of wheat (*Triticum aestivum* cvs. Saratovskaya 29 (S 29), Saratovskaya 57 (S 57), and Mironovskaya 808 (M 808)) were surface sterilized using 2% chloramine B. The seeds were germinated at 22-24°C for 3-5 days in sterile cuvettes on filter paper moistened with distilled water.

For long-term experiments, the seeds were placed in sterile test tubes of coarse river sand (particle size 3-4 mm) with a mineral medium added (Barbieri *et al.*, 1986). Wheat lectin was isolated from root extracts of 3-to 4-day-old seedlings (Iosipenko *et al.*, 1992), or a commercial sample of wheat germ agglutinin (WGA) was utilized.

The ability of the lectin to interact with *Azospirillum* cells was determined by a dot-blot analysis using wheat lectin labelled with colloidal gold (Bogatyrev *et al.*, 1992). Protein estimation was carried out according to Scopes (1987).

Inoculation of 7- to 15-day-old wheat seedlings was done with *Azospirillum* cells grown overnight in Döbereiner's medium (1976) at a rate of 10^6-10^7 cells per plant.

Isolation of the extracellular fluid from the wheat seedling roots was performed using the method of Parent and Asselin (1983).

The activities of exoglycolytic enzymes were measured by the degradation of 4-nitrophenyl-glycosides (30°C, 30 min, pH 4.5) and expressed in nkat/ml. The activities of endoglycosidases were evaluated by the rate of chitin and laminarin degradation. Reducing carbohydrates were determined according to Nelson (1944).

The content of hydroxyproline-rich proteins (HPRPs) in plant cells was estimated following acid hydrolysis of the roots and cell wall preparations with an AAA-339 amino acid analyzer (Czechoslovakia).

The concentration of indole-3-acetic acid (IAA) in the culture fluid of *A. brasilense* was determined according to Tang and Bonner (1947).

The mathematical processing of the data obtained in a multifactor experiment (Maksimov, 1980) was performed using a library of mathematical and statistical procedures SSP.

Results

Studies of wheat lectin effects on the vital functions in *Azospirillum*

The root lectin isolated by the authors from wheat roots was identical to WGA in basic properties. This result is consistent with the earlier data of Raikhel *et al.* (1984). The lectin bound well to the carbohydrate receptors of *Azospirillum* cell surface (Konnova *et al.*, 1994), and bacterial strains isolated from wheat rhizo--sphere possessed a high affinity for wheat lectin (Fig.1). As distinct from *Azospirillum* type strain Sp 7, strains Sp 245 and Sp 75 isolated from wheat roots gave a clearer picture of interaction with wheat lectin labelled with colloidal gold. The binding of lectin to the carbohydrate receptors of *Azospirillum* cell surface brought about physiological and biochemical alterations in the microbial population (Antonyuk *et al.*, 1993; Iosipenko *et al.*, 1994). Considering the fact that microbial production of such a phytohormone as IAA is an essential factor affecting plant growth and development, we attempted at tracing the lectin-induced changes in IAA synthesis depending on the presence of ammonium ions in the culture medium. A multifactor mathematical analysis showed that at concentrations of 0.5-2.0 mg/L, wheat lectin was stimulatory to IAA synthesis in *Azospirillum* cells grown on N-free medium. IAA production decreased in the presence of ammonium chloride in the culture medium (Fig. 2). The maximum concentration of IAA on N-free medium was 173 mg/L; adding ammonium chloride reduced this value to as little as 130 mg/L. Thus, we have demonstrated a positive effect of wheat lectin on IAA production by *Azospirillum*.

Studies by Stinissen (1985) reported on the effects of the abscisic acid (ABA) and IAA

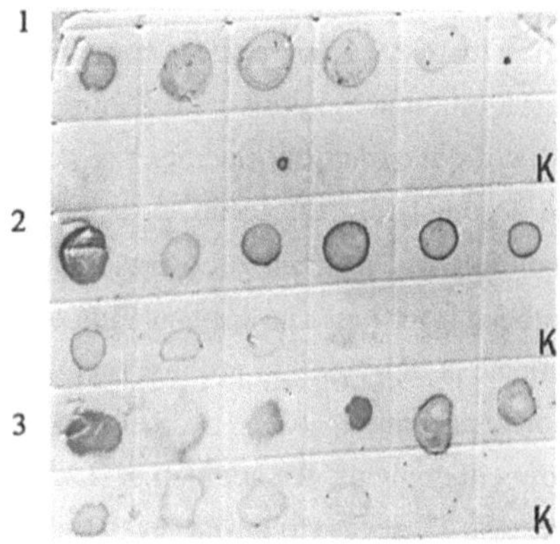

Fig.1. Interaction of colloidal gold-labelled (0.01 mg/ml) wheat lectin with *A. brasilense* Sp 7 (1); Sp 75 (2); Sp 245 (3) (initial concentration $4 \cdot 10^9$ cells/ml followed by 2-fold dilution); K - control (colloidal gold-labelled lectin without bacteria added).

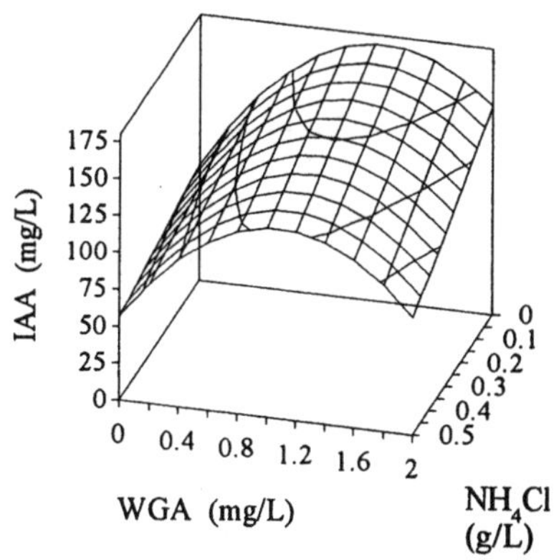

Fig. 2. Effect of WGA and NH_4Cl on IAA production by *Azospirillum brasilense* Sp 245.

phytohormones on lectin biosynthesis in the roots of young wheat seedlings. The phytohormones produced and excreted by associative microorganisms into the environment may also influence the physiological processes in the plant cell, thereby affecting the level of the root lectin content.

Studies of changes in lectin activity and in wheat seedling root development induced by phytohormones and *Azospirillum* inoculation.

Under natural conditions, the concentration of IAA in the plant-bacterium system ranges from unities to tens of micrograms per millilitre. Zimmer and Bothe (1988) reported that IAA concentration in the growth medium of *Azospirillum* was in the range of 15 -100 mM, which does not contradict our findings.

The accumulation of lectin in the wheat root seedlings occurs much more rapidly under conditions of drought or osmotic shock making the plant most sensitive to a potential microbial infection. The accumulation of ABA is induced under the same conditions. This led Cammue *et al.* (1990) to propose that the regulation of lectin synthesis and accumulation is of a hormonal nature.

In our experiments, the addition of IAA and ABA to the growth medium of wheat (cv. Saratovskaya 29) seeds followed by 3-and 7-day incubations brought about substantial changes in the development of seedling roots and in the level of lectin activity, which was judged by the hemagglutination reaction (RHA) of the root extracts (Table 1). At a concentration of 10^{-5} M, IAA inhibited root growth and development, yet it caused an increase in lectin activity. Addition of IAA at a concentration of 10^{-10} M stimulated the growth of lateral roots, increased root weight, and decreased specific lectin activity. Adding ABA at a concentration of 10^{-5} M also retarded root growth and development, yet lectin activity increased even more. Adding ABA at a concentration of 10^{-7} M yielded negligible changes. These included a slight decrease in the root length with the root weight being normal, and in lectin activity that differed little from that of the control.

Table 1. Effect of phytohormones on growth parameters and lectin activity of 3-day-old wheat seedlings (10 plants)

Treatments	Parameters		
	root length, cm	root weight, g	titer RHA/g fresh weight
Control	2.84	0.38	168
IAA 10^{-5}M	0.30	0.12	1066
IAA 10^{-10}M	3.13	0.46	69
ABA 10^{-5}M	0.81	0.10	5120
ABA 10^{-7}M	1.76	0.40	160

Thus, lectin content in seedling roots increased with decelerating root growth and development. This deceleration was caused by definite concentrations of the phytohormones exogenously supplied to the growth medium.

Inoculation with strain Sp 245 and with its mutants in IAA synthesis (Sp 245.90 and Sp 245.63) yielded data to suggest that the response of the root system of 14-day-old wheat seedlings may be explained to a large degree by IAA synthesis (Table 2).

Inoculating the seedlings with strain Sp 245 resulted in increased wet root weight (mainly due to increased length of lateral roots), and in decreased specific lectin activity. Inoculation with the IAA-overproducing strain Sp 245.90 led to decreased length and numbers of lateral roots, to decreased root weight, and to increased specific lectin activity. The changes observed following inoculation with strain Sp 245.63, that had lost its ability to synthesize IAA, were insignificant.

We conclude that phytohormones either exogenously supplied or produced by *Azospirillum* have outstanding and not always unequivocal effects on the morphological and physiological processes in the plant root system. Our data are in good agreement with the earlier conclusions (Barbieri *et al.*, 1986; Barbieri and Galli, 1993) indicating that the root growth-beneficial effect of inoculation is due to microbial synthesis of IAA.

Table 2. Effect of inoculation of 14-day-old wheat seedlings with *Azospirillum* exhibiting altered IAA synthesis on specific lectin activity and root growth (average in 10 plants)

Treatments	Parameters	
	root weight, g	titer RHA/g fresh weight
Control	0.10	160
A.brasilense Sp 245	0.13	123
A.brasilense Sp 245.90 (IAA^{++})	0.08	200
A.brasilense Sp 245.63 (IAA^{-})	0.09	177

Studies of accumulation of extracellular proteins in wheat seedling roots upon *Azospirillum* inoculation.

The question now arises as to whether *Azospirillum* induces any biochemical responses in the colonized root tissues of wheat.

The earlier studies of Patriquin and Döbereiner (1978), and Levanony *et al.*, (1989) which have addressed plant root colonization by *Azospirillum* demonstrate that bacteria colonize the surface of roots and penetrate via the epidermis to fill the intercellular spaces in the outer cortical layers, where they exist during the life span of plant tissues.

On the basis of this information, attention has been given to the apoplast, the intercellular compartment of wheat roots. We now turn to consider the effect of *Azospirillum* inoculation on the spectrum of glycolytic enzyme activities and on HPRP accumulation in the plant root cell walls.

These biochemical parameters were chosen to estimate the degree of pronouncement of: i) the mechanism of partner recognition; and ii) plant adaptation to a microbiological infection. They have been amply documented for the early steps of the pathogenesis (De Wit and Joasten, 1989; Mazau *et al.*, 1986). The induction of HPRP synthesis was also seen during the symbiotic process at the stage of nodule formation, presumably reflecting the adaptive morphological and physiological processes in cells surrounding the interaction zone (Benhamou *et al.*, 1991). The present findings indicate that these reactions reveal themselves in the case of associative interactions, too.

The first step of our work was to examine the effect of *Azospirillum* inoculation on the level exo- and endoglycolytic enzyme activities in the wheat root apoplast (Fig. 3).

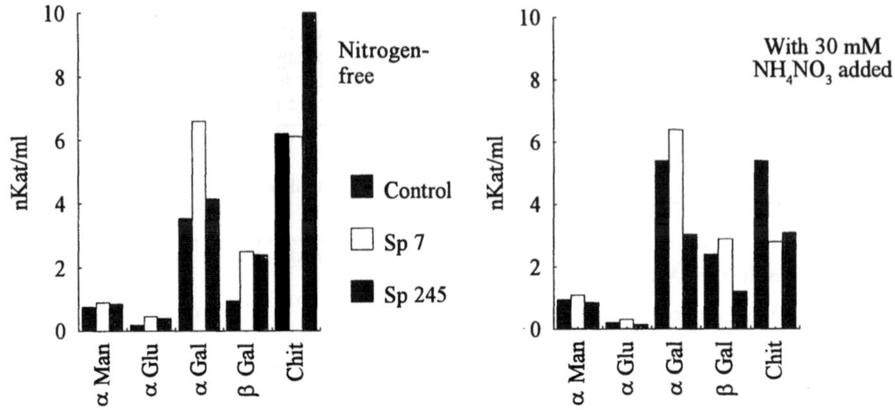

Fig. 3. The effect of *A. brasilense* inoculation on the activities of glycosidases in the wheat root apoplast within 3 days after inoculation.

Inoculation of the roots of 7- to 10-day-old wheat plants led to substantial changes in the spectrum of exoglycolytic activities in the root apoplast. The enzymes studied were α-mannosidase, α- and β-glucosidases, and α- and β-galactosidases. Exo-β-glucosidase and α-galactosidase exhibited the highest activities; however, α- and β-galactosidases were the most sensitive to bacterial infection.

It is significant that application of different *Azospirillum* strains yielded different results. Thus, following inoculation with strains Sp 245 and Sp 7, α-galactosidase activity increased by 17% and 85%, respectively, and β-galactosidase activity increased by 152% and 163%, respectively. *Azospirillum* inoculation also resulted in a 2-fold increase in the activity of α-glucosidase; however, the activity level of this enzyme in the apoplast was very low. Inoculating the plants on a medium supplemented with 30 mM NH_4NO_3 slightly changed the spectrum of glycolytic activities. In this case, the activity level of glycosidases in sterile plants was higher than that seen on N-free medium. On the other hand, inoculation with strain Sp 245 decreased the activities of all enzymes.

As distinct from exoglycosidases being more sensitive to Sp 7, the activity of chitinase in the root apoplast displayed a higher sensitivity to strain Sp 245. The increase in chitinase activity on N-free medium was 65% of the control. By contrast, inoculation with strains Sp 7 and Sp 245 in the presence of nitrogen caused a decrease in chitinase activity of 51% and 57%, respectively.

An explanation for the observed differences may be that the establishment of a successful association would only occur on N-free medium. Correspondingly, the system of plant recognition is adequately "switched on" under these conditions only.

We explored the effect on HPRP accumulation of *Azospirillum* strains exhibiting variation in

Table 3. HPRP content in cell walls of wheat seedling roots upon *Azospirillum* inoculation

Treatments	48 h nM/Hpr mg/dry wt	48 h %	72 h nM/Hpr mg/dry wt	72 h %
Control	0.668	100	0.664	100
A.brasilense Sp 245	1.057	156	1.319	199
A.brasilense Sp 7	0.654	98.3	0.751	113
A.brasilense Sp 75	0.642	96	0.912	137
A.lipoferum Sp 43	0.653	98.3	0.713	107

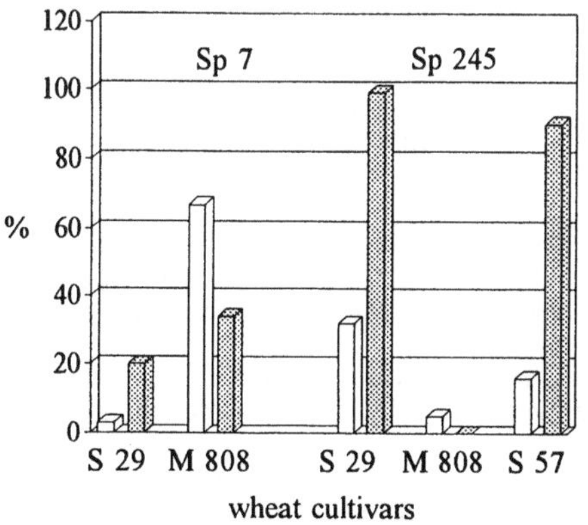

Fig. 4. HPRP accumulation in different wheat cultivars upon *A. brasilense* inoculation
□ - 48 h ▨ - 72 h

the specificity for wheat. The results listed in Table 3 indicate that unlike less specific strain Sp 7 and *A.lipoferum* Sp 43, strains Sp 245 and Sp 7 isolated from wheat rhizosphere evoke the most pronounced plant response. Attempts at studying cultivar-specific HPRP accumulation revealed increased HPRP content in all the cultivars tested of both spring and winter wheats (Fig. 4). This increase was 67-102% of the control. The results demonstrate that HPRP accumulation in the cultivars differed both in the induction level and in strain specificity. For the winter wheat Mironovskaya 808, the maximum content of HPRPs was achieved within 40 h after inoculation, while in spring wheat the accumulation occurred more evenly, with the maximal values being obtainable within 72 h.

For the strain specificity, it may be noted that inoculation with strain Sp 245 had a greater effect on spring wheat than it did on winter wheat, whereas strain Sp 7 had a more marked effect on the winter cultivar Mironovskaya 808. The data we present attest to the efficient physiological adaptation of root cells to the colonization by *Azospirillum*.

In summary, cereal infection by *Azospirillum* has been shown here to induce substantial physiological and biochemical changes both in the bacterium and in the plant. More work is still needed to get a clearer understanding of the nature of these changes.

Acknowledgements

The research described in this paper was sponsored by the Russian Foundation of Basic Research under grant no. 93-04-20680. It is a pleasure to thank Dr. E. Katzy for the gift of strains Sp 245.90 and Sp 245.63. Thanks also go to D. Tychinin for his assistance in preparing this manuscript.

References

Antonyuk LP, Fomina OR, Galkin MA, Ignatov VV (1993) The effect of wheat germ agglutinin on dinitrogen fixation, glutamine synthetase activity and ammonia excretion in *Azospirillum brasilense* Sp 245. FEMS Microbiol Lett 110:285-290

Barbieri P, Galli E (1993) Effect on wheat root development of inoculation with an *Azospirillum brasilense* mutant with altered indole-3-acetic acid production. Res Microbiol 144:69-75

Barbieri P, Zanelli T, Galli E, Zanetti G (1986) Wheat inoculation with *Azospirillum brasilense* Sp 6 and some mutants altered in nitrogen fixation and indole-3-acetic acid production. FEMS Microbiol Lett 36:87-90

Benhamou N, Lafontane PJ, Mazau D, Esquerre-Tugaye MT (1991) Differentional accumulation of hydroxyproline-rich glycoproteins in bean root nodule cells infected with a wild-type strain or C_4-dicarboxylic acid mutant *Rhizobium leguminosarum bv.phaseoli*. Planta 184:457-467

Bogatyrev VA, Dykman LA, Matora LYu, Schwartsburd BI (1992) The serotyping of *Azospirillum* spp. by cell-gold immunoblotting. FEMS Microbiol Lett 96:115-118

Cammue BPA, Broekaert WF, Peumans WJ (1990) Wheat germ agglutinin in wheat seedling roots: induction by elicitors and fungi. Plant Cell Rep 9:264-267

De Wit JGM, Joasten HAJ (1989) Identification of several pathogenesis related proteins in tomato leaves inoculated with *Cladosporium fulvum* as 1,3-b-glucanases and chitinases. Plant Physiol 89:945-951

Döbereiner J, Day JM (1976) Associated symbiosis in tropical grasses: characterisation of microorganisms and dinitrogen fixing sites. In: Proceedings of the 1st International Symposium on Nitrogen Fixation. Newton WH, Nymans CJ (eds). Washington State University Press, Pullman, Wash. pp 518-538

Iosipenko AD, Sergeeva EI, Antonyuk LP, Ignatov VV (1994) The effect of wheat lectin on indole-3-acetic acid synthesis by *Azospirillum brasilense* Sp 245. Doklady Rossiiskoy Akademii Nauk 336 4:559-561 (in Russian)

Iosipenko OA, Selivanov NYu, Evtushenko IYa, Stadnik GI, Ignatov VV (1992) Changes of lectin isoform content during wheat germination. In: Lectins: Biology, Biochemistry and Chemical Biochemistry. Kallikorm A, Bøg-Hansen TC (eds). Sigma Chemical Company, S. Louis

Konnova SA, Makarov OE, Skvortsov IM, Ignatov VV (1994) Isolation, fractionation and some properties of polysaccharides produced in a bound form by *Azospirillum brasilense* and their possible involvement in *Azospirillum*-wheat interactions. FEMS Microbiol Lett 118:93-100

Levanony H, Bashan Y, Romano B, Klein E (1989) Ultrastructural localization and identification of *Azospirillum brasilense* Cd on and within wheat root by immuno-gold labeling. Plant and Soil 117:207-218

Maksimov VN (1980) The multifactor experiment in biology. MGU Moscow (In Russian)

Mazau D, Rumeau D, Esquerre-Tugaye MT (1986) Biochemical study of hydroxyproline-rich glycoproteins in plant-pathogen interactions. In: Biology and molecular biology of plant-patogen interactions NATO ASI. Bailey JA (ed). Series H Vol 1. Springer-Verlag New York

Nelson NJ (1944) New method for colorimetric determination of sugars. J Biol Chem 153:375-375

Parent J-G, Asselin A (1983) Detection of pathogenesis-related proteins (PR or b) and of other proteins in the intercellular fluid of hypersensitive plants infected with tobacco mosaic virus. Can J Bot 62:564-569

Patriquin DG, Döbereiner J (1978) Light microscopy observations of tetrazolium reducing bacteria in the endorhizosphere of maize and other grasses in Brazil. Can J Microbiol 24:734-742

Raikhel NV, Mishkind ML, Palevitz BA (1984) Characterization of a wheat germ agglutinin-like lectin from adult wheat plants. Planta 162:55-61

Scopes RK (1987) Protein purification: principles and practice. Springer-Verlag New York

Stinissen HM, Chrispeels MJ, Peumans WJ (1985) Biosynthesis of lectin in roots of germinating and adult cereal plants. Planta 164:278-286

Tang YW, Bonner J (1947) The enzymatic inactivation of indoleacetic acid. I. Some characteristics of the enzyme contained in pea seedlings. Arch Biochem 13:11-25

Zaat SAJ, Van Brussel AAN, Tak T, Pees E, Lugtenberg BJJ (1989) Aminoethoxyvinylglycine restores normal nodulation by *Rhizobium leguminosarum* biovar *viciae* on *Vicia sativa* by suppressing the «Thick and short root» phenotype. Planta 177:141-150

Zimmer W, Bothe H (1988) The phytohormonal interaction between *Azospirillum* and wheat. Plant and Soil 110:239-247

Azospirillum brasilense Exopolysaccharide Complexes, their Possible Involvement in Bacteria-Wheat Roots Interactions and the Suggested Nature of these Interactions

Igor Skvortsov, Svetlana Konnova, Oleg Makarov, Raisa Prokhorova and Vladimir Ignatov
Institute of Biochemistry and Physiology of Plants and Microorganisms
Russian Academy of Sciences
pr. Entuziastov 13
410015 Saratov
Russia

Abstract

Two exopolysaccharide complexes hereafter termed lipopolysaccharide-protein complex and polysaccharide-lipid complex were isolated by means of washing-off with physiological solution from the cell surface of *Azospirillum brasilense* Sp107, Sp245, S17, 8, 15, 55, 75 and 80 of a complex biopolymer mixture followed by separation. On exposure to wheat seedling root hairs (*Triticum aestivum* cv. Saratovskaya 29) LPPCs and PSLCs were shown to cause root hair deformations similar to morphological changes of root hairs under the effect of bacterial cells. The data indicating the ability of complexes to bind to WGA, localized on wheat seedling roots were obtained by means of studying LPPCs and PSLCs inhibition in agglutination test under the effect of WGA. PSs containing GlcN, specifically interacting with WGA, were isolated from PSLCs.

Key words: *Azospirillum brasilense*, wheat, root hairs, deformation, exopolysaccharide complex, polysaccharide, wheat germ agglutinin, interaction

Abbreviations: HMWC - high molecular weight compounds; LMWC - low molecular weight compounds; EPS - exopolysaccharide; LPPC - lipopolysaccharide-protein complex; PSLC - polysaccharide-lipid complex; LPS - lipopolysaccharide; PS - polysaccharide; KDO - 2-keto-3-desoxyoctonate; WGA - wheat germ agglutinin; Glc - glucose; Gal - galactose, GalA - galacturonic acid; Man - mannose; Rha - rhamnose, Fuc - fucose; Xyl - xylose; GlcN - glucosamine; GlcNAc - N-acetyl-D-glucosamine; GalN - galactosamine

Introduction

It is well known that bacteria of the genus *Azospirillum* are capable of interaction with roots both of cereals (Michiels *et al.*, 1989; Bashan and Levanony, 1990) and non-cereal plants (Bashan and Levanony, 1990; Bashan *et al.*, 1989). The acceleration of plant growth and the increase of plant yield is an external and desirable in practice manifestation of *Azopsirillum brasilense*-plant interactions. Invading rhizosphere and rhizoplane of plants, or root tissues, bacterial cells may be effected on macroassociant by its exudates. As an interaction test we used an easily observed phenomenon of wheat root hair deformation as effected by *A.brasilense* (Jain and Patriquin, 1984).

The process of interaction between azospirilla and surface of roots and root hairs is a very complex

NATO ASI Series, Vol. G 37
Azospirillum VI and Related Microorganisms
Edited by I. Fendrik et al.

process, involving typically various in its nature, physico-chemical and biological processes, wherein both LMWC and HMWC might be involved. The paper (Jain and Patriquin, 1985) provides us with an example of one LMWC, namely, indole-3-acetic acid, effect on morphology of wheat root hairs produced by *A.brasilense*.

In the work presented our attention was focused on the HMWC, localized on the cell surface of azospirilla, which can also spread in the environment, excreting in the process of cells growth of a liquid medium.

Rhizobial EPSs are known to stimulate morphological changes in root hairs of legumes, similar to the ones, induced by the whole bacterial cells (Halverson and Stacey,1986). In this connection we set the goal to isolate HMWC, relatively loosely bound to the cell surface, to get the provisional information on their chemical nature and to investigate its effect on morphology of wheat root hairs. Besides, we tried to figure out whether the process of interaction of the LPPC and PSLC with root hairs has specific characteristic and to ascertain the elements of their structure, causing their biological effect.

Material and methods

Bacterial strains and growth conditions

A.brasilense bacterial strains used in this study and their origin are presented in Table 1.

Isolation, analysis and physico-chemical properties of LPPCs, PSLCs, and PSs

LPPCs and PSLCs from *A.brasilense* strains Sp7, Sp245, Sp246, S17, 15 and 75 were obtained from a material washed out from the cell surface and purified by means of gel filtration and ion-exchange chromatography techniques according to (Konnova *et al.*, 1994).

Modified LPPCs and PSLCs were obtained from LPPC and PSLC mixtures of strains Sp107 and Sp246 with the help of mild hydrolysis, followed by treatment of hydrolysis products as described in (Konnova *et al.*, 1994).

PS-1, PS-2, PS-3, PS-4, and PS-5 were isolated from PSLC of *A.brasilense* strains Sp7, Sp245, S17, 15, and 75, whereas PS-6 and PS-7 - were isolated from slightly modified PSLCs of *A.brasilense* Sp107 and Sp246, according to method described in (Konnova *et al.*, 1994).

Analysis of LPPCs, PSLCs and PSs

The preparations obtained were tested for carbohydrates, proteins, nucleic acids, KDO and phosphorus by methods described in (Dubois *et al.*, 1956; Lowry *et al.*, 1951; Spirin, 1958; Karkhanis *et al.*, 1978; Berenblum and Chain, 1938), respectively. Glucose, human serum albumin and Na_3PO_4x 12 $H_2$0 were used as standards in determination of carbohydrates, proteins and phosphorus.

When constructing elution profiles by carbohydrates, optical density of reaction products of fractions

Tabl.1. Strains of *A. brasilense* used in the work, sources of isolation,origin

Strain	Host plant	Source	Reference
Sp7 (ATCC 29145)	*Digitaria*	D.Jannssens, Rijksuniversiteit, Gent.Lab.Voor.Microbiol., Gent, Belgium	Tarrand *et al.*, 1978
Sp 107	Wheat	J.Döbereiner, Empresa Brasileira de Pesquisa Agropecuaria, Rio de Janeiro, Brazil	De Polli *et al.*, 1980
Sp 245	Wheat	J.Döbereiner, Brazil	Baldani *et al.*, 1983
Sp 246	Wheat	J.Döbereiner, Brazil	Baldani *et al.*, 1986
S 17	*Pearl millet*	A.N.Lahiri, Division of Soil-Water-Plant Relationship, Central Arid Zone Institute, Jodhpur, India	
8	*Zerna inermis L.*	Culture Collection Institute of Biochemistry & Physiology of Plants & Microorganisms, Saratov, Russia	Pozdnyakova *et al.*, 1988
15	*Dactilis glomerata L.*	Idem	Ibid
55	Wheat	Idem	Ibid
75	Wheat	Idem	Ibid
80	Wheat	Idem	Ibid

with phenol and H_2SO_4 was measured at 490 nm, whereas by proteins - optical density was measured at 280 nm.

Monosaccharide composition of the PSs was determined as describer previously (Konnova *et al.*, 1994).

<u>Studying physico-chemical properties of LPPCs, PSLCs and PSs</u>

Physico-chemical properties of LPPC, PSLC and PS were defined according to (Konnova *et al.*, 1994).

LPPCs, PSLCs, PSs concentrations required for 50% inhibition of the agglutination test of trypsinized rabbit erythrocytes with WGA, were conducted in accordance with the well known method (Beeley,1985).

Direct effect of PSs and WGA was observed with the help of an interfacial ring test (Kato *et al.*, 1980). Aqueous solutions both of WGA (0.25%) and PS (initial concentration 0.25%) were used to test PS titers. The following controls were performed in parallel: (i) interaction of WGA with dextrans of Mr 500 and 20 kDa (Pharmacia); (ii) interaction of PS with gelatin solution (3%); (iii) interaction of PS with WGA preliminary inhibited by GlcNAc (0.3 M).

Seeds and terms of germination

Seeds of common wheat (*Triticum aestivum*, cv.Saratovskaya 29) reproduced in 1990 were obtained from the Institute for South-East Agriculture and used in the work presented.

Seeds were thoroughly rinsed in water, treated with 70% ethanol for 1 min, followed by 5-times rinsing with sterile water. Then seeds were treated with 9% sodium hypochlorite and again 6 times rinsed with sterile water. Seeds sterility was tested on Petri dishes with LB medium, incubated for 48h at 30^0 C.

Corn seeds thus prepared were placed in sterile Petri dishes on moistened filters and germinated for 24h at 25^0 C in the darkness.

Studying root hair deformations of wheat seedlings induced by LPPCs and PSLCs

Fahraeus slide technique (Fahraeus, 1957) was used to observe root hair deformations. Seedlings with 10-15 mm roots were placed on a glass slide, roots were sealed with 6-7 drops of mineral medium, containing 0.4% of agar, cooled to 30^0 C, and then covered with a cover slide. After solidification of agar, preparations were placed into tubes filled with mineral medium (Fahraeus, 1957), used as control, and into tubes filled with the same medium enriched by LPPCs and PSLCs (from 0.1 to 0.4 mg/ml). Concentrations of the latter, wherein root hair deformations were not observed, were sorted out experimentally. 12 - 15 plants were taken for each experiment. A set of 4 parallel experiments was carried out. 24-hours-old seedlings under study were incubated with LPPCs and PSLCs for 24 hours in the darkness and then examined through a light microscope at 100-fold magnification. The number of root hair deformations was counted at every cm along the length of the root.

Various types of deformations were taken into account, namely, formation of forks with branchings various and equal in length, wavy appearance, thickenings on the ends of roots or lengthwise the root, root hair curlings, curves, etc. For statistic calculations of the results of the measurements, total number of measurements in all experiments was taken.

Results

Analysis of LPPCs and PSLCs, their activity in inhibition test resulted in agglutination of erythrocytes

Table 2 displays data on the LPPC and PSLC content, their activity in inhibition of hemagglutination test of rabbit erythrocytes with WGA.

All LPPCs were shown to contain carbohydrates from 18.1 (strain 75) to 33.8% (strain S17), proteins

Tabl. 2. Composition of LPPCs and PSLCs isolated from *A.brasilense* strains and data on their interaction with WGA

Strain	Complex	Content, % (per weight)				Concentration needed for 50% inhibition, of WGA, mg/ml
		Carbohyd-rates	Proteins	Nucleic acid	KDO	
Sp 7	LPPC	27.0	19.3	0.1	3.0	5.0
	PSLC	50.0	8.0	0.8	0.3	5.4
Sp 245	LPPC	24.8	22.1	not found	1.1	4.0
	PSLC	57.0	12.0	1.7	0.5	4.2
S 17	LPPC	33.8	27.6	not found	2.0	3.0
	PSLC	48.4	12.3	6.7	0.7	4.3
8	LPPC	23.9	23.0	0.2	3.0	9.5
	PSLC	48.1	4.2	0.5	0.5	4.7
15	LPPC	28.5	17.9	1.1	2.0	no inhibition up to 8
	PSLC	63.2	7.6	5.0	0.5	no inhibition up to 8
55	LPPC	22.4	20.5	0.5	1.5	7.7
	PSLC	47.8	15.2	0.8	0.3	5.2
75	LPPC	18.1	28.1	0.1	1.8	5.1
	PSLC	46.3	15.3	0.8	0.6	7.7
80	LPPC	25.3	18.8	0.3	3.0	4.2
	PSLC	45.8	15.9	0.5	0.6	3.6

- from 17.9 (strain 15) to 37.6% (strain S17), KDO - from 1.1 (strain Sp245) to 30% (strains Sp7, 8, 80), and minor amounts of nucleic acids (up to 1.1%) and phosphorus (up to 0.5%).

Compared to the LPPCs, the PSLCs of the same strain contained more carbohydrates, varying from 45.8 (strain 80) to 63.2 (strain 15) and less protein - from 4.2 (strain 8) to 15.9 (strain 80) and KDO - from 0.3 (strains Sp7 and 55) up to 0.7 (strain S17). Nucleic acid impurities in the preparations of the PSLC were higher than that of the LPPC, phosphorus content was lower, varying from 0.1 to 0.3%.

The LPPCs and PSLCs preparations with one exception of low-active LPPC and PSLC from the strain 15, demonstrated similar activity in a 50% inhibition of the agglutination test of trypsinized rabbit erythrocytes.Concentrations, inducing the above effect,vary within the range of 3.6-9.5 mg/ml.

<u>The effect of LPPCs and PSLCs on deformation of wheat root hairs</u>

LPPCs and PSLCs, in a concentration range from 0.1 to 0.4 mg/ml were active with respect to induce

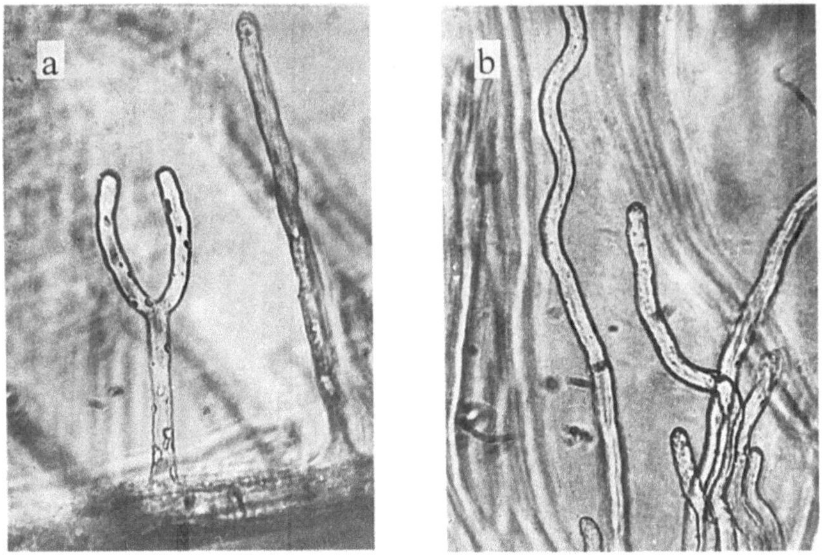

Fig. 1. Light photomicrograph of root hair deformations in 2-days-old wheat seedlings (*Triticum aestivum* cv. Saratovskaya 29) treated with PSLC from *A.brasilense* Sp245. (a) Equally branched root hair («tuning fork»); (b) A wavy appearance.

stimulation of root hair deformations. These concentrations caused various types of root hair deformations, including symmetric U-type branchings («tuning forks») and unsymmetric root hairs branchings, as well as other deformations. Examples of root hair deformations as affected by LPPCs and PSLCs are presented in Fig.1.

Quantitative data on the effect of the LPPCs and PSLCs isolated from various *A.brasilense* strains are listed in Table 3. LPPCs and PSLCs activities of each strain were of the same order.

Reliable effects ($P < 0.01$) in the order of increasing activity were observed for the LPPCs isolated from 15, 80, Sp7, 8 and 75 strains and for the PSLCs isolated from strains 55, 8, 80, Sp7, 75 and 15. In the set under consideration high activity was registered for LPPC isolated from strain 8 (15.3 $\pm$ 4.8), PSLC from strain 15 (17.0 $\pm$ 2.9) and both complexes from strain 75 (16.0 $\pm$ 3.0 and 16.9 $\pm$ 2.9). Strain Sp245 revealed low activity. Both complexes isolated from strain S17 and LPPC from Sp245 and 55 strains did not show reliable activity.

Monosaccharide content and some properties of PSs isolated from PSLCs

All the PSs isolated from PSLCs were acidic, formed transparent aqueous solutions

Tabl. 3. Effect of LPPCs and PSLCs of *A.brasilense* on the frequency of deformed root hairs in wheat (cv. Saratovskaya 29)

Strain	Complex	The sum of different kinds of root hair derormations/cm of root (a number of measuruments)				P
		Treated		Untreated		
Sp7	LPPC	10.9 ± 2.1	(43)	1.4 ± 0.8	(23)	< 0.01
	PSLC	10.9 ± 3.0	(45)	1.3 ± 1.0	(11)	< 0.01
Sp245	LPPC	1.8 ± 0.5	(34)	1.1 ± 0.4	(10)	> 0.05
	PSLC	5.5 ± 1.0	(20)	3.1 ± 1.3	(10)	< 0.05
S17	LPPC	1.3 ± 0.2	(30)	1.4 ± 2.2	(7)	> 0.05
	PSLC	4.2 ± 1.7	(24)	2.4 ± 1.6	(7)	> 0.05
8	LPPC	15.3 ± 4.8	(45)	2.2 ± 1.1	(18)	< 0.01
	PSLC	5.9 ± 1.2	(27)	2.4 ± 0.3	(15)	< 0.01
15	LPPC	6.7 ± 1.4	(39)	2.7 ± 0.7	(22)	< 0.01
	PSLC	17.0 ± 2.9	(42)	4.9 ± 1.6	(24)	< 0.01
55	LPPC	3.4 ± 0.8	(43)	2.4 ± 1.0	(25)	> 0.05
	PSLC	5.0 ± 1.6	(19)	1.6 ± 1.2	(9)	< 0.01
75	LPPC	16.0 ± 3.0	(22)	2.4 ± 1.2	(7)	< 0.01
	PSLC	16.9 ± 2.9	(20)	5.2 ± 2.9	(11)	< 0.01
80	LPPC	7.5 ± 1.1	(34)	3.4 ± 0.5	(18)	< 0.01
	PSLC	6.4 ± 0.8	(33)	2.7 ± 0.9	(14)	< 0.01

Confidence intervals are given for 95% reliability

and had molecular mass ranging from 17 to 500 kDa. Some PSs (PS-1, PS-2, PS-3, PS-4) were positive in charge of specific rotation, the other ones (PS-6, PS-7) were negative (see, table 4). The PSs contained Rha, GalA which taken together were quantitatively predominant in all PSs, with the exception of PS-3. All PSs were found to contain GlcN, though in low amounts. Most PSs were found to contain Gal and GalN. Besides, PS-1 contained Fuc, and PS-3 - Man.

PSs differed essentially in inhibition test of the agglutination of erythrocytes, and on the basis of increasing activity were aligned in the following row, beginning from inactive PS-4, PS-1, PS-3, PS-5, PS-6, PS-2, PS-7.

The activity of all PSs in reaction of direct interaction with WGA, was expressed in quantities of one order (Table 4), PS-5 and PS-7 and PS-2, PS-3 manifested themselves somewhat more active.

Table 4. Physico-chemical properties and monosaccharide composition of polysaccharides isolated from polysaccharide-lipid complexes of surface substances of *Azospirillum brasilense* cells

Strain	PS	$[\alpha]_D$,° /C,mg/ml in water/	Mr, kDa	C*, mM	PS**, titer	Monosaccharides, % ***						
						Fuc	Rha	Gal	Man	GalA	GlcN	GalN
Sp 7	PS-1	+ 45.9 /0.015/	20	0.28	8	10.5±0.5	51.5±3.8	17.6±0.6		16.6±0.5	1.5	tr.
Sp 245	PS-2	+ 81.2 /0.0175/	500	0.008	16		77.2±1.2	tr.		1.5±0.2	tr.	7.4
S 17	PS-3	+ 37.3 /0.014/	20	0.16	16		4.1±0.5	11.0±1.5	63.0±2.1	11.9±0.1	5.6	1.2
15	PS-4	+ 90.5 /0.013/	17	no inhibit.	4		84.4±1.2			13±1.3	tr.	tr.
75	PS-5	-56.5 /0.013/	500	0.06	32		72.1±0.8	24.4±2.0		2.9±0.1	1.2	4.4
Sp 107	PS-6	-5.8 /0.0265/	500	0.014	8		72.1±1.9			8.1±0.8	1.1	5.4
Sp 246	PS-7	-75.0 /0.0095/	200	0.005	32		82.9±1.8			7.0±0.4	4	tr.

*Concentration needed for 50% inhibition of WGA; ** Titer of PS solution (0.25%) in the precipitation reaction with WGA is given; *** Confidence intervals are given for 95% reliability. PS-5 contains xylose, 3.0. Amounts less than 1% are considered as traces (tr.)

Discussion

The work presented deals with the isolation of HMWC, localized on a cell periphery and loosely bound to the surface. These substances were arbitrarily called LPPC and PSLC (Konnova *et al.*, 1994) until their nature is revealed in details. They were weakly bound to the cell surface, as it was shown earlier (Konnova *et al.*,1994), and in the process of cultivation on a liquid synthetic medium, they passed from the cell surface into the environment culture liquid as a result of partial dissolving.

LPPC and PSLC obtained by means of washing-off with physiological solution from the cell surface, separated from culture liquid (Konnova *et al.*, 1994) were used for experiments. Complexes isolated from cell surface were proposed to be more natural. That was the reason why the LPPC and PSLC, presented in a culture liquid, were not used in the experiments.

In the process of gel-filtration on a Sepharose CL-4B column all LPPCs were eluted with void volume as turbid solutions. Hence it follows that in aqueous solutions LPPCs form aggregated particles with Mr about 5 x 10^6 Da. In all cases proteins were the abundant (17.9 - 28.1%) components of the LPPCs. LPPCs also contained carbohydrates and lipids (Konnova *et al.*, 1994), KDO and nucleic acid and phosphorus impurities. It is well known that KDO is an integral part of LPS of gram-negative microorganisms (Rietschel *et al.*, 1987) and the presence of the above component in 1.1-3.0% gives all the reason to believe that LPS was a constituent part of the LPPCs.

When gel filtered, PSLCs were eluted following LPPCs as turbid solutions. PSLCs particles were also aggregated, although mass of the aggregated particles was lower as compared with the LPPCs. Physico-chemical properties of the PSLCs solutions were, probably, conditioned by the fact that both hydrophylic - polysaccharide and hydrophobic - lipids components composed these complexes (Konnova *et al.*, 1994). PSLCs had higher PS content (45.8-63.2%) than LPPCs (18.1 - 33.8%). PSLC were found to involve less protein and KDO, than LPPC. The LPPC were supposed to incorporate small quantities of the LPPC, undetectable by the used methods.

The reliable ($P < 0.01$) increase in the number of root hair deformations per 1 cm of wheat root (cv.Saratovskaya 29) was induced by most preparations of LPPCs and PSLCs, as it is evident from Table 3. LPPCs and PSLCs from Sp245, 55, 80 strains associated with wheat appeared to be less active as compared to the complexes isolated from Sp7, 15, and 8 strains associated with wild cereals, except for strain 75. LPPCs and PSLCs from *A.brasilense* 75 strain, isolated from wheat seedlings (cv.Saratovskaya 29) stand out by their activity among the set of complexes (Pozdnyakova *et al.*, 1988). The revealed fact suggests the idea of *Azospirillum* strain specificity in the particular example cited above and correlates well with the previously described observations, that testify for the existence of such specificity in azospirilla (Bashan and Levanony, 1990; Jain and Patriquin, 1984; Jain and Patriquin, 1985; Baldani *et al.*,1983).

One of the possible mechanisms of *Azospirillum*-cereal root hair interaction can be presumably obtained with the help of macroassociant lectins.

Interaction of *Azospirillum* cell surface with various lectins was shown (Yagoda-Shagam *et al.*, 1988). WGA was among the lectins attached to the bacterial cell surface. Since WGA and WGA-like proteins were detected on the surface of wheat and roots of other cereals (Mishkind *et.al.*, 1980; Mishkind *et.al.*, 1983; Stinissen *et.al.*, 1985), they were of particular importance (Del Gallo *et.al.*, 1989; Del Gallo and Haegi, 1990; Haegi and Del Gallo, 1991).

Most of the *A.brasilense* strains under study are associated with wheat (Table 1). Type strain *A.brasilense* Sp7 isolated from roots of *Digitaria* has definite affinity to wheat, being capable of adsorbtion to wheat roots (Bastarrachea *et al.*, 1988; Michiels *et al.*, 1991).

Based on all this, as well as on our own results,we tried to find out the role of one of the possible ways for realization of these interactions. GlcN specifically interacts with WGA. This hapten has been detected in the composition of the complexes. That is why affinity of LPPCs and PSLCs preparations to WGA was investigated. From Table 2 it follows that both LPPCs and PSLCs from the strains of azospirilla under study (except for strain 15) exhibited to WGA affinity of the same order. Agglutinating erythrocytes themselves, LPPC and PSLC, isolated from strain 15, were incapable of lectin activity inhibition.

Taking into account the results obtained and based on Dazzo's lectin hypothesis (Dazzo and Hubbel, 1975; Dazzo, 1984), LPPCs and PSLCs complexes isolated from the cell surface may be proposed to play an important part in early stages of bacterial recognition and attachment to plant root surface. As PSs are incorporated into complexes, we decided to find out whether they are an active «starting point» of these complexes.

PSs, isolated from PSLCs of Sp7, Sp245, S17, 75, Sp107 and Sp246 were taken as examples. Agglutination test of trypsinized rabbit erythrocytes as affected by WGA (Table 4) appeared to be inhibited by the corresponding PS-1, PS-2, PS-3, PS-5, PS-6, and PS-7. Taking into account WGA specificity, it may be concluded that GlcN comprising PSs, was in N-acetylated form and determined PSs activity. All the PSs, except for PS-4 isolated from strain 15, inhibited WGA. 50% inhibition of WGA activity required PSs in 1-9.8 mg/ml (0.005-0.28 mM) and free GlcNAc required in 14 mg/ml (63 mM). Thus, PSs, except for PS-4, were more active as compared to GlcNAc, despite low GlcN content in PS (0.8 - 5.6%). This statement may be considered as a manifestation of high PSs affinity to WGA.

In precipitation test WGA pre-inhibition by GlcNAc completely suppressed the formation of precipitin rings for PS-2, PS-3 and PS-7, isolated from the PSLCs isolated from *A.brasilense* Sp245, S17, Sp246 strains and partly for PS-1, PS-4 and PS-6 from the PSLCs of Sp7, 15 and Sp107 strains. The data obtained permit one to suggest specific interaction in the first case and the existence of unspecific ways of WGA-PS interaction besides specific interactionin the second case.

Low GlcN content in most PS preparations could hardly be explained based on regularity of structures of microbial PSs. The above monosaccharide residue is probably bound to the main PS chain in an irregular way. Moreover it is not improbable that PSs under study are a mixture of two or more PSs, with similar physical properties, and they could not be separated by means of the applied methods. In one of the proposed minor PSs, the GlcNAc content , may be particularly more abundant.

Acknowledgements

This work was partly supported by grant from Russian Fund of Fundamental Investigations (project N 94-04-11206-a). We are grateful to Dr. D.Janssens (Belgium), Dr. J.Döbereiner (Brazil), Dr. A.N.Lahiri (India), Dr.V.V.Nikiforov and Dr. L.S.Fedorova (this institute) for providing *A.brasilense* strains, Dr. G.I. Stadnik (this institute for quantitative analysis of amino sugars and G.G. Masevkina (this institute) for help in the preparation of the manuscript.

References

Baldani VLD, Alvarez MAdeB, Baldani JI, Döbereiner J (1986) Establishment of inoculated *Azospirillum* spp. in the rhizosphere and in roots of field grown wheat and sorghum. Plant Soil 90: 35-46

Baldani VLD, Baldani JI, Döbereiner J (1983) Effects of *Azospirillum* inoculation on root infection and nitrogen incorporation in wheat. Can J Microbiol 29: 924-929

Bashan Y, Levanony H (1990) Current status of *Azospirillum* inoculation technology: *Azospirillum* as a challenge of agriculture. Can J Microbiol 36: 591-608

Bashan Y, Ream Y, Levanony H, Sade A (1989) Nonspecific responses in plant growth, yield, and root colonization of noncereal crop plants to inoculation with *Azospirillum brasilense* Cd. Can J Bot 67: 1317-1324

Bastarrachea F, Zamudio M, Rivas R (1988) Non-encapsulated mutants of *Azospirillum brasilense* and *Azospirillum lipoferum*. Can J Microbiol 34: 24-29

Beeley JG (1985) Glycoproteins and proteoglycan techniques. Elsevier Amsterdam New York Oxford: 327-333

Berenblum I, Chain E (1938) An improved method for the colorimetric determination of phosphate. Biochem J 32: 295-298

Dazzo FB Bacterial adhesion to plant root surfaces. Marshall KC (ed) (1984) Microbial adhesion and aggregation. Springer Berlin Heidelberg New York Tokyo: 85-93

Dazzo FB, Hubbell DH (1975) Cross-reactive antigens and lectin as determinants of symbiotic specificity in the *Rhizobium*-clover association. Appl Microbiol 30: 1017-1033

Del Gallo M, Haegi A (1990) Characterization and quantification of exocellular polysaccharides in *Azospirillum brasilense* and *Azospirillum lipoferum*. Symbiosis 9: 155-161

Del Gallo M, Negi M, Neyra CA (1989) Calcofluor - and lectin-binding exocellular polysaccharides of *Azospirillum brasilense* and *Azospirillum lipoferum*. J Bacteriol 171: 3504-3510

De-Polli H, Bohlool BB, Döbereiner J (1980) Serological differentiation of *Azospirillum* species belonging to different host-plant specificity groups. Arch Microbiol 126: 217-222

Dubois M, Gilles KA, Hamilton JK, Rebers PA, Smith F (1956) Colorimetric method for determination of sugar and related substances. Analyt Chem 28: 350-356

Fahraeus G (1957) The infection of clover root hairs by nodule bacteria studied by simple glass slide technique. J Gen Microbiol 16: 374-381

Haegi A, Del Gallo M *Azospirillum*-plant interaction: a biochemical approach.Polsinelli M, Materassi R, Vincenzini M (eds) (1991) Nitrogen fixation. Kluwer Acad Publ Dordrecht Boston London: 147-153

Halverson LJ, Stacey G (1986) Signal exchange in plant-microbe interactions. Microbiol Rev 50: 193-225

Jain DK, Patriquin DG (1984) Root hair deformation, bacterial attachment, and plant growth in wheat-*Azospirillum* associations. Appl Environ Microbiol 48: 1208-1213

Jain DK, Patriquin DG (1985) Characterisation of a substance produced by *Azospirillum* which causes branching of wheat root hairs. Can J Microbiol 31: 206-210

Karkhanis YD, Zeltner JY, Jackson JJ, Carlo DJ (1978) A new and improved microassay to determine 2-keto-3-deoxyoctonate in lipopolysaccharide of gram-negative bacteria. Anal Biochem 85: 595-601

Kato G, Maruyama Y, Nakamura M (1980) Role of bacterial polysaccharides in the adsorption process of the *Rhizobium*-pea symbiosis. Agric Biol Chem 44: 2843-2855

Konnova SA, Makarov OE, Skvortsov IM, Ignatov VV (1994) Isolation, fractionation and some properties of polysaccharides produced in a bound form by *Azospirillum brasilense* and their possible involvement in *Azospirillum*-wheat root interactions. FEMS Microbiol Letters 118: 93-99

Lowry OH, Rosebrough NJ, Farr AL, Randall RJ (1951) Protein measurement with the Folin phenol reagent. J Biol Chem 193: 265-275

Matveev VYu, Petrova LP, Zhuravleva EA, Panasenko VI (1987) Peculiarities of dissociation in *Azospirillum brasilense* Sp7 cultures. Molecul Gen Microbiol Virusol (Russ) 8: 16-18

Michiels KW, Croes CL, Vanderleyden J (1991) Two different modes of attachment of *Azospirillum brasilense* Sp7 to wheat roots. J Gen Microbiol 137: 2241-2246

Michiels K, Vanderleyden J, Van Gool A (1989) *Azospirillum*-plant root associations: A review. Biol Fertil Soils 8: 356-368

Mishkind M, Keegstra K, Palevitz BA (1980) Distribution of wheat germ agglutinin in young wheat plants. Plant Physiol 66: 950-955

Mishkind ML, Palevitz BA, Raikhel NV, Keegstra K (1983) Localization of wheat germ agglutinin-like lectins in various species of the *Gramineae*. Science 220: 1290-1292

Pozdnyakova LI, Kanevskaya SV, Levanova GF, Barysheva NN, Pilipenko TYu, Bogatyrev VA, Fedorova LS (1988) Taxonomic studies of *Azospirillum* isolated from cereals in the Saratov region. Microbiol (Russ) 57: 275-278

Rietchel ETh, Brade L, Schade U, Seydel U, Zahringer U, Kusumoto S, Brade H Bacterial endotoxins: properties and structure of biologically active domains. Schrinner E, Richmond MH, Seibert G, Schwarz U (eds) (1987) Surface structures of microorganisms and their interactions with the mammalian host. VCH: 1-41

Spirin AS (1958) Spectrofotometric determination of the total amount of nucleic acids. Biokhimia(Russ) 23: 656-662

Stinissen HM, Chrispeels MJ, Peumans WJ (1985) Biosynthesis of lectin in roots of germinating and adult cereal plant. Planta 164: 278-286

Tarrand JJ, Krieg NR, Dobereiner J (1978) A taxonomic study of the *Spirillum lipoferum* group, with descriptions of a new genus, *Azospirillum* gen nov. and two species, *Azospirillum lipoferum* (Beijerinck) comb.nov. and *Azospirillum brasilense* sp.nov. Can J Microbiol 24: 967-980

Yagoda-Shagam J, Barton LL, Reed WP, Chiovetti R (1988) Fluorescein isothiocyanate-labelled lectin analysis of the surface of the nitrogen-fixing bacterium *Azospirillum brasilense* by flow cytometry. Appl Environ Microbiol 54: 1831-1837

Continuous Culture Application in Physiological Investigations on Diazotrophic Bacteria

Joerg Ueckert and Istvan Fendrik
Institute of Biophysics
University of Hannover
Herrenhaeuser Str. 2
30419 Hannover
Germany

Abstract

Benefits of continuous culture application in the bioreactor for research on diazotrophic bacteria are outlined. Topics include determination of basic growth parameters and enzymatic activities as well as assessment of oxygen-induced metabolic effects. Immobilization of bacteria and creation of spatial heterogeneity allows modelling of the rhizosphere in the lab scale. Finally potential industrial applications are pointed out.

Introduction

Research on diazotrophic plant-associated bacteria has increased significantly after its potential value for agriculture became apparent. Field studies assessing the beneficial impact on plant growth after inoculation with rhizosphere-bacteria (PGPR) often showed variabilities which have limited commercial use of PGPRs. Basic requirement for model studies is therefore determination of the physiological state under conditions assumingly present in the habitat of interest. Bacterial metabolism is commonly studied by biochemical means or batch cultures, of which results are affected by nutrient depletion and toxic products. An important tool to master the setbacks of traditional approaches and to model rhizosphere conditions is continuous fementor cultivation. Experimental parameters can be maintained constant at a predetermined level (steady state), according to environmental conditions to be simulated. Bioreactor designs can especially be modified to meet experimental requirements, e.g. immobilization of bacteria in the rhizosphere (Ueckert et al., 1992). The chemostat, which is widely used for continuous culture studies is briefly characterized as follows. Fresh nutrient solution is supplied at a constant flow rate (D). The bacterial culture grows constantly at the same rate ($\mu = D$; $\mu < \mu_{max}$) and growth is limited only by a selectable substrate, usually the

NATO ASI Series, Vol. G 37
Azospirillum VI and Related Microorganisms
Edited by I. Fendrik et al.

Table 1. Continuous culture studies of diazotrophic bacteria (Overview)

Species	**Cultivation**	**Investigations**	**References**
Acetobacter diazotrophicus	minimal medium	glucose metabolism, gluconic acid product.	Attwood et al., 1991
Arthrobacter spp.	batch and cont. culture	pO_2 - effects on growth	Cacciari et al., 1979
Arthrobacter giacom. *Azospirillum brasilense*	single and mixed binary culture	nitrogenase activity, competition studies	Cacciari et al., 1986
Arthrobacter fluorescens	batch and continuous culture	regulation of alanine dehydrogenase	Cacciari et al., 1994
Azotobacter spp.	batch and continuous culture	hydrogenase effects on diazotrophic growth	Aguilar et al., 1985
Azospirillum spp.	suspension culture	pH-, temp.- effects on Vitamin B production	Dahm et al., 1993
Azospirillum spp.	suspension culture	µmax and Ks	Fritzsche et al., 1991*b*
Azospirillum spp. and Kallar grass isolates	suspension culture	pO_2 - effects on growth and PHB content	Hurek et al., 1987*a*
Azospirillum brasilense strain Cd	suspension culture	diff. pO_2 on carotinoids and enzymatic activity	Nur et al., 1982
Azosp. brasilense Sp7 and rice isolates	medium with and without NH_4Cl	nitrogenase activity	Fritzsche et al., 1990
Azospirillum brasilense Sp7	batch and continuous culture	carotinoid effects on O_2-tolerance of N_2ase	Hartmann et al., 1988
Azosp. brasilense Sp7	suspension culture	µmax , Ks , PHB, N_2ase	Kloss et al., 1983
Azospirillum brasilense Sp7 and Sp245	suspended and immobilized single cultures	nitrate reductase activity	Ueckert et al., 1991
Azospirillum brasilense Sp7 and Sp245	suspended and immobilized single cultures	pH - effects on nitrate reductase activity	Ueckert et al., 1994
Azospirillum brasilense Sp7	batch and continuous culture	denitrification, nitrite resp., nitrogen fixation	Zimmer et al., 1984
Azospirillum spp. and Kallar grass isolates	medium with NH_4Cl present	pO_2 - effects on growth and PHB content	Hurek et al., 1987*b*
Azospirillum spp. and rice isolates	mixed suspension cultures	enrichment and competition studies	Kloss et al., 1986
Kallar grass isolates	suspended,immobilized single and mixed culture	growth, N_2ase activity, competition studies	Ueckert et al., 1992
Rice root isolates	single, mixed, +N -N, cultures in a Gradostat	NH_4 -,O_2- growth effects, competition studies	Fritzsche et al., 1991*a*

carbon source. Parameters like temperature, dissolved oxygen levels, pH and the stirring rate, which determines the oxygen transfer rate in suspended culture, can be adjusted and varied. Substancial physiological data like the maximal growth rate (μ_{max}) or the substrate affinity (K_s) under different conditions can easily be determined.

Determination of basic growth parameters

According to Lynch (1982) rhizosphere organisms are carbon limited in their natural habitat. The substrate affinity to the carbon source can be determined and plays an important role in establishment of each strain in its ecological niche. Strains with a high affinity (low K_s-value) have a competitive advantage in the habitat. Determination of μ_{max} is less important for modelling, since competition in natural habitats occurs at submaximal growth rates.

Kloss (1983) employed a bioreactor as one of the first for physiological characterization of diazotrophic *Azospirillum brasilense* Sp7 and determined μ_{max}, K_s, nitrogenase activity, PHB formation and protein exudation. Fritzsche and coworkers (1991*b*) compared basic growth parameters of several diazotrophic isolates obtained from rice and wheat. In N-free media maximum growth rates of microaerobic diazotrophs were within a small range of 0.11 h^{-1} to 0.21 h^{-1}. K_S- values, in contrast, showed 40fold differences under optimal conditions. Even within species remarkable variations were obvious, of which knowledge will help to choose suitable strains for field inoculation studies.

Modelling the rhizosphere

Especially in the rhizosphere bacteria grow on available surfaces. This has been demonstrated for instance by Reinhold and Hurek (1989) for the Kallar grass association. To simulate aggregation and attached growth in the lab scale, fermentors have been specifically redesigned to allow growth of bacteria entrapped in alginate gel beads. The gel matrix enables gases, metabolites and substrates to diffuse in and out without limitations. The medium surrounding the carriers could be maintained cell-free.

Entrapment of *Azospirillum brasilense* strains Sp7 and Sp245 in gel beads showed differences in nitrate reductase activity by a factor of three between strains (Ueckert et al., 1991) and caused in comparison to suspended culture an increase in enzymatic activity by 40 % (Ueckert et al., 1994). The pH - range of maximum enzymatic activity could be extended, indicating a protective effect of aggregated growth in the rhizosphere.

Competition studies with diazotrophic Kallar grass isolates in mixed entrapment cultures in gel beads indicated population shifts due to immobilization in comparison to suspended culture. In addition ammonia accumulation in the supernatant has been found to be at least twice as high for immobilized cells (Ueckert et al., 1992).

Significant changes in bacterial metabolism and in competitive capacity after immobilization suggest a more close simulation of natural habitats for experiments in the lab scale. This will allow improved assessment of mechanisms in diazotrophic associations or of potential agricultural benefits after inoculation.

To model the influence of spatial heterogeneity of solutes and gases in the rhizosphere, Fritzsche et al. (1991*a*) constructed a five-vessel gradostat (multistage chemostat) and studied pO_2- and Δ NH_4Cl effects on pure and mixed cultures of diazotrophic and non-fixing bacteria. The diazotrophic isolates could successfully compete with the non-diazotrophic ones, even in presence of ammonia in the culture medium. This approach led to a proposal for nitrogen-transfer to non-fixing organisms.

Mixed culture- and competition-studies

The natural habitat is characterized as well by cell to cell contact and microbial competition and antagonism. In contrast to immobilized mixed culture approaches of previous chapter, most mixed culture- and competition-studies are usually performed in suspension.

Cacciari et al. (1986) studied the effects of dilution rate and carbon-limitation on the interaction between *Azospirillum brasilense* and *Arthrobacter giacomelloi* in single and binary continuous culture in nitrogen-free medium with succinate as C-source. It turned out that the ratio of organisms varied with varying growth rate. This binary continuous culture device for coinoculation proved to be a suitable tool to evaluate the extend to which N_2ase-activity and survival of *Azospirillum* can be influenced by the presence of organisms competing for the same limiting nutrient.

Kloss et al. (1986) used a similar approach to study competitive interactions of diazotrophic microorganisms. Enrichment of strains from soil as well as successful competition among *Azospirillum brasilense, A. lipoferum* and rice soil isolates largely depended on dilution rates and substrate concentrations applied, and basically followed the models of Monod (1949) and Veldkamp & Jannasch (1972).

Oxygen-effects and -tolerances

Oxygen-sensitivity of nitrogenase has produced several papers, stressing the fact that oxygen gradients occur around the rhizodermis. The oxygen-status of the rhizosphere can change dramatically due to flooding.

Oxygen tolerance studies under different culture conditions revealed the narrow O_2-range especially of microaerophilic diazotrophic bacteria grown on N_2 and differences among species or even strains.

Hurek et al. (1987*a*) studied O_2-tolerances of rhizoplane-isolates *Azospirillum lipoferum* Rp5 and *A. halopraeferens* as well as of two endorhizosphere-isolates from Kallar grass *Leptochloa fusca* L. Kunth, grown with N_2 in continuous culture. It has been found that oxygen tolerance of strains from Kallar grass was root zone specific. Strains predominating in the endorhizosphere still grew under nitrogen at pO_2 - levels already inhibitory to the rhizoplane-isolates. Hurek postulated that differences in oxygen tolerance profiles of isolates reflect adaptions to the microenvironment provided by the host plant.

Continuous culture studies on the same strains, but grown in presence of ammonia in the medium at different O_2 concentrations (Hurek, 1987*b*) showed no significant metabolic changes and could not explain different colonization patterns in the Kallar grass rhizosphere.

Different oxygen partial pressures were adjusted to elucidate parameters growth, respiration rate and nitrogen fixation of aerobic *Arthrobacter fluorescens* in batch and continuous culture (Cacciari et al., 1979). The pO_2 optima could be determined and compared with those of *Azotobacter* spp.

Growth-studies with carotinoid-overproducing mutants of *Azospirillum brasilense* Sp7 in batch and continuous culture under nitrogen-deficient conditions revealed a rise in carotinoid-production by increasing O_2-concentrations (Hartmann and Hurek, 1988). It was concluded that high levels of carotinoids slightly enhanced the oxygen tolerance, but could not extend the optimum pO_2 for nitrogen fixation to higher O_2-concentrations.

In contrast to the previous study Nur et al. (1982) studied effects of different dissolved oxygen tensions (DOT) in continuous culture on carotinoid production and enzymatic activity of *Azospirillum brasilense* Cd, but in presence of NH_4Cl in the medium. The highest carotinoid content has been observed at the highest DOT adjusted (0.18 atm). Succinate-oxidase activity increased consistantly with increasing DOT, while specific superoxide-dismutase activity increased sharply above 0.16 atm oxygen.

Studies on enzymatic activity

Though this topic is already part of the previous chapter, in this part the special suitability of continuous culture for determination of enzymatic activity is outlined. It has been found that under nitrogen fixation conditions most of the cellular N is present in form of protein, since no substantial amounts of combined N are released by bacteria (Fritzsche et al., 1991*b*). Protein content (and yield) serve so as a simple measure for quantification of nitrogenase activity.
The most striking study on nitrogenase activity and continuous culture has been published by Fritzsche and Niemann (1990). With *Azospirillum brasilense* Sp7 and two rice isolates and by using N-balancing and ^{15}N dilution techniques, simultaneous growth on ammonium chloride and N_2 as nitrogen source has been proven to occur. It could be shown that fixed nitrogen actually became part of the biomass. In conclusion it could be stated, that nitrogen fixation does not depend on the total absence of other N-sources.
Zimmer and coworkers (1984) focused on denitrification and nitrite respiration of *Azospirillum brasilense* Sp7 in anaerobic batch and continuous culture. Nitrate reductase activity was always present under anaerobic conditions. Nitrite formed was at high concentrations and pH critical for cell growth. Nitrite was quantitatively converted to nitrogen, without formation of free ammonia, nitric oxide or nitrous oxide. It has been suggested that *Azospirillum brasilense* has a cytochrome c, d-type of dissimilatory nitrite reductase.
The influence of hydrogenase in aerobic *Azotobacter chroococcum* under nitrogen-fixing, carbon-limiting conditions in continuous and batch culture has been investigated by Aguilar et al. (1985). The experiments with Hup^+ (uptake hydrogenase) parent and Hup^- mutant strain suggested a beneficial effect of hydrogenase upon initiation of diazotrophic growth under restricted C- and energy-source.
Diazotrophic *Arthrobacter fluorescens* possesses the same mechanisms for ammonium assimilation as other N_2-fixing bacteria, but lacks a functional glutamate dehydrogenase (GDH). The formation of alanine from pyruvate and ammonium is catalyzed by alanine dehydrogenase. The study of Cacciari et al. (1994) examined the regulation of alanine dehydrogenase by ammonium and glucose of *Arthrobacter fluorescens* in batch and continuous culture. The results indicated that alanine dehydrogenase and glutamine synthetase activities were inversely related. In carbon-limited chemostat culture glucose limitation had a negative influence on dehydrogenase activity.

Potential commercial use of diazotrophic bacteria

Besides research diazotrophic bacteria have a potential to be used in bioreactors for industrial production, which is outlined in two examples.

Three *Azospirillum* strains were isolated by Dahm et al. (1993) from coniferous ectomycorrhizae and sporocarps of ectomycorrhizal fungi and subsequently grown in media with different pH (5.5 - 7.0) at different temperatures (10 - 26°C) for production of B-group vitamins (e.g. thiamin, biotin, nicotinic acid, riboflavin or panthotetic acid). Content and composition of vitamins in culture liquids depended on growth temperature, medium pH and the strain studied. No biotin has been detected, but thiamin was synthesized independently of temperature in the largest quantities at pH 5.5 and was not detected at all at high pH.

The second study describes the glucose metabolism of *Acetobacter diazotrophicus*. This diazotrophic, acid-tolerant acetic acid bacterium has been grown in glucose-limited minimal media in a chemostat at pH 3.5 and exhibited high rates of gluconic acid formation (Attwood et al., 1991). Enzyme activities in cell-free extracts suggested that glucose metabolism in *A. diazotrophicus* proceeds exclusicely via the hexose monophosphate pathway. Due to the fact that glucose-oxidation was less sensitive to low pH than glucose oxidation by *Gluconobacter oxydans* and not affected by high gluconic acid concentrations, the authors suggest consideration of *Acetobacter diazotrophicus* for industrial production of gluconic acid.

References

Aguilar OM, Yates MG and Postgate JR (1985) The beneficial effect of hydrogenase in *Azotobacter chroococcum* under nitrogen-fixing, carbon-limiting conditions in continuous and batch cultures. J Gen Microbiol 131:3141-3145

Attwood MM, Dijken-van JP and Pronk JT (1991) Glucose metabolism and gluconic acid production by *Acetobacter diazotrophicus*. J Ferment Bioeng 72:101-105

Cacciari I, Lippi D and Bordeleau M (1979) Effect of oxygen on batch and continuous cultures of a nitrogen fixing *Arthrobacter* sp. Can J Microbiol 25:746-751

Cacciari I, Del Gal, Ippoliti S, Pietrosanti T and Pietrosanti W (1986) Growth and survival of Azospirillum brasilense and *Arthrobacter giacomelloi* in binary continuous culture. Plant and Soil 90:107-116

Cacciari I, Lippi D and Pietrosanti T (1994) Regulation by ammonium and glucose of *Arthrobacter fluorescens* alanine dehydrogenase. Arch Microbiol 161:11-16

Dahm H, Rozycki H, Strzelczyk E and Li CY (1993) Production of B-group vitamins by *Azospirillum* spp. grown in media of different temperatures. Zentrbl Mikrob 148:195-203

Fritzsche C and Niemann E-G (1990) Nitrogen fixation in continuous culture with NH_4Cl-containing media. Appl Env Microbiol 56:1160-1161

Fritzsche C, Huckfeldt K and Niemann E-G (1991*a*) Ecophysiology of associative nitrogen fixation in a rhizosphere model in pure and mixed culture. FEMS Microb Ecol 1991:279-292

Fritzsche C, Ueckert J and Niemann E-G (1991*b*) Growth parameters of microaerobic diazotrophic rhizobacteria determined in continuous culture. In: Nitrogen Fixation. R. Polsinelli (ed). Kluwer Academic Publisher London

Hurek T, Reinhold B, Fendrik I and Niemann E-G (1987*a*) Root-zone-specific oxygen tolerance of *Azospirillum* spp. and diazotrophic rods closely associated with Kallar Grass. Appl Env Microbiol 53:163-169

Hurek T, Reinhold B and Niemann E-G (1987*b*) Effect of oxygen on ammonia-grown continuous cultures of *Azospirillum* spp. and diazotrophic rods closely associated with Kallar grass. Can J Microbiol 33:919-922

Hartmann A and Hurek T (1988) Effect of carotinoid overproduction on oxygen tolerance of nitrogen fixation in *Azospirillum brasilense* Sp7. J Gen Microbiol 134:2449-2455

Kloss M, Iwannek K-H and Fendrik I (1983) Physiological properties of *Azospirillum brasilense* SP7 in a malate limited chemostat. J Gen Appl Microbiol 29:447-457

Kloss M, Iwannek K-H, Fendrik I and Niemann E-G (1986) Enrichment of diazotrophic bacteria from rice soil in continuous culture. Plant and Soil 90:151-164

Lynch JM (1982) The rhizosphere. Burns G. and Slater J.H. Blackwell Scientific Publications Oxford

Monod J (1949) The growth of bacterial cultures. Ann Rev Microbiol 3:371-394

Nur I, Okon Y and Henis Y (1982) Effect of dissolved oxigen tension on production of carotenoids, poly-beta-hydroxybutyrate succinate-oxidase and superoxide-dismutase by *Azospirillum brasilense* Cd grown in continuous culture. J Gen Microbiol 128:2937-2943

Reinhold B and Hurek T (1989) Location of diazotrophs in the root interior with special attention to the kallar grass association. In: Skinner F.A., et al. Kluwer Academic Publishers Amsterdam

Ueckert J, Doebereiner J, Fendrik I and Niemann E-G (1991) Nitrate reductase activity of *Azospirillum brasilense* Sp7 and Sp245 V- and C-forms in continuous culture. pp. 249-253. In: Nitrogen Fixation. Polsinalli M. et al. (eds). Kluwer Academic Publisher London

Ueckert J, Niemann EG and Fendrik I (1992) Mixed continuous suspended and immobilized culture of diazotrophic isolates from root-free soil and the endorhizosphere of *Leptochloa fusca* L. Kunth. Symbiosis 13:75-84

Ueckert J, Niemann EG and Fendrik I (1994) pH-dependent activity changes of nitrate reductase upon immobilization of *Azospirillum brasilense* in continuous culture. pp. 351-354. In: Nitrogen fixation with non-legumes. NA Hegazi (ed). AUC Press Cairo

Veldkamp H and Jannasch HW (1972) Mixed culture studies with the chemostat. J Appl Chem Biotechnol 22:105-123

Zimmer W, Stephan MP and Bothe H (1984) Denitrification by *Azospirillum brasilense* Sp7. Arch Microbiol 138:206-211

Changes in Membrane Potential upon Chemotactic Stimulation of *Azospirillum brasilense*

Igor B. Zhulin*, Lisa E. Sarmiento, and Barry L. Taylor
Department of Microbiology and Molecular Genetics
Loma Linda University
Loma Linda CA 92350
U. S. A.

Introduction

Azospirillum brasilense is capable of aerotaxis (Barak *et al.*, 1982), redox taxis (Grishanin *et al.*, 1991), and chemotaxis towards different organic compounds (Barak *et al.*, 1983). The aerotactic response in *A. brasilense* was shown to be dependent on electron flow through the redox chain and consequent changes in the proton motive force (pmf) (Grishanin *et al.*, 1991). The functioning redox chain was shown to be responsible for producing several types of behavioral responses in bacteria including aerotaxis, phototaxis and some cases of chemotaxis. The proton motive force generated by electron transport is thought to be a parameter measured by the cell in order to produce these tactic responses (Taylor, 1983). Since *A. brasilense* lacks dedicated chemoreceptors or methyl-accepting chemotaxis proteins (Zhulin and Armitage, 1993), we propose that the main chemotaxis pathway in this bacterium might be dependent on pmf-sensing. In this paper, we present data supporting this proposal. Changes in chemotactic behavior of *A. brasilense* coincided with changes in membrane potential, the main component of the proton motive force.

*Permanent address: Laboratory of Bioenergetics, Institute of Biochemistry and Physiology of Plants and Microorganisms, Russian Academy of Sciences, Saratov, 410015 Russia

NATO ASI Series, Vol. G 37
Azospirillum VI and Related Microorganisms
Edited by I. Fendrik et al.

Materials and methods

Bacteria. A. brasilense Sp7 (ATCC 29145), a wild-type strain, was used throughout the study.

Growth conditions. A. brasilense was grown on defined malate medium at 30^{0}C and prepared for chemotaxis experiments as described previously (Grishanin *et al.*, 1991; Zhulin and Armitage, 1993).

Negative chemotaxis in a spatial assay. The chemical-in-plate method of Tso and Adler (1974) was used to determine negative chemotaxis in a spatial gradient of pentachlorophenol (PCP).

Temporal assay for chemotaxis. All assays were conducted using a Leitz dark-field videomicroscope. Frame-by-frame analysis of video recordings was carried out as previously described (Zhulin and Armitage, 1993).

Measurement of membrane potential. Tetraphenyl phosphonium ion (TPP^+) is a lipophilic cation that freely passes through the cytoplasmic membrane and equilibrates according to the membrane potential. The TPP^+ ion distributes between intra- and extracellular spaces according to the Nernst equation:

$$\Delta\Psi = (RT/zF)\ \ln(C_{out}/C_{in})$$

where $\Delta\Psi$ is the membrane potential with respect to the outside of the cells. C_{out} and C_{in} are the concentrations of TPP^+ in the external solution and inside the cells, respectively, and z is the valency (1 for TPP^+). R, T and F have their usual thermodynamic definitions. C_{in} was calculated from the decrease of the TPP^+ concentration in the external medium measured with a TPP^+-selective electrode. The TPP^+-selective electrode was constructed by the method of Kamo *et al.* (1979). All measurements were carried out at 30^{0}C and pH 7.5.

The cells were made permeable to TPP^+ by treatment with EDTA as previously described (Shioi *et al.*, 1988). In calculating membrane potential, a value of 5×10^{-15} l was used for the intracellular cell volume. This value was determined using assumptions suggested by Goodsell (1991).

Abbreviations: pmf, proton motive force; TPP^+, tetraphenyl phosphonium; EDTA, ethylenediaminetetraacetic acid; PCP, pentachlorophenol.

Results

Chemotaxis of A. brasilense *away from a protonophore.* Pentachlorophenol (PCP) is a lipophilic carrier which transports protons into the cell and therefore decreases the proton motive force. Fig. 1 shows negative chemotaxis of *A. brasilense* cells exposed to a spatial gradient of PCP.

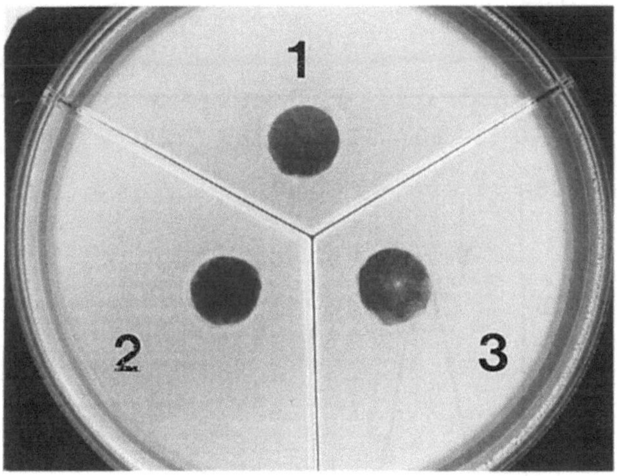

Fig. 1. Negative chemotaxis of *A. brasilense* in a spatial gradient of PCP. Concentration of PCP: 1, no PCP; 2, 10 μM; 3, 100 μM. 2 hours incubation.

PCP diffused from 3% agar into a bacterial suspension that was placed in a hole in the agar and formed a spatial gradient. Bacteria escaping from the repellent concentrated in the center of the hole and became visible as a white dot on a black background.
A temporal gradient assay was used to quantitate the negative chemotactic response. Fig. 2 shows the effect of PCP on cell speed and reversal frequency of an *A. brasilense* population in a temporal gradient. The cell speed gradually decreased upon addition of 20 μM PCP, indicating negative chemokinesis, with almost all the cells becoming non-motile after 5 min incubation (Fig. 2A). A negative chemotactic response was seen within the first minute of incubation with PCP, when reversal frequency increased from 0.33 to 0.62 s^{-1} (Fig. 2B).

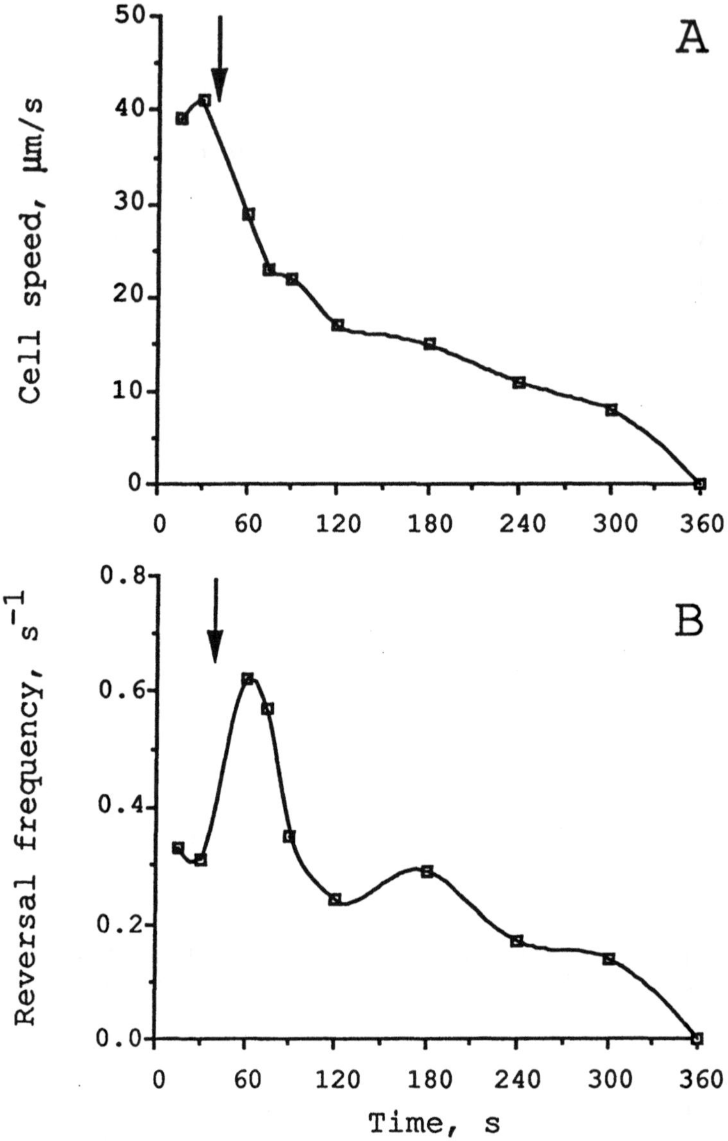

Fig. 2. Effect of PCP addition on the speed (A) and reversal frequency (B) of free-swimming *A. brasilense* cells. Concentration of PCP, 20 µM. Arrows show addition of PCP.

Chemotaxis of A. brasilense *towards malate.* Both positive chemotaxis and chemokinesis of *A. brasilense* towards malate have been previously demonstrated in both spatial and temporal gradients (Zhulin and Armitage, 1993). We found that the peak concentration for the positive chemotactic response of *A. brasilense* towards malate was 5-10 mM when measured in a temporal gradient assay (data not shown).

Changes in membrane potential upon stimulation of A. brasilense *with an attractant, a non-chemoeffector, and a repellent.* Fig. 3 shows the effect of an attractant (malate), a non-chemoeffector (glycerol), and a repellent (PCP) on membrane potential in starved cells of *A. brasilense*. No changes in membrane potential were observed when a non-chemoeffector, glycerol was added to either starved cells ($\Delta\Psi$ = -190 mV) (data not shown) or cells stimulated with malate ($\Delta\Psi$ = - 90 mV) (Fig. 3). Addition of an attractant, malate, to starved cells caused a decrease in membrane potential of up to -90 mV (Fig. 3). Similar results were obtained with another strong attractant, succinate (data not shown). In non-starved cells, neither glycerol nor malate caused any changes in membrane potential. The value of membrane potential in non-starved cells was about -90 mV.

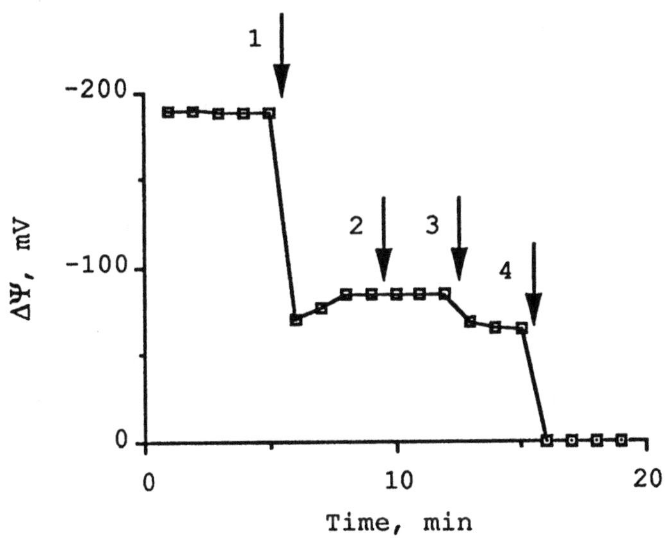

Fig. 3. Effect of an attractant, a non-chemoeffector, and a repellent on membrane potential in starved *A. brasilense* cells. Cells were incubated in HEPES buffer without any carbon source for 30 min. Arrows indicate addition of stimuli: 1, 10 mM malate; 2, 10 mM glycerol; 3, 20 μM PCP; 4, 30 μM PCP.

PCP caused a decrease in membrane potential of up to 60-70 mV but only in a very narrow concentration range, 10-25 μM. Membrane potential totally collapsed after addition of 30 μM PCP. PCP caused a negative chemotactic response over the 10-25 μM concentration range, with all the cells becoming non-motile at 30 μM PCP.

Discussion

A. brasilense showed positive chemotaxis and positive chemokinesis towards malate and negative chemotaxis and negative chemokinesis away from pentachlorophenol. The chemokinetic response in bacteria is known to be pmf-dependent, since the bacterial flagellar motor is powered by pmf (Manson *et al.*, 1977; Glagolev and Skulachev, 1978). The data presented here indicates that the chemotactic response in *A. brasilense* might also be pmf-dependent.

Membrane potential in neutrophilic bacteria is the main component of the proton motive force at pH 7.0-7.5 (Kashket, 1985), therefore changes in membrane potential indicate similar changes in pmf. Both chemoeffectors caused changes in membrane potential while a non-chemoeffector, glycerol did not. Chemoeffectors changed membrane potential and caused a chemotactic response at the same concentration range. In the presence of exogenous substrate (non-starved cells), bacteria had a membrane potential of -90 mV and did not respond to an attractant. In starved cells, the membrane potential was -190 mV and addition of an attractant decreased it to approx. -90 mV. Addition of a repellent decreased membrane potential to below -90 mV in both starved and non-starved cells. It is possible that the value of approx. -90 mV is optimal for membrane potential in *A. brasilense*. Compounds that change pmf towards optimum appear to be attractants. Compounds that change pmf away from optimum are repellents. Non-chemoeffectors do not alter pmf. More compounds should be tested in order to prove this hypothesis.

Our findings along with the recent discovery that pmf-sensing is involved in phototaxis of *Halobacterium salinarium* (Bibikov *et al.*, 1993) support the hypothesis that direct sensing of the proton motive force is an independent pathway in bacterial signal transduction (Taylor *et al.*, 1979; Glagolev, 1980; Laszlo and Taylor, 1981; Sherman *et al.*, 1981; Taylor, 1983; Glagolev, 1984; Shioi and Taylor, 1984; Shioi *et al.*, 1988).

We are grateful to M. S. Johnson for reading the manuscript and helpful discussion.
This work was supported by Public Health Service grant GM29481 (to B. L. T.) from the National Institute of General Medical Sciences.

References

Barak R, Nur I, Okon Y (1983) Detection of chemotaxis in *Azospirillum brasilense*. J Appl Bacteriol 53:399-403
Barak R, Nur I, Okon Y, Henis Y (1982) Aerotactic response of *Azospirillum brasilense*. J Bacteriol 152:643-649
Bibikov SI, Grishanin RN, Kaulen AD, Marwan W, Oesterhelt D, Skulachev VP (1993) Bacteriorhodopsin is involved in halobacterial photoreception. Proc Natl Acad Sci USA 90:9446-9450
Glagolev AN (1980) Reception of the energy level in bacterial taxis. J Theor Biol 82:171-185
Glagolev AN (1984) Bacterial $\Delta\mu H^+$-sensing. Trends Biochem Sci 9:397-400
Glagolev AN, Skulachev VP (1978) The proton pump is a molecular engine of motile bacteria. Nature 272:280-282
Goodsell DS (1991) Inside a living cell. Trends Biochem Sci 16:203-206
Grishanin RN, Chalmina II, Zhulin IB (1991) Behaviour of *Azospirillum brasilense* in a spatial gradient of oxygen and a 'redox' gradient of an artificial electron acceptor. J Gen Microbiol 137:2781-2785
Kamo N, Muratsugu M, Hongoh R, Kobatake Y (1979) Membrane potential of mitochondria measured with an electrode sensitive to tetraphenyl phosphonium and relationship between proton electrochemical potential and phosphorylation potential in steady state. J Membrane Biol 49:105-121
Kashket ER (1985) The proton motive force in bacteria: a critical assessment of methods. Annu Rev Microbiol 39:219-242
Laszlo DJ, Taylor BL (1981) Aerotaxis in *Salmonella typhimurium*: role of electron transport. J Bacteriol 145:990-1001
Manson MD, Tedesco P, Berg HC, Harold FM, van der Drift C (1977) A proton motive force drives bacterial flagella. Proc Natl Acad Sci USA 74:3060-3064
Sherman MYu, Glagolev AN (1981) Sensing of the proton motive force in *Escherichia coli* chemotaxis. FEMS Microbiol Lett 12:121-124
Shioi J, Taylor BL (1984) Oxygen taxis and proton motive force in *Salmonella typhimurium*. J Biol Chem 259:10983-10988
Shioi J, Tribhuwan RC, Berg ST, Taylor BL (1988) Signal transduction in chemotaxis to oxygen in *Escherichia coli* and *Salmonella typhimurium*. J Bacteriol 170:5507-5511
Taylor BL (1983) Role of proton motive force in sensory transduction in bacteria. Annu Rev Microbiol 37:551- 573
Taylor BL, Miller JB, Warrick HM, Koshland DE, Jr (1979) Electron acceptor taxis and blue light effect on bacterial chemotaxis. J Bacteriol 140:567-573
Tso WW, Adler J (1974) Negative chemotaxis in *Escherichia coli*. J Bacteriol 118:560-576
Zhulin IB, Armitage JP (1993) Motility, chemokinesis, and methylation-independent chemotaxis in *Azospirillum brasilense*. J Bacteriol 175:952-958

Physiological aspects of phytohormone production by *Azospirillum brasilense* Sp 7

Alexander Iosipenko and Vladimir Ignatov
Institute of Biochemistry and Physiology of Plants and Microorganisms
Russian Academy of Sciences
pr. Entuziastov 13 Saratov 410015
Russia

Abstract

The culture fluid of *A. brasilense* Sp 7 was found to contain 15 mg/L of indole-3-acetic acid and 267 μg/L of abscisic acid. In the course of 10 days of growth of *A. brasilense* Sp 7 on root exudate medium one can trace well-defined dynamics of cytokinin activity within a period of 4 to 6 days of culture growth. It was shown that zeatin and zeatin-riboside were absent in the culture fluid of *A. brasilense* Sp 7. *Azospirillum brasilense* strains isolated from cereals from the Saratov Region, when grown in a tryptophan-supplemented medium for 3 days, produced indole-3-acetic acid in amounts of 72.6 to 108.9 μg/mg dry matter of intact cells. No close correlation was shown between the biosynthetic activity of indole-3-acetic acid and N_2-fixation activity in 14 *Azospirillum* strains studied. A mathematical model was developed for indole-3-acetic acid synthesis by *A. brasilense* Sp 7 in relation to pH values, tryptophan concentration and temperature.

Introducton

One of the major points of the theory of associativeness is the study of trophical interactions in a plant-microbe system. Phytohormone synthesis by rhizosphere bacteria, along with nitrogen fixation, is a determining factor in successful realization of plant growth and development that result finally in higher crop yields. Nitrogen-fixing bacteria of the genus *Azospirillum* are potentially applicable as biofertilizers owing to their ability to excrete phytohormone-like substances such as auxins (Tien *et al.*, 1979; Falik *et al.*, 1989; Zimmer and Bothe, 1988; Horemans and Vlassak, 1985; Bar and Okon, 1993), cytokinins (Tien *et al.*, 1979; Conard and Stegemann, 1991), and abscisic acid (Omay *et al.*, 1993). The main objective of the present work was to elucidate the peculiarities of phytohormone synthesis by *Azospirillum*.

Materials and methods

The *Azospirillum* strains used were *A. brasilense* type strain Sp 7 (ATCC 29145) obtained from the Type Culture Collection, Institute of Agricultural Biotechnology, Russian Academy of Agricultural Sciences (Moscow); and 14 wild strains isolated from local cereals in the Genosystematics Laboratory of our institute. Bacteria were cultivated either in a synthetic malate medium (Döbereiner

NATO ASI Series, Vol. G 37
Azospirillum VI and Related Microorganisms
Edited by I. Fendrik et al.

and Day, 1976) supplemented with NH_4Cl (0.5 g/L), or in a RE (root exudates) medium proposed by the authors. Cells were grown to stationary phase on a shaker at 30°C. The activities of auxins were bioassayed with wheat (cv Saratovskaya 42) coleoptile segments (Nikolaeva and Daletskaya, 1963). The activity of abscisic acid (ABA) was determined with a bioassay based on germination of mustard seeds (variety Sareptskaya) (Nikolaeva, 1979). Cytokinin activity was bioassayed with *Amaranthus caudatus L.* (Lough and Jameson, 1990). Thin-layer chromatography (TLC) of the ethyl acetate extracts was performed on Silufol UV-254 plates with the solvent ethyl acetate-chloroform-acetic acid (100:100:1). Indole-3-acetic acid (IAA) was freed from accompanying substances on a Sephadex LH-20 column (68×2 cm). Elution was performed with 50% ethanol (pH 5.0) at a flow rate of 20 ml/h. N-butanol extracts were chromatographed on Silufol UV-254 plates with the solvent system n-butanol-ammonia solution-water (86:5:9). N-butanol extracts were purified on a Sephadex LH-20 column (68×2 cm). Elution was performed with 35% ethanol (pH 7.0) at a flow rate of 20 ml/h. IAA was identified: i) from ultraviolet (UV) spectra obtained on a Specord M-40 spectrophotometer in ethanol; and ii) from infrared (IR) spectra obtained on a Specord M-80 spectrophotometer in a KBr pellet. IAA content in the samples obtained was determined spectrophotometrically with the Salkowski reagent (Tang and Bonner, 1947). The amounts of IAA, ABA, zeatin and zeatin-riboside were determined by gas-liquid chromatography (GLC) (Negretskii *et al.*, 1984) on a Chrom-4 instrument (Czechoslovakia) equipped with a flame ionization detector. Low resolution mass-spectra were obtained on a mass-spectrometer (LKB-2091, Sweden) with the evaporator held at 50° to 200°C, and ionizing voltage of 70 eV (Palni *et al.*, 1986). All the calculations were performed on a computer using a library of mathematical and statistical procedures SSP (Maksimov, 1980).

Results

Bioassays of plant growth regulating substances present in extracts of the supernatant fluids

Plant growth regulating substances contained in the ethyl acetate extract of *A. brasilense* Sp 7 culture supernatant fluid were bioassayed. Assaying the ethyl acetate extract of the supernatant fluid showed it to contain substances exhibiting growth promoting and inhibiting activities. Histograms in Fig. 1 show that the substance eluted from the spot on TLC at Rf 0.3 to 0.4 increases wheat coleoptile length by 58% compared with the control, although its stimulating effect on mustard seed germination is only weak (17%). These findings are in favour of the assumption that the substance corresponding to the Rf values 0.3-0.4 is a bacterial auxin. The substance eluted from TLC at the Rf values 0.2-0.3 suppresses mustard seed germination by 49%, and decreases wheat coleoptile length by 26%, thereby exhibiting a strong inhibitory effect on the plants. The substance takes up a position corresponding to the Rf values of ABA (Sigma), and therefore we can assume it to be abscisic acid. By this means, we showed that the supernatant fluid of *A. brasilense* Sp 7 contained 15 mg/L of IAA and 267 μg/L of ABA.

Assaying the n-butanol extract of the supernatant fluid showed it to contain cytokinin-like sub-

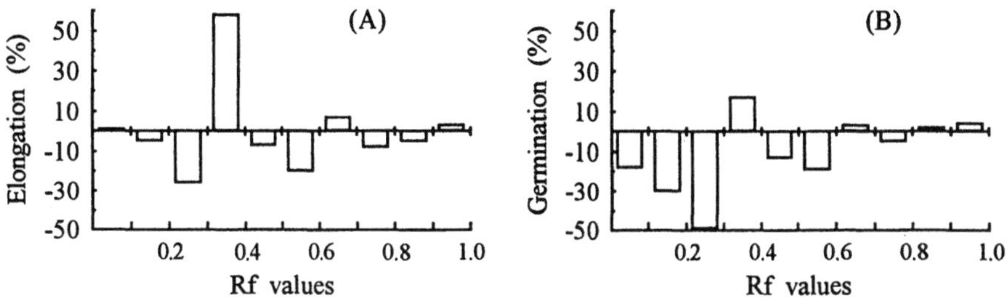

Fig. 1. Histograms of the biological effect produced by the ethyl acetate extract of *A. brasilense* Sp 7 culture supernatant fluid: (A) increase in length of wheat coleoptile segments; (B) mustard seed germination

stances. A 6-day-old sample showed cytokinin activity equivalent to 3 μg/L of zeatin-riboside and 19 μg/L of zeatin. A 10-day-old culture displayed cytokinin activity equivalent to 1 μg/L of zeatin-riboside. Thus, one can trace well-defined dynamics of cytokinin activity over 10 days of growth of *A.brasilense* Sp7 on RE medium. Cytokinin activity has its maximum in a period of 4 to 6 days of culture growth.

Identification of substances present in the extracts of the supernatant fluid of *A. brasilense* Sp 7

Zones on the thin-layer chromatogram at Rf 0.0 to 0.4 contained substances of an indole nature, including IAA at Rf 0.34, which corresponds in position to the Rf value of authentic IAA (Sigma). GLC of the fraction taken at Rf 0.2 to 0.4 on the TLC is shown in Fig. 2. It can be seen that retention time of authentic IAA and ABA is equal to that of IAA and ABA from the culture supernatant fluid. UV absorption spectra of the ethanol solution of authentic IAA and those of the ethanol solution of a promoting substance isolated from the culture fluid using Sephadex LH-20 have their absorption maxima at 280 nm. The IR spectrum of the sample isolated from the culture fluid of *A. brasilense* Sp 7 was identical to that of authentic IAA.

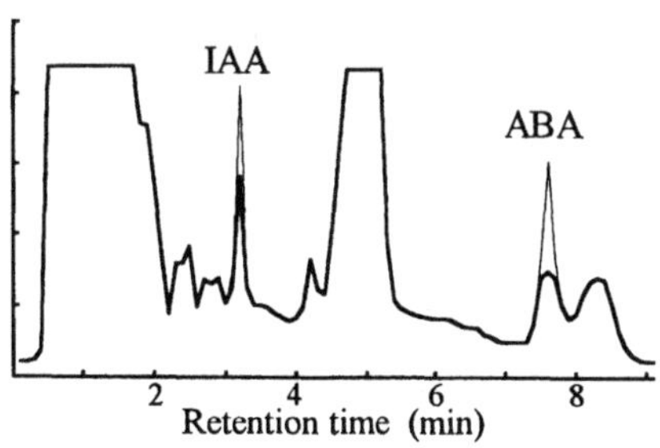

Fig. 2. Gas liquid chromatography (—) of the IAA and ABA fractions of a solution of the *A. brasilense* Sp 7 culture after ethyl acetate extraction and TLC separation at Rf 0.2 to 0.4; (—) cochromatography with authentic IAA and ABA

To identify substances contained in the n-butanol extract of the culture fluid of *A. brasilense* Sp 7, the samples were chromatographed on Silufol UV-254 plates with the solvent n-butanol-ammonia solution-water. Spots on the chromatograms at Rf values 0.2 to 0.3 and 0.4 to 0.5 corresponding to zeatin-riboside and zeatin were scraped off from the TL plates, eluted with ethanol and evaporated to dryness under vacuum at 60°C. By this means samples R_1 (Rf 0.2-0.3) and R_2 (Rf 0.4-0.5) were prepared, which were then examined by chromatography and by chromatomass spectrometry of trimethylsilyl derivatives. Data obtained suggest that the culture fluid of *A. brasilense* Sp 7 is free from such cytokinins as zeatin and zeatin-riboside.

Peculiarities of IAA synthesis in *A.brasilense* Sp 7

DL-tryptophan-supplemented medium containing carbon and nitrogen sources in reasonable quantities permitted us to reveal the dynamics of IAA synthesis by *A. brasilense* Sp 7. The increase in the amount of IAA produced into the culture fluid coincides in time with the attainment by the culture of the stationary phase of growth (after 24 h). The amount of IAA continues to increase up to the 3rd day of cultivation and is kept at a high level up to the 7th day of cultivation. Thus, at the stationary phase of culture growth when cell specialization mechanisms are initiated, the activation of IAA synthesis from tryptophan takes place.

One of the main questions in the diagnosis of the functional state of partners in the microorganism-plant system is to elucidate the peculiarities of IAA synthesis by *A. brasilense* Sp 7 under the simultaneous action of different tryptophan concentrations, pH and temperature values.

We have performed a multifactor analysis of the joint action of tryptophan (at concentrations of 20, 100, and 180 mg/L), pH (5, 7, and 9), and temperature (20°C and 30°C) on IAA synthesis by *A. brasilense* Sp 7. One of the requirements of the regression analysis is that dispersions of separate experiments must be homogeneous, i.e. their differences should be statistically insignificant. To test this hypothesis, we used Cohren's and Fischer's criteria. Our calculated values of the criteria were less than those adduced by the above authors at a 5% significance level and with the appropriate number of degrees of freedom. Therefore, we have accepted the hypothesis for the dispersions' homogeneity. Regression equations, which take into account only significant coefficients and adequately describe experimental results, have the form (in natural variables' values):

$$Y = -197.4 + 0.14\times Trp + 61.9\times pH - 4.2\times pH^2 \quad \text{for } 20°C$$

$$Y = -81.0 + 0.21\times Trp + 28.6\times pH - 1.9\times pH^2 \quad \text{for } 30°C$$

Equations obtained show (Fig. 3) that the range of factors studied involves optimal pH values, and pH optimum lies close to the experiment's center. The optimal pH values are 7.3 at 20°C and 7.6 at 30°C. The root-mean-square effect of the Trp factor is statistically insignificant, and it is practically impossible to determine its optimal values. This means that at concentrations of 20 to 180 mg/L, tryptophan content in the medium does not limit IAA synthesis. It is suggested that the optimum zone of the Trp factor lies in the higher concentrations range.

Determination of the IAA synthesizing activity in 14 *Azospirillum* strains

After cultivating of *Azospirillum* in a liquid medium with tryptophan for 3 days, the acidic ethyl acetate extracts of the culture supernatant fluid of all strains were separated on Silufol UV-254 plates with the solvent ethyl acetate-chloroform-acetic acid. Spots at the Rf value corresponding to that of the authentic IAA (Sigma) were scraped off from the plates and eluted with ethanol. IAA content in the samples obtained was determined with the Salkowski reagent. Table 1 presents data on the absolute and specific IAA biosynthetic activity and specific N_2-fixation activity of *A. brasilense*. The strains under study varied greatly in the intensity of IAA accumulation in the

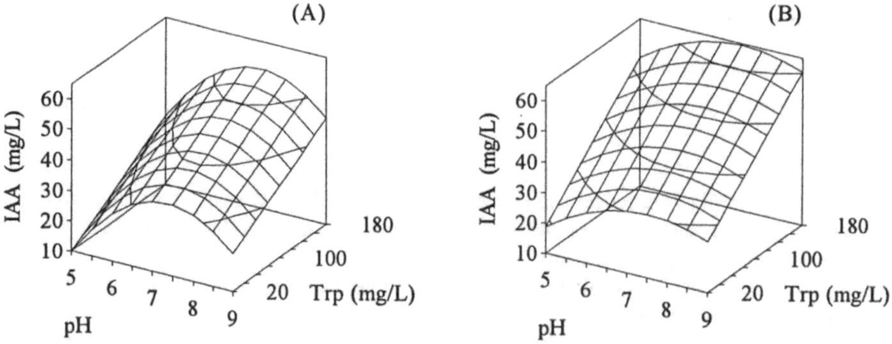

Fig. 3. Surface of response of IAA production by *A. brasilense* Sp 7 to the change in pH values, tryptophan concentrations, and temperature (A) at 20°C; (B) at 30°C

culture medium. With probability of 0.95, the arithmetic average of the sample X=90.7 lay in the range of $72.6 \leq X \leq 108.9$ µg IAA/mg dry matter of intact cells. Strain 80 appeared to be most active and produced 185.2 µg IAA/mg dry matter of intact cells. Of interest were strains SR 14 and SR 75 which actively produced IAA and actively fixed nitrogen. Calculation of the correlation between IAA biosynthetic activity and N_2-fixation activity in 15 *Azospirillum* strains showed that the variation of both criteria in the sample occured independently - R=0.07.

Table 1. Production of IAA by *A. brasilense* isolated from cereals of the Saratov Region

Strains	IAA concent. in the culture fluid (mg/L)	mkg IAA/mg dry matter of intact cells	N -fixation* nM ethylene per mg dry matter per h	Objects of isolation*	Sources of isolation*
SR 80	10	185.2	305.2	Wheat Saratovskaya 49	seedlings
SR 14	22.5	164.2	596	Festuca pratensis Huds	roots
SR 75	9.5	163.8	467.4	Wheat Saratovskaya 29	seedlings
SR 37	39	150	236.2	Elytrigia intermedia	seedlings
SR 8	44.5	130.8	260.4	Zerna inermis Leyss	roots
Sp 7	17	125.9	346.4	Digitaria decumbens	roots
SR 88	19	116.6	428.3	Wheat Kharkovskaya 46	seedlings
SR 32	17	99.4	21.4	Zerna inermis Leyss	roots
SR 100	16	84.2	350.4	Rye Odesskaya	seedlings
SR 115	13	81.2	489.8	Sorghum Saratovskoe	seedlings
SR 103	12	80	356.3	Oat Ruslan	roots
SR 15	19.5	79.3	1102.5	Dactylis glomerata L.	roots
SR 109	13	76.4	571.7	Corn Zubovidnaya	seedlings
SR 111	13	72.2	424.4	Millet Saratovskaya 853	seedlings
SR 7	5	38.5	306.3	Dactylis glomerata L.	roots

*- data obtained in the Laboratory of Genosystematics (Pozdnyakova *et al.*, 1988)

Acknowledgements

This work was supported in part by grant no. 93-04-20680 from the Russian Foundation of Basic Research. We are grateful to S.V.Kanevskaya for her help in the preparation of the manuscript.

References

Alekseeva IU (1975) Theoretical and experimental studies of the laws of error distribution, their classification and methods of estimating their parameters. Author's abstract of Cand. Tech. Sci. thesis, Leningrad 20 p (In Russian)

Bar T, Okon Y (1993) Tryptophan conversion to indole-3-acetic acid via indole-3-acetamide in *Azospirillum brasilense* Sp7. Can J Microbiol 39:81-86

Conard K, Stegemann B (1991) Does *Azospirillum* influence plant growth by cytokinins and auxins? Wennigsen Deister Germany Sept 4-7, Hannover 15-16 (Abstract)

Crozier A, Arruda P, Jasmin JM, Monteiro AMT, Sandberg G (1988) Analysis of indole-3-acetic acid and related indoles in culture medium from *Azospirillum lipoferum* and *Azospirillum brasilense*. Appl Environ Microbiol 54:2833-2837

Döbereiner J, Day JM (1976) Associated symbiosis in tropical grasses: characterisation of microorganisms and dinitrogen fixing sites. In: Proceedings of the 1st International Symposium on Nitrogen Fixation. Newton WH, Nymans CJ (eds) Washington State University Press, Pullman, Wash. pp 518-538

Fallik E, Okon Y, Epstein E, Goldman A, Fischer M (1989) Identification and quantification of IAA and IBA in *Azospirillum brasilense* - inoculated maize roots. Soil Biol Biochem 21:147-153

Hald A (1956) Mathematical statistics with technical application. Moscow 642 p (In Russian)

Horemans S, Vlassak K (1985) Production of indole-3-acetic acid by *Azospirillum brasilense*. In: *Azospirillum* III: Genetics, Physiology, Ecology: Proceedings of the 3rd Bayreuth *Azospirillum* Workshop. Klingmüller W (ed) Springer-Verlag, Berlin pp 98-108

Lough TJ, Jameson PE (1990) Comparative effects of four naturally-occurring cytokinins in the *Amaranthus* bioassay. J Plant Physiol 136:638-640

Maksimov VN (1980) The multifactor experiment in biology. Moscow 280 p (In Russian)

Negretskii VA, Verenchikov SP, Lozhnikova VN (1984) Identification and isolation of cytokinins from tobacco plants. Agrokhimiya 1:89-92 (In Russian)

Nikolaeva MG, Daletskaya TV (1963) Studies on physiologically active substances in dormant seeds. Proceedings of the Botany Institute, USSR Academy of Sciences 16:49-63 (In Russian)

Nikolaeva MG (ed) (1979) Methods of determining phytohormones and phenols in seeds. Leningrad 78 p (In Russian)

Omay SH, Schmidt WA, Martin P (1993) Indoleacetic acid production by the rhizosphere bacterium *Azospirillum brasilense* Cd under *in vitro* conditions. Can J Microbiol 39:187-192

Palni LMS, Tay SAB, MacLeod JK (1986) GC-MS methods for cytokinins and metabolites. Gas Chromatogr. Mass Spectrom., Berlin pp 214-253

Pozdnyakova LI, Kanevskaya SV, Levanova GF, Barysheva NN, Pilipenko TYu, Bogatyrev VA, Fedorova LS (1988) Taxonomic studies on azospirilla isolated from cereals of the Saratov region. Mikrobiologiya 2:275-278 (In Russian)

Tang YW, Bonner J (1947) The enzymatic inactivation of indoleacetic acid. I.Some characteristics of the enzyme contained in pea seedlings. Arch Biochem 13:11-25

Tien TM, Gaskins MH, Hubbell DH (1979) Plant growth substances produced by *Azospirillum brasilense* and their effect on the growth of pearl millet (*Pennisetum americanum L.*). Appl Environ Microbiol 37:1016-1024

Zimmer W, Bothe H (1988) The phytohormonal interaction between *Azospirillum* and wheat. Plant and Soil 110:239-247

Laccase activity in *Azospirillum lipoferum* : characteristics, melanization and Tn*5* mutants.

D. Faure, C. Jacoud, M.L. Bouillant and R. Bally
Laboratoire d'Ecologie Microbienne du Sol URA CNRS 1450
Université Claude Bernard - Lyon I
43, boulevard du 11 novembre 1918
69622 Villeurbanne Cedex, France

Abstract

In non-motile *Azospirillum lipoferum* 4T and *A. lipoferum* 4Bp, a laccase activity was characterized by several substrates and inhibitors. This laccase activity was required for the production of a brown-dark pigment, as melanin. Following random Tn*5* mutagenesis in *A. lipoferum* 4T, several mutants affected in melanization and laccase activity were obtained. These mutations were located on the chromosome or on the cryptic 300 MDa plasmid of *A. lipoferum* 4T.

Key Words:
laccase, tyrosinase, polyphenol oxidase, melanin, Tn*5* mutant, plasmid, motility.

Abbreviations:
SHAM: Salicylhydroxamic acid
ABTS: 2-2'azino-bis-(3-ethulbenzthiazoline-6-sulfonic acid)
CTAB: hexadecyltrimethylammonium bromide

Introduction

Laccases can oxidize, demethylate, polymerize or depolymerize phenolic compounds. Laccases have been associated with lignification in plants (Bao et al., 1993) and with different biological mechanisms in fungi, such as phytopathogenicity or resistance to environmental or microbial stress (Bell et al., 1986). The laccases oxidize *p*-and *o*-diphenols whereas tyrosinases oxidize *o*-diphenols and monophenols. Laccases and tyrosinases are both classified in the polyphenoloxidase (PPO) group. PPOs are widespread in nature, laccases are mostly present in various fungi and higher plants, while the occurrence of tyrosinases in organisms is almost universal (Mayer et Harel, 1979). In bacteria, tyrosinase is the best known PPO type enzyme,

NATO ASI Series, Vol. G 37
Azospirillum VI and Related Microorganisms
Edited by I. Fendrik et al.

and was first described in *Streptomyces* species (Lerch et al., 1972). In many soil diazotrophs, tyrosinase activity or melanin synthesis has also been reported such as in *Azospirillum brasilense* (Sadasivan et al., 1987), in *Azotobacter chroococcum* (Shivprasad et al., 1989), and in several *Rhizobium* strains (Cubo et al, 1988). Both laccase activity and melanization could potentially contribute to adaptative mechanisms of *A. lipoferum* 4T in rhizosphere of rice, such as O_2 or toxic phenol scavanging. These hypothesis were also discussed with respect to melanin-producing *Rhizobium* and *Azotobacter* strains (Hynes et al, 1988; Shivprasad, 1989).

In this work, we reported on the identification of laccase activity in *Azospirillum lipoferum* 4T and *A. lipoferum* 4Bp, a spontaneous non-motile form derived from *A. lipoferum* 4B (Givaudan et al, 1993). *A. lipoferum* 4B and *A. lipoferum* 4T were isolated from the same rhizosphere of rice (Bally et al, 1983). In *A. lipoferum* 4T, we also investigated the expression of laccase activity and its role in melanization. Several *A. lipoferum* 4T Tn*5* mutants allowed the localization of several loci which were required for melanization and laccase activity (Faure et al, 1994).

Materials and Methods

A. lipoferum strains were grown at 28°C on TY medium, on Nfb medium, on Nfb medium supplemented with 20 mM NH_4Cl (NN), or on potato dextrose agar (PDA; Difco). *E. coli* S17.1, harboring the plasmid pSUP102-Gm-Tn*5*-B20, was grown on Luria broth (LB) at 37°C. Concentrations of antibiotics (µg/mL) used were: kanamycin (Km) 40; colistin methane sulfonate (Col) 50; streptomycin (Sm) 100. Laccase activity was revealed with specific substrate, syringaldazine (Givaudan et al., 1993), and melanization was observed as previously described (Faure et al, 1994). DNA isolation, restriction analysis and DNA transfer were performed by conventional techniques, and plasmid DNA isolation and hybridizations were previously described (Bally et Givaudan, 1988).

Results and discussion

By selected substrates and inhibitors, laccase activities could be characterized and ditinguished from *o*-PPO activities, in purified enzyme preparations, as well as in crude extracts (Mayer et Harel, 1979). As characteristic substrates of the plant and fungal laccases, catechol and hydroquinone (and their derivatives), but not tyrosinase, were used by *A. lipoferum* laccase activity. 2-6 methoxyphenol and several methoxy substitued monophenols, such as lignin componants, were described as attacked by plant, fungal (Mayer, 1987) and bacterial laccase activities (see Jacoud et al, this volume). A specific laccase substrate in absence of H_2O_2, syringaldazine (Harkin and Obst, 1973), which was usually used for measuring specific laccase activity, and others laccase substrates, such as *p*-phenylenediamine and *p*-aminophenol were also oxidized by *A. lipoferum* laccase activity. Several chemicals, which were not laccase substrates on their own, could be affected by fungal or bacterial laccase activities, in presence of a coupling reagent, as ABTS (Givaudan et al., 1993). From *o*- and *p*-diphenols to the compounds oxidized in presence of ABTS, the range of substrates which were utilized by plant, fungal and bacterial laccase activities was similar and large, suggesting an analogous functional mechanism.

In impure or purified enzyme preparations, the sensitivity to several specific inhibitors was the second characteristic of tyrosinase and laccases (Mayer, 1987; Walker et al., 1980). Both tyrosinases and laccases, as well as *A. lipoferum* laccase activity, were inhibited by azide, cyanide, EDTA and copper chelators. However, *A. lipoferum* laccase activity was affected by CTAB and N-hydroxyglycine, as laccase inhibitors. Among specific tyrosinase inhibitors, SHAM and tropolone which were selective at low concentrations, did not decreased *A. lipoferum* laccase activity. By the way of comparison between *A. lipoferum* and *P. oryzae* laccase activities, the sensitivity to several inhibitors was tightly similar, renforcing the hypothesis of an analogous active site (Givaudan et al., 1993; Jacoud et al., this volume). In this topic, biochemical and genetic investigations of *A. lipoferum* laccase activity were started.

Like several *Rhizobium* strains, *A. lipoferum* 4T produced a brown pigment only in old colonies grown on rich medium such as TY (Cubo et al., 1988). No melanization was obtained on minimal media (Nfb or NN) or on PDA which induces the synthesis of a pink pigment as previously described (Krieg et al., 1984). Unlike the secondary metabolites as melanin or pink pigment, laccase activity was revealed on both minimal media (Nfb, NN) and rich media (TY, PDA) (Faure et al., 1994). If the polyphenoloxidase activity is needed for melanization, it is not sufficient, because synthesis of melanin precursors is also necessary, as for example for the biosynthesis of different fungal melanins (Bell et al., 1986). In poor medium, melanin precursors are absent or not in contact with laccase activity.

Ten Tn*5* mutants which failed to produce melanin or to exhibit laccase activity were isolated. These mutants could be classified into two groups: the first characterized by an alteration of melanization and of specific laccase activity (class 1), the second by the absence of both functions (class 2). The mutants of class 1 expressed laccase activity at 54 to 6 % that of the wild type. In these mutants, the decreased expression levels of laccase activity were correlated with red brown to light brown pigmentations obtained on TY medium. These loci could affect laccase activity at transcriptional or post-transcriptional levels, and would not in the structural gene of the laccase activity. Indeed, in *Streptomyces antibioticus* and in *Rhizobium*, both mutations in the structural gene of tyrosinase or in a positive regulatory gene confered *mel*$^-$

phenotype (Bortakur et al., 1987; Mercado-Blanco et al., 1993). In *A. lipoferum* 4T class 2 mutants (*mel*⁻, *laccase*⁻), Tn*5* insertions could affect different loci strictly required for melanization and laccase activity, as laccase structural gene or their regulatory factors.

All class 1 mutations were presumably located on the chromosome, whereas, in class 2, most mutations were on the same 300 MDa plasmid. In *Rhizobium* species, the role of both symbiotic and non-symbiotic plasmids in the melanization process has been demonstrated (Bortakur et al., 1987; Mercado-Blanco et al., 1993). In various bacterial genera, the genes required for biosynthesis of secondary metabolites, as antibiotics or pigments, were also clustered on plasmids (Martin and Liras, 1989). In *A. lipoferum* and in *A. brasilense* the known plasmidic functions have been investigated and found to be borne either on the 150 MDa plasmid or the 90 MDa plasmid (Bally et al., 1988; Michiels et al., 1989; Onyeocha et al., 1990). These plasmids have large DNA homologies among *A. lipoferum* or *A. brasilense* strains (Givaudan et al., 1991; Vieille et al., 1989). In this work, no insertion was obtained on 150 MDa plasmid of *A. lipoferum* 4T, but on two cryptic plasmids of 270 and 300 MDa, respectively.

References

Bally, R., and A. Givaudan. 1988. Mobilization and transfer of *Azospirillum lipoferum* plasmid by the Tn*5*-*mob* transposon into a plasmid-free *Agrobacterium tumefaciens* strain. Can. J. Microbiol. 34: 1354-1357.

Bally, R., D. Thomas-Bauzon, T. Heulin, J. Balandreau, C. Richard, and J.De Ley. 1983. Determination of the most frequent N_2 fixing bacteria in the rice rhizosphere. Can. J. Microbiol. 29: 881-887.

Bao, W., D. M. O'Malley, R. Whetten, and R. R. Sederoff. 1993. A laccase associated with lignification in loblolly pine xylem. Science 260: 672-674.

Bell, A.A., and M.H. Wheeler. 1986. Biosynthesis and functions of fungal melanins. Ann. Rev. Phytopathol. 24:411-451.

Bortakur, D., J.W. Lamb, and A.W.B. Johnston. 1987. Identification of two classes of *Rhizobium phaseoli* genes required for melanin synthesis, one of which is required for nitrogen fixation and activates the transcription of other. Mol. Gen. Genet. 207: 155-160.

Cubo,.M.T., A.M. Buendia-Claveria, J.E. Beringer, and J.E. Ruiz-Sainz. 1988. Melanin production by *Rhizobium* strains. Appl. Environ. Microbiol. 54: 1812-1817.

Faure, D., M.L. Bouillant and R. Bally. 1994. Isolation of *Azospirillum lipoferum* Tn*5* mutants affected in melanization and laccase activity. Appl. Env. Microbiol. 60: 3413-3415.

Givaudan, A., and R. Bally. 1991. Similarity between large plasmids of *Azospirillum lipoferum*. FEMS Lett. 78: 245-252.

Givaudan, A., A. Effosse, D. Faure, P. Potier, M.L. Bouillant, and R. Bally. 1993. Polyphenoloxidase in *Azospirillum lipoferum* isolated from rice rhizosphere: evidence for laccase activity in non-motile strains of *Azospirillum lipoferum*. FEMS Microbiol. Lett. 108: 205-210.

Harkin, J.M., and J.R. Obst. 1973. Syringaldazine, an effective reagent for detecting laccase and peroxidase in fungi. Experientia. 29: 381-508.

Hynes, M.F., K. Brucksch, and U. Priefer. 1988. Melanin production encoded by a cryptic plasmid in a *Rhizobium leguminosarum* strain. Arch. Microbiol. 150: 326-332.

Krieg, N.R., and J. Döberener. 1984. The genus *Azospirillum*. *In* N.R. Krieg and J.G. Holt (ed.), Bergey's manual of methods for general microbiology. The Williams and Wilkins Co, p.94-104.

Lerch, K., and L. Ettlinger. 1972. Purification and characterization of a tyrosinase from *Streptomyces glaucescens*. Eur. J. Biochem. 31: 427-437.

Martin, J.F., and P. Liras. 1989. Organization and expression of genes involved in the biosynthesis of antibiotics and other secondary metabolites. Annu. Rev. Microbiol. 43: 173-206.

Mayer, A. M., and E. Harel. 1979. Polyphenol oxidases in plants. Phytochemistry 18: 193-215.

Mayer, A. M. 1987. Polyphenol oxidases in plants- recent progress. Phytochemistry 26: 11-20.

Mercado-Blanco J., F. Garcia, M. Fernandez-Lopez, and J. Olivares. 1993. Melanin production by *Rhizobium meliloti* GR4 is linked to nonsymbiotic plasmid pRmeGR4: cloning, sequencing, and expression of the tyrosinase gene *mepA*. J. Bacteriol. 175: 5403-5410.

Michiels, K., P. De Troch, I. Onyeocha, A. Van Gool, C. Elmerich, and J. Vanderleyden. 1989. Plasmid localization and mapping of two *Azospirillum brasilense* loci that affect exopolysaccharide synthesis. Plasmid 21: 142-146.

Onyeocha, I., C. Vieille, W. Zimmer, B. Baca, M. Flores, P. Palacios and C. Elmerich. 1990. Physical map and properties of a 90 MDa plasmid of *Azospirillum brasilense* Sp7. Plasmid 23: 169-182

Sadasivan, L.N., and Neyra, C.A. 1987. Cyst production and brown pigment formation in aging cultures of *Azospirillum brasilense* ATCC 29145. J. Bacteriol. 169: 1670-1677.

Shivprasad, S., and W.J. Page. 1989. Catechol formation and melanization by Na^+-dependent *Azotobacter chroococcum*: a protective mechanism for aeroadaptation ? Appl. Environ. Microbiol. 55:1811-1817

Vieille, C., I. Onyeocha, M. Galimand, and C. Elmerich. 1989. Homology between plasmids of *Azospirillum brasilense* and *Azospirillum lipoferum* . *In* Nitrogen fixation with non-legumes, Skinner F.A, R. M. Boddey and I. Fendrik (eds), Kluwer Academic publishers, pp. 165-172.

Walker, J.R.L., and R. F. Mccallion. 1980. The selective inhibition of ortho- and para-diphenoloxidases. Phytochem. 19: 373-377.

Wheat Lectin Possibly Serves as a Signal Molecule in the *Azospirillum* - Wheat Association

Lyudmila Antonyuk, Oksana Fomina, Alla Kalinina, Sergey Semenov,
Marina Nesmeyanova† and Vladimir Ignatov
Institute of Biochemistry and Physiology of Microorganisms
Russian Academy of Sciences†
Institute of Biochemistry and Physiology of Plants and Microorganisms
Russian Academy of Sciences
pr. Entuziastov 13 Saratov 410015
Russia

Abstract

Addition of a wheat lectin (wheat germ agglutinin, WGA) to the *Azospirillum brasilense* Sp 245 culture was shown to induce not only the previously reported enhancement of N_2-fixation, ammonia excretion, and indole-3-acetic acid production, but also protein biosynthesis in the cell. Also, variation in the acidic phospholipid proportion in the membrane was noted. The proposals that the observed effects could occur naturally and that WGA might serve as a signal molecule in the *Azospirillum*-wheat association are discussed.

Introduction

In studying the molecular mechanisms of interaction of the plant root system with the plant-growth promoting bacteria, several questions still need to be answered; among them are the signal exchange in the plant-bacterium association and the subsequent metabolic alterations in the macro- and micropartners. In this connection, the role of plant lectins has been addressed (Bashan and Levanony 1988; Del Gallo *et al.*, 1989). It has been well documented that the ability of rhizobia to bind to root lectin is an important determinant of bacterial attachment to legume roots and successful colonization (Diaz *et al.*, 1989). Furthermore, plant lectins are involved in the partners' interactions in the plant-pathogen system (Lyubimova *et al.*, 1990). Wheat lectin was originally found in embryos and later in the caps and tips of adventitious roots. It is considered to be accessible to bacteria (Mishkind *et al.*, 1982). WGA was also reported to bind to the cell surface of *A. brasilense* and *A. lipoferum* (Yagoda-Shagam *et al.*, 1988; Del Gallo *et al.*, 1989).

Our previous research has established that the binding of WGA to *A.brasilense* Sp 245, a natural wheat colonizer, greatly increases glutamine synthetase (GS) activity, N_2-fixation, and ammonia excretion in the bacterium (Antonyuk *et al.*, 1993). From the experiments performed, however, it was unclear whether WGA binding to *Azospirillum* surface resulted in an increased

NATO ASI Series, Vol. G 37
Azospirillum VI and Related Microorganisms
Edited by I. Fendrik et al.

biosynthesis of the enzymes or simply triggered their activation. We proposed that WGA caused the enhancement of biosynthetic processes in the bacterium as it did in human lymphocytes (Udey *et al.*, 1980). Another proposal was that during transmembrane signaling there occured alterations in *Azospirillum* membrane. This study has been designed to shed some light upon these questions.

Materials and Methods

For all experiments, *A.brasilense* Sp 245 was grown for 12 h on N-free medium as described previously (Antonyuk *et al.*, 1993) and WGA concentration used was 0.5 μg/ml. For preparation of the cell extracts, the culture of *A. brasilense* was divided into two parts equal in volume and in cell number, one of which was incubated with WGA (0.5 μg/ml, 1 h). Then both cultures were thoroughly washed off. The cells were disrupted by sonication, the cell debris was removed by centrifugation, and the supernatants were assayed by SDS-PAGE according to Laemmli (1970) and immunoelectrophoretically according to Weeke (1973). Protein was estimated by the method of Bradford (1976). The cell number and GS activity were determined as described previously (Antonyuk *et al.*, 1993). Extraction and quantification of phospholipids were carried out according to Ames (1968). Phospholipids were chromatographed for separation on silica-gel plates (Merk, Germany) with the solvent system chloroform - methanol - water - ammonia (60 : 37.5 : 3.38 : 0.62, v/v).

Results and Discussion

Enhancement of GS biosynthesis in *A. brasilense* sp 245 under the effect of WGA

To answer the question of whether GS activity increases at the expense of biosynthesis or of activation of the enzyme, we used rocket immunoelectrophoresis of the above extracts with antibodies against the electroforetically homogeneous GS from the same strain (Fig.1). In the

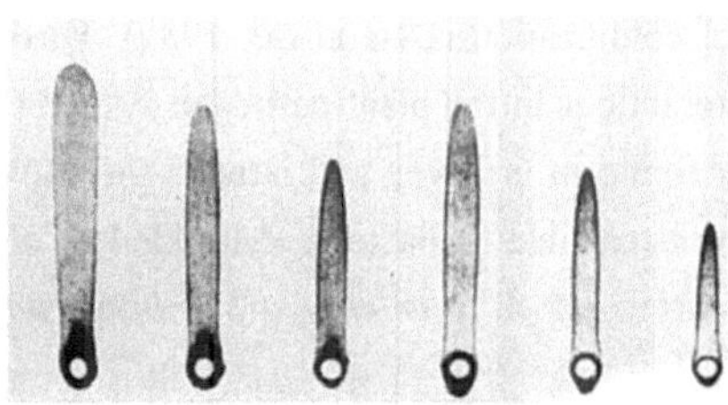

A B C D E F

Fig.1. Rocket immunoelectrophoresis of the extracts of WGA-incubated cells (A,B,C) and non-incubated cells (D,E,F) of *A. brasilense* Sp 245. Extracts were undiluted (A,D), two-fold diluted (B,E) and four-fold diluted (C,F).

case of WGA-incubated cells, the rockets were higher (Fig. 1; A,B,C) than the corresponding control rockets (Fig. 1; D,E,F). Thus, the first extract contained more molecules than the second extract. It turned out that non-diluted extract of lectin-incubated cells and two-fold diluted extract of cells which were not exposed to WGA (Fig. 1; B,D) were the same height (27,5 mm), and hence, the number of enzyme molecules approximately doubled when *Azospirillum* cells were incubated with the lectin. This is in good agreement with the results of determination of GS activities in the extracts. Consequently, binding of WGA to the cell surface of *A. brasilense* Sp 245 causes enhancement of GS biosynthesis rather than activation of the enzyme.

Effect of WGA on the biosynthesis of other proteins

Protein estimation in the cell extracts revealed that the extract of lectin-incubated cells contained 3 times more protein than the alternative extract (380 µg/ml against 120 µg/ml). Both extracts were next subjected to SDS-PAGE (Fig. 2.). In the control electrophoregramm (Fig. 2, C) one can count 26 protein bands against 64 bands which were exhibited when cells were exposed to WGA (Fig. 2, B and A). Thirty-eight new proteins made their appearance.

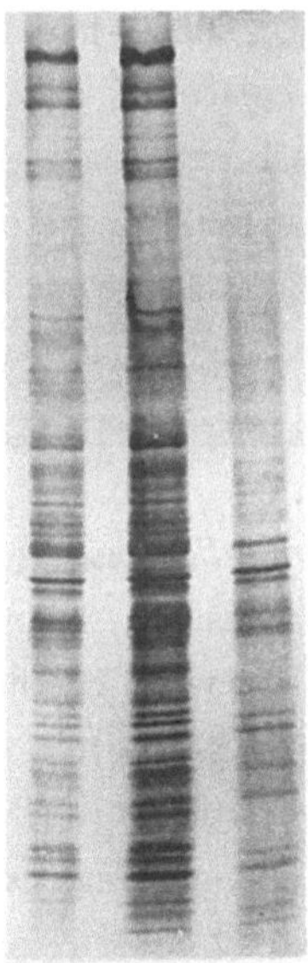

A B C

Fig. 2. SDS-PAGE of the extracts of WGA-incubated cells (A,B) and non-incubated cells (C); A - the extract was diluted two-fold.

Careful visual comparison of the protein profiles revealed that none of the protein bands of the control disappeared in the alternative. Furthermore, several proteins present in both protein profiles had a different intensity, more intensive in the case of incubation with lectin and less intensive in the control. From these results, it may be concluded that under the influence of the lectin there occur increases in the biosynthesis of both presynthesized and new proteins in *Azospirillum brasilense* Sp 245. This, in turn, counts in favour of the expression of rather a large number of genes as a bacterial response to WGA.

Effect of WGA on the membrane of *A.brasilense* Sp 245

The manifold response of *A. brasilense* Sp 245 to the binding of WGA implies transmembrane signaling which is liable to result in a membrane modification. Phospholipid content was chosen to test the membrane modification under the action of lectin, since in bacteria signals from the environment are used to cause variation in the acidic phospholipid proportion (Nesmeyanova, 1990). The membranes of *A. brasilense* Sp 245 were shown to contain cardiolipin (CL), phosphatidylethanolamine (PE), and phosphatidylglycerol (PG) at a 8.2 : 76.6 : 15.2 percental ratio. Two-dimensional thin-layer chromatography revealed two unidentified phospholipids (not shown). Cultivation of *A.brasilense* Sp 245 in the presence of WGA did not influence PE content, but reduced CL content to as little as 3.4% and increased PG content to as much as 19.9 %. The combined content of acidic phospholipids, CL and PG, remained virtually invariant, whereas the PG : CL ratio increased more than 3-fold. It can be inferred that during the binding of the wheat lectin to the bacterial cell surface there may occur acceleration of the acidic phospholipid exchange and in particular, of their interconversion. The acidic phospholipids probably mediate the transmembrane signaling.

What are the chances for wheat lectin to serve as a signal molecule in the *Azospirillum* - wheat association ?

We propose that the activation of *Azospirillum* discovered in laboratory experiments may take place in the *Azospirillum*-wheat association, i.e., WGA is likely to serve as a signal molecule to change the *Azospirillum* metabolism in a wheat-beneficial way. Several lines of circumstantial evidence indicate that this lectin can play such a role. Firstly, WGA enhances two processes that are most significant for plant growth promotion, indole-3-acetic acid production (Iosipenko *et al.*, 1994) and N_2-fixation followed by ammonia excretion (Antonyuk *et al.*, 1993). The activation is well marked (2-, 5.3-and 12.9-fold respectively). It is significant that all the effects described, namely, enhancement of N_2-fixation, ammonia excretion, GS activity, indole-3-acetic acid production, protein biosynthesis and the PG : CL ratio in *Azospirillum* membrane have been shown up at low WGA concentrations (10^{-8}-10^{-9}M). It should be remembered that this concentration is typical of hormones and other signal molecules. Furthermore, wheat lectin can act as a hormone on fat cells (Cuatrecasas and Tell 1973), and this study revealed that the binding of the lectin to *Azospirillum* cell surface brought about transmembrane signaling and expression of a number of genes. And finally, confrontation of the data on lectin distribution in wheat with those on *Azospirillum* localization allows one to hope that the interaction of WGA with the bacterium takes place in the association. A detailed study of the former question conducted by Mishkind *et al.*, (1980, 1982) revealed that in embryos WGA is limited to the regions that establish contact with the soil during germination of the grain. A small population of cells in embryos (the radicle, the first adventitious roots, the coleoptile and scutellum) possesses almost all the lectin present in the grain, and consequently, its concentration is rather high (Mishkind *et al.*, 1982). In adult plant the lectin is located in the caps and tips of adventitious roots. The root cap cells are easily sloughed off and eroded which results in the availability of the lectin to bacteria. The lectin at the root surface can be eluted by washing intact roots in buffer (Mishkind *et al.*, 1980). It was found in wheat roots during the first month of plant growth (Mishkind *et al.*, 1980), and in roots of two-month-old wheat plants (Mishkind *et al.*, 1982). *A.brasilense* was also shown to colonize the apex part of the wheat root (the tip, the elongation, and the root hair zones) (Levanony *et al.*, 1989). Further experiments are necessary to either support or refute this working hypothesis.

Acknowledgements

This work was supported in part by grant no. 93-04-20680 from the Russian Foundation of Basic Research. We thank Dr. L. Matora for providing the antibodies and D. Tychinin for his help in the preparation of the typescript.

References

Antonyuk LP, Fomina OR, Galkin MA, Ignatov VV (1993) The effect of wheat germ agglutinin on dinitrogen fixation, glutamine synthetase activity and ammonia excretion in *Azospirillum brasilense* Sp 245. FEMS Microbiol Lett 110:285-290

Ames GF (1968) Lipids of *Salmonella typhimurium* and *Escherichia coli*: Structure and Metabolism. J Bacteriol 95:833-843

Bashan Y, Levanony H (1988) Migration, colonization and adsorption of *Azospirillum brasilense* to wheat roots. In Freed D, Bog-Hansen TC (eds): "Lectins: Biology, Biochemistry, Clinical Biochemistry", vol 6:69-84 Sigma Chemical Co. St. Louis Mo

Bradford MN (1976) A rapid and sensitive method for the quantification of microgram quantities of protein. Anal Biochem 72:248-254

Cuatrecasas P, Tell GPE (1973) Insulin-like activity of conconovalin A and wheat germ agglutinin - direct interactions with insulin receptors. Proc Nat Acad Sci USA 70:485-489

Del Gallo M, Negi M, Neyra CA (1989) Calcofluor- and lectin-binding exocellular polysaccharides of *Azospirillum brasilense* and *Azospirillum lipoferum*. J Bacteriol 171:3504-3510

Diaz CL, Melchers LS, Hooykaas PJJ, Lugtenberg BJJ, Kijne JW (1989) Root lectin as a determinant of host-plant specificity in the *Rhizobium*-legume symbiosis. Nature 338:579-581

Iosipenko AD, Sergeeva EI, Antonyuk LP, Ignatov VV (1994) Effect of wheat lectin on the synthesis of indole-3-acetic acid by *Azospirillum brasilense* Sp 245. Doklady Rossiiskoy Akademii Nauk 336:559-561 (in Russian)

Laemmli UK (1970) Cleavage of structural proteins during the assembly of the head of bacteriophage T4. Nature 277:680-685

Levanony H, Bashan Y, Romano B, Klein E (1989) Ultrastructural localization and identification of *Azospirillum brasilense* Cd on and within wheat root by immuno-gold labeling. Plant and Soil 117:207-218

Lyubimova NV, Kashulin PA, Lakhtin VM, Shcherbukhin VD, Merzeyak MN (1990) Lectin-carbohydrate interaction and its role in the cell-cell recognition in the system potato tuber. potato Blight fungus. In Bog-Hansen TC (ed): "Lectins: Biology, Biochemistry, Clinical Biochemistry", vol 7:273-285, Valter de Gryuter Berlin

Mishkind M, Keegstra K, Palevitz BA (1980). Distribution of wheat germ agglutinin in young wheat plants. Plant Physiol 66:950-955

Mishkind M, Raikhel NV, Palevitz BA, Keegstra K (1982) Immunocytochemical localization of wheat germ agglutinin in wheat. J Cell Biol 92:753-764

Nesmeyanova MA (1990) Molecular mechanisms of protein secretion in bacteria. VINITI Moscow (in Russian)

Udey MS, Chaplen DD, Wedner HJ, Parker CW (1980) Early activation events in lectin-stimulated human lymphocytes: evidence that wheat germ agglutinin and mytogenic lectins cause similar early changes in limphocyte metabolism. J Immunol 125:1544-1550

Weeke B (1973) Rocket immunoelectrophoresis. In Axelsen NH, Kroll J, Weeke B (eds):"A Manual of Quantitative Immunoelectrophoresis": 43-57, Universitetforlaget Oslo

Yagoda-Shagam J, Barton LL, Reed WP, Chiovetti R (1988) Fluorescein isothiocyanate labeled lectin analysis of the surface of the nitrogen-fixing bacterium *Azospirillum brasilense* by flow cytometry. Appl. Environ. Microbiol. 54: 1831-1837

Biochemical Characterization of IAA Mutants of Azospirillum brasilense Sp7 Induced by Transposon Tn5 Mutagenesis

G. Gastelúm-Reynoso, C. Cruz-Rufz, Y.G. Xochihua-Corona, J. Luzuriaga-Galicia and B.E. Baca.
Department of Biochemica and Genetics Microbiology
Universidad Autónoma de Puebla, Instituto de Ciencias
Apartado Postal 1622, C.P. 72,000, Puebla Pue. México

Abstract

Azospirillum brasilense Sp7 mutants impaired in IAA production were isolated by Tn5 mutagenesis, Km^r-Sm^r exconjugants were obtained at frequencies ranging from 10^{-6} to 10^{-5}. Among 3000 mutants tested, 0.3 % with IAA changed production were found. Supernatants from IAA-mutants were analysed by HPLC. Chromatograms acid extracts showed ILA, IBA, IAA, and anthranilic acid. Whereas in neutral extracts only IEtOH was detected. According to phenotype the mutants were clustered in three classes: Mutants underproducing IAA, those excreting high amount of IAA-Trp dependent, some of them showing AAT2 overproduction; others showing Trp-independent IAA-overproduction, suggesting to be impaired on the control of Trp pathway.

Introduction

Azospirillum is a diazotroph associated with roots of grasses and cereals (Bashan et al., 1990). The excretion of phytohormones such as cytokinins, indole-3-acetic acid (IAA) and gibberelins have been proposed involved in promote plant growth. The production of phytohormones by the microorganism seems play an essential role in the *Azospirillum*-plant interaction. Severals pathways have been reported for the conversion of tryptophan (Trp) to IAA by bacteria. Recently Prinsen et al., (1993) have provide evidence for multiple IAA-Trp and non-Trp dependent pathways in *Azospirillum.* In order to identify genetic loci involved in IAA production, we have generated and isolated Tn5-IAA mutants, analyzed their metabolite excretion and enzymatic activities involved in IAA biosynthesis.

Materials and Methods

Bacterial conjugations.

A. brasilense Sp7 (ATCC) Sm^r was grown in medium LA previously described (Vázquez-Cruz et al., 1992), supplemented with Sm (100 µg/ml). The mobilizing strain *E. coli* S17.1 harboring pSUP5011 (Simon et al., 1983), was grown in LB medium supplemented with Km (20µg/ml), incubated overnight 30° C. For bacterial conjugations the cultures were diluted tenfold in the same medium and growth for 6 h to logarithm phase. Donor and recipient were concentrated by centrifugation in Eppendorf tube (1.5 ml). A mixture of cells were plated onto LA medium added with 1.5 % (w/v) of yeast extract. The plates were incubated overnight at 30° C for 24 h. Then the bacteria were

NATO ASI Series, Vol. G 37
Azospirillum VI and Related Microorganisms
Edited by I. Fendrik et al.

resuspended in 5 ml saline, 100 µl was plated onto LA medium with Sm (100µg/ml) and Km (60 µg/ml). The Tn5 mutants were submitted to a rapid screening method. The Salkowski reaction was used to measure the amount of indole compounds excreted by *Azospirillum*-Tn5 induced mutants.

Extraction of indole compounds, identification and quantification by HPLC.

To quantify the indole compounds synthesized by *Azospirillum* Tn5 -IAA mutants. The strains were grown in the medium described by Jain & Patriquin (1985). When indicated the cultures were supplemented with Trp (100 µg/ml). *Azospirillum* strains were grown in 250 ml flaks containing 130 ml of medium for 3 days at 30° C, with shaking (160 rpm). Bacterial cultures were centrifuged 10,000 x g for 30 min at 4° C, and extracted with ethyl acetate Tien et al., (1979). High pressure liquid chromatograms were produced by injecting 20 µl of the ethyl acetate extract into 5 µm reverse phase column (Beckman, Ultrasphere ODS, 4.6 x 250 mm), in a Beckman HPLC with an ultraviolet detector absorbing at 280 nm. The isocratic solvent used was methanol 40 %: glacial acetic acid 1 % in water for 2 min. After that a gradient of methanol 100 to 0 % was applied for 2 min. Anthranilic acid (ANT), Trp and Indole (IND) were eluted using a isocratic reverse phase chromatography with acetonitrile 40 %- glacial acetic acid 1 % in water for 18 min at a flow rate of 1 ml/min, with the ultraviolet detector absorbing at 240 nm. Retention times for peaks were compared with those of authentics standards. Quantification was done by comparison of peaks area.

Enzymatic determinations.

Aromatic amino transferases (AAT1, AAT2) and tryptophan 2 monooxygenase (2TMO) were detected by enzyme activity assays on a non denaturing polyacrylamide gel electrophoresis (PAGE). Gels were stained at room temperature in the dark for 2 TMO as described Sánchez-Alonso et al., (1991), and for AATs as previously was described by Baca et al., (1994).

Results

A suicide plasmid mutagenesis procedure was used Tn5 induced mutants, exconjugants were obtained at frequencies ranging from 10^{-6} to 10^{-5} in different experiments. IAA production was determined with Salkowski reagent as a rapid screening. IAA mutants were found at a frequency of 0.9 % among 3000 Km^r strains containing Tn5 insertions. Both IAA underproducers and IAA-overproducers could be identified when the IAA excretion was compared with that of the wild-type. All IAA mutants exhibited normal growth, as determined by culture absorbance determinations, at 540 nm.

Among 28 IAA-mutants we found 8 overproducers Trp-dependent phenotype and 20 strains showed an IAA underproducer Trp-dependent phenotype with 32-68 % IAA production less than the wild-type.

Table 1.-
Indole Compounds Determination by HPLC from IAA-Tn5 Mutants of *A. brasilense* Sp7.
Class A Underproducers.

STRAIN	IAA	ILA	IPA μg/ml	IBA	ANT	IEtOH	IAN
Sp7	4.5	1.1	0.3	0.8	nd[a]	7.6	nd
7227	0.8	0.5	0	0.2	4.5	0	nd
7035	3.96	0.3	0	0.3	nd	1.56	nd
7908	2.8	1.7	0	0.3			nd
7021	0	0.76	0	0	0.4	0	nd
7914	0	1	0	0	0.3	0.9	nd
(Medium Jain & Patriquin + L-Trp 100 μg/ml)							
Sp7	16	1.1	0.3	0.8	nd	17.6	nd
7227	5.8	3.5	0	0.2	2.3	0	nd
7035	1.3	0.6	1.2	0.3	nd	1.56	nd
7908	11.2	1.2	0	3.4			
7021	4.1	1.6	0	0.9	0.6	0	nd
7914	2.8	0	0	0	0.3	1.5	nd

Data represent the means of the three experimental determinations with standards errors, after 72 h culture; [a] nd = not detected.

Taking in account the results obtained with the wild-type, we compared the indole compounds, and amount excreted by different Tn5-induced mutants. Accumulation of intermediate metabolites (ANT and IEtOH), higher or less amount of IAA than wild-type strain were found in some cultures supernatants. As regarding to growth conditions, addition of Trp or not to culture medium, IAA-mutants were clustered in three classes, as is shown in Tables 1, 2 and 3. On the other hand enzymatic activities of AAT1 and AAT2 were detected on cell free extracts by PAGE non-denaturing test, but 2 TMO activy was never detected.

1).- Mutants class A (Table 1). Mutants excreting very low amount of IAA (underproducers). Cells cultured without Trp IAA production ranged from 0.08 to 3.9 µg/ml; in case Trp was added to medium IAA production was 1.3 to 11.2 µg/ml. Cell-free extract of IAA 7914 lost AAT2 (Fig 1).

2).- Mutants class B (Table 2). Mutants excreting less IAA than wild-type cultured without Trp 0.03 to 4.2 µg/ml, or showing overproduction of IAA 21.6 to 90 µg/ml, in case Trp was supplied to medium. Cell-free extract of IAA-7921 displayed overproduction of AAT2 (Fig. 2).

3).- Mutants class C (Table 3). Mutants showing IAA overproduction Trp independent phenotype: medium not supplemented with Trp IAA excretion ranged from 4.9 to 51.2 µg/ml, however Trp was added to medium IAA production ranged from 27.4 to 112 µg/ml. Mutant IAA-7031 was also resistant to F-Trp and excreted very high amount of anthranilic acid (53 µg/ml) in both culture conditions.

Table 2.-
Indole Compounds Determination by HPLC from IAA-Tn5 Mutants of *A. brasilense* Sp7.
Class B Underproducer less Trp and Overproducer plus Trp

STRAIN	IAA	ILA	IPA µg/ml	IBA	ANT	IEtOH	IAN
Sp7	4.5	1.1	0.3	0.8	nd[a]	7.6	nd
7787	0.13	0	0	0.28	0.1	0	nd
7880	0.03	0.2	0.04	0.32	0.3	0	nd
7922	2	0	0	0.1	0.1	0	nd
7936	1.1	0.3	0.4	0.7	nd	0	nd
7614	0.1	0.2	0	0.25	nd	0	nd
7349	4.2	0.6	0	0.21			nd
(Medium Jain & Patriquin + L-Trp 100 µg/ml)							
Sp7	16	1.1	0.3	0.8	nd	17.6	nd
7787	90	2.6	0	0.3	4.3	2.1	nd
7880	51.2	1.9	0.4	0.8	0.9	5	nd
7922	66	0.7	0	1.1	0.1	3.6	nd
7936	21.6	1.9	0.4	0	nd	2	nd
7614	26.1	2.2	0.5	1	nd	5.9	nd
7349	78.9	1.6	0	0.35			

Data represent the means of the three experimental determinations with standards errors, after 72 h culture; [a] nd = not detected.

Table 3.-
Indole Compounds Determination by HPLC from IAA-Tn5 Mutants of *A. brasilense* Sp7.
Class Overproducer Trp-independent

STRAIN	IAA	ILA	IPA µg/ml	IBA	ANT	IEtOH	IAN
Sp7	4.5	1.1	0.3	0.8	nd[a]	7.6	nd
7031	51.2	2.3	0	0.8	53	3.5	nd
7022	30.1	0.9	0.5	0.3	0.5	0.6	nd
7034	47.6	1.6	0	1.3	3.2	0.6	nd
7708	16.1	1.2	0	1.2	0.3	0	nd
7848	15.5	0	0.16	0	0.1	0.3	nd
7921	11.6	2.6	0	0.6	0.1	0.6	nd
7934	4.7	0.26	0	3.8	0.8	3.8	nd
IAA	ILA	IPA	IBA	ANT	IEtOH	IAN	
(Medium Jain & Patriquin + LTrp 100 µg/ml)							
Sp7	16	1.1	0.3	0.8	nd[a]	17.6	nd
7031	55.6	2.3	0	0.8	53	1.7	nd
7022	27.4	1	0.5	3.7	0.4	1.1	nd
7034	104	1.7	0	0	2.5	4.1	nd
7708	100	1.4	0	0.9	0.5	2.5	nd
7848	112	1.5	0	0	4.3	0.1	nd
7921	63.7	2.6	0	0.9	0.3	2.6	nd
7934	101	0.26	0	3.4	0.4	1.8	nd

Data represent the means of the three experimental determinations with standards errors, after 72 h culture; [a] nd = not detected.

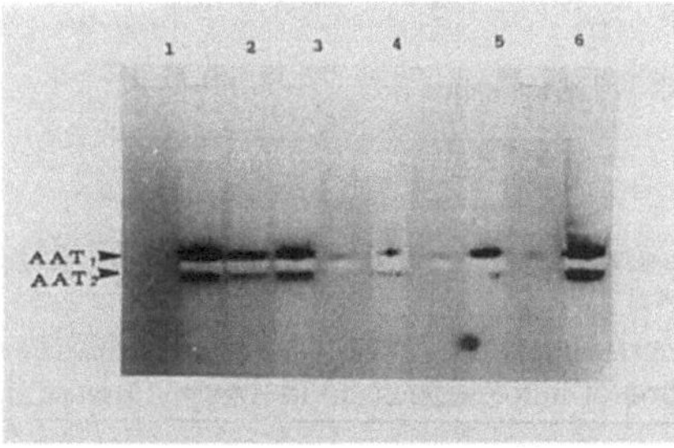

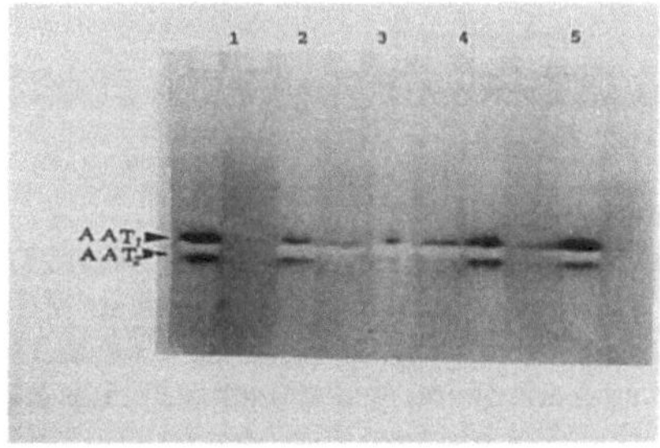

Fig. 1. Non-denaturing PAGE of AATs. Lane 1.- IAA-7034; 2.- IAA-7921; 3.- Sp7; 4.- IAA-7022; 5.- IAA-7914, and 6.- IAA-7035. Gel was stained with Trp as amino-donor

Fig. 2. Non-denaturing PAGE of AATs. Lane 1.- Sp7; 2.- IAA-7787; 4.- IAA-7848; 5.- IAA-7880. Gel was stained with Trp as amino-donor.

Discussion

We have succesfully used a suicide plasmid technique to isolate a series of Tn5-induced IAA mutants from A. *brasilense* Sp7, that produce IAA 1.3 to 150-fold less than wild-type and overproduce 1.7 to 14.15 fold more than the parenteral strain. As others (Abdel-Salam et al., 1987; Ruckdäschel and Klingmüller, 1992), no completely negative mutant could be obtained, indicating that the mutagenesis of one gene is not sufficient for the isolation of a minus mutant: Indeed, it has been proposed that at least three biosynthetic pathways are present in A. *brasilense*, (Prinsen et al., 1993).

In culture supernatant intermediates ILA and IEtOH were detected. This results are in agreement with those obtained for A. *lipoferum* by Ruckdäschel and Klingmüller (1992). Data obtained on the supernatants from bacterial cultures growing with L-Trp excess, indicate the occurrence of the IPyA pathway (Costacurta et al., 1994), where AAT1 and AAT2 (Baca et al., 1994) are responsible for IPyA production, and probably explain the failure to isolate a IAA deficient mutant. Mutants class A culturing on media supplemented or not with Trp excreted from 1.3 to 5.6-fold levels lower that wild-type. One of this, mutant IAA 7914 lost AAT2.

Mutants class B IAA levels excreted to media ranged from 1.07 to 150-fold lower that values reached with Trp added to medium. On the other hand Trp supplementation resulted 1.8 to 5.6-fold increase of IAA excretion. Mutant IAA-7034 showed an increase level of AAT2 enzymatic activity.

Finally mutants class C IAA levels excreted to media ranged from 1.04 to 14.2 times more than those excreted by parenteral strain and this are independent from Trp supplementation. Mutant IAA7031was resistant to 5FTrp, and excreted high level of anthranilic acid suggesting to be altered in regulation control of Trp pathway (Katzy et al., 1992; Zimmer et al., 1991).

Acknowledgements:

This work was partially supported by a CONACyT F193-N9204 Grant

References

Abdel Salam MS and Klingmüller W (1987) Transposon mutagenesis in *Azospirillum lipoferum:* isolation of indoleacetic acid mutants MGG 210:165-170

Baca BE, Soto-Urzúa L, Xochihua-Corona YG and Cuervo-García, A. (1994) Characterization of two aromatic amino acid aminotransferases and production of indoleacetic acid in *Azospirillum* strains Soil Biol Biochem 26:57-63

Bashan Y and Levanony H (1990) Current status of *Azospirillum* inoculation technology: *Azospirillum* as a challange for agriculture Can J Microbiol 36:591-608

Costacurta A, Keijers V and Vanderleyden J (1994) Molecular cloning and sequence analysis of an *Azospirillum brasilense* indole-3-pyruvate decarboxilase gene MGG 243:463-472

Jain DK and Patriquin DG (1985) Characterization of a substance produced by *Azospirillum* which causes branching of wheat roots hairs Can. J. Microbiol 31:206-210

Katzy E, Brodnikova N, Egorenkov D, Panasenko V and Ignatov V (1992) Anthranilate and indol-3-acetate production in *Azospirillum brasilense* Sp245 mutants derivatives Symbiosis 13:147-150

Prinsen E, Costacurta A, Michiels K, Vanderleyden J and Van Onckelen H (1993) *Azospirillum brasilense* indole-3-acetic acid biosynthesis evidence for a non-tryptophan dependent pathway MPMI 6:609-615

Ruckdäschel E and Klingmüller W (1992) Analysis of IAA biosynthesis in *Azospirillum lipoferum* and Tn5 induced mutants Symbiosis 13: 123-131

Sánchez-Alonso MGP, Vázquez-Cruz C, Martínez-Morales LJ, Fluren-Martínez BE and Baca BE (1991) Actividades enzimáticas responsables de la síntesis de la auxina ácido indol ace tico y su relación con los plásmidos presentes en las cepas de *Azospirillum* spp Rev Lat-a mer Microbiol 33:25-34

Tien TM, Gaskins MH and Hubbel DH (1979) Plant growth substances produced by *Azospirillum* and their effect on the growth of pearl millet (*Pennisetum americanum* L) Appl Environ Microbiol 39:219-226

Vázquez-Cruz C, Tapia-Hernández AR, Mascarúa-Esparza MA, Caballero-Mellado J and Baca BE (1992) Plasmid profile modification after elimination of bacteriocin activity in *Azospirillum brasilense* strains Microbios 69:195-204

Zimmer W, Aparicio C and Elmerich C (1991) Relations between tryptophan biosynthesis and indole-3-acetic acid production in *Azospirillum*: identification and sequencing of a *trp*GDC cluster MGG 229:41-51

Influence of Asparagine and Aspartic Acid on Growth of *Azospirillum*.

Sini K.,[1] F. Flouri and C. Balis
Laboratory of General and Agricultural Microbiology
Agricultural University of Athens
Iera Odos 75, 118 55 Athens
Greece.

Abstract

The effects of amino acids on growth of *Azospirillum* have not been studied in detail. Amino acids, as constituents of root exudates, may influence both growth and nitrogenase activity of the *Azospirillum*-root association. The present study was undertaken to assess the effect of asparagine and aspartic acid on growth of *A. brasilense* and *A. lipoferum* and to examine the possibility of using this character as an additional taxonomic characteristic. The results indicate that *A. lipoferum* grows luxuriously on asparagine and aspartic acid whereas *A. brasilense* grows poorly or not at all. Utilization of asparagine as the sole source of carbon offers an additional characteristic for the taxonomic separation of these two species.

Introduction

Organic acids support vigorous growth and nitrogen fixation of all *Azospirillum* species, but the organisms differ widely in utilizing various sugars (Krieg and Döbereiner, 1984). The effects of amino acids on growth of *Azospirillum* have not been studied in detail (Rao and Venkateswarly, 1982). In studies concerning other nitrogen-fixing microorganisms, it has been well documented that amino acids influence nitrogen fixation activity (Tubb, 1976, Yoch and Pengra, 1966). Amino acids, as constituents of root exudates, may influence both growth and nitrogenase activity of the *Azospirillum*-root association. Hartmann *et al* (1988) found that the utilization of amino acids for growth and their effects on nitrogen fixation differs among several strains of *Azospirillum* spp.

Results from experiments dealing with the interactions between *Azospirillum* and *Phialophora radicicola* showed that growth of *A. brasilense* was poor in the *Phialophora* medium whereas *A. lipoferum* grew very well. The content carbon sources of *Phialophora* medium is asparagine and glucose.

So, the present study was undertaken to assess the effect of asparagine and aspartic acid on growth of *A. brasilense* and *A. lipoferum* and to examine the possibility of using this character as an additional taxonomic characteristic.

Materials and Methods

A. brasilense Sp.7 (29145) and *A. lipoferum* Sp.59 (29707), type strains of their respective

[1]Present adress: Instituti Studimit Tokave, Tirana, Albania.

NATO ASI Series, Vol. G 37
Azospirillum VI and Related Microorganisms
Edited by I. Fendrik et al.

species, were obtained from the American Type Culture Collection, Rockvill, Md. and maintained in a nitrogen-free malate medium. The other strains were isolated from soil samples from Experimental Field of Agricultutal University of Athens.

Azo medium (Krieg and Döbereiner 1984) consisted of (in grams per litre): succinic acid, 5.0; K_2HPO_4, 0.5; $FeSO_4 7H_2O$, 0.01; $Na_2MoO_4 2H_2O$, 0.002; $MgSO_4 7H_2O$, 0,2; NaCl, 0.1; $CaCl_2 2H_2O$, 0.026; $(NH_4)_2SO_4$ 1.0. The pH was adjusted to 7.0 with KOH. The carbon sources used were succinic acid, 5.0g/l, asparagine, 2.0g/l and 6.36g/l and aspartic acid, 2.0g/l and 5.63g/l. Inoculation was with 0.1 ml of 24-h-old culture grown in 10 ml of Azo-succinic broth. The cultures were incubated at 27°C for 48 hours on a reciprocal shaker. Turbitity of the growing cultures was measured in a Spectronic 20 Spectrophotometer at 2h intervals.

Growth parameters were calculated from the experimental data by the following equation:

$$y=\frac{A}{1+e^{KA(T-t)}}$$

where: A is the maximal value which the population can reach, T is the time at which the population attains the A/2 level (inflection point) and KA is the growth rate (=μ).

Results

Influence of asparagine and aspartic acid on growth of *A. brasilense* and *A. lipoferum*

A. brasilense could not grow or grew poorly in the presence of asparagine and aspartic acid. In contrast, *A. lipoferum* could use asparagine and aspartic acid as carbon sources effectively. As it is shown in Table 1, the values of growth rates did not differ significantly between succinic acid and asparagine or aspartic acid. Also, the values of A, were higher when *A. lipoferum* grew in asparagine and aspartic acid, especially in low concentrations, than in succinic acid.

Utilization of asparagine as an additional taxonomic characteristic

On the basis of their difference between type cultures of *A. brasilense and A. lipoferum* in utilization of asparagine, it was considered interesting to examine if this property can be used for the identified strains of our collection. Data presented in Table 2 indicate that strains characterized as *A. lipoferum* use also asparagine as carbon source. Contrary to that, strains characterized as *A. brasilense* can not grow in this medium.

Table 1. Effects of asparagine and aspartic acid on the growth of *Azospirillum* spp.

Treatment	Nutrient medium	Parameters of the kinetics growth curve			
		KA(=μ)	A	T	r^2
A. brasilense	succinic acid 5.00 g/l	0.24	1.27	1.93	0.99
	asparagine 2.00 g/l	-	-	-	-
	asparagine 6.36 g/l	0.08	0.31	24.67	0.99
	aspartic acid 2.00 g/l	0.13	0.40	28.50	0.99
	aspartic acid 5.63 g/l	0.21	0.08	5.20	0.95
A. lipoferum	succinic acid 5.00 g/l	0.28	0.34	17.80	0.99
	asparagine 2.00 g/l	0.24	0.32	21.20	0.99
	asparagine 6.36 g/l	0.22	1.51	31.00	0.99
	aspartic acid 2.00 g/l	0.11	1.20	27.30	0.99
	aspartic acid 5.63 g/l	0.18	0.91	25.70	0.99

Table 2. Influence of asparagine on the growth of *Azospirillum* spp.

Strain	Species	Growth capability in asparagine
Sp7 (ATCC 29145)	A. brasilense	-
Sp59 (ATCC 29707)	A. lipoferum	+
A	A. lipoferum	+
B	A. brasilense	-
C	A. brasilense	-
D	A. brasilense	-
E	A. brasilense	-
4a1	A. brasilense	-
4a2	A. brasilense	-
4a3	A. brasilense	-
H2	A. brasilense	-
H3	A. brasilense	-
H2/N2	A. lipoferum	+
T5	A. brasilense	-

Discussion

The utilization of substrates available in the rhizosphere is a prerequisite for growth, colonization of the root and finally nitrogen fixation of *Azospirillum* spp. The utilization of various organic acids and sugars by *Azospirillum* is well documented (Reinhold *et al*, 1987; Tarrand *et al* 1978) and has been mainly used for taxonomic purposes. Hartmann *et al* (1988) studied the influence of amino acids on growth and nitrogen fixation of *Azospirillum* spp. and found that *A. lipoferum* and *A. amazonense* use many amino acids as the sole sources of carbon, nitrogen and energy more readily than do *A. brasilense* and *A. halopraeferens*. Also they propose the utilization of histidine as an additional taxonomic characteristic and for selective isolation of *A. lipoferum* Our results agree with theirs as we have found that strains of *A. brasilense* are unable to grow on asparagine and grow poorly on aspartic acid.

References

Hartmann A, Fu H and Burris RH, (1988) Influence of amino acids on nitrogen fixation ability and growth of *Azospirillum* spp. Applied and Environmental Microbiology 54:87-93.

Krieg NR, and Döbereiner J, (1984) Genus *Azospirillum* in Krieg NR, and Holt JG, (eds) Bergey's Manual of Systematic Bacteriology, Vol. 1, pp. 94-104, Williams and Wilkins, Baltimore.

Rao AV, and Venkateswarly B, (1982) Associative symbiosis of *Azospirillum lipoferrum* with dicotyledonous succulent plants on the Indian desert. Canadian Journal of Microbiology 28:778-782.

Reinhold B, Hurek T, Fendrik I, Pot B, Gillis M, Kerster K, Thielemans S, de Ley J, (1987) *Azospirillum halopraeferens* sp. nov., a nitrogen-fixing organism associated with roots of kallar grass. (*Leptochloa fusca* (Linn.). International Journal of Systematic Bacteriology 37:43-51

Tarrand JJ, Krieg NR, Döbereiner J, (1978) A taxonomic study of the *Spirillum lipoferum* group, with descriptions of a new genus, *Azospririllum* gen. nov and two species, *Azospirillum lipoferum* (Beijerinck) comb. nov. and *Azospirillum brasilense* sp. nov. Canadian Journal of Microbiology 24:967-980

Tubb RS, (1976) Regulation of nitrogen fixation in *Rhizobium* species. Applied and Environmental Microbiology 32, 483-488

Yoch DC, Pengra RM, (1966) Effect of aminoacids on the nitrogenase system of *Klebsiella pneumoniae*, Journal of Bacteriology 92:618-622

Inoculation with *Azospirillum brasilense* Cd Affects the Root System Development of *Sorghum bicolor*

Barbieri P., Trambaioli C.[1], Zanetti G.[1], Galli E.
Dipartimento di Genetica e Biologia dei Microrganismi
Università degli Studi di Milano
Via Celoria, 26 - 20133 Milano (Italia)

Abstract

Inoculation of seedlings of *Sorghum bicolor* with *Azospirillum brasilense* Cd led to morphological, physiological, and biochemical modifications of the root system. In *Azospirillum* inoculated plants the main root was shortened, but the total root surface, namely the number and length of lateral roots and the density of the root hairs, was significantly increased in respect to control roots. These modifications were affected by the plant age and by inoculum concentration. The levels in the roots of glutamate dehydrogenase and glutamate synthase were not substantially varied by *Azospirillum* inoculation. In the inoculated roots the level of free indoleacetic acid was lower than in the control roots.

Introduction

Azospirillum brasilense colonizes the root system of a number of plant species and many authors reported that it promotes plant growth (for reviews see: Bashan & Levanony, 1990; Okon, 1994). Inoculated plants often show an improved mineral and water uptake that could be related to the enhancement of root development (Kapulnik et al., 1985; Levanony & Bashan, 1989; Sarig et al., 1988). However, plant-*Azospirillum* association does not lead to the formation of special structures in the roots (i.e. nodules). The aim of this work was to study morphological, physiological and biochemical modifications of the root system of *Sorghum bicolor* induced by *Azospirillum* and possibly single out some parameters that could be used as markers of the association.

---

[1] Dipartimento di Fisiologia e Biochimica Generali

NATO ASI Series, Vol. G 37
Azospirillum VI and Related Microorganisms
Edited by I. Fendrik et al.

Materials and Methods

Preparation of *Azospirillum* suspensions for inoculation experiments and germination of seeds of *S. bicolor* were carried out as previously described (Barbieri et al. 1986). Unless otherwise indicated, the seedlings were inoculated two days after germination with 2-3x10^7 cell/ml in a hydroponic axenic system with nitrogen- and carbon-free medium (NFb) (Barbieri et al. 1986); each system allowed the growth of six plants and was considered as one replicate. After 15 days of incubation the fresh and dry weight of inoculated and control roots were measured and the total root surface was evaluated by the titration method described by Carley & Watson (1966) and expressed as ml of NaOH (0.1 N) per plant. Results were analyzed by ANOVA.
Protein concentration was measured according to Bradford (1976). Glutamate dehydrogenase (Kanamori et al., 1972) and glutamate synthase (Lea et al., 1990) were assayed essentially as reported. IAA was measured by immunological methods (Mertens et al., 1985) and by GC-MS (Law et al., 1990).

Results and Discussion

Visual comparison of the root systems of inoculated and uninoculated sorghum seedlings grown for 24 or 48 hours in Petri dishes, showed that the main root of *Azospirillum* inoculated plants was shortened, but the density and length of lateral roots were increased in respect to control plants. Inoculated roots also showed a remarkable increase in root hairs, especially in the root tip zone.
Although some differences in root development between inoculated and uninoculated plants were visually observed, we did not obtain statistically significant data. On the contrary, in plants inoculated in the hydroponic system the total root surface and the root water content were significantly increased in respect to control roots (uninoculated or inoculated with dead cells) (Table 1).
The modifications of the root system induced by the inoculation were not affected by the growth phase of the cells used as inoculum neither by the carbon source used to grow the bacteria. Although the carbon source affects the composition of *Azospirillum* exopolysaccharides (Del Gallo & Fendrik, 1994), it does not seem to have any effect on root morphology, at least in 15 days grown plants; in fact, we did not find any difference between plants inoculated with *Azospirillum* cells grown in the presence of fructose or malate.

Table 1. Effects of inoculation on root surface and root water content

Treatment	Root surface (ml NaOH/plant) (P=0.0133)	Water content (g H_2O/ g dry weight) (P=0.0001)
Uninoculated	1.25 (0.14) a	20.66 (0.92) a
Dead cells	1.18 (0.19) a	19.34 (0.44) a
Viable cells	1.99 (0.16) b	27.16 (0.78) b

Data represent means of 4 replicates. Internal standard error is given in parentheses. Numbers followed by the same letter in each column do not differ significantly.

On the contrary the root system development was affected by the cell concentration of the inoculant: a concentration of 10^6 cell/ml poorly affected the root growth, while in plants inoculated with 10^8 cell/ml the root system appeared to be inhibited in its growth, although both the root surface and the water content were still higher than those of the control roots (Table 2).

Table 2. Influence of inoculum concentration

Inoculum concentration (cells/ml)	Root surface (ml NaOH/plant) (P=0.0033)	Water content (g H_2O/ g dry weight) (P=0.0013)
Uninoculated	1.07 (0.05) a	18.54 (0.31) a
10^6	1.11 (0.11) a	24.69 (1.52) b
10^7	1.47 (0.05) b	24.87 (2.91) b
10^8	1.49 (0.09) b	37.72 (2.65) c

Data represent means of 4 replicates. Internal standard error is given in parentheses. Numbers followed by the same letter in each column do not differ significantly.

The age of the plants at the time of inoculation was also critical: both the root surface and the water content decreased in a linear way in plants inoculated after 3 and 5 days from germination, and roots of plants inoculated after 7 days from germination were undistiguishable from the control ones (Table 3).

Table 3. Influence of plant age

Days after germination	Root surface (ml NaOH/plant) (P=0.0001)	Water content (g H_2O/ g dry weight) (P=0.0049)
2	1.83 (0.10) a	24.65 (1.06) a
3	1.74 (0.17) a	28.79 (4.73) a
5	1.42 (0.09) ab	19.64 (0.79) ab
7	1.07 (0.13) b	17.40 (0.93) b
Uninoculated	1.01 (0.06) b	17.31 (0.30) b

Data represent means of 4 replicates. Internal standard error is given in parentheses. Numbers followed by the same letter in each column do not differ significantly.

Due to the nitrogen-fixing ability of *A. brasilense*, it was expected an influence by the inoculation on the plant enzymes involved in nitrogen metabolism. The activity levels of glutamate dehydrogenase and ferredoxin-dependent glutamate synthase, enzymes of the two alternative metabolic routes for ammonia assimilation, were measured in crude extracts of the roots after 15 days of growth in the presence and absence of *Azospirillum* and were found very similar both in the control and inoculated roots.

Alternatively, the beneficial effect of *Azospirillum* inoculation on the plant could be due to indole-3-acetic acid (IAA) production by the microorganism. The levels of free IAA in the roots of seedlings grown for 15 days in the presence and absence of *Azospirillum* were determined using commercial monoclonal antibodies (Idetek, Inc.) To avoid interferences, after addition of [^{3}H]-IAA as a tracer, the methanolic extracts were filtered twice through SepPak C18 (Waters) at two different pH's, subjected to methylation and finally, Met-IAA was separated by HPLC on a C18 column. In Table 4 the values of free IAA found in roots by ELISA are reported on a fresh weight basis. Clearly, the inoculated roots showed a lower content of IAA. This could be due in part to the increase in water content of the inoculated roots (Table 1).

To verify that the values of IAA determined were not grossly underestimated, due to low reproducibility of the data, IAA was also quantified using gas-chromatografy coupled to mass spectrometry (GC-MS) and the results confirmed the data obtained by ELISA.

Table 4. Levels of IAA in roots of *S. bicolor* seedlings grown for 15 days

Treatment	IAA (pmol/gFW) ELISA	IAA (pmol/gFW) GS-MS
Uninoculated	243	199
Inoculated	103	113

Thus, it seems that inoculation with *Azospirillum* would rather decrease the amount of free IAA in the roots. Further studies are in progress to evaluate the amounts of IAA in the roots at shorter times of inoculation.

On the whole our data demonstrate that inoculation with *A. brasilense* Cd affects the root development of *S. bicolor* at morphological, physiological and biochemical levels; studies are in progress with soil microorganisms other than *Azospirillum* to test if these changes are highly specific of the *Azospirillum*-sorghum association.

Research supported by the Consiglio Nazionale delle Ricerche, Italy, special project RAISA, sub-project n. 2.

References

Barbieri P, Zanelli T, Galli E, Zanetti G (1986) Wheat inoculation with *Azospirillum brasilense* Sp6 and some mutants altered in nitrogen fixation and indole-3-acetic acid production. FEMS Microbiol. Letters 36:87-90.

Bashan Y, Levanony H (1990) Current status of *Azospirillum* inoculation technology: *Azospirillum* as a challenge for agriculture. Can. J. Microbiol. 36:591-608.

Bradford MM (1976) A rapid and sensitive method for the quantitation of microgram quantities of protein utilizing the principle of protein-dye binding. Anal. Biochem. 72:248-254.

Carley HE, Watson RD (1966) A new gravimetric method for estimating root-surface area. Soil Science 102:289-291.

Del Gallo M, Fendrik I (1994) The rhizosphere and *Azospirillum*. In "*Azospirillum*-plant associations" (Okon Y, ed) CRC Press, Boca Raton USA. pp. 57-75.

Kanamori T, Shigeki K, Eiichi T (1972) Inducibile formation of glutamate dehydrogenase in rice roots by the addition of ammonia to the media. Physiol. Plant. 26:1-6.

Kapulnik Y, Gafny R, Okon Y (1985) Effect of *Azospirillum* spp. inoculation on root development and NO_3^- uptake in wheat (*Triticum aestivum* cv. Miriam) in hydroponic system. Can. J. Bot. 63:627-631.

Law DM, Davies PJ (1990) Comparative indole-3-acetic acid levels in the slender pea and other pea phenotypes. Plant Physiol. 93:1539-1543.

Lea PJ, Blackwell RD, Chen FL, Hecht U (1990) Enzymes in ammonia assimilation. Meth. Plant Biochem. 3:257-276.

Levanony H, Bashan Y (1989) Enhanacement of cell division in wheat root tips and growth of root elongation zone induced by *Azospirillum brasilense* Cd. Can. J. Bot. 67:2213-2216.

Mertens R, Eberle J, Arnscheidt A, Ledebur A, Weiler EW (1985) Monoclonal antibodies to plant growth regulators. II. Indole-3-acetic acid. Planta 166:389-393.

Okon Y (ed) (1994) *Azospirillum*-plant associations. CRC Press, Boca Raton USA.

Sarig S, Blum A, Okon Y (1988) Improvement of water status and yield of field-grown grain sorghum (*Sorghum bicolor*) by inoculation with *Azospirillum brasilense*. Agric. Sci. 110:271-277.

Laccase Activity in *Azospirillum lipoferum*: Substrates and Inhibitors

Colette Jacoud, Denis Faure, Aline Effosse, Patrick Wadoux[1] and Marie L. Bouillant
Laboratoire d'Ecologie Microbienne du Sol URA CNRS 1450
Université Claude Bernard - LYON 1
43, boulevard du 11 novembre 1918
69622 Villeurbanne Cedex, France

Abstract

Azospirillum lipoferum 4B and non-motile *A. lipoferum* 4T have been isolated from rice rhizosphere. *A. lipoferum* 4T and *A. lipoferum* 4Bp, a non-motile form of *A. lipoferum* 4B, exhibit a laccase activity (Bally R. *et al.* 1983). Different substrates and inhibitors, which are characteristic of both laccase and tyrosinase activities, have been screened. Oxidation products and enzymatic activity measurements have been carried out using spectrophotometry and HPLC. *p*- and *o*-diphenols, *p* and *o*-aminophenols and other fungal laccase substrates were oxidized by *A. lipoferum 4Bp* extracts. Both fungal and *Azospirillum* laccase activities oxidized aldehydic derivatives and syringic acid. These compounds are known to be precursors in lignins biosyntheses. Tyrosinase inhibitors, like kojic acid or salicylhydroxamic acid, were not effective on *Azospirillum* laccase activity.

Key Words:
laccase, polyphenoloxidase substrates, polyphenoloxidase inhibitors, tyrosinase, syringaldehyde, HPLC, spectrophotometry.

Abbreviations:
PPO: Polyphenoloxidase
SHAM: Salicylhydroxamic acid
DOPA: Dihydroxyphenylalanin
ABTS: 2,2'-azino-bis(3-ethylbenz-thiazoline-6-sulfonic acid)
DMBQ: 2,6-dimethoxy-*p*-benzoquinone
HPLC: High Performance Liquid Chromatography

---

[1] LIPHA, 10, avenue de Lattre de Tassigny, 69330 Meyzieu, France

NATO ASI Series, Vol. G 37
Azospirillum VI and Related Microorganisms
Edited by I. Fendrik et al.

Introduction

In bacteria *Azospirillum lipoferum* isolated from rice rhizosphere, motile and non motile forms were observed (Bally *et al.* 1983). Furthermore, it was found that non motile forms could be distinguished from motile ones by a brown pigment formation, qualified as melanins (Givaudan *et al.* 1991). Traditionnally, melanins are considered to be formed by the oxidative polymerization of phenolic compounds. It was demonstrated that non motile forms of *A. lipoferum* exhibit a laccase activity (Givaudan *et al.* 1993) probably implicated in a phenomenon of polymerization. Laccases are present in many fungi and sometimes in higher plants but none has yet been described in bacteria.
With a view of comparing this bacterial laccase activity with fungal ones, different substrates and inhibitors, which are characteristic of both laccase and tyrosinase activities have been screened.

Materials and Methods

Substrates analysis

Acellular extract from *A. lipoferum* 4Bp and commercial *Pyricularia oryzae* laccase preparation were used in assays and acellular extract from *A. lipoferum* 4B was used as control.
Each chemical solution containing either laccase or tyrosinase substrates (Table 1) were added together with their buffer into the enzyme solution.
After a few hours of incubation, a spectral analysis of reaction mixture was performed from 200 to 700 nm.
Compounds were considered as substrates of *A. lipoferum* 4Bp laccase activity when the spectrum obtained was:
1° different from the spectra obtained
-without enzyme and
-with an acellular extract from the laccase$^-$ *A. lipoferum* 4B
2° similar to the spectra obtained with *P. oryzae* laccase.

Inhibition measurements:

Commercial *Agaricus bisporus* tyrosinase and *P. oryzae* laccase preparation and acellular extracts of *A. lipoferum* 4Bp were used.
Selective substrates of *o*-polyphenoloxydase (*o*-aminophenol) and *p*-polyphenoloxydase (*p*-aminophenol and syringaldazine) were used to test the effects of:
- specific tyrosinase inhibitors (kojic acid, SHAM, tropolone)

- specific laccase inhibitor (N hydroxyglycine)

Measurement of the PPO activities of the solutions containing:

1° the enzyme preparation in buffer and

2° substrate with or without inhibitor,

have been carried out using a microplate reader at 540 nm (syringaldazine) and 450 nm (*o* and *p*-aminophenols).

Results

Substrates of bacterial laccase activity (Table 1)

All *o*- and *p*-PPO substrates tested were used by both *A. lipoferum* and *P. oryzae* laccase activities. On the other hand, characteristic tyrosinase substrates and *m*-diphenols were not oxidized by both bacterial and fungal laccase activites.

Table 1: Substrates of *P. oryzae* and *A. lipoferum* 4Bp laccase activities: in the presence of *P. oryzae* or *A. lipoferum* 4Bp, the spectral profile (200-700 nm) of chemicals was modified (+) or not (-).

Substrate	Characteristic activity	*A. lipoferum* 4Bp extract	*P. oryzae* laccase
L-Tyrosine	tyrosinase	-	-
p-coumaric acid		-	-
catechol	catecholoxidase	+	+
gallic acid		+	+
L-Dopa		+	+
o-aminophenol		+	+
hydroquinone	laccase	+	+
p-aminophenol		+	+
syringic acid	laccase	+	+
syringic aldehyde		+	+
syringaldazine	laccase	+	+
ABTS		+	+
resorcinol		-	-
5-pentadecylresorcinol		-	-

Most of the substrates tested are naturally occuring compounds:

- L-tyrosine, DOPA and catechol are intermediate products in melanins biosynthesis.
- *p*-coumaric acid is a precursor of lignin monomer.

- gallic acid can be polymerised into hydrolysable tannins.
- resorcinolic compounds are very toxic. They have been described in particular in root exudates (Bouillant *et al.* 1994).
- both syringaldehyde and syringic acid are lignin degradation products. Syringaldehyde is oxidized in syringic acid by a laccase activity and syringic acid is also converted by a laccase activity into DMBQ. This compound is involved in haustorium development of *Striga asiatica*, a parasitic plant of Sorgho (Chang and Lynn, 1986).

Effects of PPO's inhibitors on bacterial laccase activity (Table 2)

Specific tyrosinase inhibitors like kojic acid, SHAM and tropolone, and a specific laccase inhibitor, N-hydroxyglycine, allowed extensive characterization of *A. lipoferum* laccase activity.

Table 2: Inhibitors effects (+) or not (-) on the expression of tyrosinase and laccase activities.

	o-PPO activity			*p*-PPO activity	
inhibitors	*A.bisporus*	*P.oryzae*	*A.lipoferum*	*P.oryzae*	*A.lipoferum*
kojic acid	+++	-	-	+	++
SHAM	+++	-	-	-	-
tropolone	+++	-	-	-	-
N-hydroxyglycine	-	++	-	+++	++

For example, an increase of SHAM concentration did not affect the expression of both *A. lipoferum* and *P. oryzae o*-PPO activities while *o*- PPO activity of *A. bisporus* significantly decreased. Both *A. lipoferum* and *P. oryzae p*-PPO activities were not affected when the substrate was *p*-aminophenol. These *p*-PPO activities were more sensitive to SHAM when the substrate was syringaldazine.

Specific laccase inhibitor, N-hydroxyglycine, is effective on both *P. oryzae* and *A. lipoferum* but not on *A. bisporus* PPO activities.

Conclusion

Azospirillum lipoferum laccase activity oxidize the same substrates as *Pyricularia oryzae* laccase. Moreover, this laccase activity is sensitive to fungal laccase inhibitors. This suggests a strong functional similarity between fungal and bacterial laccase activities.

To date, the function of laccase activity in *A. lipoferum* is unknown. But the ability to oxidize naturally occuring phenolic compounds could be related to the competitivity of *Azospirillum* in the rhizosphere.

References

Bally R., Thomas-Bauzon D., Heulin T., Richard C. and De Ley J. 1983. Determination of the most frequent N_2-fixing bacteria in a rice rhizosphere. Can. J. Microbiol. 29: 881-887.

Bouillant M.L., Jacoud C., Favre-Bonvin J. and Bally R. 1994. Identification of 5-(12-heptadecenyl)-resorcinol in rice root exudates. Phytochem. 35: 769-771.

Chang M. and Lynn D.G. 1986. The haustorium and the chemistry of host recognition in parasitic angiosperms. J. Chem. Ecol. 12: 561-579.

Givaudan A., Effosse A. and Bally R. 1991. Melanin production by *Azospirillum lipoferum* strains. In: Nitrogen Fixation (Polsinelli M., Materassi R. and Vincenzini M. Eds), pp 311-312, Kluver Academic Publishers.

Givaudan A., Effosse A., Faure D., Potier P., Bouillant M.L. and Bally R. 1993. Polyphenoloxidase in *Azospirillum lipoferum* isolated from rice rhizosphere: evidence for laccase activity in non-motile strains of *Azospirillum lipoferum*. FEMS Microbiol. Lett. 108: 205-210.

Conversion of Tryptophan, Indole-3-Pyruvic Acid, Indole-3-Lactic Acid and Indole to Indole-3-Acetic Acid by *Azospirillum brasilense* Sp7

Tami Bar and Yaacov Okon
Department of Plant Pathology & Microbiology
Faculty of Agriculture,
The Hebrew University of Jerusalem,
Rehovot 76100, Israel

Abstract

The free-living nitrogen-fixing soil bacterium *Azospirillum brasilense* Sp7 produces indole-3-acetic acid (IAA) from tryptophan (Trp). Ammonia and oxygen are involved in the processes of IAA production and nitrogen fixation. Oxygen is required during the conversion of Trp to IAA via indole-3-acetamide (IAM). One ammonia molecule is released during the conversion of Trp to IAA. Under microaerophilic conditions, the cells were found to preferentially catalyze the enzymatic conversion of the tryptophan deaminated compound indole-3-pyruvic acid (IPyA) to IAA. Addition of ammonia caused a decrease in the conversion of Trp and IPyA but not indole-3-lactic acid (ILA) to IAA. Indole-3-pyruvic acid was found in the growth medium after addition of either Trp or ILA, indicating the function of an aminotransferase pathway for IAA synthesis from Trp. Three tryptophan-aminotransferases were observed on nondenaturing gels. No tryptophan-aminotransferase activity was observed upon addition of IPyA to the bacterial culture. Addition of 0.5 mM indole to the growth medium resulted in accumulation of IAA by an enzymatic process. Holotryptophanase activity was detected in cell-free extracts of *A. brasilense* Sp7 grown in a medium amended with indole. Addition of Trp to *A. brasilense* Sp7 culture as the sole N source partially inhibited nitrogenase activity. Study of the factors affecting both IAA production and nitrogen fixation is important for better understanding of the bacterial-plant association.

Introduction

Phytohormone production and nitrogen fixation by soil microorganisms are important processes governing beneficial bacteria-plant associations, such as symbiosis and associative symbiosis (Dreyfus et al., 1988; Loper and Schroth, 1986). The phytohormones produced by

NATO ASI Series, Vol. G 37
Azospirillum VI and Related Microorganisms
Edited by I. Fendrik et al.

microorganisms may modify the plant's own content of growth regulators, resulting in plant growth stimulation.

Free-living diazotrophic soil bacteria of the genus *Azospirillum* are potentially applicable as plant growth-promoting rhizobacteria, due to their ability to fix nitrogen and excrete phytohormones, such as IAA (Fallik et al., 1989; Omay et al, 1992), gibberrellins (Bottini et al., 1989) and cytokinin-like substances (Tien et. al., 1979). These abilities may be partly responsible for the close association of *Azospirillum* with the roots of many economically important cereals (Dobereiner and Pedrosa, 1987; Okon, 1985).

The IAA biosynthesis pathway involving tryptophan (Trp) as a precursor and indole-3-acetamide (IAM) as an intermediate was found in several bacteria including *Agrobacterium tumefaciens,* (Thomashow et al., 1984), *Bradyrhizobium japonicum* (Sekine et al., 1989), *Pseudomonas syringae pv. savastanoi* (Hutcheson and Kosuge, 1985; Yamada et al., 1985), pathogenic strains of *Erwinia herbicola* (Clark et al., 1993) and recently in *Azospirillum brasilense* (Bar and Okon, 1993). In the IAM pathway, the conversion of Trp to IAM is catalyzed by the enzyme tryptophan-2-monoxygenase (TMO) followed by conversion of IAM to IAA, which is catalyzed by IAM hydrolase. The genetic determinants for these enzymatic conversions are *iaa*M and *iaa*H, respectively (Yamada et al., 1985). Oxygen is required for the first step, releasing one molecule of water and one molecule of CO_2 (Hutcheson and Kosuge, 1985). A tryptamine-dependent pathway, in which the first step involves decarboxylation has also been suggested for *Azospirillum brasilense* (Hartmann et al., 1983).

Four aromatic amino acid aminotransferases (AAT) that can catalyze the deamination of Trp *in vitro* were found in *A. lipoferum*. Amino acid aminotransferase activity banding patterns were not influenced by different N sources, but one of these enzymes did not appear when Trp was missing from the growth medium (Ruckdaschel et al., 1988). Two aromatic aminotransferases were observed in *A. brasilense* strain UAP-14. The two enzymes could be distinguished by their electrophoretic mobility, they were constitutively produced and none of them was repressed by tyrosine or phenylalanine and tyrosine (Sanchez-Alonso et al., 1992). Three deaminated Trp compounds, indole-3-lactic acid (ILA), indole-3-methanol and indole-3-ethanol, have been identified in *A. brasilense* 703 Ebc culture medium (Crozier et al., 1988).

Indole-3-pyruvic acid (IPyA) was found in extracts of *Rhizobium* and *Bradyrhizobium* (Badenoch-Jones et al., 1982; Kaneshiro et al., 1983). The presence of IPyA indicates another pathway for IAA biosynthesis, not *via* IAM. This pathway combines deamination of Trp in the first step to form IPyA, without oxygen consumption (Perez-Galdona et al., 1989). Reduction of IPyA to ILA, and lactate oxidation by lactate dehydrogenase was observed in *R. meliloti* S22 (Trinchant and Rigaud, 1974).

The deamination of Trp to IPyA in bacteria, and its subsequent conversion to IAA, is in agreement with the biosynthesis of IAA from Trp via IPyA that occurs in most higher plants

(Cohen and Bialek, 1984). IPyA, ILA and Trp may therefore play a role in the regulation of IAA biosynthesis in the plant (Ernstsen et al., 1986).

An interrelationship between IAA production and nitrogen fixation in *Rhizobium* is implied by the observation that Trp inhibits acetylene reduction in *R. japonicum* tan-coloured mutants (Kaneshiro et al., 1983). Indole-acetic acid production by *Rhizobium* increases when 2-oxoglutaric acid is added (Garcia-Rodriguez et al., 1982), implying that a reduction of ammonia content in the cell enhances IAA production. Nitrogen fixation in *A. brasilense* Sp7 is inhibited by oxygen and ammonia and influenced by addition of amino acids to the medium (Hartmann and Burris, 1987).

In this report, IAA production, from Trp and IPyA, and nitrogen fixation under various conditions were investigated in order to study possible relationships between the two processes of IAA production and nitrogen fixation.

Materials and Methods

Bacteria and growth conditions

Azospirillum brasilense Sp7 (ATCC 29145) streptomycin resistant cells were grown in flasks in a shaking water bath (unless otherwise stated), under batch culture conditions (100 rpm, 30°C) in the medium described by Zimmer et al. (1988). Initial pH was adjusted to 6.8. NH_4 Cl was added to a final concentration of 5 mM.

Determination of indole derivatives by high-performance liquid chromatography (HPLC).

Culture broth was centrifuged at 13,000 g for 5 min. The cell-free supernatant was filtered through a 0.2 µm sterile disposable filter (Schleicher and Schuell). In order to minimize losses of indole derivatives (Zimmer et al., 1988), to avoid non-enzymatic conversion of Trp and other indole derivatives to IAA and to prevent the non-enzymatic oxidation of IAA, no medium extraction was made (Ernstsen et al., 1986).

Filtered supernatant (10 µl) was injected into a Hewlett Packard 1090 HPLC at room temperature. Indole derivatives were analyzed by isocratic reverse phase chromatography on a Merck 50943 LiChrospher 100RP-18 column (0.8 cm i.d.; 12.5 cm long; 5 µm particle size). The column was equilibrated with methanol: glacial acetic acid (1%) in water (40:60, v/v) to prevent non-enzymatic conversion of IPyA to IAA (Ernstsen et. al., 1986; Frankenberger and Poth, 1987), since IPyA is unstable in a basic solvent system (Kaneshiro et al., 1983). Flow rate was adjusted to 0.5 $ml \cdot min^{-1}$. Substances were quantified by integrating the areas under the peak (U.V. detector at 278 nm) using an Hewlett Packard 3394A integrator and indole derivatives (Merck) as standards. Indole-3-acetic acid concentration was expressed per $O.D_{450}$

of the examined suspension. High performance liquid chromatography sensitivity limit was 2 μmol of the indole derivatives.

Addition of indole compounds

Indole, Trp, IPyA or ILA were added to 48 h (for indole) or 72 h cultures in 100 ml flasks containing 50 ml growth medium. Variation in culture density was expected because of the different growth conditions (microaerophilic without ammonia supplementation vs. aerobic culture with 5 mM NH_4Cl). Because IAA production per cell was the matter of interest, IAA production was expressed as μmol per unit cell suspension (O.D. $_{450}$). Indole compounds were added to a final concentration of 0.5 mM. After the indole compound was added, growth rate and protein synthesis were examined by monitoring ^{35}S-methionine incorporation (10 $\mu Ci \cdot ml^{-1}$ for 30 min at 30°C, 100 rpm), in order to detect possible toxicity (Bar and Okon, 1992). No inhibition of growth or protein synthesis was observed in any of the experiments. Indole-3-acetic acid formation was determined by HPLC, as detailed above. All cultures were incubated in the dark to minimize the instability of indole compounds (Frankenberger and Poth, 1987). Microaerophilic conditions were obtained in bottles sealed with rubber stoppers by replacing air with N_2 3-4 h before addition of the indole compounds. Spontaneous conversion to IAA was determined under the same conditions as described above. Indole-3-acetaldehyde was not investigated because of the inherent instability of this compound.

Aromatic amino acid aminotransferase (AAT) assay.

Amino acid aminotransferase in bacterial culture was determined on a nondenaturating gel as described by Kittell et al. (1989), except for the lack of agarose in the staining reaction mixture, 18 h after the addition of an indole compound to the growth medium. Dark stained protein bands indicating Trp-aminotransferase activity did not appear in the control experiments (to which no indole compounds were added), thus confirming the specificity of the assay for AAT. Holotryptophanase activity was tested by adding 0.01 mM of the chromogenic substrate SOPC (S-O-nitrophenyl-L-cysteine) to a cell free extract and by measuring the absorbance peak formed at 420 nm. Since authentic S-O-nitrophenyl, that is formed by holotryptophanase activity from SOPC, could not be purchased or synthesized it was not possible to determine its actual concentration. Specific activity was expressed as (O.D. $_{420}$ (mg protein $\cdot$ ml^{-1}) $^{-1}$ $\cdot$ h^{-1}) (Behbahni-Nejad et al., 1983).

Nitrogen fixation assay

Nitrogenase activity was measured by the acetylene reduction activity (ARA) assay. The effect of Trp and 2-oxoglutaric acid on ARA was measured 3 and 6 h after the addition of the compounds to washed *A. brasilense* cells after overnight static incubation.

Immunoblot assay

Culture samples (1 ml) were washed with TE buffer (50 mM Tris · Cl, 20 mM EDTA (pH 8), seven hours after the addition of Trp (0.5 mM) or ammonia (5 mM), resuspended in 100 µl SDS-lysis buffer (10% (v/v) glycerol, 2.5% (w/v) SDS, 1 mM EDTA, 10 mM Tris base, 0.01% (w/v) bromophenol blue and 5% (v/v) 2-mercaptoethanol (pH 8)) and boiled for 5 min. The supernatant was transferred to a new Ependorf tube after 5 min of centrifugation at 15,000 g. Protein extract was electrophoresed on a polyacrylamide gel (9x12x0.15 cm, 12% separating gel, 3% stacking gel, 0.1% SDS) according to Laemmli (1970) in a Bio Rad cooling system at a constant 250V. Proteins were electroblotted to a nitrocellulose membrane (Schleicher and Schuell, 0.45 µm). The membrane was transferred to blocking solution (20 mM Tris base, 500 mM NaCl, 10% (v/v) 1% skimmed milk, adjusted to pH 7.5 with HCl) for 3 h, followed by 3 h in a solution containing anti-dinitrogenase reductase antibodies (diluted 1:500) of a *Klebsiella pneumoniae* batch αHL (kindly supplied by Prof. Ada Zamir, Weizmann Institute of Science, Israel) (Berman et al., 1985). Each membrane was washed three times for 20 min with blocking solution and then incubated for 1 h in ^{125}I-labelled protein A (Miles Yeda). After the incubation, the membranes were washed again three times each for 20 min in blocking solution containing 0.05% (v/v) Tween-20 to minimize background, air-dried and finally autoradiographed for 24 h at -70°C using Kodak X-OMAT film.

All the experiments were carried out with two to three cultures of independent origin. Four samples were taken from each culture to determine O.D., IAA content, or nitrogen fixation.

Results

Conversion of Trp, IPyA and ILA to IAA under various growth conditions

The effect of NH_4Cl supplement and oxygen on IAA production by *A. brasilense* Sp7 was examined 7 h after the addition of Trp, IPyA or ILA to batch cultures (Table 1).

Conversion of Trp to IAA was higher but not significantly under aerobic as compared to microaerophilic conditions, in the presence of 5 mM NH_4^+ by a factor of 2.1 and without NH_4^+ by a factor of 1.15. Addition of NH_4^+ reduced IAA production from Trp by a factor of 4.8 under aerobic conditions and of 8.8 under microaerophilic conditions (Table 1).

The production of IAA after addition of IPyA to the medium was in general much higher as compared to that produced from Trp. Production of IAA from IPyA was stimulated by microaerophilic conditions by a factor of 5.3 with NH_4^+ and of 7.0 without NH_4^+. In the presence of NH_4^+ IAA production from IPyA was reduced by a factor of 4.6 under aerobic and 6.2 under microaerophilic conditions (Table 1).

Spontaneous conversion of IPyA to IAA was monitored (Table 1). Three hours after IPyA addition measurements of IAA concentration in the medium remained constant for 24 h. No spontaneous conversion of Trp or ILA to IAA could be detected.

Conversion of ILA to IAA was higher under aerobic as compared to microaerophilic conditions in the presence of 5 mM NH_4^+ by a factor of 11.75. The same trend was observed when cultures were incubated without NH_4Cl: under aerobic conditions 43 µmol IAA per unit cell suspension ($O.D._{450}$) was accumulated in the medium, while no IAA was found under microaeophilic conditions. Addition of NH_4^+ increased IAA production from ILA by a factor of 2.2 under aerobic conditions, and the same trend was observed under microaerophilic conditions (Table 1).

TABLE 1: Effect of ammonium chloride (5 mM) and oxygen on the production of IAA in *Azospirillum brasilense* Sp7 after addition of Trp, IPyA or ILA (0.5 mM) and spontaneous conversion to IAA.

Growth Conditions	Trp addition NH_4^+	Trp addition No NH_4^+	IPyA addition NH_4^+	IPyA addition No NH_4^+	ILA addition NH_4^+	ILA addition No NH_4^+
			IAA production (µmol)			
+[a]	64±8.3[b]	305±118.9	148±4.4	686±34.3	94±2.7	43±1
-	30±0.6	265±7.9	780±15.6	4,819±240.9	8±0.3	n.d.
+	n.d.[d]	n.d.	85±0[c]	97±1.5[c]	n.d.	n.d.
-	n.d.	n.d.	37±0.9[c]	n.d.	n.d.	n.d.

[a] (+) aerobic, (-) microaerophilic

[b] IAA (µmol IAA per unit cell suspension, $O.D._{450}$) was analyzed by HPLC 7 h after addition of Trp, IPyA or ILA.

[c] Spontaneous conversion to IAA (µmol), without bacteria in the growth medium.

[d] n.d. not detected.

AAT assay

Four polypeptide bands exhibiting aminotransferase activity were observed by AAT assay. When Trp was omitted from the staining reaction mixture, three of the bands were eliminated (Fig. 1, part 2). No change in Trp-aminotransferase banding pattern was observed 18 h after the addition of tyrosine, Trp or IAM (Fig. 1, part 1, lanes a, c and g). Addition of Trp to an ammonia-free culture resulted in the same Trp-aminotransferase banding pattern (Fig. 1, part 1, lane d). This pattern was also exhibited by bacterial culture alone (Fig. 1, part 1, lane b). The latter result indicates, despite the changes in the second band intensity, that these Trp-aminotransferases are not induced by Trp. IPyA addition to bacterial culture with or without ammonia eliminated all of the bands (Fig. 1, part1, lanes e and f, respectively). The addition of indole caused inhibition of Trp-aminotransferase activity (Fig. 1, part 1, lane h).

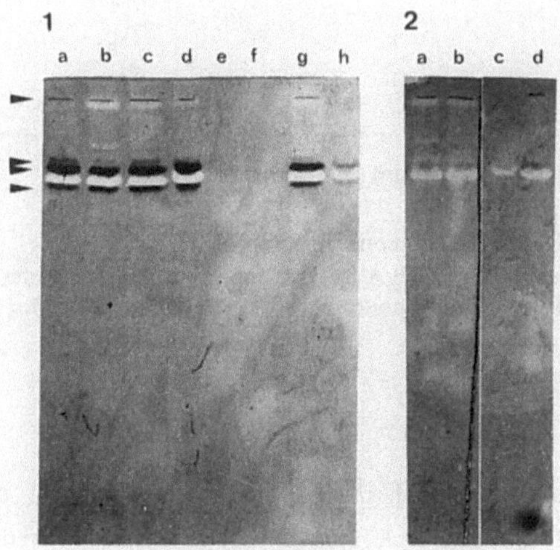

Fig. 1: Tryptophan-aminotransferase activity of *Azospirillum brasilense* strain Sp7.
Protein extract from bacterial cells at log-phase was loaded on a nondenaturing gel. Proteins with aminotransferase activity appear as dark bands. Gels were stained with or without Trp in the staining reaction mixture (parts 1 and 2 of the photograph, respectively). Part 1, lane b - no addition to the bacterial culture, lanes a, c, d, e, f, g and h - tyrosine, Trp, Trp without ammonia supplementation, indole-pyruvic acid, IPyA without ammonia supplementation, IAM and indole addition, respectively. Part 2, lane a - no addition to the bacterial culture, lanes b, c and d - with the addition of Trp, indole and tyrosine, respectively.

Determination of IPyA in the growth medium

Under microaerophilic conditions, i.e., the preferred condition for IPyA conversion to IAA, IPyA concentration in the medium was lower than under aerobic conditions. Similarly, IPyA concentration in the medium after the addition of Trp, was lower when ammonia was omitted (Table 2).

TABLE 2: Determination of IPyA in the growth medium of *Azospirillum brasilense* Sp7

NH_4^+ supplementation [a]	Growth conditions [b]	Addition of [c]	IPyA [d]
+	+	ILA	6±0.24
+	-	ILA	0
+	+	Trp	2±0.3
+	-	Trp	less than 0.5
+	+	IPyA	100±8
+	-	IPyA	45±6.7
+	+	Trp	11±1.5
-	+	Trp	less than 0.5

[a] (+) 5 mM NH_4^+ , (-) without ammonium chloride supplement.
[b] (+) aerobic (-) microaerophilic
[c] addition of Trp, ILA or IPyA to a final concentration of 0.5 mM.
[d] the exact spontaneous conversion of IPyA to IAA could not be determined, therefore, IPyA is expressed as percent of IPyA addition under aerobic conditions (100%).

Biosynthesis of IAA from Indole

Addition of 0.5 mM indole to a 48 h culture (O.D.$_{450}$ 0.65) of *A. brasilense* caused accumulation of IAA (17±2.3 μM IAA after 24 h) in the growth medium. No IAA accumulated after addition of indole to dead cells.

The possibility of conversion of indole to IAA in *Azospirillum* by tryptophanase was tested. It was found that after addition of 0.5 mM (final concentration) indole to the cell free extract, a holotryptophanase specific activity of 0.194 (O.D. 420 · (mg protein · $ml^{-1})^{-1}$ $\cdot h^{-1}$) could be obtained whereas with 1.6 mM (final concentration) indole the specific activity of holotryptophanase was 0.021. No activity could be detected after addition of Trp alone.

Effect of Trp on nitrogen fixation by A. brasilense Sp7

Nitrogenase activity was measured 3-6 h after the addition of Trp and/or 2-oxoglutaric acid (Table 3). Acetylene reduction activity was inhibited by 58% after Trp addition. Addition of 10 mM 2-oxoglutaric acid increased activity by 113% and 2-oxoglutaric acid with Trp by 33%.

TABLE 3: Effect of Trp and/or 2-oxoglutaric acid on acetylene reduction activity (ARA) by *A. brasilense* Sp 7.

Addition of	ARA (% of control)
None	100
Trp[a]	42
2-oxoglutaric acid[b]	213
Trp[a] and 2-oxoglutaric acid[b]	133

[a] 0.5 mM Trp
[b] 10 mM 2-oxoglutaric acid disodium salt
100% ARA determined as 194 nmole $C_2H_4 \cdot h^{-1} \cdot$ O.D. $_{450}$ unit^{-1}

Western blot analysis

The possibility of nitrogenase inhibition by the ammonia released from Trp during IAA biosynthesis, was studied and compared to the inhibition of nitrogenase by ammonia added directly to the medium. Seven hours after the addition of Trp or ammonia to the culture, Western blot analysis with *Klebsiella pneumoniae* anti-dinitrogenase reductase antibodies revealed a single polypeptide with a molecular weight of approximately 35.4 KDa interacting with the antibodies. The amount of this polypeptide was significantly reduced by the addition of Trp and disappeared after addition of 5 mM NH_4^+ (Fig. 2).

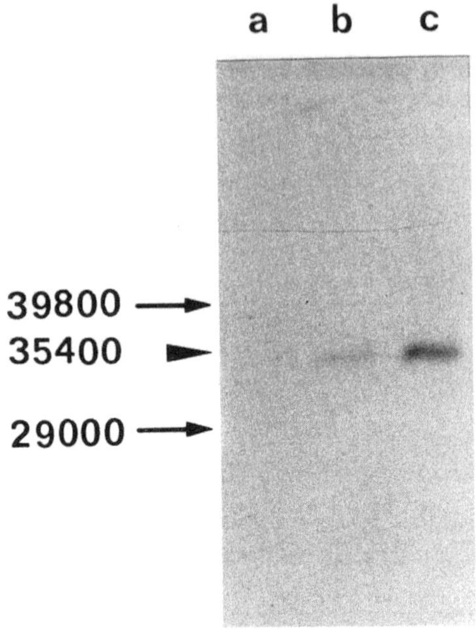

Fig. 2: Western blot autoradiograph of *Azospirillum brasilense* Sp7 extract.
7h after Trp or NH_4^+ addition, the extracts were probed with antidinitrogenase reductase antibodies. Lane a, NH_4^+ addition; lane b, Trp addition and lane c, untreated. The numbers on the left indicate the molecular weights in KDa.

Discussion

It has been reported that *A. brasilense* in batch cultures amended with tryptophan and NH_4^+ accumulated more IAA at the stationary growth phase, compared to cultures growing on N_2 (Hartmann et al., 1983; Omay et al., 1992). In our work after addition of Trp or IPyA to a 72 h culture with NH_4^+ we observed reduced production of IAA as compared to cultures growing on N_2. We have observed that the production of IAA after a pulse of Trp to 48 and 72 h cultures was linear versus time (Bar and Okon, 1982). Detection of IAA after addition of precursors to already grown culture is a more accurate way of determining IAA production rates by *A. brasilense* cultures.

In the absence of oxygen, IAA production from either ILA or Trp was reduced . On the other hand, IPyA conversion to IAA was five to seven fold higher under microaerophilic conditions than under aerobic conditions, contrary to its tendency during spontaneous conversion. Therefore, the existence of various biosynthetic pathways with different regulatory strategies in *A. brasilense* Sp7 for IAA production is suggested, based on the differences in IAA production under aerobic vs. microaerophilic conditions.

A reduction in the spontaneous conversion of IPyA to IAA has been reported in the presence of antioxidants. The maximum reported spontaneous conversion without antioxidants in acidic systems (pH 3.5) after 49 h was 6% (Ernstsen et al., 1986).

The presence of IPyA in the growth medium after addition of Trp or ILA, when no spontaneous conversion of Trp or ILA to IAA occurred, indicates that IAA synthesis can occur *via* IPyA.

The diversity of pathways for synthesis of IAA in *A. brasilense* is further complicated by the observation in this work that IAA production was stimulated in the presence of indole possibly *via* tryptophanase activity.

Formation of IAA from indole could be the basis of enzymatic changes of heterocyclic compounds and/or indication of tryptophanase activity (Behbahni-Nejad, et al., 1987).

Inhibition of tryptophan aminotransferase activity by high indole concentration indicates possible control of IAA synthesis from Trp. The use of Trp as a sole carbon source in *A. brasilense* may indicate activity of tryptophanase, as it was observed in *Klebsiella aerogenes* (Paris and Magasanik, 1981). We detected tryptophanase activity in *A. brasilense* after addition of indole to the growth medium. No activity was found after addition of Trp alone.

The high amount of IPyA conversion to IAA, particularly under microaerophilic conditions, indicates the advantage of these conditions in *A. brasilense* for the catalysis of IPyA conversion to IAA. Conversion of both Trp and IPyA to IAA was reduced in the presence of ammonia under all tested conditions (Table 1). Western blot with dinitrogenase reductase antibodies revealed nitrogenase repression caused by Trp, which may suggest the possible inhibition of nitrogenase activity by ammonia released during the conversion of Trp to IAA. Several amino acids are shown to influence nitrogenase activity in the stationary growth phase of a culture of *A. brasilense* Sp7 (Hartmann et al., 1988).

Nitrogenase activity was higher when 2-oxoglutaric acid was added alone or with Trp. This, as well as the fact that *A. brasilense* can grow in the presence of Trp as a sole N source (Ruckdaschel et al., 1988) suggests that Trp, as a N-source, could affect both nitrogen fixation and IAA production. From the data presented in this work it is suggested that, limited air and conditions for nitrogen fixation, which usually occur in soil could increase IAA production by *A. brasilense*.

Acknowledgement

This work was supported by a grant from the Minerva-Stipendien-Komitte, Deutsche-Israelische Wissenschaftliche Zusammenarbeit.

References

Badenoch-Jones J, Summons RE, Entsch B, Rolfe BG, Parker CW and Letham DS (1982) Mass spectrometric identification of indole compounds produced by *Rhizobium* strains. Biomed Mass Spectron 9:429-437.

Bar T and Okon Y (1992) Induction of indole-3-acetic acid synthesis and possible toxicity of tryptophan in *Azospirillum brasilense* Sp7. Symbiosis 13:191-198.

Bar T and Okon Y (1993) Tryptophan conversion to indole-3-acetic acid *via* indole-3-acetamide in *Azospirillum brasilense* Sp7. Can J Microbiol 39:81-86.

Behbahni-Nejad I, Dye JL and Suelter CH (1987) Tryptophanase from *Escherichia coli* B/It7-A. Method in Enzymol 142:414-422.

Berman J., Gershoni JM and Zamir A (1985) Expression of nitrogen fixation genes in foreign hosts. J Biol Chem 260:5240-5243.

Bottini R, Fulchieri M, Pearce D and Pharis, RP (1989) Identification of gibberellins A1, A3 and iso-A3 in culture of *Azospirillum lipoferum.* Plant Physiol 89:1-3.

Clark E, Manulis S, Ophir Y, Barasch I and Gafni Y (1993) Cloning and characterization of iaaM and iaaH from *Erwinia herbicola* pathovar *ghysophilae*. Phytopathol 83:234-240.

Cohen JD and Bialek K (1984) The biosynthesis of indole-3-acetic acid in higher plants. In: The Biosynthesis and Metabolism of Plant Hormones , pp. 165-181, A. Crozier and JR Hillman (eds.). Cambridge University Press, Cambridge, Mass.

Crozier A, Arruda P, Jasmin JM, Monterio AM and Sandberg G (1988) Analysis of indole-3-acetic acid and related indoles in culture medium from *Azospirillum lipoferum* and *Azospirillum brasilense*. Appl Environ Microbiol 54:2833-2837.

Dobereiner J and Pedrosa FO (1987) Nitrogen-fixing bacteria in nonleguminous crop plants. In: Brock Springer Series in Contemporary Biology, pp. 1-155, Sciencia Tech Publishers: Madison Wis. Springer Berlin.

Dreyfus BL, Diem HG and Dommergues YR (1988) Future directions for biological nitrogen fixation research. Plant Soil 108:191-199.

Ernstsen A, Sandberg G and Crozier A (1986) Effects of sodium diethyldithiocarbamate, solvent, temperature and plant extracts on the stability of indoles. Physiol Plant 68:519-522.

Fallik E, Okon Y, Epstein E, Goldman A and Fischer M (1989) Identification and quantification of IAA and IBA in *Azospirillum brasilense* - inoculated maize roots. Soil Biol Biochem 21:147-153.

Frankenberger WT Jr and Poth M (1987) Determination of substituted indole derivatives by ion suppression-reverse-phase high-performance liquid chromatography. Anal Biochem 165:300-308.

Garcia-Rodriguez T, Gutierrez-Navarro AM, Garcia R and Perez Silva J (1982) Indole acetic acid production by *Rhizobium*: effect of 2-ketoglutaric acid. Soil Biol Biochem 14:153-155.

Hartmann A and Burris RH (1987) Regulation of nitrogenase activity by oxygen in *Azospirillum brasilense* and *Azospirillum lipoferum.* J Bacteriol 169:944-948.

Hartmann A, Fu H and Burris RH (1988) Influence of amino acids on nitrogen fixation ability and growth of *Azospirillum* spp. Appl Environ Microbiol 54:87-93.

Hartmann A, Singh M and Klingmuller W (1983) Isolation and characterization of *Azospirillum* mutants excreting high amounts of indoleacetic acid. Can J Microbiol 29:916-923.

Hutcheson SW and Kosuge T (1985) Regulation of 3-indoleacetic acid production in *Pseudomonas syringae* pv. *savastanoi.* Purification and properties of tryptophan 2-monoxygenase. J Biol Chem 260:6281-6287.

Kaneshiro T, Slodki ME and Plattner RD (1983) Tryptophan catabolism to indolepyruvic and indoleacetic acids by *Rhizobium japonicum* L-259 mutants. Curr Microbiol 8:301-306.

Kittell BL, Helinski DR and Ditta GS (1989) Aromatic aminotransferase activity and indoleacetic acid production in *Rhizobium meliloti.* J Bacteriol 171:5458-5466.

Loper JE and Schroth MN (1986) Influence of bacterial source of indole-3-acetic acid on root elongation of sugar beet. Phytopathology 76:386-389.

Okon Y (1985) *Azospirillum* as a potential inoculant for agriculture. Trends Biotechnol 3:223-228.

Omay SH, Schmidt WA and Martin P (1992) Indoleacetic acid production by the rhizosphere bacterium *Azospirillum brasilense* Cd under in vitro conditions. Can J Microbiol 39:187-192.

Paris CG and Magasanik B (1981) Tryptophan metabolism in *Klebsiella aerogenes:* regulation of the utilization of aromatic amino acid as source of nitrogen. J Bacteriol 145:257-265.

Perez-Galdona R, Corzo J, Leon-Barrios M and Gutierrez-Navarro AM (1989) Aromatic amino acid aminotransferases in *Rhizobium leguminosarum* biovar *trifolii.* Their role in indoleacetic acid synthesis. Soil Biol Biochem 21:573-579.

Ruckdaschel E. Kittell BL, Helinski DR and Klingmuller W (1988) Aromatic amino acid aminotransferases of *Azospirillum lipoferum* and their possible involvement in IAA biosynthesis. In: *Azospirillum* IV: Genetics, Physiology, Ecology. pp. 49-53, W Klingmuller (ed.) Springer Verlag, Berlin and Heidelberg.

Sanchez-Alonso MPG, Varzquez-Cruz C, Martinez-Morales LU, Fluren-Martinez BE and Baca BE (1992) Enzymatic activities involved in auxin indoleacetic acid (IAA) biosynthesis and their relationship with plasmids harboured in *Azospirillum* strains. Rev Lat Amer Microbiol 33:25-34.

Sekine M, Watanabe K and Syono (1989) Molecular cloning of a gene for indole-3-acetamide hydrolase from *Bradyrhizobium japonicum*.. J Bacteriol 171:1718-1724.

Thomashow LS, Reeves S and Thomashow MF (1984) Crown gall oncogenesis: evidence that a T-DNA gene from the *Agrobacterium* Ti plasmid pTiA6 encodes an enzyme that catalyzes synthesis of indoleacetic acid. Proc Nat Acad Sci USA 81:5071-5075.

Tien TM, Gaskins MH and Hubbell DH (1979) Plant growth substances produced by *Azospirillum brasilense* and their effect on the growth of pearl millet (*Pennisetum americanum* L.) Appl Environ Microbiol 37:1016-1024.

Trinchant JC and Rigaud J (1974) Lactate dehydrogenase from *Rhizobium* . Purification and role in indole metabolism. Physiol Plant 32:394-399.

Yamada T, Palm, CJ, Brooks B and Kosuge T. (1985) Nucleotide sequences of the *Pseudomonas savastanoi* indoleacetic acid genes show homology with *Agrobacterium tumefaciens* T-DNA. Proc Natl Acad Sci USA 82:6522-6526.

Zimmer W, Roeben K and Bothe H (1988) An alternative explanation for plant growth promotion by bacteria of the genus *Azospirillum.* Planta 176:333-342.

Transformation of Aniline and its Chlorinated Derivatives by *Azotobacter agilis*

Stefan Russel, Stanisław Muszyński*
Department of Agricultural Microbiology
Warsaw Agricultural University
02-528 Warsaw
ul.Rakowiecka 26/30
Poland

Abstract

Aniline and its different chlorinated derivatives were metabolized in a liquid growth medium by a *Azotobacter agilis* which was isolated from soil. After one week of incubation unsubstituted aniline was completely disappeared. From monochloroanilines, 4-chloroaniline was transformed in 85% where as at the same time *Azotobacter agilis* metabolized about 50% of dichlorinated anilines.

When ^{14}C-ring-labeled 4-chloroaniline was used as a substrate in the growth medium after 7 days of incubation only about 5% of total radioactivity was not detected but the most of radioactivity after extraction with ethyl-ether in organic phase was found.

Clear evidence of the ability of the *Azotobacter agilis* to metabolize 4-chloroaniline was obtained by TLC analysis. Several metabolites of 4-chloroanilne was detected by TLC method.

4-chloroacetanilide and 4-chloropropionanilide were identified as a main products. This identification was confirmed by melting point, UV and IR spectra and mass-spectra of isolated metabolites and chemically synthesized 4-chloroacetanilide and 4-chloropropionanilide.

Using gas-chromatography it was possible to find in *Azotobacter agilis* culture medium amended with 4-chloroaniline two additional metabolites: 4,4'- dichloroazoxybenzene and 4,4'-dichloroazobenzene.

Introduction

Many agriculturally important pesticides are derivatives of substituted anilines (Cantier, Bastide, Yan and Coste, 1988; Rouchaud, Rouchaud, Benoit, Ceustermans, Gillet and Maraite, 1989).These compounds release aniline moieties on breakdown in microbial culture media and natural environment (Bollag, 1974).

Our knowledge of the further transformation of these intermediary products is still fragmentary. So far, three major biological transformation reactions of substituted anilines have been

---

* Department of Plant Breeding and Seed Science, Warsaw Agricultural University, 02-528 Warsaw, ul. Rakowiecka 26/30, Poland

NATO ASI Series, Vol. G 37
Azospirillum VI and Related Microorganisms
Edited by I. Fendrik et al.

described: (a) dimerization to an azo-derivative or to a diphenylamine; (b) acylation resulting in a formylated acetylated aniline; and (c) oxidation of the amine to a nitro group (Bollag, 1974; Minard, Russel and Bollag, 1977).

The present study describes the transformation onf aniline and its chlorinated derivatives in *Azotobacter agilis* culture medium.

Materials and methods

The strain of *Azotobacter agilis* used in this study was isolated from silt loam soil.

The bacterium was grown aerobically in nitrogen-free medium (Burk and Lineweaver, 1930) at 28°C. Aerobic conditions were provided by shaking the cultures in Erlenmeyer flasks on a rotary shaker (150 oscillations/min).

The various anilines were dissolved in absolute ethanol, sterilized by Millipore filters, and added to the growth medium at the final concentration 20ppm.

The following chemicals were used: 4-chloroaniline - (ring-UL-^{14}C) sulfate (specific activity 4.5 µCi/mM), aniline, 2-chloroaniline, 3-chloroaniline, 4-chloroaniline, 2,3-dichloroaniline, 2,4-dichloroaniline, 2,5-dichloroaniline and 2,6-dichloroaniline were purchased from Aldrich Chemical Co., Inc., Milwaukee, Wis., USA. The purity of ^{14}C-4-chloroaniline was verified by cochromatography on TLC with different solvent systems.

The transformation of anilines was measured initially by colorimetric method and subsequently by thin-layer chromatography analysis (Bollag and Russel, 1976). For thin-layer chromatography, the culture medium was centrifuged and the supernatant was extracted with an equal volume of diethyl ether. After concentration, the ether extract was applied on a silica gel F-254 thin-layer plate of a thickness of 0.25mm (Brinkmann Instruments, Inc., Westbury, N.Y., USA). The following solvent system was employed: etherhexane (4:1, v/v). Spots on plates were detected with ultraviolet light at 254nm and by spraying with a p-aminobenzaldehyde reagent.

Radioactivity was determined with Beckman liquid scintillation counter. The radioactive samples were measured in a scintillation solution composed of 8g Omnifluor and 2% Bis-MSB, 60g naphthalene, and 100ml of methanol in 1 liter of dioxane. Radioactive zones from thin-layer plates were counted after scraping the entire area into the scintillation solution and adding Cab-0-Sil (thixotropic suspension powder) at a concentration of 6%.

Mass spectra were taken with Varian MAT CH 5 mass spectrometer at an ionization potential of 70eV using the direct insertion probe. Ultraviolet spectra were determined with Beckman spectrometer, and infrared spectra were obtained with a Perkin-Elmer spectrometer using a KBr disk technique.

Gas chromatography was carred out with Beckman gas chromatograph.
Radioactivity on TLC plates was also measured using DC-Scanner, Modell LB 2723.

Results

Among the bacteria isolated from silt loam soil, we found one strain that was active in transformation of different anilines. This strain was identified as *Azotobacter agilis*.

Table 1. Transformation of aniline and its chlorinated derivatives in, *Azotobacter agilis* culture medium

Compound	Optical density at 550nm	Degradation of substrate in %	Number of metabolites
Aniline	0.42	100	6
2-chloroaniline	0.38	73	5
3-chloroaniline	0.35	78	5
4-chloroaniline	0.40	85	5
2,3-dichloroaniline	0.29	51	4
2,4-dichloroaniline	0.29	56	4
2,5-dichloroaniline	0.32	53	5
2,6-dichloroaniline	0.30	50	4

After 7 days of incubation, the isolated strain completely transformed unsubstituted aniline. From monochloroanilines, 4-chloroaniline was transformed in 85%, but dichlorinated anilines were degraded in about 50% (Table 1).

In Fig. 1. The distribution of radioactivity during growth of *Azotobacter agilis* in culture medium amended with 20ppm of C^{14} -labeled 4-chloroaniline between aqueous and organic phases (after extraction with ether) is shown. During growth of bacterium the total radioactivity was almost not changed, but after 7 days of incubation, about 85-90% of radioactive material in ether phase was found.

Clear evidence of the ability of the *Azotobacter agilis* to metabolize 4-chloroaniline was obtained by thin-layer chromatographic analysis of the ether extract from the culture medium (Fig. 2). 4-chloroaniline was almost completely transformed, and formation of metabolites, which were named M-1, M-2 and M-3, could be observed.

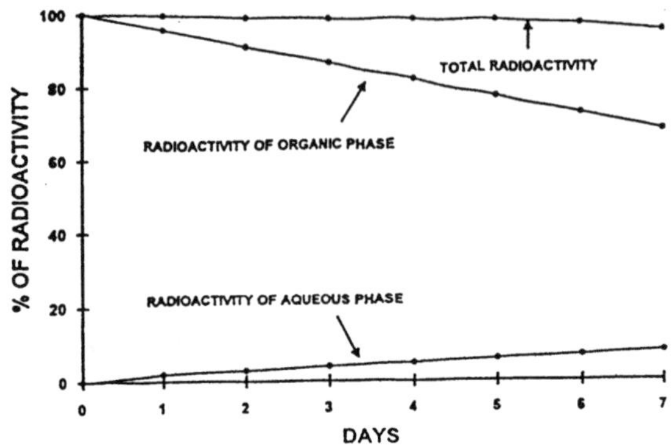

Fig. 1. Distribution of radioactivity during growth of *Azotobacter agilis* after extraction of C^{14}-labeled 4-chloroaniline with ethyl-ether

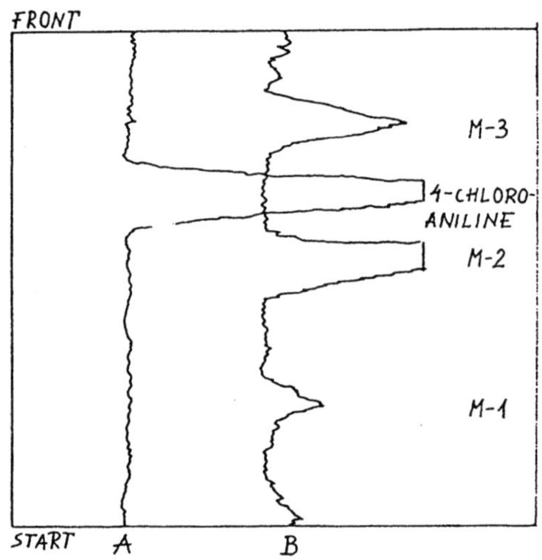

Fig. 2. Metabolism of ^{14}C-labeled 4-chloroaniline by *Azotobacter agilis* as indicated by TLC analysis using radioactivity scanner. (A) - control medium with 20ppm of 4-chloroaniline, (B) - culture medium with the same amounts of substrate

The metabolism of 4-chloroaniline during growth of *Azotobacter agilis* was also sudied by gas chromatography (Fig. 3). Using this method it was possible to discover two additional metabolites, named as M-4 and M-5.

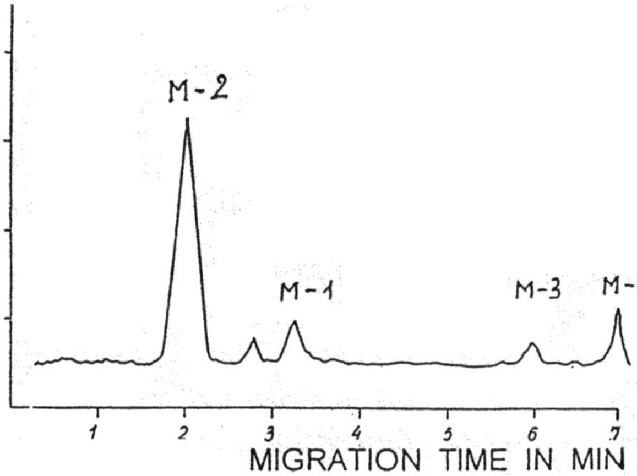

Fig. 3. Gas chromatography of ether extract of *Azotobacter agilis* culture medium amended with 20ppm of 4-chloroaniline after 7 days of incubation

Attempts were made to isolate and characterize the metabolites produced from 4-chloroaniline in *Azotobacter agilis* culture medium, but only enough material could be obtained from M-1 and M-2 for identification (Table 2).

Table 2. Some characteristics of metabolites M-1 and M-2 produced in *Azotobacter agilis* culture medium

Metabolite	Identified as	R_f value (ether:hexane=4:1)	Melting point in °C	UV nm	Mass spectrometry	
					Mol. Ion	Fragments
M-1	4-chloroacetanilide	0.21	177°	248	169	153,127,75
M-2	4-chloropropionanilide	0.53		242	183	127,75,43

Based on chromatographic characteristics, melting point, UV- and mass-spectra, metabolite M-1 was identified as 4-chloroacetanilide, and metabolite M-2 as 4-chloropropionanilide.

The infrared spectra of chemically synthesized 4-chloroacetanilide and 4-chloropropionanilide and the isolated metabolites were identical (Fig.4 and Fig. 5).

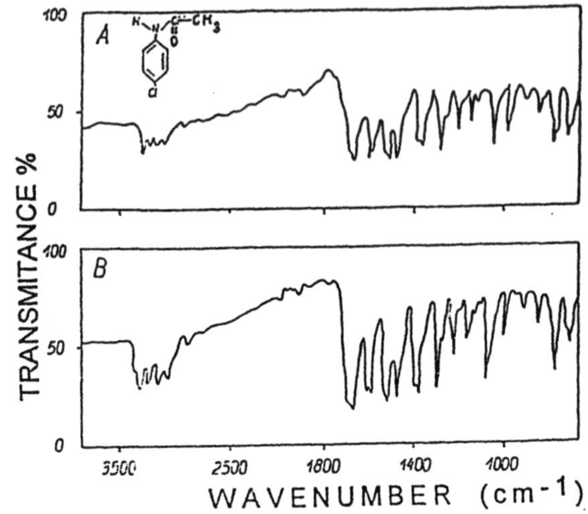

Fig.4. The infrared spectra of 4-chloroacetanilide (A) and metabolite M-1 (B)

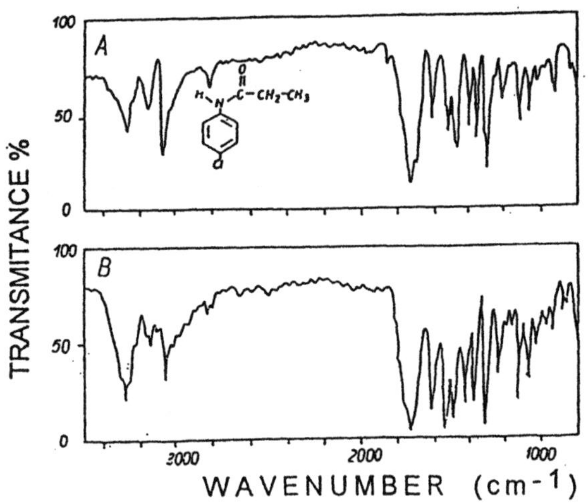

Fig. 5 The infrared spectra of 4-chloropropionanilide (A) and metabolite M-2 (B)

The results of acylation of various anilines in *Azotobacter agilis* culture medium were collected in Table 3.

Table 3. Acetylation and propionylation of aniline and its chlorinated derivatives by *Azotobacter agilis*

Compound	Content of 4-chloroacetanilide in % of orginal radioactivity	Content of 4-chloropropionanilide in % of orginal radioactivity
Aniline	25	41
2-chloroaniline	29	35
3-chloroaniline	23	39
4-chloroaniline	21	33
2,3-dichloroaniline	15	21
2,4-dichloroaniline	13	28
2,5-dichloroaniline	10	30
2,6-dichloroaniline	12	26

From Table 3 it appears that *Azotobacter agilis* transformed all investigated anilines to acetylated and propionylated prodacts. In all cases higher amounts of propionanilides than acetanilides were observed.

Discussion

Acylation is a major methabolic pathway of foreign anilines as it has been demonstrated especially with mammalian tissues (Weber, 1973). However, there is a little knowledge on the mechanism of acylation of halogenated anilines by microorganisms. Various anilides were isolated from microbial culture media, but no evidence of the active enzymes in this process was given (Bollag and Russel, 1976).

In present study, the *Azotobacter agilis*, which was isoalted from soil, transformed aniline and its chlorinated derivatives to the two acylated products, acetanilides and propionanilides. At this process several other unidentified metabolites were also found.

As far as we are aware, the propionylation of halogenated anilines has not been reported with microbial cultures. Since our investigation provides clear evidence that *Azotobacter* species produces the propionylation of anilines, one could suppose that other microorganisms are also able of modifying amino groups in such a way.

References

Bollag J.-M. (1974) Microbial transformation of pesticides. Advan. Appl. Microbiol. 18:75-130

Bollag J.-M., Russel S. (1976) Formylation and acetylation of 4-chloroaniline by a Streptomyces sp. Acta Microbiol. Pol. 26:59-64

Burk D., Lineweaver H. (1930) The influence of fixed nitrogen on Azotobacter. J. Bacteriol. 19:389-391

Cantier J.M., Bastide J., Yan Z., Coste C. (1988) Degradation of carbetamide and its chief metabolites in alkaline soil. Pesticide Sci. 23:327-335

Minard R.D., Russel S., Bollag J.-M. (1977) Chemical transformation of 4-chloroaniline to a triazene in a bacterial culture medium. J. Agr. Food Chem. 25:841-844

Rouchaud J., Rouchaud P., Benoit F., Ceustermans N., Gillet J., Maraite H. (1989) Plant and soil metabolism of the herbicide chlorpropham in field grown crops of lettuce, scorzonera and Italian chicory. Soils and Fertilizers 52:113-120

Weber W.W. (1973) Acetylation of drugs. Metabolic conjugation and metabolic hydrolysis. Academic Press New York

Reduction of 4-chloronitrobenzene by *Azospirillum lipoferum*

Stefan Russel, Stanisław Muszyński*
Department of Agricultural Microbiology
Warsaw Agricultural University
02-528 Warsaw
ul. Rakowiecka 26/30
Poland

Abstract

The strain of *Azospirillum lipoferum* which reduces 4-chloronitrobenzene under aerobic and anaerobic conditions was isolated from the soil. In experiments with ^{14}C-ring-labeled 4-chloronitrobenzene, approximately 40% of the radioactive material disappeared from the growth medium after incubation under anaerobiosis within 48 hours, but under aerobic conditions only small decrease of radioactivity in the growth medium was observed. The major two metabolites produced from 4-chloronitrobenzene in a *Azospirillum* culture medium were isolated and characterized. These metabolites appear to be 4-chloroacetanilide and 1,3-bis(p-chlorophenyl)triazene.

Introduction

The nitroaromatic compounds, such as nitrobenzenes, are used in the manufacture of pesticides, dyes, explosives and industrial solvents. The microbial degradation of these substances is of environmental importance (Bollag, 1974).

In natural environment nitrobenzenes and their different derivatives can be transformed by various groups of microorganisms (Hallas and Alexander, 1983; Patil and Shinde, 1988). The most characteristic microbiological transformation reaction of nitroaromatic pollutants is reduction of nitro to a amine group (Bollag, 1974) resulting in aniline formation.

Evidences exist that the genus *Azospirillum* is engaged in denitrification processes (Neyra, Döbereiner, Lalande and Knowles, 1977; Neuer, Kronenber and Bothe, 1985). In denitrification, nitrate is reduced to nitrite and further to N_20, N_2 or ammonia (Ingraham, 1981). *Azospirillum*

----

* Department of Plant Breeding and Seed Science, Warsaw Agricultural University, 02-528 Warsaw, ul. Rakowiecka 26/30, Poland

NATO ASI Series, Vol. G 37
Azospirillum VI and Related Microorganisms
Edited by I. Fendrik et al.

can even grow in liquid culture anaerobically with nitrate as respiratory electron acceptor (Bothe, Klein, Stephan and Döbereiner, 1981).

The present communication describe the reduction of 4-chloronitrobenzene by *Azospirillum lipoferum* under aerobic and anaerobic conditions in culture medium with nitrate as nitrogen source.

Materials and methods

The strain of soil bacterium identified as *Azospirillum* was cultivated in a Czapek-Dox medium (Minard, Russel and Bollag, 1977). After autoclaving sterilized 4-chloronitrobenzene was added.

The 4-chloronitrobenzene was purchased from Buchler, Amersham, Germany and recrystalized several times from hexane; its purity was determined by TLC, melting point determination, and mass spectral analysis. Uniformly ^{14}C-labeled 4-chloronitrobenzene was obtained from Buchler, Amersham, Germany and had a specific activity 103.3mCi/mM. The labeled material was diluted with sufficient unlabeled compound to reach the desired concentration. The 4-chloronitrobenzene was introduced into 50ml of medium in 0.2ml of ethanol solution after sterilization by membrane filtration (0.22µm pore size Milipore filter).

To obtain anaerobic conditions, the flasks were flushed with nitrogen until all air was removed (Bollag and Nash, 1974) and incubated at 28°C.

For experiments designed to study chemical transformation, a modified Czapek-Dox medium was used. Instead of nitrate, nitrite was added.

Nitrite was measured colorimetrically (American Public Health Association, 1971).

For all routine analysis, 5ml of growth medium or chemical reaction mixture was extracted with an equal volume of diethyl ether. The organic phase was used for measurements of radioactivity and TLC analysis.

Radioactivity was determined with Beckman liquid scintillation counter.

Precoated thin-layer plates (silica gel F-254) with fluorescent indicator (layer thickness, 0.25mm) were used. The following solvent system was employed: ether-hexane (4:1, v/v).

For routine TLC analysis, 1.0ml of ether extract was concentrated to 0.1ml and spotted on plates. All compounds were made visible with ultraviolet light at 254nm. The 4-chloroaniline was futher determined by spraying the plates with an aromatic amines specific reagent composed of 1g of p-dimethylaminobenzaldehyde dissolved in 180ml of 1-butanol, 30ml ethanol and 3ml of HCl.

For determination of 4-chloroaniline formation in the culture medium, a colorimetric method was applied using p-di-methylaminobenzaldehyde-reagent (Bollag and Russel, 1976).

Mass spectra were taken with Varian MAT CH 5 mass spectrometer at ionization potential of 70eV using direct insertion probe. Ultraviolet spectra were determined with Beckman spectrometer.

Results

During growth of *Azospirillum lipoferum* in a Czapek-Dox medium with 20ppm of 4-chloronitrobenzene under aerobic and anaerobic conditions, 4-chloroaniline was produced (Fig. 1)

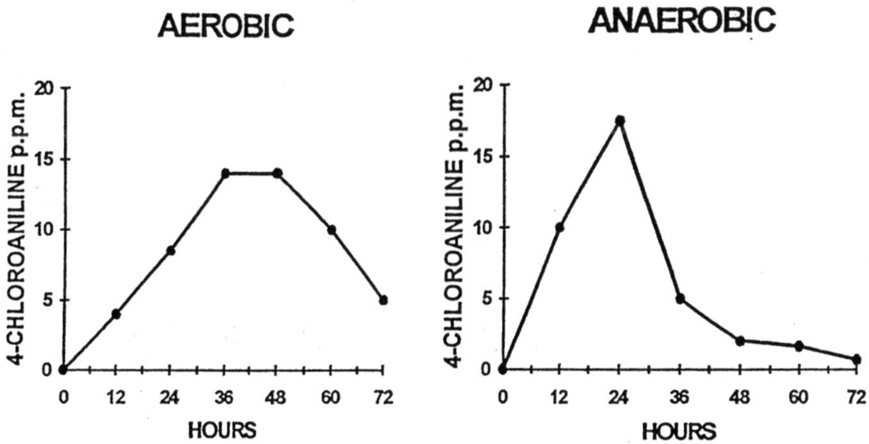

Fig.1 Reduction 4-chloronitrobenzene to 4-chloroaniline and products reacting with the applied colorimetric method by a *Azospirillum lipoferum* under aerobic and anaerobic conditions

When ^{14}C-ring-labeled 4-chloronitrobenzene was used as substrate in the growth medium of *Azospirillum lipoferum*, no apparent decrease of radioactivity could be detected after 7 days under aerobic conditions. However; incubation of the bacterial culture during the same time period under anaerobic conditions resulted in the loss of approximately 40% of the radioactive material (Fig.2). When nitrogen was continuously flushed through the anaerobic culture and subsequently passed through a n-butanol, it was possible to trap approximately 40% of radiolabeled material. No radioactivity was found when 5% NaOH was used.

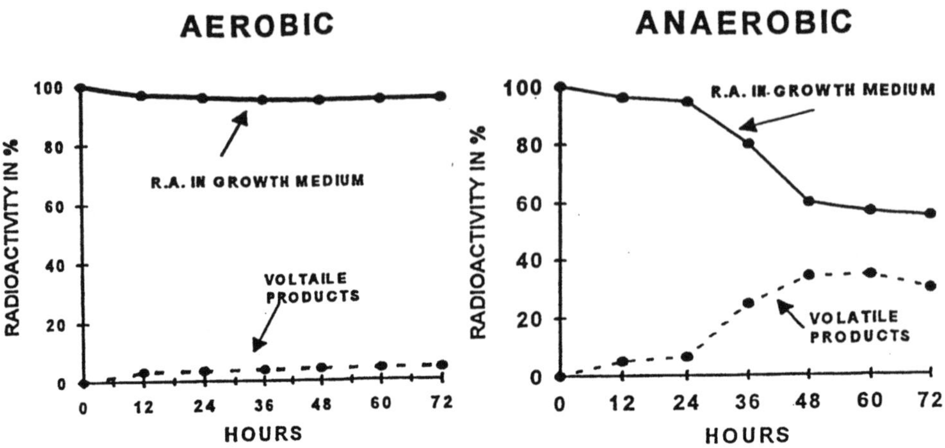

Fig. 2. Transformation of ^{14}C-ring-labeled 4-chloronitrobenzene by a *Azospirillum lipoferum* under aerobic and anaerobic conditions, as indicated by disappearance of radioactivity from the culture medium and trapping of volatile metabolites in an alcohol solution

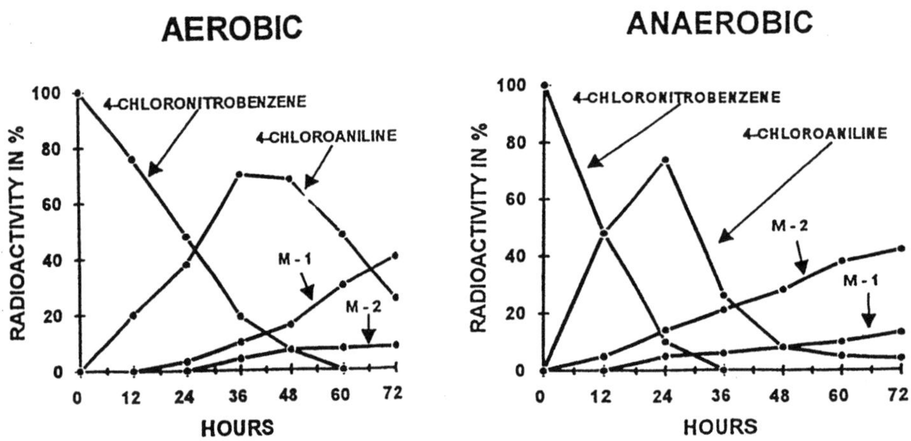

Fig.3. Metabolism of 4-chloronitrobenzene as indicated by TLC of the ether extract from aerobic and anaerobic cultures of a *Azospirillum lipoferum*

Metabolism of ^{14}C-labeled 4-chloronitrobenzene during growth of a *Azospirillum lipoferum* was also followed by TLC analysis of the ether extract from culture medium. The radioactive spots on plates were removed for counting of radioactivity and for further characterization and identification (Fig.3). Figure 3 shows that under aerobic conditions 4-chloronitrobenzene was completly metabolized after 60 hours of incubation. The same effect was seen under anaerobic conditions after 36 hours. Concurrently to the degradation of 4-chloronitrobenzene some metabolites were produced. Especially, the accumulation of two metabolites, named M-1 and M-2, was observed.

Figure 4 shows the changes of pH and N-NO_2^- formation in a *Azospirillum lipoferum* culture medium under aerobic and anaerobic conditions.

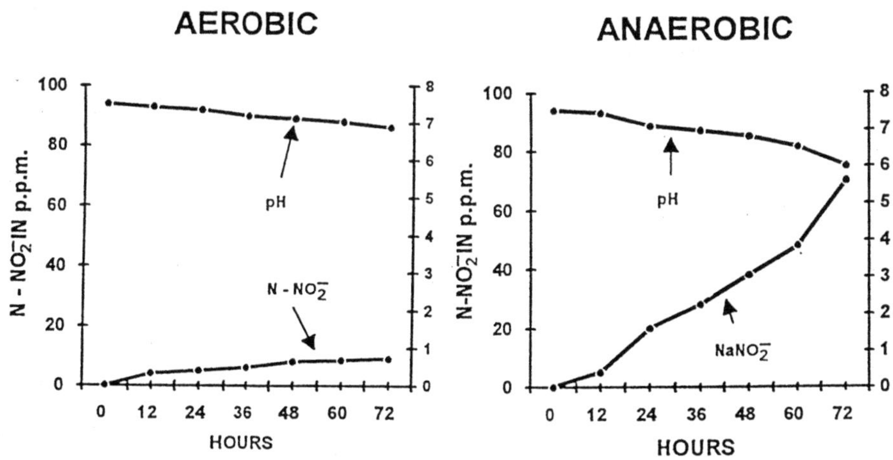

Fig. 4. Changes of pH and N-NO_2^- formation in a *Azospirillum lipoferum* culture medium under aerobic and anaerobic conditions

During aerobic growth of a *Azospirillum lipoferum* only small changes of pH and traced amounts of nitrites production were found. Under anaerobic conditions significant changes of pH and formation of N-NO_2^- could be detected.

The major two metabolites, M-1 and M-2, produced from 4-chloronitrobenzene in a *Azospirillum* culture medium were isolated and characterized (Table 1). The metabolites M-1 and M-2 apear to be 4-chloroacetanilide and 1,3-bis(p-chlorophenyl)triazene, respectively.

Table 1. Some characteristic of 4-chloronitrobenzene metabolites produced in the culture medium of a *Azospirillum lipoferum*

Metabolite	Identified as	R_f value (ether:hexane=4:1)	Melting point in C°	UV nm	Mass Spectrometry Mol. Ion	Mass Spectrometry Fragments
M-1	4-chloroacetanilide	0.21	177	248	169	157,127,75
M-2	1,3-bis(p-chlorophenyl) triazene	0.74	128	-	265	237,167,75

The chemical production of 1,3-bis(p-chlorophenyl)triazene from 4-chloroaniline when to growth medium instead of nitrate 100ppm of nitrite and 20ppm of 4-chloroaniline were added but pH was adjusted to 5.9, was shown in Table 2.

Table 2. Formation of 1,3-bis(p-chlorophenyl)triazene during anaerobic incubation (24 hours) 20ppm of ^{14}C-labeled-4-chloroaniline and 100ppm of N-NO_2^- in the sterile growth medium (pH 5.9)

4-chloroaniline ppm	N-NO_2^- ppm	Transformation of substrate	Loss of radioactivity	Formation of triazene
0	100	-	-	-
20	100	96	41	45

It is an interesting observation, that at a concentration of 20ppm of 4-chloroaniline, in the chemical reaction mixture about 45% of substrate was converted to the triazene. In this reaction mixture a loss of approximately 41% of radioactivity occurred.

Discussion

Pesticides are exposed to numerous biological, chemical and physical factors, any of which can cause their transformation. It is often not easy to distinguish which factor is responsible for a given structural transformation since all three factors are so closely interdependent.

In the present investigation there appears to be experimental evidence for such a secondary effect stemming from two biologically mediated activities: (1) transformation of nitrobenzene to aniline and (2) the lowering of the pH of the medium and formation of nitrite from nitrate. This leads to the proper physico-chemical conditions which cause the conversion of 4-chloro-aniline to 1,3-bis(p-chlorophenyl)triazene.

A number of condensation products arising from substituted anilines have been found in soils and in microbial culture media receiving high dosages of anilines or aniline-based pesticides (Bollag, 1974). Most reports describe the isolation of azobenzenes, but in soil it was also found that diphenylamine (Briggs and Ogilvie, 1971) can be formed. The actual cause of condensation or polymerization could not unequivocally be elaborated.

Plimmer, Kearney, Chisaka, Yount and Klingebiel (1970), who isolated 1,3-bis(3,4-dichlorophenyl)triazene from propanil in soils, proposed that soil nitrite reacts with 3,4-dichloroaniline to form an intermediate diazonium cation, which subsequently couples with free 3,4-dichloroaniline resulting in the formation of.the triazene.

Although the formation of a triazene derivative from an aniline compound under described conditions is a well-known chemical reaction, the occurrence of this reaction under specific natural conditions must always be considered a possible dangerous environmental factor since triazene are known for their multiple effects.

References

American Public Health Association (1971) "Standard Method for Examination of Water and Wastewater". American Public Health Association, 13th ed., New York

Bollag J.-M. (1974) Microbial transformation of pesticides. Adv. Appl. Microbiol. 18:75-130

Bollag J.-M., Nash C.L. (1974) Effect of chemical structure of phenylureas and anilines on the denitrification process. Bull. Environ. Contam. Toxicol. 12:241-249

Bollag J.-M. (1977) Formylation and acetylation of 4-chloroaniline by a *Streptomyces* sp. Acta Microbiol. Pol. 26:59-64

Bothe H., Klein B., Stephan M.P., Döbereiner J. (1981) Transformations of inorganic nitrogen by *Azospirillum spp.* Arch. Microbiol. 130:96-100

Briggs G.G., Ogilvie S.Y. (1971) Metabolism of 3-chloro-4-methoxyaniline and some N-acyl derivatives in soil. Pest. Sci. 2:165-172

Hallas L.E., Alexander M. (1983) Microbial degradation of nitro-aromatic compounds in sewage effluent. Appl. Environ. Microbiol. 45:1234-1241

Ingraham J.L. (1981) Microbiology and genetics of denitrifiers. In: Delwich C.C. Ed. Denitrification, nitrification and atmospheric nitrous oxide. Wiley, New York, pp 45-65

Minard R.D., Russel S., Bollag J.-M. (1977) Chemical transformation of 4-chloroaniline to traiazene in a bacterial culture medium. J. Agr. Food Chem. 25:841-844

Neur G., Kronenberg A., Bothe H. Denitrification and nitrogen foxation by *Azospirillum*. III. Properties of a wheat-*Azospirillum* association. Arch. Microbiol. 141:364-370

Neyra C.A., Döbereiner J., Lalande R., Knowles R. (1977) Denitrification by N_2-fixing *Spirillum lipoferum*. Can. J. Microbiol. 23:300-305

Patil S.S., Shinde V.M. (1988) Biodegradation studies of aniline plant waste water by gas chromatography. Environ. Sci. Technol. 22:1160-1165

Plimmer J.R., Kearney P.C., Chisaka H., Yount J.B., Klingebiel U. (19,70) 1,3-bis(3,4-dichlorophenyl)-triazene from propanil in soil. J.Agr.Food Chem. 18:859-863

Immunological Properties of *Azospirillum* Cell Surface: the Structure of Carbohydrate Antigens and Evaluation of their Involvement in Bacteria-Plant Contact Interactions

Larisa Matora, Galina Solovova, Oksana Serebrennikova, Nikolai Selivanov, & Sergei Shchyogolev
Institute of Biochemistry & Physiology
of Plants & Microorganisms
Russian Academy of Sciences
Prosp. Entuziastov 13
410015 Saratov
Russia

Abstract

Immunization of rabbits with glutar aldehyde-treated intact *Azospirillum* cells results in the production of antibodies specific mainly with regard to the cell O-antigens. EPS, CPS, and O-specific polysaccharides of *A. brasilense* Sp 7 are antigenically identical. Pre-treatment of the wheat seedling roots and a carrot cell suspension with LPS of strain Sp 245 promotes significantly bacterial adhesion to wheat roots and *Azospirillum*-induced agglutination of carrot cells. It also leads to an increased synthesis of the major proteins in the wheat root apoplast, which is comparable with the plant response to inoculation with whole cells. This may be explained by an active effect of *Azospirillum* LPS on the receptor systems of plant cells.

Introduction

The carbohydrate components of the outer membranes of Gram-negative bacteria were recognized as very important structural elements. Data available particularly for enterobacteria indicated that bacterial O-antigens may serve as suitable molecular markers (Sela, 1975) to draw intraspecific classification schemes. A key role of lipopolysaccharides (LPS) of soil microorganisms in the host plant recognition by the bacterium was observed for instance in the *Rhizobium*-legume symbiosis (Kato *et al.*, 1979; Wolpert and Albertshen, 1976). In this case, however, no direct involvement of the LPS in the process of bacterial attachment to plant tissues was revealed (Smit *et al.*, 1989). On the other hand, bacterial LPS are known to participate actively in the contact interactions of some pathogenic agrobacteria with plant cells (Whatley *et al.*, 1976; Matthysse *et al.*, 1978).

The aims of the current work were: i) to expand our knowledge of the serological activity, specificity, and structure of O-antigenic determinants of soil associative bacteria of the genus *Azospirillum*; and ii) to obtain data characterizing the involvement of *Azospirillum* LPS in the bacterium-cereal interaction as compared with similar results reported for the *Rhizobium*-legume symbiosis and phytopathogenesis. We hope these may be rather interesting from the

NATO ASI Series, Vol. G 37
Azospirillum VI and Related Microorganisms
Edited by I. Fendrik et al.

viewpoint of the known analogies between the associativeness, symbiosis, and phyto-pathogenicity phenomena (Elmerich *et al.* 1987; Okon *et al.*, 1988).

Materials and Methods

Azospirillum type strains were grown on a synthetic nutrient medium (von Bulow and Dobereiner, 1975). This study utilized exopolysaccharide (EPS) and capsular polysaccharide (CPS) preparations, and a preparation of LPS-protein complex (LPPC) isolated from *A. brasilense* strain Sp 7 (S-type) by Konnova *et al.* (1994). Also, we used LPS preparations obtained by proteinase treatment of whole cells (Hitchcock and Brown, 1983), and crude extracts of the outer *Azospirillum* membranes isolated following the method described by Poxton *et al.* (1985).

To obtain antibodies specific to the carbohydrate antigens of the bacterial cell surface, we compared the possibilities of the conventional strategy (Levanony and Bashan, 1989) (immunization of rabbits with intact bacterial cells) with a modified procedure involving immunization of the animals with whole cells treated with 2% glutaraldehyde solution (Bogatyrev *et al.*, 1992). For immunization, we used cell suspensions at a concentration of 10^9 cells per ml. Animals were immunized 6 times daily by intravenous injections of 0.4 to 1.6 ml of cell suspension. After a week, the inoculations were repeated 4 times daily using 2 ml of the suspension at a time. Bleeding was made within 6 days after the last injection.

Immunodiffusion, as well as immunoelectrophoresis and its variations, were performed according to a well known description (Axelsen and Krell, 1977). Estimation of the relative quantity of the cells attached to wheat seedling roots during their 2-h incubation with a bacterial suspension (calculated on unit mass of root) was done by using a radioisotopic analysis (Solovova *et al.*, 1994). Experiments on agglutination of a carrot cell suspension induced by bacterial adhesion were carried out according to a standard procedure (Matthysse *et al.*, 1981). Seed sterilization and germination, seedling inoculation, and isolation of extracellular proteins from wheat seedling roots were performed as described by Ignatov *et al.* (1994). When treating the seedlings with an LPS preparation from *A. brasilense* Sp 245, the sample of LPS was suspended in water and added to the plant growth medium to final concentrations of 0.6 to 1 mg/ml.

Results and Discussion

It was found that pre-treatment of whole cells with glutar aldehyde permits one to obtain antisera containing antibodies which are specific predominantly with regard to the carbohydrate components of the outer membrane of *Azospirillum*. Note that in this case agglutination titres are higher than those for antisera obtained using untreated cells (1: 20 000 instead of 1: 5 000). Thus, such a modification of the immunization procedure resulted in producing monospecific antibodies against *Azospirillum* O-antigens without heat treatment of cells or additional purifications of antisera. This, in particular, was revealed by two-dimensional immunoelectrophoresis with the separation of cell membrane preparation of *A. brasilense* Sp 7 in poly-acrylamide gel (Bogatyrev *et al.*, 1992). The overwhelming bulk of the antibodies obtained interacted mainly with the fractions possessed by the cell surface polysaccharide components. These fractions were identified as O-antigens on the basis of special elecrophoretical experiments with LPPC preparations. Fig. 1 reports results of a serological analysis of preparations of the outer membranes of five *Azospirillum* strains with antibodies against strain Sp 7. These antibodies were obtained by using both the conventional (immunization with untreated cells with Freund's adjuvant) and the modified procedures. Only in the latter case, a clear cross reaction which was comparable with a direct reaction was observed for *A. brasilense* Sp 7 and *A. brasilense* Cd. This finding is in good agreement with the earlier data reported for these two closely related strains (Schloter *et al.*, 1992). The antibodies obtained were thus shown to be strain-specific.

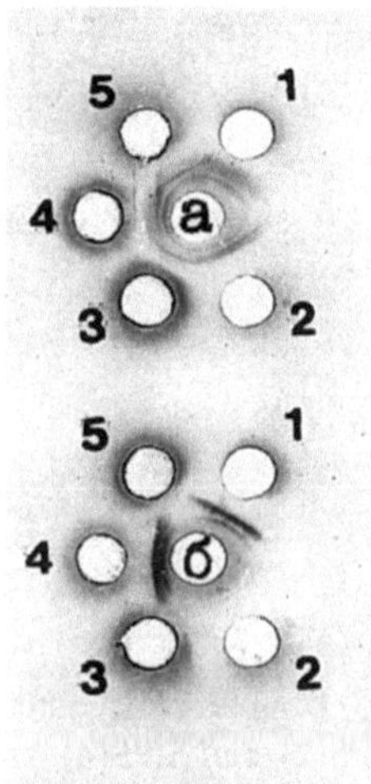

Fig. 1 Immunodiffusion of preparations of the outer membranes of *Azospirillum* strains: Cd(1), Sp 245 (2), JM 125 A2(3), Sp 7(4), and Sp 107(5) with antibodies to Sp 7 obtained using conventional (a) and modified (b) procedures.

The polysaccharides isolated from capsular material were compared with EPS and LPS from *A. brasilense* Sp 7 (S-type) with respect to their antigenic composition. The data yielded by double immunodiffusion showed the antigenic determinants of these polysaccharide preparations to be identical (Fig. 2). This was supported by double cross immunoelectrophoresis data (Fig. 3). Hence,

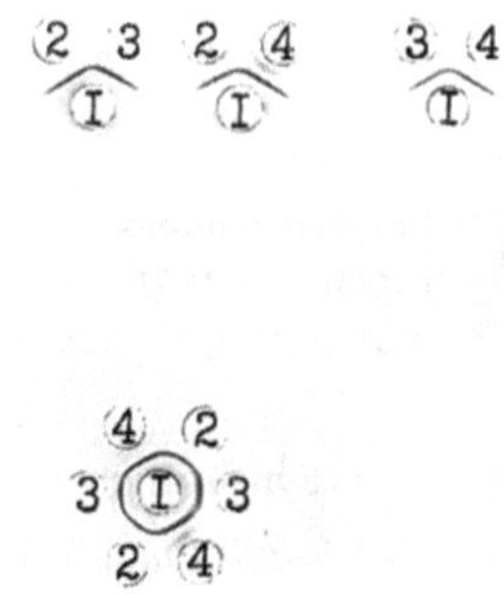

Fig. 2 Immumodiffusion of EPS (2), CPS(3), and LPS (4) of *A. brasilense* Sp 7 with the antibodies obtained (1).

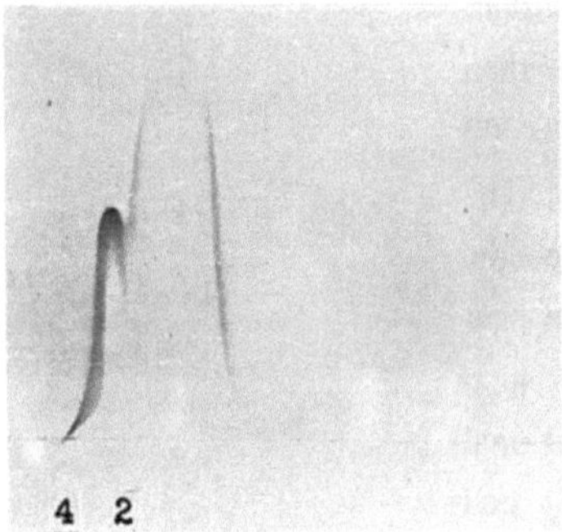

Fig. 3 Double cross immunoelectrophoresis of EPS(2) and LPS(4) preparations of *A. brasilense* Sp 7.

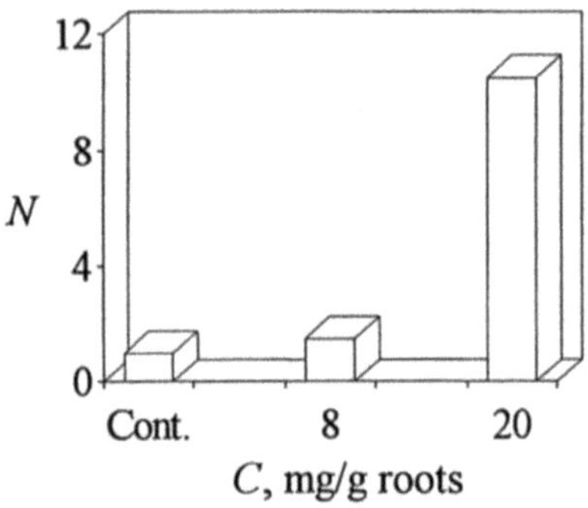

Fig. 4 Effect of pre-treatment of wheat (cv. Saratovskaya 29) roots with an LPS preparation of *A. brasilense* Sp 245 on the relative number *N* of bacteria attached to plant roots (from 1 ml suspension per gram roots), *C* - concentration of LPS.

the somatic antigen has immunodeterminant sites similar both to the CPS and to the EPS. According to the results reported by Konnova *et al.* (1990), the above biopolymers have the same monosaccharide composition. Thus, it may be concluded that the S-type of *A. brasilense* Sp 7 has no specific capsular material. The CPS and EPS of the microorganism involved are likely to be O-specific LPS fragments which are excreted by cells into the culture medium.

The effect of treating plant material with preparations of the isolated *Azospirillum* LPS on the resuls of bacterium-plant interactions was investigated using strain Sp 245 as an example with taking account of the data on: 1) *Azospirillum* adhesion to wheat roots; 2) plant responses to bacterial inoculaion which were revealed by the peculiarities of the synthesis of specific polypeptides in the wheat root apoplast; and 3) *Azospirillum*-induced agglutination of carrot cell suspensions. It was found that in all cases such a pre-treatment promotes significantly the effect of *Azospirillum*-plant contact interactions. For example, the amount of the cells of *A. brasilense* Sp 245 attached to wheat roots significantly increased in response to such a treatment (Fig. 4). Pre-incubation of roots with an LPS solution increased the effect of bacterial adhesion 1.5-fold (with 8 mg/g root), and even 10-fold (with 20 mg/g root). An analogous effect was observed for *Azospirillum*-plant contact interactions in cultures of pre-treated carrot cells. Also, it was established (Fig. 5) that pre-treatment of wheat roots with LPS from *A. brasilense* Sp 245 induced plant response (an enhanced synthesis of the major proteins in the root apoplast) comparable with that to inoculation with a cell suspension of this strain. Note that in the latter case there was also an increase in glycosidase activity in the root apoplast (Ignatov *et al.*, 1994).

Thus, it is clear that *Azospirillum* LPS take part in the adhesion process immediately in contrast to the LPS of *Rhizobium* (Smit *et al.*, 1989). However, unlike *Agrobacterium* LPS (Whatley *et al.*, 1976; Matthysse *et al.*, 1978) LPS of azospirilla do not inhibit effects of bacterium-plant contact interactions. It is apparent that the role of *Azospirillum* LPS is not confined to their involvement in bacterial attachment to plant tissues. Presumably, LPS also have an active effect on the receptor systems of plant cells, which specifically leads to the promotion of induction of the above plant metabolites. In addition, our studies have demonstrated that the antigenic properties of bacterial LPS, which were revealed by means of the modified immunization procedure, provide a basis for the creation of a reliable system for the monitoring, identification and immunological classification of soil bacteria of the genus *Azospirillum*.

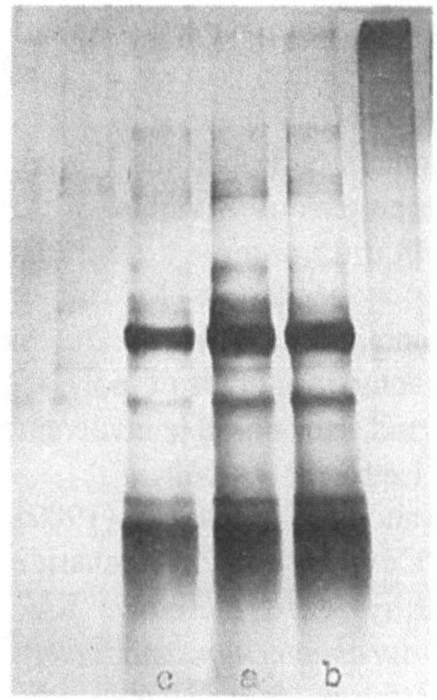

Fig. 5 Effects of inoculation of wheat (cv. Saratovskaya 29) seedlings with strain Sp 245 (a) and of their pretreatment with LPS of this strain (b) on protein accumulation in the apoplast compared with the control (untreated seedlings) (c).

Acknowledgments

Part of this work was supported by a grant from the Foundation for International Projects in Life Sciences and Biotechnology (Russian Ministry of Science and Technical Policy). The authors thank Prof. I. Skvortsov for kindly providing the samples of *Azospirillum* polysaccharide containing polymers, Mrs. T. Naumova for preparing carrot cell cultures, and Mr. D. Tychinin for his invaluable assistance in preparing the manuscript.

References

Axelsen N, Krell N (eds) (1977) A guide to quantitative immunoelectrophoresis. Mir Moscow

Bogatyrev VA, Dykman LA, Matora LYu, Schwartsburd BI (1992) The serotyping of *Azospirillum* spp. by cell-gold immunoblotting. FEMS Microbiol Lett 96: 115-118

Bulow von YFW, Dobereiner J (1975) Potential for nitrogen fixation in maize genotypes in Brazil. J Proc Nat Acad Sci USA 76: 2389-2393

Elmerich C, Bozouklian H, Vieille C, *et al.* (1987) *Azospirillum*: genetics of nitrogen fixation and interaction with plants. Philos Trans Royal Soc London Ser B 313: 183-192

Hitchcock PJ, Brown TM (1983) Morphological heterogeneity among *Salmonella* lipopolysaccharide chemotypes in silver-stain polyacrylamide gels. J Bacteriol 154: 269-277

Ignatov VV, Stadnik GI, Iosipenko OA, Selivanov NYu, Iosipenko AD, Sergeeva EI (1994)

Interactions between partners in the association "wheat-*Azospirillum brasilense* Sp 245" This book
Kato G, Maruyama Y, Natamura M (1979) Role of lectin and lipopolysaccharides in the recognition process of specific legume-*Rhizobium* symbiosis. Agric Biol Chem 43: 1085-1092
Konnova S, Baberdina I, Makarov O, Skvortsov I, Ignatov V (1990) Polysaccharide-lipid complex of S-variant of *Azospirillum brasilense* Sp 7 of the surface of cells and culture liquid: complex degradation, fractionation and monosaccharide composition of polysaccharides. Mikrob Zh 52: 40-45
Konnova SA, Makarov OE, Skvortsov IM, Ignatov VV (1994) Isolation, fractionation and some properties of polysaccharides produced in a bound form by *Azospirillum brasilense* and their possible involvement in *Azospirillum*-wheat root interactions. FEMS Microbiol Lett 118: 93-100
Levanony H, Bashan Y (1989) Localization of specific antigens of *Azospirillum brasilense* Cd in its exopolysaccharide by immuno-gold staining. Curr Microbiol 18: 145-149
Matthysse AG, Holmes KV, Gurlitz RHG (1981) Elaboration of cellulose fibrils by *Agrobacterium tumefaciens* during attachment to carrot cells. J Bacteriol 145: 583-595
Matthysse AG, Wyman PM, Holmes KV (1978) Plasmid-dependent attachment of *Agrobacterium tumefaciens* to plant tissue culture cells. Infect Immun 22: 516-522
Okon Y, Fallik E, Sarig E, *et al.* (1988) Plant growth promoting effects of *Azospirillum*. Nitrogen fixation: hundred years after. Gustav Fisher Stuttgart 741-746
Poxton J, Bell G, Barclay G (1985) The association on SDS-polyacrylamide gels of lipopolysaccharide and outer membrane proteins of *Pseudomonas aeruginosa* as revealed by monoclonal antibodies and Western blotting. FEMS Microbiol Lett 27: 247- 251
Schloter M, Bode W, Hartmann A (1992) Characterization of monoclonal antibodies against cell surface structures of *Azospirillum brasilense* Sp 7 using ELISA Techniques. Symbiosis 13: 37-45
Sela M (ed) (1975) The Antigens, Vol. 3. Acad. Press New York
Smit G, Kijne JW, Lugtenberg BJJ (1989) Roles of flagella, lipopolysaccharide and Ca^{2+}-dependent cell surface protein in attachement of *Rhizobium leguminosarum* biovar *viciae* to pea root hair tips. J Bacteriol 171: 569-572
Solovova GK, Velikov VA, Krivopalov YuV, Chumakov MI (1994) Short-term attachmemt of bacteria from the *Rhizobiaceae* family to the roots of cereals. This book
Whatley MH, Bodwin JS, Lippincott BB, Lippincott JA (1976) Role for *Agrobacterium* cell envelope lipopolysaccharide in infection site attachment. Infect Immun 1976 13: 1080-1083
Wolpert JS, Albertshen P (1976) Host-symbiont interactions. 1. The lectins of legumes interact with the O-antigen containing lipopolysaccharides of their symbiont *Rhizobium*. Biochem Biophys Res Comm 70: 729-737

Role of Choline and Glycine Betaine in *Azospirillum brasilense* Cd Under Salinity Stress

Anil Kumar Tripathi, Bal Mukund Mishra and Michael Schloter[1]

Laboratory of Bacterial Genetics

School of Biotechnology, Faculty of Science

Banaras Hindu University, Varanasi-221 005, India.

Abstract

Glycine betaine has been reported to act as osmoprotectant in *Azospirillum brasilense* Sp7 but choline failed to produce a similar effect. A plasmid (pFF600) containing *bet*ABT operon of *E. coli* on a broad host range vector was conjugatively mobilized into *A. brasilense* Cd. The transconjugants failed to use choline as osmoprotectant. The comparison of growth and acetylene reducing activity of parent and transconjugant revealed that neither choline nor glycine betaine acted as osmoprotectant. Acetylene reducing activity was proportionally inhibited with increasing salinity stress. Glycine betaine but not choline inhibited the growth completely at 20mM concentration in the absence of salinity stress. Analysis of ^{35}S-methionine labeled proteins after salinity upshock revealed the inhibition of at least four proteins and induction/enhancement of about six proteins. The protein profile revealed a ca. 31 kDa protein which was induced under salinity stress but was inhibited by glycine betaine. Western blotting of total and periplasmic proteins and immunological detection with anti-glycine betaine binding protein antibody of *E.coli* revealed cross reaction with a 31 kDa protein and also with a protein of 18 kDa. The 31 kDa cross reacting band was present with or without salinity stress with or without choline or glycine betaine. These observations suggest that osmoregulation in *A. brasilense* Cd differs from *A. brasilense* Sp7.

Introduction

Bacteria have evolved mechanisms to sense changes in the environmental osmolarity and to adapt rapidly to such adverse conditions. One of the mechanisms of osmotic adaptation is to accumulate osmoprotective compounds such as trehalose, proline, glutamate, glycine betaine and proline betaine by synthesis or by uptake from the external environment. The intracellular accumulation of glycine betaine is of central importance for osmoadaptation and

[1] GSF - Forschungszentrum fur Umwelt und Gesundheit, Institut fur Bodenökologie, Ingolstaedter Land Str. 1, Neuherberg, D-85758 Oberschleissheim, Germany.

NATO ASI Series, Vol. G 37
Azospirillum VI and Related Microorganisms
Edited by I. Fendrik et al.

has been investigated in greater detail in *E. coli* (Lucht and Bremer, 1994). Under osmotic stress, *E. coli* accumulates glycine betaine from the external environment by a high affinity uptake system, ProU or by enzymatic conversion of choline into glycine betaine via *bet*ABT. Both, proU and *bet*ABT operons are induced by osmotic stress. Neither choline nor glycine betaine is utilized as a source of carbon and/or nitrogen for the growth of *E. coli* . In contrast to *E. coli* , glycine betaine can function as carbon, nitrogen and energy source as well as an osmoprotectant in *Rhizobium meliloti* (Smith et a., 1988).

Like other bacteria, *Azospirillum* spp. also accumulated compatible solutes such as trehalose, glutamate, proline and glycine betaine under salinity stress (Hartmann, 1988). Osmotolerance declined in the order *A. halopraeferens, A. irakense, A. brasilense, A. lipoferum* and *A. amazonense*. In *A. halopraeferens* and *A. brasilense*, glycine betaine stimulated growth and nitrogen fixation under salt stress. *A. halopraeferens* could convert choline into glycine betaine but *A.brasilense* and *A.lipoferum* failed to do so (Hartmann et al., 1991). *A. brasilense* Sp7 has been investigated in detail with respect to the role of glycine betaine in osmotolerance and the physiological mechanisms of glycine betaine uptake (Riou and Le Rudulier, 1990; Riou et al., 1991). The aim of the present study was to investigate the role of choline and glycine betaine in salinity stress response of *A. brasilense* strain Cd (ATCC 29710).

Materials and Methods

Bacterial strains and Growth Conditions - A spontaneous rifampicin (100μg/ml) resistant mutant of *A. brasilense* strain Cd (ATCC 29710) was used in this study. This strain was maintained in Luria broth supplemented with rifampicin. Cultures were grown at 30°C, aerobically with shaking and microaerobically without shaking in the semisolid minimal medium. The K minimal medium contained malate (1 % w/v) as carbon source and NH_4Cl (20 mM) as nitrogen source (Franche & Elmerich, 1981). *Escherichia coli* strain S17-1 (pFF600) was grown in LB medium supplemented with kanamycin (20μg/ml).

Acetylene Reducing Activity - It was measured according to Singh and Klingmüller (1986).

Conjugative Mobilization of pFF600 - The plasmid, pFF600, containing all the known bet genes of *E. coli*, cloned in a *SalI* site in the broad host range vector pRK293 (Andresen et al.,

1988) was transformed into *E. coli* strain S17-1. *A. brasilense* was then mated with E. *coli* S17-1 (pFF600) in an Eppendorf tube (Singh & Klingmüller, 1986). Exconjugants were selected on LB plates containing rifampicin and kanamycin. Presence of pFF600 in the exconjugants was confirmed by lysis of cells according to Kado & Liu (1981) and resolution of DNA by electrophoresis in a 0.7% agarose gel.

Labelling of Salinity Shock Induced Proteins - Cells grown for 24h in LB medium were harvested, resuspended in K minimal medium and grown for 12h at 37°C in an orbital shaker at 120rpm. Sodium chloride was then added to a final concentration of 300mM for creating salt stress condition. Where indicated, chloramphenicol (25µg/ml), choline and glycine betaine (1mM each) were also added with NaCl. Cultures were then incubated for 1 h. Proteins were radiolabelled with ^{35}S-methionine (added at 0.0375mCi/ml) during last 30min of exposure to NaCl stress. Cells were lysed by adding cracking buffer, boiling for 15min and then returning to ice followed by centrifugation. Supernatants were used for resolving proteins on 10% discontinuous SDS-PAGE at 100V for 2h followed by another 2h at 200V and finally autoradiographed according to Apte and Bhagwat (1989).

Immunological Identification of glycine betaine binding protein (GBBP) - *A. brasilense* was grown in K minimal medium for 36h with shaking at 30°C. Equal amount of cells (measured by O.D.560) were taken for extraction of total proteins by the lysis of cells in cracking buffer as described above. Periplasmic proteins were extracted by chloroform shock procedure (Ames et al., 1984). The proteins were resolved in 10% dicontinuous SDS-PAGE and electroblotted onto Immun-Lite (BioRad) membrane in a Biometra Fast blot system as per manufacturer's recommendations. The blotting was then completed in 1h at 5W. The blot was then dried in an incubator at 60°C for 30min. After blocking and washing steps, the blot was treated with the immunoadsorbed anti- GBBP antibody (Barron et al., 1987) which was kindly provided by M. Villarejo, University of California, Davis, USA. Immunoreactive bands were visualized by using horseradish peroxidase conjugate and 4-chloro-1-naphtol for detection.

Results and Discussion

Increasing concentrations of NaCl proportionally inhibited the growth and acetylene reduction in *A. brasilense* Cd. Inhibition of acetylene reducing ability was more pronounced than the growth in medium supplemented with NH_4Cl. Supplementation of 350mM NaCl completely inhibited the acetylene reducing ability.

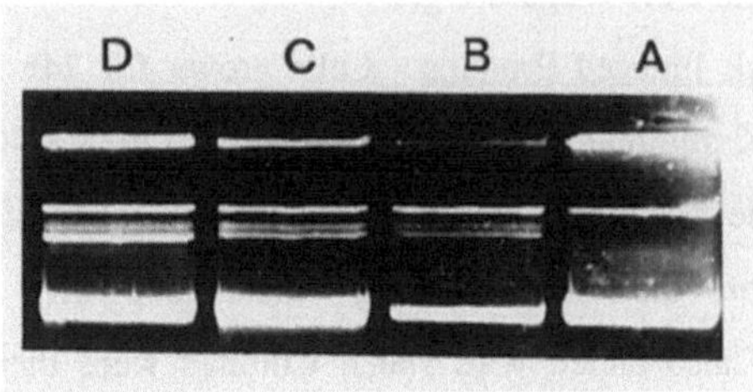

Figure 1 - Agarose gel electrophoresis of the plasmids of *A. brasilense* (lane A) and transconjugants (lanes B,C &D) showing presence of pFF600

Treatments	Acetylene Reducin Ability	
	A.brasilense	*A. brasilense* (pFF600)
1. Control	42 %	41 %
2. 300 mM Nacl	2 %	2 %
3. 300 mM Nacl + 1mM glycine betaine	1.6 %	1.8 %
4. 300 mM Nacl + 1mM choline	1.5 %	1.6 %
5. 1 mM glycine betaine	42 %	41 %
6. 10 mM glycine betaine	39 %	40 %
7. 20 mM glycine betaine	0	0
8. 1mM choline	43 %	42 %
9. 10 mM choline	44 %	41 %
10. 20 mM choline	45 %	40 %

Table 1 - Effect of choline and glycine betaine on acetylene reducing ability of *A. brasilense* and its transconjugant containing pFF600

In order to assess the ability of the cloned *bet*ABT operon of *E. coli* to confer the ability to convert choline into glycine betaine on *A. brasilense*, a plasmid (pFF600) harbouring all the *bet* genes of *E. coli* on a broad host range vector was conjugatively mobilized into *A. brasilense* Cd. The physical presence of the transferred plasmid was ensured in the exconjugants (Figure 1) before using them for osmoregulatory experiments. Table 1 shows

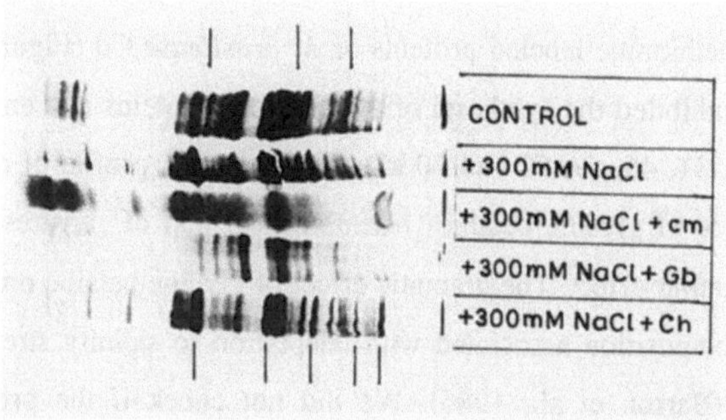

Fig 2 - Autoradiogram of the ^{35}S - methionine labelled proteins of *Azospirillum brasilense* after salinity upshock. Lines on the right of the picture indicate induced / enhanced proteins and those on the left indicate inhibited proteins (Cm - chloramphenicol, Gb - glycine betaine, Ch - choline)

the effect of choline and glycine betaine on the acetylene reducing ability of *A. brasilense* Cd and its pFF600 bearing exconjugant. Acetylene reducing ability of *A. brasilense* Cd and *A. brasilense* Cd (pFF600) was drastically inhibited by 300mM NaCl. Supplementation of 1 mM glycine betaine or choline failed to restore the NaCl mediated inhibition of nitrogen fixation ability. Acetylene reducing ability, without salinity stress, was not affected by 10mM glycine betaine. But, supplementation of 20mM glycine betaine completely inhibited the pellicle formation and the acetylene reducing ability. Choline did not show any positive effect in restoration of acetylene reducing ability under salinity stress in *A. brasilense* Cd (pFF600). With 20mM choline supplementation, the uninhibited growth of *A. brasilense* Cd (pFF600) indicated that choline was not converted to glycine betaine. Although the osmoprotective effect of glycine betaine is reported to result from high intracellular accumulation of glycine betaine, *A. brasilense* Cd was inhibited by 20mM levels of glycine betaine. In *A. brasilense* Sp7, the intracellular concentrations of glycine betaine have been reported to reach up to 480mM under salinity stress (Riou and Le Rudulier, 1990).

The analysis of ^{35}S-methionine labeled proteins of *A. brasilense* Cd (Figure 2) revealed that the salinity upshock inhibited the synthesis of at least four proteins and enhanced/induced at least six proteins (21, 31, 45, 59, 62 and 70 kDa). The protein profile of cells shocked with salinity in the presence of glycine betaine resembled to that of unstressed cells. Choline failed to produce a similar effect. The dramatic effect of glycine betaine on reversal of all the changes in protein composition associated with adaptation to salinity stress have also been observed in *E. coli* (Barron et al., 1986). We did not check if the proteins, induced in response to salinity upshock, persisted at elevated levels until a downshift. Such an experiment would indicate if these proteins are synthesized transiently as in heat shock response or they continue to be synthesized at elevated levels until the salinity stress persisted. Most notable of the induced proteins in our experiment were 21, 31 and 45 kDa proteins which on the basis of their sizes belong to *kdp*C, *pro*X and *pro*V gene products of *E. coli.* Appearance of 31 kDa protein under salinity stress but its disappearance with glycine betaine supplementation that this protein might be the glycine betaine binding protein (GBBP) which is known to be located in periplasm in *E. coli* and *S. typhimurium* (Lucht and Bremer, 1994). In order to identify and localize GBBP, the western blots containing total proteins and periplasmic proteins were used for immunological detection with E. coli anti GBBP - antibody which showed cross-reacting bands in all the lanes (Figure 3). A cross-reacting band of the same size (31 kDa) was present in all the treatments i.e. with or without salinity stress with or without choline or glycine betaine. Interestingly, another smaller protein cross-reacted with anti-GBBP antibody in the protein extract from the unstressed cells and was also periplasmically located. Similar experiments with *A. brasilense* Sp7 did not reveal this new cross-reacting protein (Riou et al., 1991).

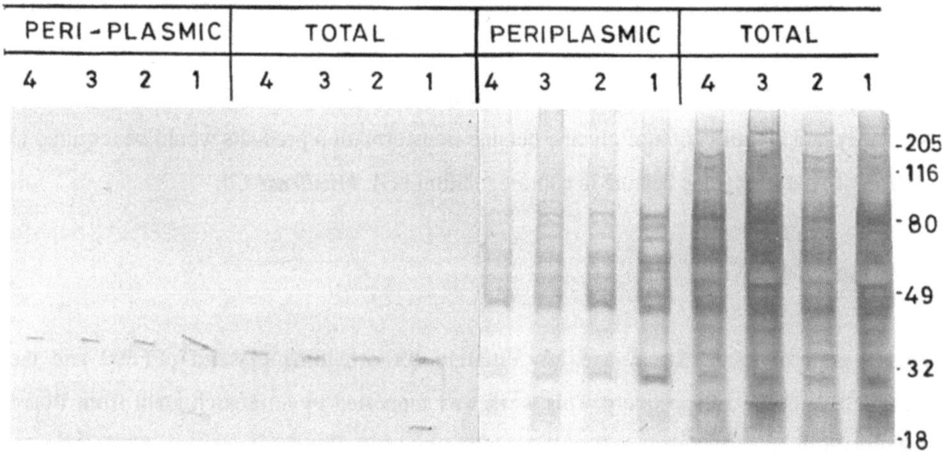

Figure 3 - Immunodetection of total and periplasmic proteins with anti - glycine betaine binding protein antibody. (Lane 1 - proteins extracted from untreated cells, lane 2 - proteins extracted from cells grown in 200mM. NaCl stress, Lane 3-proteins extracted from cells grown in 200mM NaCl + 1 mM glycine betaine, Lane 4- proteins extracted from cells grown in 200mM NaCl + 1 mM choline). The picture on the right shows the profile of the blotted proteins and picture on the left shows the cross reacting bands.

From the observations reported above it appears that *A. brasilense* Cd and *A. brasilense* Sp7 differ in their osmoregulatory characteristics with reference to the role of glycine betaine in osmoregulation. Studies in *E. coli* show that the high-affinity glycine betaine transport system (ProV) requires concerted action of at least four proteins (OmpC, ProX, ProW and ProV) under salinity stress (Lucht and Bremer, 1994; Dattananda and Gowrishankar, 1989). The western blotting experiments indicated that ProX (the glycine betaine binding protein) is present in *A. brasilense* but does not seem to be osmoregulated. The inability of glycine betaine to restore growth and nitrogen fixation ability in *A. brasilense* Cd seemed to indicate for a defect in glycine betaine uptake. But, the effect of glycine betaine on salinity stress induced proteins and inhibition of growth and acetylene reduction by 20mM glycine betaine indicated against this possibility. It, therefore, appears that glycine betaine is taken up but is quickly metabolized into some derivative which was toxic for the growth of *A. brasilense* Cd.

Further experiments on glycine betaine uptake, stability of induced proteins at elevated osmolarity and identification of glycine betaine transformation products would be required to explain the role of glycine betaine in osmoregulation in *A. brasilense* Cd.

Acknowledement

We are grateful to A. Strom and M. Villarejo for providing plasmid pFF600 and the anti-GBBP antibody, respectively. This work was supported by a research grant from Board of Research in Nuclear Sciences (Bombay, India) to AKT. Special thanks are due to Mrs. B. Gubitz for typing the manuscript.

References

Ames GFL, Prody C, Kustu S (1984) Simple, rapid and quantitative release of periplasmic proteins by chloroform. J Bacteriol 160: 1181-1183

Andresen PA, Kaasen I, Styrvold OB, Boulnois G, Strom AR (1988) Molecular cloning, physical mapping and expression of the bet genes groverning the osmoregulatory choline-glycine betaine pathway of *Escherichia coli.* J Gen Microbiol 134: 1737-1746

Apte SK, BhagwatAA (1989) Salinity-stress-induced proteins in two nitrogen-fixing *Anabaena* strains differentially to salt. J Bacteriol 171: 909-915

Barron A, Jung JU, Villarejo M (1987) Purification and characterization of a glycine betaine binding protein from *Escherichia coli.* J Biol Chem 262: 11841-11846

Barron A, May G, Bremer E, Villarejo M (1986) Regulation of envelope protein composition during adaptation to osmotic stress in *Escherichia coli.* J Bacteriol 167:433-438

Dattananda CS, Gowrishankar J (1989) Osmoregulation in *Escherichia coli*: Complementation analysis and gene-protein relationships in the proU locus. J Bacteriol 171: 1915-1922

Franche C, Elmerich C (1981) Physiological properties and plasmid content of several strains of *Azospirillum brasilense* and *Azospirillum lipoferum.* Ann de Microbiol 132A: 3-18

Hartmann A (1988) Osmoregulatory properties of *Azospirillum* spp. in *Azospirillum* IV: Genetics, Physiology, Ecology. Klingmüller W (ed). Springer Verlag Berlin 122-130

Hartmann A, Prabhu SR, Galinski EA (1991) Osmotolerance of diazotrophic rhizosphere bacteria. Plant & Soil 137: 105-109

Kado CS, Liu ST (1981) Rapid procedure for detection and isolation of large and small plasmids. J Bacteriol 145: 1365-1373

Lucht JM, Bremer E (1994) Adaptation of *Escherichia coli* to high osmolarity environments: Osmoregulation of the high-affinity glycine betaine transport system ProU. FEMS Microbiol Rev 14: 3-20

Riou N, Poggi MC, Le Rudulier D (1991) Characterization of an osmoregulated periplasmic glycine betaine binding protein in *Azospirillum brasilense* Sp7 Biochimie 73: 1187-1193.

Riou N, Le Rudulier D (1990) Osmoregulation in *Azospirillum brasilense*: glycine betaine transport enhances growth and nitrogen fixation under salt stress. J Gen Microbiol 136: 1455-1461

Singh M, Klingmüller W (1986) Transposon mutagenesis in *Azospirillum brasilense*: isolation of auxotrophic and Nif⁻ mutants and molecular cloning of the mutagenized nif DNA. Mol Gen Genet 202: 136-142

Smith LT, Pocard JA, Bernard T, Le Rudulier D (1988) Osmotic control of glycine betaine biosynthesis and degradation in *Rhizobium meliloti*. J Bacteriol 170: 3142-3149

Bioreactor Systems for Research on Rhizosphere Bacteria

J. Ueckert, K. Huckfeldt and I. Fendrik
Institute of Biophysics
University of Hannover
Herrenhaeuser Str. 2
30419 Hannover
Germany

Abstract

Bioreactor systems for physiological investigations on plant growth promoting rhizosphere bacteria (PGPRs) have been specifically redesigned to model rhizosphere conditions at constant predetermined levels. Systems employed include a Stirred Tank- and Fluidized Bed Reactor and a Gadostat. Modelled conditions to study physiological responses are bacterial immobilization in the rhizosphere and spatial soil solute heterogeneity.

Introduction

Physiological investigations of soil bacteria in pure and mixed culture and simulation of particular characteristics of diazotrophic associations require application of bioreactors and continuous culture, which, in comparison to batch culture, can maintain constant growth conditions at predetermined levels. The suitability of commonly used bioreactors and our especially modified systems for research on diazotrophic microorganisms is outlined.

Systems for suspended cultivation

The Stirred Tank Reactor (CSTR) is appropriate for determination of basic physiological parameters like the maximal growth rate μ_{max} or the substrate affinity k_S. pH and pO_2 can be adjusted in a wide range. Rhizosphere organisms are believed to be carbon limited in their natural habitat (Lynch, 1982). The substrate affinity to a particular carbon source, which can be determined easily in this fermenter type, plays an important role in the establishment of each strain in the rhizosphere. High affinity strains (low K_S-value) have a competitve advantage in the ecosystem. In a common CSTR bacteria are freely suspended in the medium. The central stirrer allows high oxygen transfer rates (OTR), especially important for cultivation of microaerophilic associative diazotrophic bacteria like *Azospirillum* spp.

NATO ASI Series, Vol. G 37
Azospirillum VI and Related Microorganisms
Edited by I. Fendrik et al.

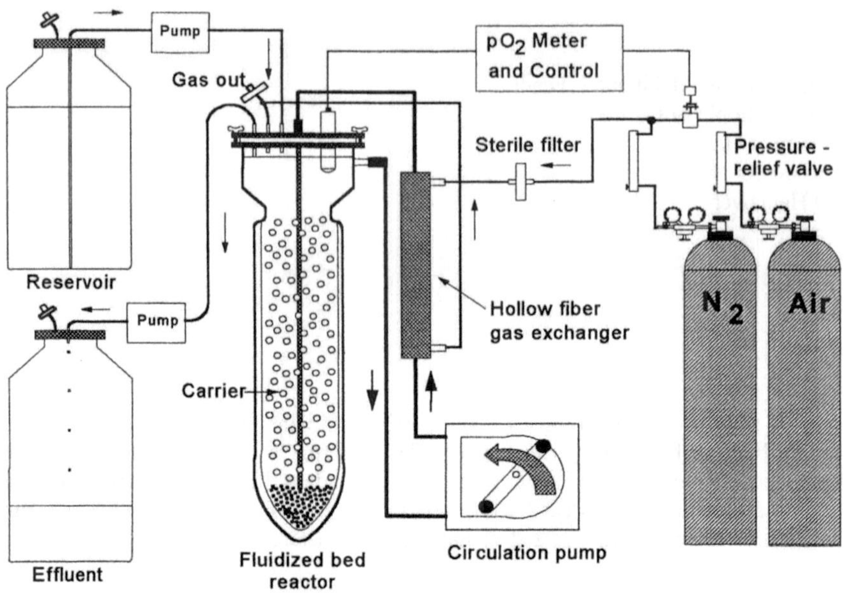

Figure 1 Fluidized Bed Reactor for carrier-bound cultivation of diazotrophic bacteria

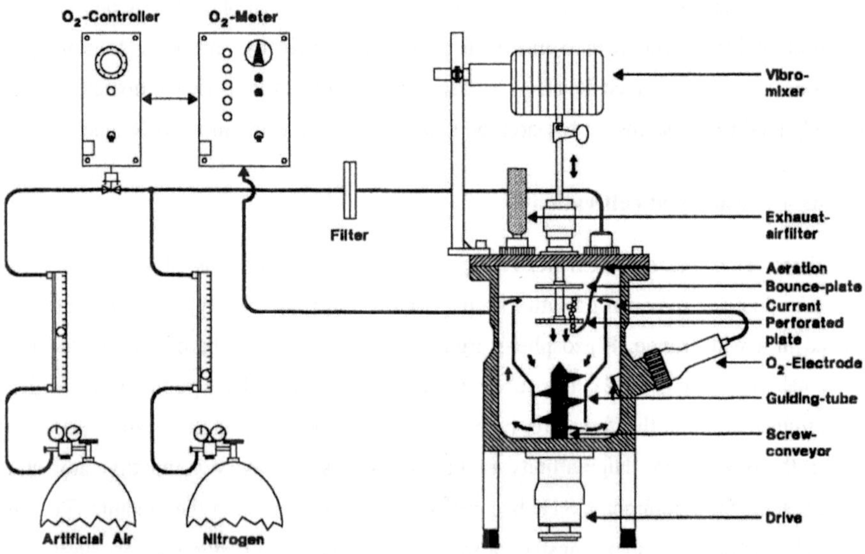

Figure 2 Modified CSTR for cultivation of alginate-entrapped microorganisms

PGPR studies on O_2-tolerances by Hurek et al. (1987), N_2-fixation in presence of ammonia (Fritzsche and Niemann, 1990) or competition studies at submaximal growth rates (Kloss et al.,1986) were only feasible using CSTRs and continuous culture.
Application of Airlift Reactors (ALR) are restricted to only aerobic or anaerobic cultivation, use of a Packed Bed Reactor (PBR) to anaerobic modes. Since most PGPRs are microaerophilic, use of ALR or PBR for modelling is limited.

Systems for immobilized cultivation

In almost all natural habitats, which are investigated at present, bacteria grow on available surfaces and not in the surrounding liquid phase (Costerton et al., 1989). This is true as well for diazotrophic associations (Reinhold et al., 1989), where bacteria are adsorbed to or entrapped in the endorhizosphere. These parameters can not be simulated by CSTRs.
For modelling particular rhizosphere conditions and subsequenly studying physiological responses we adapted different reactor types to meet experimental requirements.
To setup a model for immobilization by adsorption of bacteria to soil particels of heterogenous size or to plant root surfaces, we modified a Fluidized Bed Reactor (Bioengeneering, Switzerland) and added porous glass bead carriers with pore sizes 60-120 µm (Schott, Mainz). The carriers can float in the reactor and allow colonization by microorganisms after inoculation (Fig. 1). The typical reactor load of 250 ml glass carriers gives a colonizable surface of 22.5 m^2. As a consequence cell loads during fermentations are exceptionally high, which in comparison to CSTRs facilitates determination of responses after parametric variation. Alternatively, but less suitable, the glass particles can be replaced by alginate gel carriers. The hollow fiber gas-exchanger (Sempas, Horb) with 1 m^2 membrane surface for bubble-free aeration gives oxygen transfer rates comparable to CSTRs. This modified FBR system enables to cultivate microarophilic diazotrophs immobilized at minimal shear forces (Ueckert et al., 1994). The system can be run at dilution rates (D) smaller or higher than the specific bacterial growth rate μ_{max}. The first case results in a system with a mixture of immobilized and freely suspended cells, the second in a setup with only immobilized cells, since all suspended cells are washed out of the reactor due to $\mu_{max} < D$ (h^{-1}).
Our second promising approach for immobilizing soil microorganisms was to artificially entrap bacteria in Ca^{2+}-alginate gel beads, which is thought to simulate colonization of the endorhizosphere. The gel network allows gases, nutrients and metabolic products to diffuse both directions without limitations (Ueckert et al., 1992). For immobilized cultivation within beads a CSTR was modified (Fig. 2). To minimize shear forces and maximize OTR rates (comparable to a common CSTR) a srew-conveyor has been combined with a vibro-mixer,

which produces vertical vibrations and reduces gas bubble sizes. The model setup results in gentle circulation of shear-sensitive gel beads with entrapped microorganisms. With respect to μ_{max} and D the same applies for this modified CSTR as mentioned for the FBR. Both systems are suitable for cultivation of microaerophilic microorganisms and their special requirements (Kloss et al., 1983).

Multi-stage systems for suspended cultivation

Lovitt and Wimpenny (1982) developed a Gradostat, which is basically a multistage chemostat for suspended cultivation of soil bacteria. The objective for our setup was to simulate and study influences of spatial heterogeneity of solutes and gases as observed in the rhizosphere ecosystem.

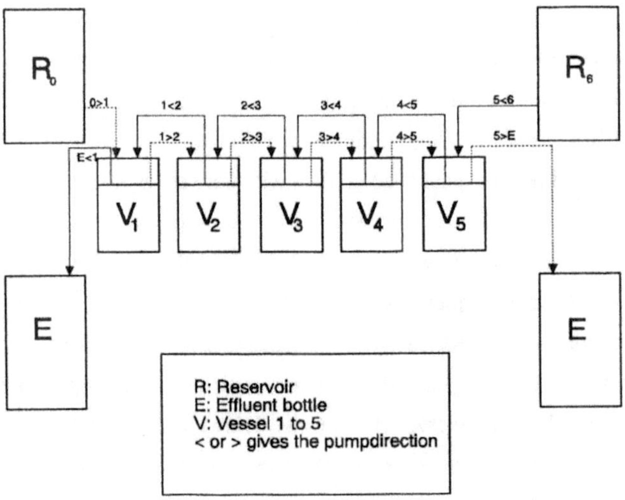

Figure 3 Gradostat composed of five vessels

The gradostat vessels are connected by peristaltic pumps (Fig. 3). The first and last vessel is connected to a reservoir. Bidirectional pumping leads to time-independent, steady-state concentrations of soluble nutrients in each vessel. The gradostat enables to establish bidirectional gradients of different solutes, e.g. of NH_4Cl-, pH- or O_2-gradients and can be utilized for pure- or mixed culture experiments (Fritzsche et al., 1991).

Due to connection of all vessels, the system allows bacteria after inoculation to multiply in the particular vessel where their specific metabolic requirements are met best.

In conclusion it can be stated that bioreactor application has a great potential for modelling parts of the complex soil ecosystem. The outlined model approaches allow to study the bacterial physiological state under conditions assumingly present in the rhizosphere of the selected habitat.

Acknowledgements
This work was supported by the Deutsche Forschungsgemeinschaft (DFG) Project FE242/3.

References

Costerton JW and Lappin-Scott HM (1989) Behavior of bacteria in biofilms. ASM News 55:650-654

Fritzsche C and Niemann EG (1990) Nitrogen fixation in continuous culture with NH_4Cl-containing media. Appl Env Microbiol 56:1160-1161

Fritzsche C, Huckfeldt K and Niemann EG (1991) Ecophysiology of associative nitrogen fixation in a rhizosphere model in pure and mixed culture. FEMS Microb Ecol 1991:279-292

Hurek T, Reinhold B, Fendrik I and Niemann EG (1987) Root-zone-specific oxygen tolerance of *Azospirillum* spp. and diazotrophic rods closely associated with Kallar Grass. Appl Env Microbiol 53:163-169

Kloss M, Iwannek KH, Fendrik I and Niemann EG (1986) Enrichment of diazotrophic bacteria from rice soil in continuous culture. Plant Soil 90:151-164

Lovitt RW and Wimpenny JWT (1982) The gradostat: a bidirectional compound chemostat and its application in microbiological research. J Gen Microbiol 127:261-268

Lynch, JM (1982) The rhizosphere. pp. 395-411. *In:* Experimental microbial ecology. RG Burns, JH Slater (eds). Oxford

Reinhold B and Hurek T (1989) Location of diazotrophs in the root interior with special attention to the kallar grass association. pp 209-221. *In:* Nitrogen fixation with-nonlegumes. FA Skinner (ed). Kluwer, Amsterdam

Ueckert J, Niemann EG and Fendrik I (1992) Mixed continuous suspended and immobilized culture of diazotrophic isolates from root-free soil and the endorhizosphere of *Leptochloa fusca* L. Kunth. Symbiosis 13:75-84

Ueckert J, Niemann EG and Fendrik I (1994) pH-dependent activity changes of nitrate reductase upon immobilization of Azospirillum brasilense in continuous culture. pp. 351-54. *In:* Nitrogen fixation with non-legumes, 6th Int. Symp. NA Hegazi (Ed). AUC Press Cairo

SECTION V:

ECOLOGY

Artificial Plant-Azotobacter Symbiosis for Atmospheric Nitrogen Fixation

István Gyurján[1], Pál Korányi[1], Éva Preininger[1], Szilárd S. Varga[1] and Gyula Paless[2]
Departments of Plant Anatomy[1] and Genetics[2]
Eötvös Lóránd University
H-1088 Budapest
Puskin u. 11-13.
Hungary

Abstract

The main aim of this work is the direct introduction of diazotrophic prokaryotes into algal cells and into the intercellular spaces of plants and the prolonged cultivation of these artificial systems. They help the better understanding of natural symbioses and the production of plants capable to fix atmospheric nitrogen. The realization of such type of symbiosis may result considerable reduction of production and utilization of nitrogen fertilizers. This would have an essential impact for the developing countries not to speak about the elimination of nitrate contamination from the soils. Our artificial alga-bacterium endosymbiosis (endocytobiosis) as well as plant-bacterium symbiosis (exocytobiosis) prove the possibility to produce stable and functioning nitrogen fixing symbiosis widely among plants. The importance of this can't be assessed yet.

Introduction

Nitrogen shortage is the main limiting factor of biomass production because only some prokaryotes can fix atmospheric nitrogen by means of their nitrogenases. These diazotrophs are sometimes abundant in terrestrial or aquatic ecosystems and their contribution to nitrogen supply is significant even in agricultural areas.

Attempts have been made at the large-scale production of some diazotrophs (e.g. *Azolla-Anabaena* system, *Rhizobia, Azospirilla*) and use them as fertilizers or inoculants on cultivated lands however, the final solution would be the transfer of nitrogen fixing ability into crop plants. Introduction and expression of *nif* genes was successful mainly in bacteria but in the case of plants there are serious limitations. An alternative solution may arise from the study of natural nitrogen fixing symbioses. Their organization ranges from loose associations to more

NATO ASI Series, Vol. G 37
Azospirillum VI and Related Microorganisms
Edited by I. Fendrik et al.

integrated partnerships and it seems possible to establish artificial symbioses between suitable diazotrophic prokaryotes and plants (Korányi et al. 1988).

Forced associations able to grow on nitrogen free media are exceptionally interesting. In the partnership of free living *Azotobacter vinelandii* with tissue culture of *Daucus carota* the nitrogen fixing bacteria were living among the callus cells (Carlson and Chaleff 1974). Partial coordination of metabolic functions was indicated by enhanced ethylene production and prolonged stability of the mixed calli. Sugarcane callus cultures were inoculated with nitrogen fixing *Azospirillum brasiliense* and nitrogenase activity was detected in the cocultures up to 18-20 months. Plantlets were regenerated from mixed calli (Berg et al. 1979, Vasil et al. 1979). Endorhizospheric partnership was established between rice and *Alcaligenes faecalis* (You and Zhou 1989), and diazotrophic cyanobacteria were introduced into tobacco shoot regenerates (Gusev at al. 1986) and into the space formed between the plasmalemma and wall of plasmolysed plant cells (Bradley 1979).

There were attempts to create artificial nitrogen fixing intracellular symbioses. Giles and Whitehead (1975, 1976) introduced *Azotobacter vinelandii* into *Rhizopogon* fungal protoplasts and nitrogenase activity was observed up to seven months. Tobacco and maize protoplasts took up diazotrophic *Gloeocapsa* cyanobacteria (Burgoon and Bottino 1976, 1977) but they harbored generally one large algal cell and the rate of uptake was less than 1%.

Stable nitrogen fixing symbioses were established between azotobacters and *Chlamydomonas reinhardtii* strains (Gyurján et al. 1984, 1986). The associative cultures have been maintained on nitrogen and carbon free medium where the energy requirement of nitrogen fixation is met by algal photosynthesis (Korányi et al. 1990, 1993). Azotobacters were chosen as diazotrophic partners because some of their species fix atmospheric nitrogen very effectively and their growth rate is similar that of the algae. They are free living bacteria forming only loose associations in nature and for this reason they may cooperate with any plant. On the other hand, they are aerobic organisms which may make possible for them to fix nitrogen even in photosynthetically active cells or tissues. The widely used and genetically well characterized unicellular green alga *C. reinhardtii* may be regarded as a prototype of plant host. The barrier of cell wall was successfully overcome by the use of wall-less mutants in the fusion experiments (Gyurján et al. 1990, 1992).

Based on the results with algal models, intercellular symbioses were established between carrot tissues and *Azotobacter* species (Korányi et al. 1993, Varga et al. 1994). In the case of multicellular plants, the introduction of microsymbionts into the intercellular spaces of host tissues may lead to viable partnerships because these spaces grow in size during differentiation and are important places of apoplasmatic transport. The effective exchange of metabolites is a prerequisite of symbiotic integration also in such plant-microbe systems. With media containing lactose, what can be utilized only by plants cells, it was possible to force azotobacters into symbiosis with carrot based on nutrient exchange between the partners.

Materials and methods

Strains and culture conditions: Cell wall-less wild type CW15 (Fig. 1a) and complementing arginine requiring 308 (CW_d arg-7-3, mt^+) and 354 (CW15 arg-7 sm_1, mt^+) strains of *C. reinhardtii* were used. Wild type cells were grown at 25°C in liquid culture (Sager and Granick 1953). Cell wall mutants auxotrophic for arginine were grown on the same minimal medium supplemented with yeast extract. Symbiotic cultures were selected and maintained on nitrogen and carbon free medium (Gyurján et al. 1984). Cultures were illuminated for 12 h daily.
The CCM 289 strain of *Azotobacter vinelandii* (Fig. 1b) and CRS-HK 5 strain of *Azomonas insignis* were maintained in nitrogen free liquid and solid cultures (Newton et al. 1953).

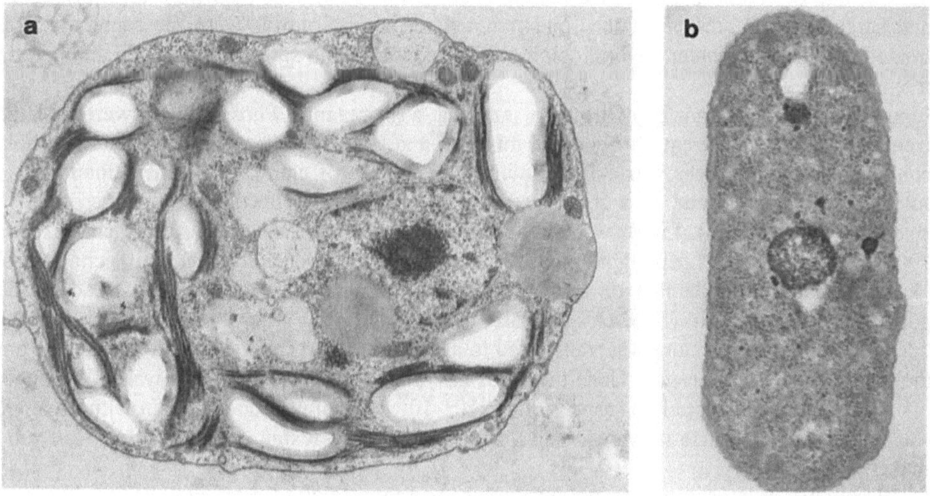

Fig. 1. Partners of alga-bacterium symbioses: a. Cell wall mutant *C. reinhardtii*. b. *A. vinelandii*.

Shoot cultures of strawberry (*Fragaria ananassa* Duch.) were maintained on MS (Murashige and Skoog 1962) basal medium supplemented with 0.5 mg/l 6-benzylaminopurine (BAP), 0.5 mg/l indole-3-butyric acid (IBA), 0.1 mg/l gibberellic acid (GA3) and 3% sucrose. Callus cultures were obtained from leaves of micropropagated plants. Cell suspension was cultured at 120 rpm agitation. Plant regeneration was induced by organogenesis. The same medium was used for callus initiation and plant regeneration consisting of MS salts and vitamins with the addition of 0.1 mg/l IBA, 1 mg/l BAP, 0.05 mg/l naphthaleneacetic acid (NAA), 0.1 mg/l 2,4-dichlorophenoxy acetic acid (2,4-D) and 3% sucrose. The pH was adjusted to 5.7. Symbioses were grown on medium containing 3% maltose at pH 7.0. Regenerated plantlets were rooted on MS medium without any growth regulator. All the cultures were incubated at 25°C in a 16 h photoperiod.

Establishment of alga-bacterium intercellular associations: Wild type algae were mixed with *A. vinelandii* cells in 100 times excess in liquid medium and the suspension was spread on the surface of the nitrogen and carbon free medium solidified with 2% agar (Gyurján et al. 1984, Korányi et al. 1990). Growing green colonies were transferred monthly. Axenic algal and bacterial cultures were used as controls.

Uptake of bacteria during PEG-induced fusion of cell wall mutants: The cells of complementing algal strains were mixed with *A. vinelandii* bacteria in 100 times excess in a minimal medium consisting of 0.2 g NH_4Cl, 0.74 g K_2HPO_4, 0.36 g KH_2PO_4, 0.02 g $MgSO_4x7H_2O$, 0.01 g $CaCl_2x2H_2O$, 2.0 g Na-acetatre and some microelements per litre, pH 7.0. 1/3 volume PEG solution was added which contained 40% (w/v) PEG-6000, 0.3 mM glucose and 66 mM $CaCl_2$, pH 7.0 (Nghia et al. 1986, 1987). The cells were carefully mixed and spread on the surface of the solid minimal medium. Controls were the same mixtures treated with distilled water. Hybrid colonies were collected and spread on nitrogen and carbon free medium.

Liposome mediated transfer of bacteria into cell wall mutants: Phospholipids were isolated from algal cells or different phospholipid mixtures and liposome kits were used. The total lipid content of separated cells was extracted with 2:1 mixture of chloroform and methanol. The solvents were evaporated in a rotary vacuum evaporator and the lipids were dissolved in chloroform (40 mg/ml). Lipids were fractionated by chromatography on silicic acid (SIL-R, 100+ mesh) column. Phospholipid mixtures were made from lecithin (40 mg/ml), cholesterol (5 mg/ml), and stearylamine or dicetyl phosphate (8 mg/ml) in chloroform. Ready-made positive and negative kits (SIGMA) were also used.

0.5 ml phospholipid mixture was transferred to a small test tube and the lipids were dried from the chloroform with nitrogen. Once the lipids had dried to a thin film on the wall of the tube, 0.5 ml *A. vinelandii* bacterium suspension ($5x10^7$-10^8 cells/ml in fusion medium without agar) was added and the bacteria were encapsulated by 3 min vortexing on a laboratory test tube shaker.

Equal volumes of liposome and algal suspensions were mixed and treated with PEG solution in 4-5 times excess. After 1 min gentle agitation the fusion products were pelleted by centrifugation and resuspended in liquid minimal medium. Growing colonies were maintained on nitrogen and carbon free medium.

Establishment of plant-bacterium partnerships: Cell suspension was mixed and cultured together with *A. insignis* for 10-14 h. The inoculated cell suspension was transferred to solidified medium for callus formation. Bacteria had to enter into the intercellular spaces of callus tissue in this step. Plants were regenerated from bacterium containing calli through organogenesis.

Microscopy: For light and transmission electron microscopy algal colonies and plant pieces were fixed in 2% glutaraldehyde for 2 h and postfixed in 1% OsO_4 for 2 h. Buffer solution was 10 mM K-Na phosphate (pH 7.2) for algae and 70 mM for plants. Samples were embedded in Durcupan ACM epoxy resin (Fluka Chemie AG). Semi-thin sections were stained with toluidine blue. Ultrathin sections were stained with uranyl and lead salts and observed in a HITACHI 7100 electron microscope.

Results

The complementing arginine auxotrophic cell wall mutants of *C. reinhardtii* behaved like plant protoplasts in somatic fusion experiments. In the presence of PEG the spherical cells aggregated quickly and fused into different forms (Fig. 2). After fusion the hybrids regenerated a thick cell wall, confirming that they are very similar to diploids. Electron microscopy provided evidence that bacteria present in the fusion mixture were taken up into the algal cells. Azotobacters are present in membrane-bound vesicles inside the host cytoplasm and the formation of vesicles was also observed. The selected intracellular symbiosis (endocytobiosis) retained its green color on the nitrogen and carbon free medium.

Liposome mediated transfer experiments provided similar results. Liposomes formed readily in dense *Azotobacter* suspensions from different phospholipid mixtures and kits. They incorporated bacteria and fused well with each other or with cell wall mutants on the effect of PEG.

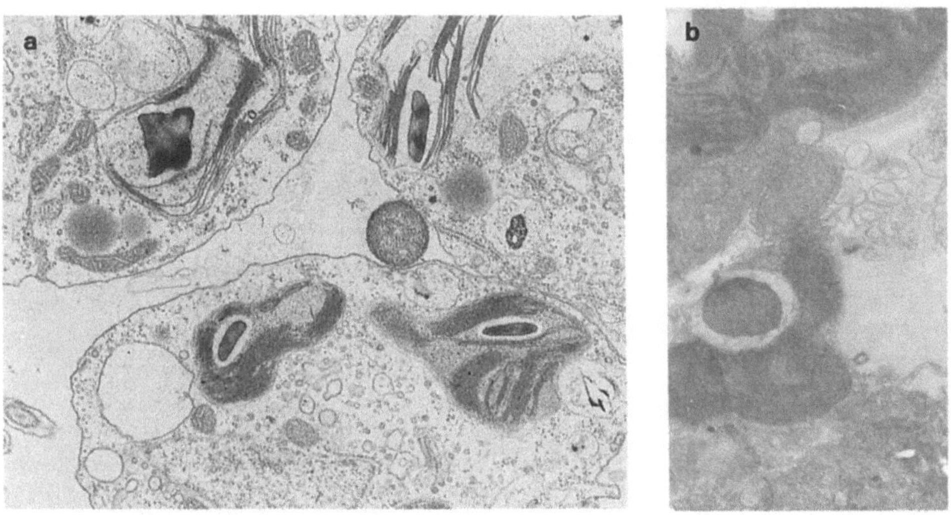

Fig. 2. PEG induced fusion of cell wall and arginine deficient algal mutants in the presence of bacteria. a. Surrounding. b. Fusion.

It is common that after prolonged cultivation of endocytobioses bacteria appear in the intercellular spaces of algal colonies and between the algal plasma membrane and cell wall or directly in the cell wall (Fig. 3). About 10% of the examined algal cells harbored endocytobionts and on average 5-10 times more azotobacters were present between the cell

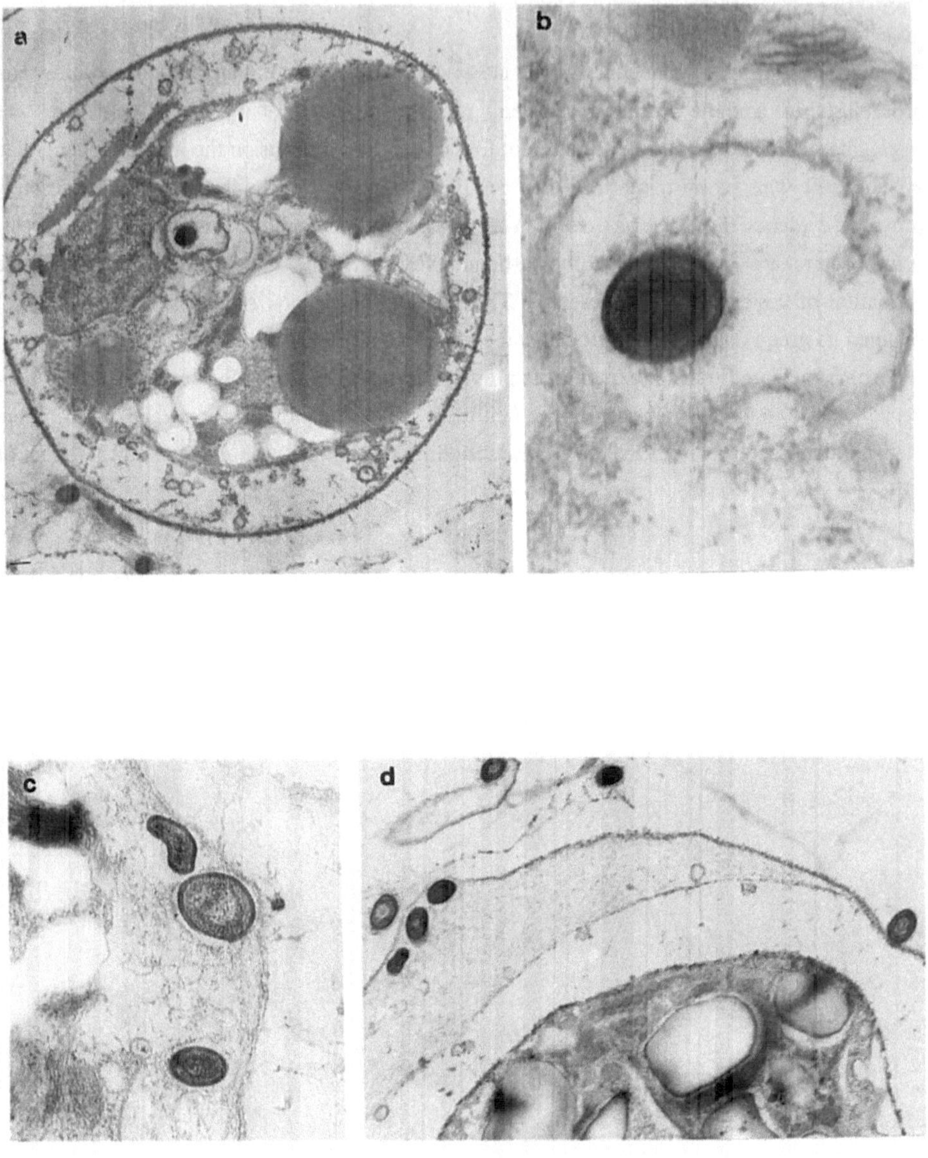

Fig. 3. Electron micrographs of *C. reinhardtii* cells harboring bacteria in the cytoplasm (a, b), between the plasma membrane and cell wall (c), and in the cell wall (d).

wall and plasma membrane than in the cytoplasm. During prolonged cultivation bacteria were also observed in the intercellular spaces of alga colonies (Fig. 4). Their number increased gradually and the former endosymbioses turned into exosymbioses.

It is interesting to mention that exosymbioses are very stable which may be a consequence of metabolic cooperation and exchange of nutrients. Photosynthetically active *C. reinhardtii* cells release glycolate, pyruvate, polysaccharides, proteins, amino acids, keto acids and vitamins into the medium. On the other hand, diazotrophic *Azotobacter* species excrete mainly nitrogen compounds, like ammonia, amino acids, vitamins and hormones. The close contact of cells in these partnerships, revealed by electron microscopy, is a prerequisite of metabolic exchange. The viability of alga-bacterium symbioses may be checked even visually because in case of nitrogen shortage yellow gametes form from the green vegetative cells. Such induction of the generative phase or gametogenesis by nitrogen deprivation is a well known phenomenon.

Based on the success with algal models, symbioses were established between strawberry and *A. insignis*. In case of a multicellular host, the introduction of bacteria into the intercellular spaces of plant tissues seemed to be an adequate solution because these spaces are important

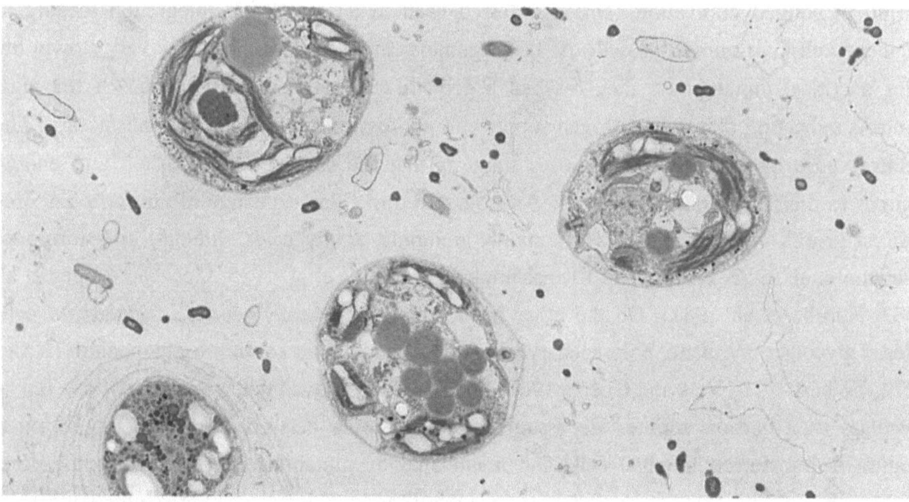

Fig. 4. Distribution of bacteria in the intercellular spaces of alga colony.

sites of apoplasmatic transport. In the artificial system bacterial metabolic dependency was ensured by the sugar source. Maltose present in the media can be utilized directly by strawberry but not by *A. insignis*. Bacteria can use only those sugars as carbon and energy source which are synthetized by the plant cells and excreted into the intercellular spaces.

Calli grown from mixed cultures contained bacteria in high number among the cells as it was shown by light and electron micrographs (Fig. 5). Selected callus regions were transferred to regeneration medium with normal level of nitrogen. Organogenesis occurred within 4-6 weeks with high efficiency (Fig. 6). Light and electron microscopic examinations proved the presence of bacteria in the intercellular spaces of regenerated plants both in lamina and petiole (Fig. 7). *A. insignis* cells were localized in the cortical parenchima and beside the tracheal elements. Rooted strawberries were planted out to a soil mixture containing low level of nitrogen where they grew vigorously and looked happy with healthy, green leaves.

Discussion

Our experimental findings are in favour of the suggestion that stable inter- and intracellular partnerships can be established between diazotrophic bacteria and plants by suitable culture, fusion and selection techniques. The *Azotobacter-Chlamydomonas* systems grow indefinitely on nitrogen and carbon free medium and may be viewed as models for the symbiotic transfer of diazotrophy into new hosts.
The metabolism of partners was not damaged but rather adapted to the mutual requirements during prolonged cultivation. This manifested itself in an increasing integration leading to balanced cell division and growth. At the beginning, these partnerships grew very slowly but after a critical initial phase they reached a growth rate comparable with that of the algal controls indicating that their nitrogen supply and photosynthetic activites are satisfactory. The autotrophic growth of these symbioses suggests that the algae provide carbon and energy sources to their symbionts and get fixed nitrogen in turn. This exchange of nutrients has only indirect proofs. Azotobacters excrete mainly ammonia, amino acids, vitamins and hormones (Newton et al. 1953, Kleiner and Kleinschmidt 1976, Kuhla et al. 1985, Gonzalez-Lopez et al. 1983, Karube et al. 1981). On the other hand, photosynthetically active *C. reinhardtii* cells release glycolate, pyruvate, polysaccharydes, proteins and amino acids into the medium (Kroes 1972, Tolbert 1979, Vela and Guerra 1966). Nghia (1986) proved that *A. vinelandii* can utilize glycolate as a carbon source. He assumed that ammonia derived from bacterial nitrogen fixation is transferred to algal cells and assimilated by glutamine synthetase. Azotobacters retained their active nitrogenase both in the neighborhood and inside of photosynthetizing algal cells.

The metabolic cooperation and exchange of nutrients may be checked visually because in the case of nitrogen shortage yellow gametes form from the green vegetative cells. If bacteria can supply enough nitrogen for the algae, they retain their green color and divide mitotically.
The close contact of partner cells in intercellular symbiosis revealed by electron microscopy may be a prerequisite of metabolite axchange and stable symbiosis formation. Extracellular bacteria appeared also in endocytobioses after some time even when they were eliminated by

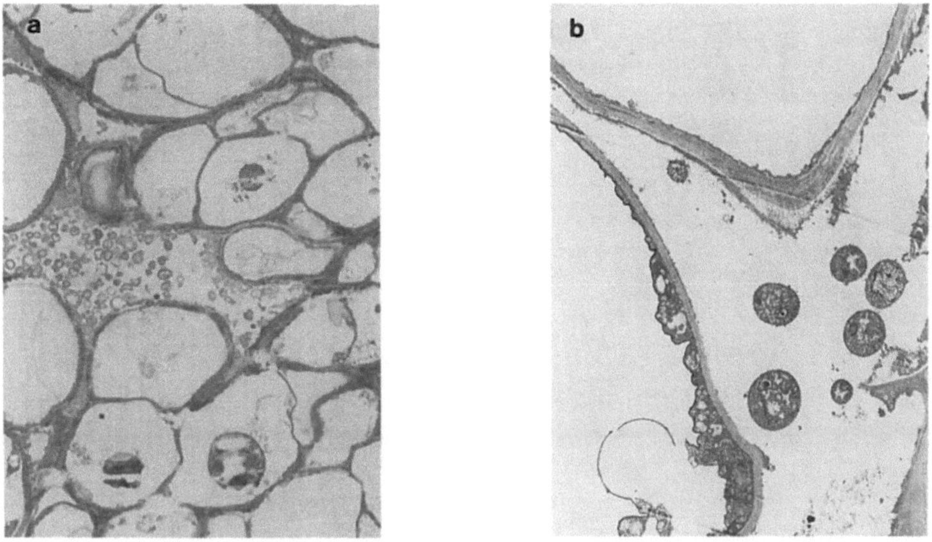

Fig. 5. Light (a) and electron (b) micrographs of *A. insignis* among the callus cells of strawberry.

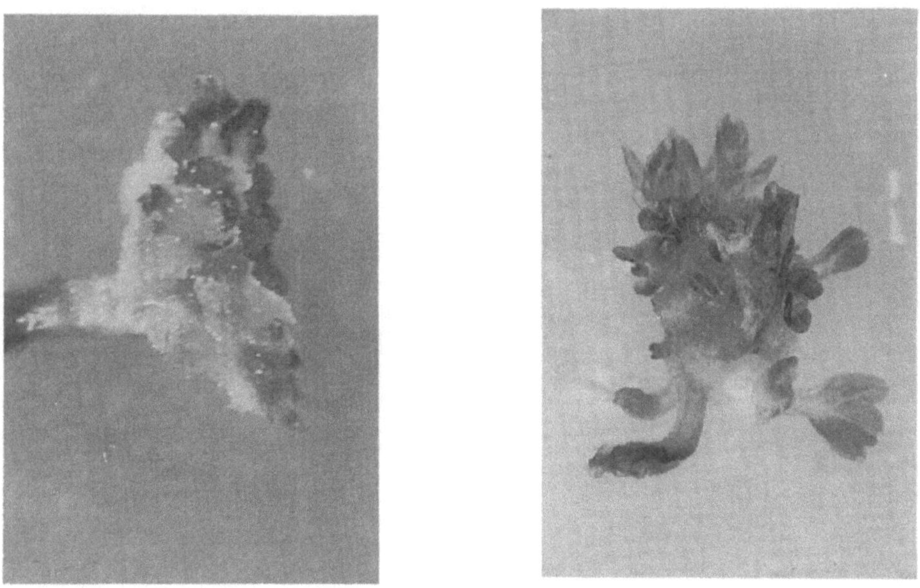

Fig. 6. Plant regeneration from bacterium containing strawberry callus.

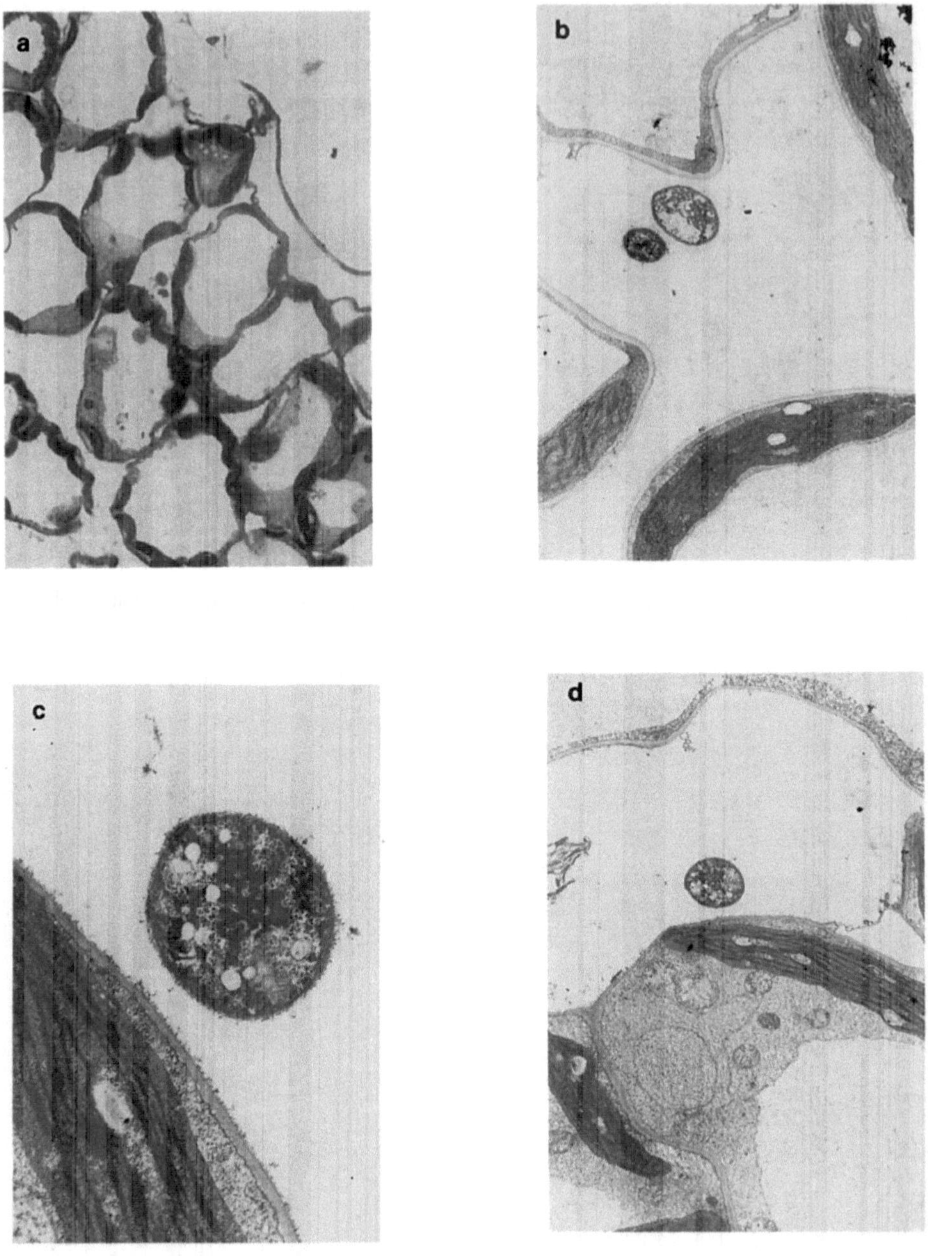

Fig. 7. Light (a) and electron (b-d) micrographs of *A. insignis* cells in the intercellular spaces of strawberry plants.

lisozyme or antibiotic treatment at the beginnig. This may be explained by the death of algae or by an active bacterial movement across the algal plasma membrane and cell wall by which the host regulates symbiont number according to the metabolic requirements. The multiplication of azotobacters even in these systems may be more frequent than the division of algae, and their excess is removed between the cell wall and plasma membrane. Exocytosis is a possible process by which they leave the host cytoplasm. Bacteria were sometimes detected directly across the algal cell wall which means that bacteria can move througt it. The inducible production of cellulases and other hydrolases by *Azotobacter* species (Samanta et al. 1989, Gyurján et al. 1992) may explain this phenomenon which was never observed in exocytobioses. This suggests that the direction of bacterial movement is outward from the host cell. Such enzymatic mechanisms may direct the exclusion of surplus azotobacters ensuring the maintenance of a balanced symbiotic partnership. A detailed description of pure endocytobiosis is not possible due to the permanent occurrence of intercellular bacteria. The establishment of a suitable selection system, based on the incorporation of resistance against different antibiotics (e.g. kanamycin) into algae, has not been successful yet.

Based on the algal models, it was possible to establish exocytobiosis between *Azomonas* and strawberry by the incorporation of bacteria into the callus tissue. The functional basis of this partnership is also the dependence of bacteria on the plant metabolic activity and vice versa. Plant cells may utilize reduced nitrogen and growth substances exuded by azotobacters into the intercellular spaces and supply the bacteria with carbon and energy source. These systems fix atmospheric nitrogen very effectively. Plant regeneration from callus was stimulated by nitrogen in the media which did not repress bacterial nitrogen fixation.

We can conclude that stable nitrogen fixing symbioses were established between diazotrophic bacteria and plants. The prolonged cultivation of these artificial symbioses suggests a successful metabolic cooperation of the partners. Intercellular spaces or algal cytoplasm seem to be good habitats for symbiotic bacteria which supply their hosts with nitrogen compounds. Intracellular bacteria of algal hosts are present in membrane-bound vesicles like true organelles.

Our artificial symbioses have both evolutional and practical significance. They provide model systems for the study of plant-microbe interactions and evolution but at the same time they may be helpful in the transfer of desired abilities into new host. The transfer of *nif* genes was implemented by the introduction of whole diazotrophic prokaryotes. In the case of higher plants, this type of transfer may be more advantageous than the direct manipulation of relevant genes.

References

Berg RH, Vasil V, Vasil IK (1979) The biology of *Azospirillum*-sugarcane association. II. Ultrastructure. Protoplasma 101: 143-163.

Bradley PM (1979) Micromanipulation of cyanelles and a cyanobacterium into higher plant cells. Physiol. Plant. 47: 293-298.

Burgoon AC, Bottino PJ (1976) Uptake of the nitrogen-fixing blue-green algae *Goeocapsa* into protoplasts of tobacco and maize. J. Heredity 67: 223-226.

Burgoon AC, Bottino PJ (1977) Uptake of the nitrogen-fixing blue-green alga *Gloeocapsa* by plant protoplasts. In: Hollaender A et al (eds) Genetic engineering for nitrogen fixation. Plenum Press New York London 213-229.

Carlson PS Chaleff RS (1974) Forced association between higher plant and bacterial cells *in vitro.* Nature 252: 393-395.

Giles KL Whitehead H (1975) The transfer of nitrogen fixing ability to a eukaryote cell. Cytobios. 14: 49-61.

Giles KL Whitehead H (1976) Uptake and continued metabolic activity of *Azotobacter* within fungal protoplasts. Science 193: 1125-1126.

Gonzalez-Lopez J Salmeron V Moreno J Ramos-Cormezana A (1983) Amino acids and vitamins produced by *Azotobacter vinelandii* ATCC 12837 in chemically-defined media and dialysed soilmedia. Soil Biol. Biochem. 15: 711-713.

Gusev MV Korzhenevskaya TG Pyvovarova LV Baulina OJ Butenko RG (1986) Introduction of a nitrogen-fixing cyanobacterium into tobacco shoot regenerates. Planta 167: 1-8.

Gyurján I Turtóczky I Tóth G Paless Gy Nghia NH (1984) Intercellular symbiosis of nitrogen-fixing bacteria and green alga. Acta Bot. Hung. 30: 249-256.

Gyurján I Nghia NH Turtóczky I Stefanovits P (1986) Photosynthesis, nitrogen fixation and enzyme activities in *Chlamydomonas-Azotobacter* symbioses. Biochem. Physiol. Pflanzen 181: 147-153.

Gyurján I Korányi P Stefanovits P Bóka K Turtóczky I (1990) Transfer of nitrogen-fixing azotobacters into *Chlamydomonas reinhardtii* cells. Endocytobiology IV. 239-244.

Gyurján I Korányi P Paless Gy (1992) Ultrastructural analysis of an artificial alga-bacterium endosymbiosis after prolonged cultivation. Symbiosis 14: 475-484.

Karube I Matsunaga T Otomine Y Suzuki S (1981) Nitrogen fixation by mobilized *Azotobacter chroococcum.* Enzyme Microb. Technol. 3: 309-312.

Kleiner D Kleinschmidt JA (1976) Selective inactivation of nitrogense in *Azotobacter vinelandii* batch cultures. J. Bacteriol. 128: 117-122.

Korányi P Kristóf Z Gyurján I (1988) Nitrogen fixation: comparative analysis of natural and artificial cytobioses. Endocyt. C. Res. 5: 47-58.

Korányi P Gyurján I Stefanovits P Bóka I Szigeti Z (1990) Characterisation of an artificial nitrogen-fixing alga-bacterium ectocytobiosis. Symbiosis 8: 175-187.

Korányi P Varga SzS Preininger É Gyurján I (1993) Symbiotic integration of diazotrophic bacteria with plant cells in artificial cytobioses. Endocytobiology V. 465-472.

Kroes HW (1972) Extracellular products from *Chlorococcum ellipsoideum* and *Chlamydomonas globosa*. Arch. Microbiol. 84: 270-272.

Kuhla J Dingler Ch Oelze J (1985) Production of extracellular nitrogen-containing components by *Azotobacter vinelandii* fixing dinitrogen in oxygen-controlled continuous culture. Arch. Microbiol. 141: 297-302.

Murashige T Skoog F (1962) A revised medium for rapid growth and bio assay with tobacco tissue cultures. Physiol. Plant. 15: 473-497.

Newton JW Wilson PW Burris RH (1953) Direct demonstration of ammonia as an intermediate in nitrogen fixation by *Azotobacter*. J. Biol. Chem. 204: 445-453.

Nghia NH Gyurján I Stefanovits P Turtóczky I (1986) Uptake of azotobacters by fusion of gametes of *Chlamydomonas reinhardtii*. Endocyt. C. Res. 3: 341-346.

Nghia NH Gyurján I (1987) Problems and perspectives in establishment of nitrogen-fixing symbioses and endosymbioses. Endocyt. C. Res. 4: 131-141.

Sager R Granick S (1953) Nutritional studies with *Chlamydomonas reinhardtii*. Ann. NJ Acad. Sci. 56: 831-838.

Samanta R Pal D Sen SP (1989) Production of hydrolases by N_2-fixing microorganisms. Biochem. Physiol. Pflanzen 185: 75-81.

Tolbert NE (1979) Glycolate metabolism by higher plants and algae. In: Gibbs M Latzkó E (eds) Encyclopedia of plant physiology. Vol. 6. Photosynthesis II. Springer-Verlag Berlin 338-352.

Varga SzS Korányi P Preininger É Gyurján I (1994) Artificial associations between *Daucus* and nitrogen-fixing *Azotobacter* cells in vitro. Physiol. Plant. 90: 786-790.

Vasil V Vasil IK Zuberer DA Hubbell DH (1979) The biology of *Azospirillum*-sugarcane association. I. Establishment of association in vitro. Z. Pflanzenphysiol. 95: 141-147.

Vela GR Guerra CN (1966) On the nature of mixed cultures of *Chlorella pyrenoidosa* tx 71105 and various bacteria. J. Gen. Microbiol. 42: 123-131.

You L Zhou F (1989) Non-nodular endorhizospheric nitrogen fixation in wetland rice. Can. J. Microbiol. 35: 403-408.

Murashige T, Skoog F (1962) A revised medium for rapid growth and bio assays with tobacco tissue cultures. Physiol Plant 15: 473–[illegible]

Newton JW, Wilson PW, Burris RH (1953) Direct demonstration of ammonia as an intermediate in nitrogen [illegible] (Azotobacter). J Biol Chem [illegible]

[illegible]

Taxonomic and Ecologic Studies of Diazotrophic Rhizosphere Bacteria Using Phylogenetic Probes*

Anton Hartmann, Jose Ivo Baldani[1], Gudrun Kirchhof, Bernhard Aßmus, Peter Hutzler[2], Nina Springer[3], Wolfgang Ludwig[3], Vera Lucia D. Baldani[1] and Johanna Döbereiner[1]

GSF-Forschungszentrum für Umwelt und Gesundheit, Institut für Bodenökologie, Neuherberg, Postfach 1129, D-85758 Oberschleißheim, Germany

Abstract

The phylogenetic relationships between diazotrophic rhizosphere bacteria were studied using 16S and 23S ribosomal RNA sequences. Certain stretches of the rRNA sequences are variable enough to provide information for the development of diagnostic probes at the species level for the five *Azospirillum* spp., *Acetobacter diazotrophicus*, *Herbaspirillum seropedica*, *H. rubrisubalbicans* and „isolate E"-bacteria. The brasilian „isolate E"- type bacteria, which reside inside roots, stems and leaves of several plants, could be grouped into the beta-subgroup of proteobacteria. According to their 16S rRNA sequences they belong to the genus *Burkholderia*. Recently isolated diazotrophic bacteria from the C4-plants *Miscanthus sinensis* and *Spartina pectinata* were identified as *Azospirillum lipoferum* and *Herbaspirillum*-like bacteria.

---

*This work was supported by the Deutsche Forschungsgemeinschaft and the brazilian german scientific & technological research cooperation (project Bra Env 34).

[1] CNPAB Seropedica, Itaguai, Rio de Janeiro, Brazil

[2] GSF-Forschungszentrum für Umwelt und Gesundheit, Institut für Pathologie, Neuherberg, Postfach 1129, D-85758 Oberschleißheim, Germany

[3] Lehrstuhl für Mikrobiologie, Technische Universität München, D-80290 München, Germany

NATO ASI Series, Vol. G 37
Azospirillum VI and Related Microorganisms
Edited by I. Fendrik et al.

Oligonucleotide probes were successfully used for a specific fluorescence-labelling of single cells *in situ. Azospirillum brasilense* strains Sp7, Sp245, and Wa3 could be identified and localized in the rhizosphere of wheat plants. Scanning confocal laser microscopy allowed to image the three-dimensional structure of colonized roots and to characterize a different localization pattern of these strains.

Introduction

Research on the phylogeny of nitrogen fixing bacteria has revealed a high diversity of diazotrophs in all groups of bacteria (Young, 1992). New diazotrophic bacteria were characterized very recently (Baldani et al., 1992, Reinhold-Hurek et al. 1993, Tran Van et al., 1994), which demonstrates that the natural diversity of diazotrophic bacteria is far from being investigated exhaustively. Especially bacteria living endophytically in different plant parts are representing an ecologically very interesting, less understood group of diazotrophic bacteria (Döbereiner et al., 1994).

Acetobacter diazotrophicus, which belongs to the alpha subgroup of proteobacteria, is found in large numbers in the roots and stems of sugar cane (Cavalcante and Döbereiner, 1988; Gillis et al., 1989). Recently a study using scanning electron microscopy revealed the localization of *Acetobacter diazkotrophicus* cells in infected sugarcane plants (James et al., 1994). *Herbaspirillum seropedicae*, described by Baldani et al. (1986), inhabits roots, stems and leaves from different plants in rather high numbers. Interestingly, it could be demonstrated that the mild plant pathogenic *Pseudomonas rubrisubalbicans* is related to *Herbaspirillum seropedicae* on the genus level (Gillis et al. 1991, Hartmann et al., 1994).

Molecular taxonomy is a powerful tool to reveal the phylogeny of organisms. The ribosomal RNA is a suitable marker molecule for evolutionary relationships and allows to derive indices of similarity from the comparison of nucleotide sequences. In addition, probes can be designed from hypervariable sequences. Oligonucleotide probes can be labelled and used as phylogenetic stains for single cells in ecological studies (De Long et al., 1989). As compared to the 16S rRNA, the 23S rRNA contains more diverse sites which are suitable for the development of diagnostic oligonucleotide probes.

The molecular taxonomy of the genus *Azospirillum* is developed rather well (Fani et al., this volume). Phylogenetic oligonucleotides were derived from a hypervariable region of the 23S rRNA sequence and allowed a species-specific identification (Kirchhof & Hartmann, 1992). Recently, species-specific oligonucleotide probes were also derived from 16S rRNA sequences (Haurat et al., 1994, Bally et al., this volume). On the basis of 16S rRNA sequences Xia et al. (1994) developed a more complete phylogenetic tree for the *Azospirillum* cluster. They confirmed the close relatedness of all five *Azospirillum* spp. and placed *Magnetospirillum magnetotacticum* and *Rhodospirillum fulvum* as closest relatives in this cluster. Surprisingly, the N_2-fixing, anoxygenic phototroph *Rhodospirillum centenum*, isolated from a hot spring (Favinger et al., 1989), was also grouped within the *Azospirillum* cluster close to *A. irakense.* In addition to ribosomal RNA sequence analysis, the DNA restriction profiles also support the subgeneric classification of *Azospirillum* (Giovanetti et al., 1990, 1992; Fani et al., 1991).

In this communication, species-specific probes for diazotrophic rhizosphere and endophytic bacteria are presented. The present state of the art in respect to practical application of these probes is summarized and examplified with the characterization of new isolates of diazotrophic bacteria from C4-plants. An example of a successful *in situ* identification and localization using fluorescence-labelled probes is given which includes the application of scanning confocal laser microscopy to improve image quality.

Materials and Methods

Bacterial strains and growth conditions

The bacteria were obtained from the german culture collection (DSM, Braunschweig) and from the collection of the CNPAB, Seropedica, Brazil. Culture conditions were applied for *Herbaspirillum* spp.as described by Baldani et al. (1986), for *Azospirillum* spp. according to Hartmann et al. (1988) and for *Acetobacter diazotrophicus* as described by Calvante & Döbereiner (1988).

Isolation of diazotrophic bacteria from the C4-plants *Miscanthus* and *Spartina*

Plant material of *Miscanthus giganteus* and *Spartina pectinata* was obtained from research plots of the Bayerische Landesanstalt für Bodenkultur und Pflanzenbau in Freising/Weihenstephan, Germany. *Miscanthus giganteus* was grown in four consecutive years with and without 150 kg N ha^{-1} a^{-1}. Roots, stems and leaves were thoroughly washed and small pieces of this material was inoculated in N-free semisolid Nfb, JNfb and LGI media (Döbereiner et al., 1994). The vials were incubated at room temperature and a small amount was transferred to fresh N-free semisolid media several times. Typical pellicles appeared in some of the vials and microscopic examination showed the presence of typical *Azospirillum lipoferum* cells and also a small, very active *Herbaspirillum*-like bacterium. All strains were able to reduce acetylene.

Probe design and dot blot hybridization

Probe development and filter hybridization were performed as described by Kirchhof and Hartmann (1992). Bacterial bulk nucleic acid (NA) was prepared according to the method of Stahl and Flesher (1987). Crude NA preparations which contain high amounts of ribonucleic acid were fixed on Zeta-probe nylon filters (BioRad, München) and immobilized by drying overnight. ^{32}P-labelled probes were used for dot blot hybridization.

Whole cell hybridization

Whole cell hybridization of single cells basically followed the protocoll of Spring et al. (1992). Root pieces from inoculated wheat plantlets were fixed with paraformaldehyde, dehydrated in an ethanol series and mounted on slides. After hybridization with fluorescein- or rhodamine-labelled oligonucleotide probes the samples were counterstained with 4,6′-diamidino-phenylindole (Porter & Feig, 1980). In addition to an *A. brasilense* probe (Kirchhof & Hartmann, 1992), the following oligonucleotide probes were used: probe EUB338, covering all bacteria (Amann et al., 1990), and probe ALF1b which is specific for alpha subgroup proteobacteria (Manz et al., 1992).

Scanning confocal laser microscopy

The samples were investigated with a Zeiss LSM 410 scanning confocal laser microscope equipped with lasers at 365, 488, and 543 nm, and filter sets suitable for detection of rhodamine- and fluorescein-coupled oligonucleotides as well as DAPI. All image processing, combining into a colored rgb-image, reconstructions of three-dimensional arrangements were performed with the standard Zeiss LSM software package.

Results and Discussion

Taxonomic studies and development of phylogenetic probes

The highly variable stretch covering helices 54-59 of the 23S rRNA can be amplified by using conservative flanking sequences as primers in the polymerase chain reaction (Höpfl et al., 1989; Ludwig et al., 1994). Table 1 summarizes the oligonucleotide sequences derived from 23S rRNA sequences using this approach. The proposed sequences were compared with data bank sequences of a broad spectrum of 23S rRNAs of other bacteria. The oligonucleotides were also tested with a large variety of bacterial nucleic acid preparations (Kirchhof, unpublished) and proved to have the proposed specificity.

Species Name	Oligonucleotide Sequence	Proteobacteria subgroup
Azospirillum amazonense	5′-GTGTGCCATGGAGGTGT-3′	alpha
Azospirillum brasilense	5′-GCCCGGCTGGGGACCC-3′	alpha
Azospirillum halopraeferens	5′-AGCGTGCTGCGGCGA-3′	alpha
Azospirillum irakense	5′-CTGAAAACCAGTATGC-3′	alpha
Azospirillum lipoferum	5′-TAGCCCCGCCTTATA-3′	alpha
Acetobacter diazotrophicus	5′-ATCCGGCTTTTGGCGCA-3′	alpha
Herbaspirillum seropedicae	5′-GTCCCGGTTTTTGCATCGA-3′	beta
Herbaspirillum rubrisubalbicans	5′-TAGTCGGTTTTTGCATCGA-3′	beta
„Isolate E"-type bacteria	5′-AAGGGTGCGAGGCCATGC-3′	beta

Table 1: Sequences of 23S rRNA derived species-specific oligonucleotide probes

The *Acetobacter diazotrophicus* probe was clearly negative with *Acetobacter aceti, Ac. hansenii, Ac. liquefaciens, Ac. pasteurianus* and *Gluconobacter oxydans.* All *Acetobacter diazotrophicus* isolates from different origins tested with this probe until now were positive. This corroborates the finding that there is limited genetic diversity in this endophytic bacterial species from plants rich in sugar (Caballero-Mellado & Martinez-Romero, 1994).

While the *Herbaspirillum seropedicae* probe was species-specific, the *H. rubrisubalbicans* probe also hybridized to NA from EF1-isolates (Kirchhof, unpublished). These mostly clinical isolates which are closely related to *H. rubrisubalbicans* are not nitrogen-fixing and differ only in a few physiological characteristics (Gillis et al. 1991).

When these oligonucleotide probes are used for a species differentiation with unknown isolates, problems may appear when other methods are used to obtain bacterial nucleic acids. Nguyen et al. (this volume) used a method which yields mostly bacterial DNA. Using DIG-labelled probes, the species-specific response was not observed in all cases. The *A. lipoferum* probe exhibited a high degree of unspecific hybridization. Probably the genome of many bacteria contains a sequence which crossreacts with the chosen 23S rRNA sequence of *A. lipoferum*. Another possible problem arrises in the non-radioactive detection system with DIG-labelled probes. NA-preparations obtained with the method of Stahl & Flesher (1987) cannot be used, because the ribonucleic acids are attacked under the alkaline conditions during the phosphatase assay used in the DIG detection system. Thus, most of the nucleic acid is lost from the filters (W. Melchinger, unpublished). Hybridizations of RNA-rich bulk nucleic acid preparations with ^{32}P-labelled probes seem to give the most reliable and sensitive results.

It also needs to be stated that the molecular taxonomic tests with species-specific oligonucleotide probes should be used together with standard taxonomic examinations. It is our experience, that thorough investigations with indicative media and culture techniques along with a careful examination of phenotypic features and physiological tests almost always led to a positive proof using the direct probing technique. The probe technique can be used as a short cut for a quick positive proof. In addition to species-specific probes, the application of oligonucleotide probes specific for the alpha-, beta- and gamma-subgroup of proteobacteria (Manz et al., 1992) corroborates the taxonomic classification. Negative results in hybridization assays could be due to an uncommon base exchange in the diagnostic oligonucleotide region of the tested strain. If negative hybridization results occur, a classical taxonomic analysis has to be performed (e.g. Krieg & Döbereiner, 1984).

„Isolate E"-type bacteria

A new type of Gram-negative, diazotrophic bacteria was isolated from the inside of sugarcane, sweet potato and rice (Olivares, 1992). These group of bacteria grew very well in semisolid, nitrogen free mannitol minimal medium, pH 4.0. They also can be obtained in medium M, containing glucose, oxalate and citrate. A series of physiological tests were performed including the API-series and carbon utilization tests using the millipore method (V. L. D. Baldani, unpublished results). The phenotypic appearance of these isolates resembles that of *Herbaspirillum* but the physiological tests suggest to group them as *Burkholderia*-like bacteria. All isolates were positive in the acetylene reduction assay. The main difference from *Burkholderia cepacia* was the lower pH-optimum (pH 4.0 - 4.5) and the higher tolerance for sucrose (up to 10%).

To evaluate the phylogenetic position, the complete sequence of the 16S ribosomal RNA of isolate PPe8 was obtained. It clearly revealed a position inside the *Burholderia*-cluster, probably representing a new *Burkholderia* species. In addition, an oligonucleotide probe for the „isolate E"-type bacteria was derived from strain PPe8 which has been isolated from sugarcane. It was tested by dot blot hybridization with nine other „isolate E"-type bacteria, *Herbaspirillum* isolates and other bacteria. The probe (Table 1) gave positive results only with the „isolate E"-strains PPe5, PPe6, Ppe7 and PPe8. This can be taken as evidence for an unrevealed diversity among the „isolate E"-type bacteria. Phylogenetic comparison with other *Burkholderia* spp. are currently under investigation.

Diazotrophs in the C4-plants *Miscanthus* and *Spartina*

Miscanthus giganteus and *Spartina pectinata* are investigated as potential energy or fiber crops in Europe. Using oligonucleotide probes, we examined the taxonomy of endophytic bacteria isolated in Germany from these plants. In dot blot hybridizations, the putative *Azospirillum lipoferum* isolates hybridized only with the *A. lipoferum* probe. However, the *Herbaspirillum*-like isolates did not hybridize with the *Herbaspirillum seropedicae* probe. It is quite possible, that this type of isolates represents a new species within the genera *Herbaspirillum* or *Burkholderia.*

In situ studies using fluorescence-labelled oligonucleotide probes

Oligonucleotide probes coupled to fluorescent dyes can be used for a specific *in situ* detection of bacteria in complex habitats (Spring et al., 1992). Using this approach, we studied the colonization of wheat roots by *Azospirillum brasilense* strains Sp7, Sp245 and Wa3 (Aßmus et al., 1995). The samples were viewed with a scanning confocal laser microscope (SCLM). This technique provides an improved image quality due to point-by point illumination of the specimen and elimination of out-of-focus signals (for review see Shotton & White, 1989). After collecting series of optical sections at different excitation wavelengths, colored images with extended focus as well as three-dimensional reconstructions of the original samples were calculated.

The species-specific probe allowed an *in situ* detection of *A. brasilense* only in root samples from monoxenic Hoagland tubes. Signal quantification with SCLM on hybridized cells revealed that this probe gave only 40% of the signal intensity as compared to probe Eub338 or ALF1b, respectively. Thus, for detection of azospirilla on soil grown roots the group-specific probe ALF1b was used. A new set of species-specific probes needs to be developed on the basis of other variable sequence stretches of the 23S rRNA which are more suitable for *in situ* application.

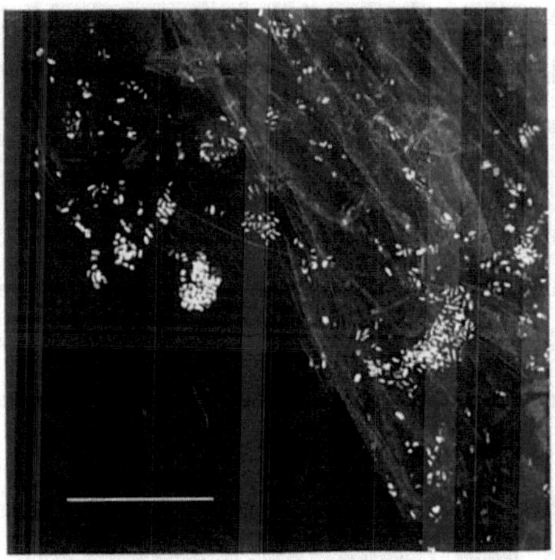

Fig. 1: A. brasilense Sp245 in the root hair zone of a Hoagland-grown wheat seedling. Staining with rhodamine-labeled probe ALF1b, excitation wavelength 543 nm. Projection of a *z*-series, 18 μm. Bar equals 20 μm.

A significant portion (10-25%) of the rhizosphere microflora gave hybridization signals with sufficient signal intensity. This provides an estimation of the physiological activity of the cells, because the content of ribosomes is supposedly high in very active cells and the penetration of probes into fixed bacteria is usually not limiting. *Azospirillum brasilense* was found on the root surface and in the root hair zone in tight layers and clumps (Fig.1). Whereas strain Sp7 was restricted to these sites, strains Sp245 and Wa3 were also detected below cells of the outer cortex (Fig. 2a). Strain Sp245 was found repeatedly in high densities in the interior of root hairs which obviously had intact cell walls (Fig. 2b). We could not yet detect strain Sp245 in the vascular system of the roots, probably because fluorescence-labelled probes may not have access to the root interior due to the tight cell walls. Improvements towards a proper sample processing should solve this problem. Strain Sp245 is known to penetrate the root cortex since it could be identified in thin sections of root vascular tissues using strain-specific monoclonal antibodies and the immunogold/silver enhancement technique (Schloter et al., 1995).

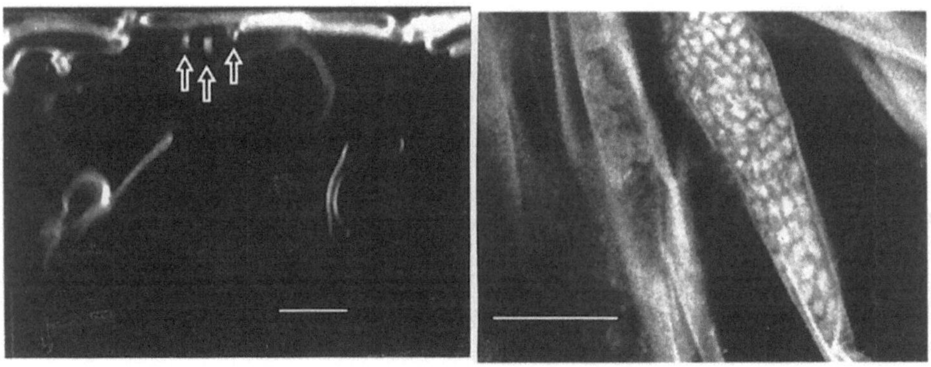

Fig 2: Colonization of the endorhizosphere of wheat seedlings by *A. brasilense*. Staining and excitation as in fig. 1. a: *z*-scan exhibiting strain Wa3 (arrows) below the outer cortex cells of a seedling grown in quartz sand. b: Root hair cell of a seedling grown in unsterile soil, tightly filled with *A. brasilense* Sp245.
Bars equal 5 μm for panel a and 10 μm for panel b, respectively.

Monoclonal or polyclonal antibodies can also be coupled to fluorescent dyes and used for an *in situ* monitoring of *Azospirillum brasilense* in the rhizosphere (Schloter et al., 1993). Whereas rRNA-targeted oligonucleotide probes can be designed at a specificity from domain down to species level, a monoclonal antibody can be even strain-specific (Schloter et al., 1992). The combination of these two labelling tools would allow the detection of a specifically introduced strain within the naturally occurring background of bacteria from the same species or genus. Research is in progress to combine the serological and oligonucleotide labelling techniques to a single protocol.

References

Amann RI, Binder BJ, Olson RJ, Chisholm SW, Devereux R, Stahl DA (1990) Combination of 16S rRNA-Targeted oligonucleotide probes with flow cytometry for analyzing mixed microbial populations. Appl Environ Microbiol 56:1919-1925

Aßmus B, Hutzler P, Kirchhof G, Amann R, Lawrence JR, Hartmann A (1995) *In situ* localization of *Azospirillum brasilense* in the rhizosphere of wheat using fluorescently labeled, rRNA-targeted oligonucleotide probes and scanning confocal laser microscopy. Appl Environ Microbiol, in press

Baldani JI, Baldani VLD, Seldin L, Döbereiner J (1986) Characterization of *Herbaspirillum seropedicae* gen. nov., sp. nov., a root-associated nitrogen-fixing bacterium. Int. J. System. Bacteriol. 36:86-93

Baldani JI, Baldani VLD, Olivares F, Döbereiner J (1992) Identification and ecology of *Herbaspirillum seropedicae* and the closely related *Pseudomonas rubrisubalbicans*. Symbiosis 13:65-73

Bally R, Haurat J, Normand P (1995) *Azospirillum* phylogeny based on 16S rRNA gene sequences. This volume

Caballero-Mellado J & Martinez-Romero E (1994) Limited genetic diversity in the endophytic sugarcane bacterium *Acetobacter diazotrophicus*. Appl Environ Microbiol 60:1532-1537

Calvante VA & Döbereiner J (1988) A new acid tolerant nitrogen-fixing bacterium associated with sugarcane. Plant Soil 108:23-31

De Long EF, Wickham GS, Pace NR (1989) Phylogenetic stains: Ribosomal RNA-based probes for the identification of single cells. Science 243:1360-1363.

Döbereiner J, Baldani VLD, Olivares F, Reis VM (1994) Endophytic diazotrophs: The key to BNF in gramineous plants. In: Proceedings of the 6th Int. Symposium on Nitrogen Fixation with Non-Legumes, Ismailia, Egypt (in press)

Fani R, Bazzicalupo M, Gallori E, Giovanetti L., Ventura S, Polsinelli M (1991) Restriction fragment length polymorphism of *Azospirillum* strains. FEMS Microbiol Lett 83:225-230

Fani R, Bandi C, Bazzicalupo M, Di Cello F, Fancelli S, Gallori E, Gerace L, Grifoni A, Damiani G (1995) Phylogenetic studies of the genus *Azospirillum*. This volume

Favinger J, Stadtwald R, Gest H (1989) *Rhodospirillum centenum* sp. nov., a thermotolerant cyt-forming anoxygenic photosynthetic bacterium. Antonie van Leeuwenhoek 55:291-296.

Gillis M, Kersters K, Hoste B, Janssens D, Kroppenstedt RM, Stephan MP, Teixeira KRS, Döbereiner J, De Ley J (1989) *Acetobacter diazotrophicus* sp. nov., a nitrogen-fixing acetic acid bacterium associated with sugarcane. Int J System Bacteriol 39:361-364

Gillis M, Döbereiner J, Pot B, Goor M, Falsen E, Hoste B, Reinhold B, Kersters K (1991) Taxonomic relationships between [*Pseudomonas*] *rubrisubalbicans*, some clinical isolates (EF group 1), *Herbaspirillum seropedicae* and [*Aquaspirillum*] *autotrophicum*, pp. 293-294. In: Nitrogen Fixation (eds: Polsinelli M, Materassi R, Vincenzini M) Kluwer Academic Publishers, Dordrecht

Giovanetti L, Ventura, S, Bazzicalupo M, Fani R and Materassi R (1990) DNA restriction fingerprint analysis of the soil bacterium *Azospirillum*. J. Gen. Microbiol. 136:1161-1166

Giovanetti L, Fedi S, Gori A, Montani P, Ventura S (1992) Identification of *Azospirillum* strains at the genome level with total DNA restriction pattern analysis. System. Appl. Microbiol. 15:37-41 (1992)

Hartmann A, Fu H, Burris RH (1988) Influence of amino acids on nitrogen fixation and growth of *Azospirillum* spp. Appl. Env. Microbiol. 54:87-93

Hartmann A, Aßmus B, Lawrence JR, Reis V, Kirchhof G (1994) Development and application of 23S rRNA-directed oligonucleotide probes for *Azospirillum* spp., *Acetobacter diazotrophicus* and *Herbaspirillum*. In: Proceedings of the 6th Int. Symposium on Nitrogen Fixation with Non-Legumes, Ismailia, Egypt (in press)

Höpfl P, Ludwig W, Schleifer KH, Larsen N (1989) The 23S ribosomal RNA higher-order structure of *Pseudomonas cepacia* and other prokaryotes. Eur. J. Biochem. 185:355-364

James EK, Reis VM, Olivarez FL, Baldani JI, Döbereiner J (1994) Infection of sugar cane by the nitrogen-fixing bacterium *Acetobacter diazotrophicus*. J Exp Botany 45:757-766

Kirchhof G & Hartmann A (1992) Development of gene probes for *Azospirillum* based on 23S-rRNA sequences. Symbiosis 13:27-35

Krieg NR, Döbereiner J (1984) Genus *Azospirillum*, pp. 94-104. In: Bergey 's Manual of Systematic Bacteriology Vol.1 (eds: Krieg NR, Holt JG) Williams & Wilkins, Baltimore, London

Ludwig W, Dorn S, Springer N, Kirchhof G, Schleifer KH (1994) PCR-based preparation of 23S rRNA-targeted group-specific polynucleotide probes. Appl. Environ. Microbiol. 60:3236-3244

Manz W, Amann R, Ludwig W, Wagner M, Schleifer K-H (1992) Phylogenetic oligodeoxynucleotide probes for the major subclasses of proteobacteria: Problems and solutions. System Appl Micrbiol 15:593-600

Nguyen ND, Kraut M, Klingmüller W (1995) Isolation and characterization of *Azospirillum* spp. from soil and rhizosphere of North-Vietnam. This volume

Olivares E (1992) Estudo da associacao entre bacterias diazotroficas e arroz. Msc Thesis, Univ. Fed. Rural Rio de Janeiro, Itaguai, Rio de Janeiro, Brazil

Porter KG & Feig YS (1980) The use of DAPI for identifying and counting aquatic microflora. Limneol Oceanogr 25:943-948

Reinhold-Hurek B, Hurek T, Gillis M, Hoste B, Vancanneyt M, Kersters K, De Ley J (1993) *Azoarcus* gen. nov., nitrogen-fixing proteobacteria associated with roots of Kallar grass (*Leptochloa fusca* (L.) Kunth), and description of two species, *Azoarcus indigens* sp. nov. and *Azoarcus communis* sp. nov. Int J System Bacteriol 43:574-584

Schloter M, Bode W, Hartmann A (1992) Characterization of monoclonal antibodies against cell surface structures of *Azospirillum brasilense* Sp7 using ELISA techniques. Symbiosis 13:37-45

Schloter M, Borlinghaus R, Bode W, Hartmann A (1993) Direct identification, and localization of *Azospirillum* in the rhizosphere of wheat using fluorescence-labelled monoclonal antibodies and confocal laser scanning microscopy. J Microscopy 171:173-177

Schloter M, Kirchhof G, Heinzmann U, Döbereiner J, Hartmann A (1995) Immunological studies of the wheat-root-colonization by the *Azospirillum brasilense* strains Sp7 and Sp245 using strain-specific monoclonal antibodies. Plant Soil, in press.

Shotton D & White N (1989) Confocal scanning microscopy: Three-dimensional biological imaging. Trends Biochem Sci 14:435-439

Spring S, Amann R, Ludwig W, Schleifer K-H, Petersen N (1992) Phylogenetic diversity and identification of nonculturable magnetotactic bacteria. System Appl Microbiol 15:116-122
Stahl & Flesher, 1987, fehlt.

Tran Van V, Gillis M, Hebbar KP, Fernadez M, Martel MH, Berge O, Meyer JM, Heulin T (1994) Isolation of a new species of nitrogen-fixing proteobacteria from the rice rhizosphere belonging to the genus *Burkholderia.* In: Proceedings of the 6th Int. Symposium on Nitrogen Fixation with Non-Legumes, Ismailia, Egypt (in press)

Xia Y, Embley TM , O'Donnell AG (1994) Phylogenetic analysis of Azospirillum by direct sequencing of PCR amplified 16S rDNA. System Appl Microbiol 17:197-201

Young JPW (1992) Phylogenetic classification of nitrogen-fixing organisms, pp. 43-86. In: Biological Nitrogen Fixation (eds: Stacey G, Burris RH, Evans HJ) Chapman and Hall, Routledge, USA

Coaggregation of Azospirillum with other Bacteria: Basis for Functional Diversity

Carlos A. Neyra, Amy Atkinson and Olubayi Olubayi
Plant Science Dept., Cook College
Rutgers University
New Brunswick, NJ 08903-0231
USA

Abstract

The phenomenon of flocculation in *Azospirilla* shows great potential in providing inoculants for plants and soils. Of particular interest is the observed flocculation of coaggregating pairs containing a mixture of two different cell types. Cocultivation of Sp7 (non-pigmented) and AB 7030 (pink) resulted in the production of light-pink flocs and a water-clear supernatant in floc media (FM). Cocultivation in FM of the pink strain *A. brasilense* Cd (ATCC 29710) and *Rhizobium leguminosarum* bv *phaseoli* (USDA 2743) also resulted in coaggregation and production of flocs containing a mixture of the two cell types. Inoculation of *Phaseolus vulgaris* with flocculated forms of *R. leguminosarum* alone or in coaggregation with *A. brasilense* resulted in a significant enhancement of nodulation and plant growth over the non-flocculated *Rhizobium.* The partnership between *Azospirillum* and *Rhizobium* is presented here to illustrate the beneficial effects of using flocculated cell forms as a delivery system for each cell type singly or in a coaggregated form.

Introduction

Many microorganisms are known to exert beneficial effects on plant growth. Among these are the well known nitrogen-fixing *Rhizobium* species, which are symbionts of leguminous plants. *Azospirillum* bacteria are well-known diazotrophs commonly found in association with the roots of grasses (Neyra and Dobereiner, 1977). *Azospirilla* are also recognized for their plant growth promoting properties (Okon and Hadar, 1987). Inoculation of seeds or soil with beneficial microorganisms for crop improvement has been practiced for a number of years (Bashan and Levanony, 1990). However, variable and inconsistent results have often been

NATO ASI Series, Vol. G 37
Azospirillum VI and Related Microorganisms
Edited by I. Fendrik et al.

obtained due to the loss of inoculant viability or variability of dosage due to changes in inoculant viability (Okon and Hadar 1987). A high quality formulation must maintain and deliver, in an economical way, a high population of bacteria (high bacterial cell titer) and exhibit prolonged shelf-life or survivability in the field. Peat-based inoculants have been the standard for the *Rhizobium* industry, but many other formulations and methods of application have been reported and proposed for commercialization (Smith, 1992).

Our laboratory is currently working on the development of novel inoculants for the delivery of *Azospirillum* and/or *Rhizobium* based on the physiological induction of massive cell aggregation or autoagglutination in a process known as flocculation (Sadasivan and Neyra, 1985; Del Gallo et al, 1989). Following transfer from a rich media (low C/N ratio) to a N-limited minimal media (high C/N ratio) there is a switch from a dispersive growth pattern to a mode of increased aggregation, clumping and flocculation. The flocculent mass sinks to the bottom of the culture medium and is readily separable from the spent culture. The separated floc comprises a mixture of vegetative and encysted cells surrounded by a polysaccharide-rich network, altogether responsible for the advantage exhibited by flocculated forms of azospirilla: extended shelf-life, survivability and adhesiveness.

The culturing of two flocculating bacterial strains of the same species (intraspecific), two different species (interspecific) or two different genera (intergeneric), result in the production of coaggregating pairs containing a mixture of the two cell types. Cocultivation of *Azospirilla* with other bacteria has already been reported in the literature (Halsall and Gibson, 1986; Kaiser, 1994), but no mention has been made about coaggregation or coflocculation. In general, there are no reports of coaggregation of rhizosphere bacteria, but it is a well-known phenomenon in the formation of methanogenic consortia (Scherer et al, 1986) and the multigeneric aggregations among oral bacteria (Kolenbrander and Andersen, 1986). We now report on the ability of several *A. brasilense* strains to form coaggregating pairs and the production of flocs containing a mixture of two cell types.The study expands into the formation of intergeneric coaggregates by mixtures of *A. brasilense* Cd and *R. leguminosarum* biovar *phaseoli*. Their positive effect on growth and nodulation of *Phaseolus vulgaris* is reported. The potential uses of multigeneric aggregations as a network system for the delivery of beneficial bacteria, will be discussed.

Materials and Methods

Bacterial strains used in this study are listed in Table 1. *Azospirillum* strains were maintained on nutrient agar-streptomycin plates (NA/Sm) and *Rhizobium* strains on yeast extract mannitol (YEM) agar plates. Studies on cell aggregation and flocculation were carried out as previously described (Sadasivan and Neyra, 1985; Del Gallo et al., 1989) with modifications according to experimental approach. The inoculum for flocculation studies was grown on nutrient broth (NB) or NB plus YEM (1:1 v/v). Cells approaching the end of log phase were harvested by centrifugation at 8000 rpm for 10 minutes and the pellet resuspended in flocculation medium as described by Del Gallo et al. (1989). Fructose (10mM) was the preferred carbon source in coaggregation studies using *A. brasilense* strains and succinate (10 mM) for experiments using *R. leguminosarum* bv *phaseoli.*

In all coaggregation studies the initial cell density was adjusted to around 10^9 cells/ml with each partner contributing an equal cell number. Floc yields were usually determined after 24 hours and evaluated in volume terms (ml flocs per liter) following settling of flocs in a graduate cylinder. Percent flocculence was determined as the difference in Absorbance (at 600 nm) between the initial (A_I) and the residual (A_R) after 24 hours as follows:

$$A_I - A_R / A_I \times 100 = \% \text{ flocculence}$$

Time course studies involved the removal of aliquots at proper time intervals during the flocculation time period and A_{600} was determined by spectrophotometry. Flocculation of *Azospirillum, Rhizobium* or *Azospirillum/Rhizobium* was observed as a decrease in optical density of the culture.

The nodulation efficiency of *R. leguminosarum* bv *phaseoli* on *Phaseolus vulgaris* plants were determined by counting the number of nodules per plant. Inocula comprising equivalent number of cells were used for either flocculated or non-flocculated treatments, applied as seed coating prior to planting. Methyl cellulose (1%) was used as an additive to improve adhesiveness of the inocula. Each treatment consisted of 10 replicates in a completely randomized design.

Table 1. Bacterial strains

Strains	Relevant properties	Reference or Source
A. brasilense		
Sp 7	Wild type strain (ATCC 29145)	Tarrand et al. (1978)
7030	NTG-derivative of Sp 7	Michiels et al. (1991)
Cd	Pink isolate (ATCC 29710)	Eskew et al. (1977)
7030 TN5-21	Cal^- $Floc^-$ mutant	Michiels et al. (1991)
R. leguminosarum bv *phaseoli* (Type I)	Strain USDA 2743. Renamed as *R. etli. sp. nov.*	Segovia et al. (1993) van Berkum et al. (1994)

Results

Coaggregation of *Azospirillum* strains

The percent flocculence and floc yields of several *A. brasilense* strains, cultivated singly or as dual mixtures, are presented in Table 2. The results indicate that the flocculence of strains Sp 7, 7030 and Cd are quite high reaching about 95% of all cells. The floc yield was in the order of 100 to 120 ml per Liter of culture and indicated that the phenomena of flocculation yields a flocculent mass with a high concentration of cells. A simple mathematical calculation will show that each ml of floc contains about 10 billion cells (high cell titer). As anticipated (Michiels et al. 1991), the Tn 5-21 mutant did not flocculate but was able to grow in the flocculation media (high C/N ratio). Thus, the inability of the Tn 5-21 mutant to produce macroaggregates and clumping is not due to nutritional limitations but possibly due to alterations in the cell surface. The lack of flocculating ability of Tn 5-21 was attributed by Michiels et al (1991), to the inability to produce a calcofluor-binding polysaccharide. The results in Table 2 also showed that Tn 5-21 was unable to form coaggregates with flocculation-competent cells from the Sp 7 strain. Accordingly, the combined flocculence was reduced by about 50 percent and was indicative of the fact that the only flocs formed arose from Sp 7 cells. The latter was further supported by the fact that the floc mass is whitish and typical of Sp 7 flocs. Tn5-21 produces a pink pellet upon centrifugation.

Cocultivation of strains Sp7 (white) and 7030 (pink) or Sp7 and Cd (pink) resulted in the production of light-pink flocs and suggested the presence of both cell types in the mixture. The coaggregation of Sp7 with either 7030 or Cd was further supported by the high flocculence value of the mixture (Table 2).

Table 2. Percent flocculence and floc yields in single or mixed cultures of different *Azospirillum brasilense* strains grown in flocculation media.

Strains	A_{600}		Flocculence	Floc Yields
	A_I	A_R	%	mL/Liter
Sp 7	0.83	0.04	95	100
7030	1.27	0.08	94	120
Cd	0.96	0.08	92	100
Tn 5-21	1.03	1.71	0	0
Sp7+7030	0.81	0.05	94	110
Sp7+Cd	1.07	0.06	94	122
Sp7+Tn 5-21	0.95	0.54	43	95

Coaggregation of *Azospirillum* and *Rhizobium*

Intergeneric coaggregation was demonstrated using mixed cultures of *A. brasilense* strain Cd and *R. leguminosarum* bv phaseoli. The mixed cultures exhibited a normal growth profile (sigmoidal) when grown on a rich media (low C/N ratio) containing NB-YEM in a 1:1 ratio (Fig. I A). Samples of a late log culture exhibited a dispersive growth pattern with a low frequency of microaggregates when observed under phase contrast microscopy. Microaggregates are formed by packages of small number of cells that can be visualized only with the help of a microscope. Transfer of the mixture to minimal media with a high C/N ratio resulted in the acceleration of cell aggregation, formation of coaggregates and production of macroscopic flocs containing a mixture of the two cell types. The change in optical density during coflocculation is shown in Fig. I B. Single cultures yielded pink flocs in the case of *Azospirillum brasilense* strain Cd and white flocs in *Rhizobium leguminosarum* bv phaseoli. The mixed cultures yielded light- pink flocs which are indicative of coaggregate formation. The coaggregated flocs concentrated about 82% of the total cell mass resulting in a floc yield of about 10 billion cells per ml of floc. The high concentration of cells per unit of floc volume is indicative of a high cell titer in flocculated inoculants.

Flocculation and Nodulation Efficiency

The ability of *Rhizobium leguminosarum* bv phaseoli to promote nodulation of *Phaseolus vulgaris* was studied using both, flocculated and non - flocculated forms. Additionally, the nodule promoting activity of *Azospirillum brasilense* strain Cd together with *Rhizobium leguminosarum* was evaluated using either flocculated and non - flocculated forms. The results of these studies are presented in Table 3. The data showed that the use of flocculated *Rhizobium* was superior to the non - flocculated form in terms of nodule numbers. The flocculated inoculant resulted in a doubling of the number of root nodules per plant on evaluations made three weeks after seeding. Notably, the addition of non - flocculated *Azospirillum* to non - flocculated *Rhizobium* also resulted in a dramatic increase in nodulation. The highest nodulation efficiency was observed, however, when coflocculated forms of *Rhizobium* and *Azospirillum* were used to inoculate the seeds (Table 3). The results support the contention that for maximum nodulation efficiency, an inoculant comprising coaggregated forms of *Rhizobium* and *Azospirillum* gives superior results. The enhancement of nodulation efficiency in response to the triple interaction between *Rhizobium, Azospirillum* and flocculation resulted in an enhancement of shoot and root growth (Table 3).

Table 3. Enhancement of nodulation efficiency and growth of *Phaseolus vulgaris* plants inoculated with flocculated *Rhizobium leguminosarum* bv phaseoli.✚

	Nodulation	Dry weights (g / plant)	
Treatments	**N° / plant**	**Shoots**	**Roots**
Non - flocculated			
Rhizobium only	93.25 ± 8.24	38.1 ± 1.2	14.3 ± 1.0
Rhizobium + Azospirillum	170.00 ± 39.31	54.1 ± 0.5	22.6 ± 0.6
Flocculated			
Rhizobium only	188.75 ± 2.75	62.4 ± 0.4	22.8 ± 0.33
Rhizobium + Azospirillum	216.25 ± 3.30	63.6 ± 0.7	30.2 ± 0.91

✚ Inocula was provided as seed coating prior to planting.

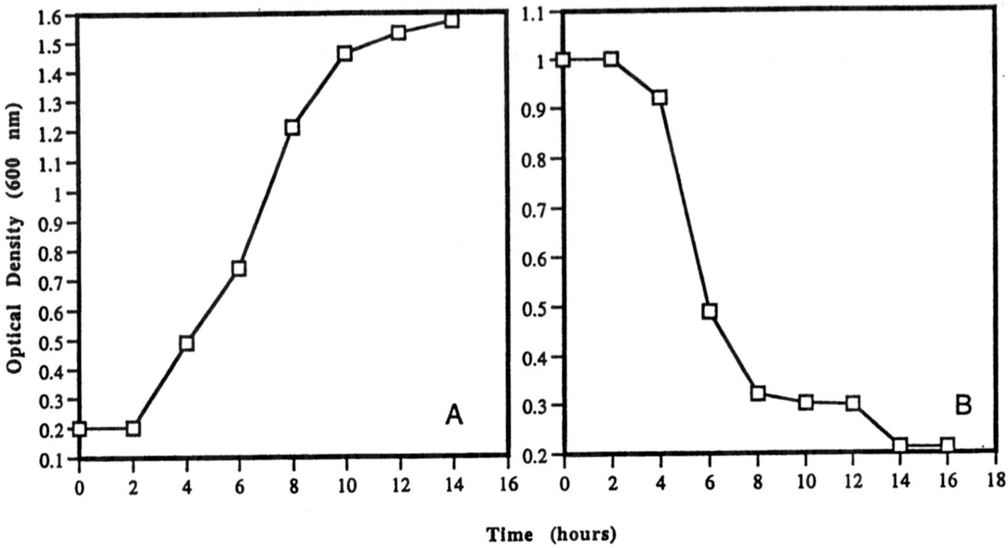

Figure I. Changes in absorbance (A_{600}) of mixed cultures of *Azospirillum brasilense* strain Cd and *Rhizobium leguminosarum* bv phaseoli on rich (A) or minimal flocculation (B) media.

Discussion and Conclusions

One of the most important biotechnological applications of bacterial attachment and cell aggregation is in the production of high quality bacterial inoculants with multiple applications in agriculture. The main research purpose of our laboratory is the production of knowledge for the development of microbial - based technologies aimed at improving the productivity of grain and forage grasses and leguminous crops. The successful use of *Azospirillum* as a Plant Growth Promoting Bacterium depends to a large extent on the availability of appropriate methods for growth (biomass), formulation and delivery of the bacterial product. The results presented in this report confirm once more that *Azospirillum* cells grown in minimal media with a high C / N ratio (Sadasivan and Neyra, 1985; Del Gallo et al. 1989) will produce both microaggregates and macroaggregates in single cultures. We also report here, for the first time, on the ability of *Azospirillum* cells to form coaggregates in mixed cultures (Table 2 and Fig. I). The observed flocculation of coaggregating pairs containing a mixture of two different cell types or strains is of major significance for the industrial production of *Azospirillum* based inoculants. First, it allows the proper combination of strains that may exhibit synergistic properties in promoting plant growth. Second, the observation of intergeneric coaggregation and coflocculation of *Azospirillum brasilense* and *Rhizobium leguminosarum* bv phaseoli suggested that the production of multiple - purpose inoculants is a real possibility, as indicated by the significant enhancement of nodulation and plant growth in *Phaseolus vulgaris* L. The interaction between *Azospirillum* and *Rhizobium* has been extensively studied but still remains a controversial topic (Bashan and Levanony, 1990). Nonetheless, we believe that the use of flocculated *Azospirillum* and *Rhizobium* cells provides a clear advantage over the use of non- flocculated cell forms.

The microbial partnership between *Azospirillum* and *Rhizobium* is presented here as a model system to illustrate the operation and benefits of using flocculated cell forms as a delivery system for each cell type singly or in coaggregated form. Our laboratory continues to search for the identification of bacterial partners capable of forming multigeneric aggregations for the production of multiple - purpose inoculants that include fungal antagonistic bacteria as well as pollutants degrading bacteria in addition to the plant growth promoting bacterial species. We hope a new generation of inoculant products, along these lines, will be forthcoming in the very near future.

Acknowledgements

This work was performed under New Jersey Agricultural Experiment Station Project 12201 supported by state and federal Hatch Act funds and is published under NJAES N° D-12201-1- 95. Additional funding was provided by a Grant from Grassland West Company, Clarkston, WA. Amy L. Atkinson and Olubayi Olubayi are doctoral students in the graduate programs of Microbiology and Molecular Genetics and Plant Biology, respectively.

References

Bashan Y and Levanony H (1990) Current status of *Azospirillum* inoculation technology: *Azospirillum* as a challenge for agriculture. Can. J. Microbiol. 36: 591 - 607.

Eskew D L, Focht D D and Ting I P (1977) Nitrogen fixation, denitrification and pleomorphic growth in a highly pigmented *Spirillum lipoferum*. Appl. Environ.Microbiol. 34: 582 - 585.

Del Gallo M, Negi M and Neyra CA (1989) Calcofluor and lectin - binding exocellular polysaccharides of *Azospirillum brasilense* and *Azospirillum lipoferum*. J. Bacteriol. 171: 3504 - 3510.

Halsall D M and Gibson A H (1986) Comparison of two *Cellulomonas* strains and their interaction with *Azospirillum brasilense* in degradation of wheat straw and associated nitrogen fixation. Appl. Environ. Microbiol. 51: 855 - 861.

Kaiser P (1995) This volume.

Kolenbranden P E and Anderson R N (1986) Multigeneric aggregations among oral bacteria: A network of independent cell - cell interactions. J. Bacteriol. 168: 851 - 859.

Michiels K, Verreth C and Vanderleyden J (1990) *Azospirillum lipoferum* and *Azospirillum brasilense* surface polysaccharide mutants that are affected in flocculation. J. Appl. Bacteriol. 69: 705 - 711.

Neyra C A and Dobereiner J (1977) Nitrogen fixation in grasses. Adv.Agron. 29:1 - 38.

Okon Y and Hadar Y (1987) Microbial inoculants as crop - yield enhancers. CRC Crit. Rev. Biotechnol. 6: 61 - 85.

Sadasivan L and Neyra C A (1985) Flocculation in *Azospirillum brasilense* and *Azospirillum lipoferum*: Exopolysaccharides and cyst formation. J. Bacteriol. 163: 716 - 723.

Scherer P A, Bochen H P, Davis J D and White D C (1986) Flocculation in methanogens, a comparative study *Methanosarcina barkeri* strains Jülich and Fusaro.32:137-144.

Segovia L, Lucas I H , Young J P W and Martinez - Romero E (1993) Phylogenetical relationships among American Rhizobium capable of nodulating bean. In New Horizons in Nitrogen Fixation. Kluver Academic Publishers, Boston, MA. pp.650.

Smith R S (1992) Legume inoculant formulation and application. Can. J. Microbiol. 38: 485 - 492.

Tarrand J J, Kreig N R and Dobereiner (1978) A taxonomic study of the *spirillum lipoferum* group, with description of a new genus, *Azospirillum* gen. nov. and two species, *Azospirillum lipoferum* (Beijerinck) comb. nov. and *Azospirillum brasilense* sp. nov. Can. J. Microbiol. 24: 967 - 980.

Van Berkum P, Navarro R B and Vargas A A T (1994). Classification of the uptake hydrogenase - positive (Hup^+) bean rhizobia as *Rhizobium tropici*. Appl. Environ. Microbiol. 60: 554 - 561.

Soil Parameters which Affect the Survival of *Azospirillum brasilense*

Yoav Bashan*, M. Esther Puente*, M. Nieves Rodriguez-Mendoza[1#], Gina Holguin*, Gerardo Toledo*, Ronald Ferrera-Cerrato#, and Sergio Pedrin@

*Department of Microbiology, and @Department of Marine Resources,

The Center for Biological Research (CIB), A.P. 128, La Paz, B.C.S. 23000, Mexico.

Abstract

The survival of *Azospirillum brasilense* strains Cd and Sp-245 was evaluated in the rhizosphere of wheat and tomato plants and in 23 plant-free soils of 23 soil types obtained from a wide range of environmental conditions in Israel and Mexico. The survival rate of *A. brasilense* was analyzed for 15 soil parameters. *A. brasilense* survived well in all the rhizospheres tested, regardless of soil type, bacterial strain, the origin of the soil, or the amount of rainfall each soil received prior to sampling. In the absence of plants, the general survival characteristics of *A. brasilense* differed and were related mainly to the geographical origin of the soil and not to prevailing environmental conditions. We propose that: (i) *A. brasilense* is a rhizosphere colonizer which survives poorly in most soils for prolonged periods. (ii) Some major physical and chemical soil parameters may affect survival of the bacteria in plantless soils.

Introduction

There is no doubt that *Azospirillum* species survive well and for prolonged periods of time in the rhizosphere of numerous plant species (Bashan and Levanony, 1990). However, survival in the soil *per se* is controversial. On one hand, several studies, mainly (Baldani et al. 1987; Döbereiner and Baldani, 1979; Döbereiner et al. 1976; Sadasivam et al. 1986; Shawky, 1989) but not exclusively (Germida, 1986) from tropical areas, have indicated that *Azospirillum* was found in nearly every sampling of soil, indicating a high capacity for survival. On the other hand, studies mainly from temperate and semi-arid zones, but also from tropical regions (Nayak et al. 1986) found that *Azospirillum* survived poorly in these soils (Albrecht et al. 1983; Bashan, 1986 a; Bashan et al. 1987; De Coninck et al. 1988; Smith et al. 1984; Vandenhove et al. 1993) and hardly lasted from one season to the next (Harris et al. 1989).

[1#]Institute of Natural Resources, Colegio de Postgraduados, Montecillo, Edo. de Mexico 56230, Mexico.

NATO ASI Series, Vol. G 37
Azospirillum VI and Related Microorganisms
Edited by I. Fendrik et al.

Overwintering of inoculated *Azospirillum* is a crucial negative factor for both the inoculation industry (one inoculation should not last for more than one season) and the growers (competition with the previous year's, inferior strains).

The main controversy originated from the diverse soil types used in the studies. No two studies used the same soil. Furthermore, these soil types were not compared with those of other studies and even with those of the same region. This controversy in the soil survival of *Azospirillum* was recognized as one of the basic unsolved questions of *Azospirillum* research (W. Klingmüller, General discussion, *Azospirillum* V workshop, 1991, Germany).

The aim of this study was to: challenge the controversy by correlating the soil parameters of 23 soil types from 4 distinct regions (arid, semi-arid and mountain soils from Israel, semi-arid and tropical soils from Central Mexico, and arid soils from Baja California, Mexico) with the survival of two common strains of *Azospirillum brasilense*, Cd and Sp-245.

Material and methods

Organisms

For all experiments we used *Azospirillum brasilense* strains Cd (ATCC 29710) and Sp-245 (Baldani et al. 1986) with the plants wheat (*Triticum aestivum*) cvs. "Deganit" (Israel) and "Morelos" (Central Mexico) and tomato (*Lycopersicon esculentum*) cv. "UC-82-L" (Baja California Sur, Mexico).

Bacterial inoculation

Bacteria were grown in one of the following growth media: Nutrient Broth (in Baja California, Mexico and Israel) or N-free medium (NFb) (in mainland Mexico) and prepared for inoculation at various concentrations as previously described (Bashan, 1986 b, Bashan et al. 1993). The final bacterial concentrations were: $1x10^6$ cfu/ml soil either in the presence or absence of plants in Israel, $1.77x10^7$ cfu/ml soil for both soil or plant inoculation in Baja California, Mexico, and $4.2x10^4$ (to inoculate plants) or $1.44x10^8$ cfu/ml (to inoculate soil) in Central Mexico. Soil was directly inoculated by applying double washed bacterial suspension to each pot. Plants were inoculated at sowing as previously described (Puente and Bashan, 1993).

Soil analyses

Twenty three soil types were collected in various regions of Israel and Mexico and kept in hermetically sealed plastic containers at 4±1°C. All soil samples were collected by commercial core samplers from the soil layer 20-30 cm deep after discarding the top soil. No attempt was made to preserve the soil intact and therefore, all samples should be considered disturbed. Several soils were from cultivated areas and the others were from uncultivated land. Soils were collected from arid (< 200 mm rainfall / year)(11 soils), semi-arid (400-600 mm rainfall /year) (5 soils), mountains (500-800 mm rainfall /year)(6 soils), and tropical (1800+ mm rainfall /year) (one soil).

The physical and chemical characteristics of each soil were determined using soil analysis standard methods: texture and organic matter (Royce, 1980); pH, water holding capacity and electric conductivity (Chapman and Pratt, 1984); nitrogen content (Volonteri, 1983); and $CaCo_3$ content (Jackson, 1976).

Soil sterilization

To avoid competition with native microorganisms and possible complications in the final analysis, all soils were sterilized by a standard tyndelization procedure (1 h at 15 lbs in an autoclave. Later, the soil was incubated for 24 h at 30±1°C. This was repeated 3 times). Preliminary comparison between sterile and non-sterile soil showed no significant difference in the level of survival for both strains Cd and Sp-245.

Plant growth conditions and inoculation

Plants were grown in 500 ml plastic pots containing 400 g of soil. All pots were disinfected with 10% NaOCl and thoroughly washed with sterile tap water before use. Inoculation of soils without plants was carried out in identical 250 ml pots. Plants were grown in a growth chamber at 25±1°C, 100 mole/m^2/sec and 60±2 % relative humidity. Plants were irrigated every week with 5-15 ml sterile, distilled water to avoid saturation as required by the different size of the growing plants. Plants were fertilized once a week with 5 ml, half-strength Hoagland's solution.

Sampling and bacterial counts from soil and root samples

Samples (2 g soil or approx 500-1000 mg (fresh weight) roots and the adhering soil particles) were taken at each sampling. Rhizosphere bacteria were considered as the bacteria that colonize the roots and the adhering soil particles which were not removed. The samples were lightly sonicated at 25 W for 5 min (Cole Parmer Series 4710, USA) and then decimally diluted in 0.06M Phosphate buffer pH 7.0. Bacteria from the soil were counted after similar treatment by conventional plate count method on Nutrient Agar (soils from Baja California, Mexico) or on N-free NFb medium (Bashan et al. 1993)(soils from Central Mexico), by indirect-ELISA (soils from Israel) (Levanony and Bashan, 1991) or by the time-limited liquid enrichment technique (Bashan et al. 1991) when the number of bacteria fell below the level of detection by the ELISA ($<10^4$ cfu/sample).

Experimental design and statistical analysis

All experiments were carried out in 3-4 replicates per treatment. The rate of growth or death of the bacterial population was calculated according to Krebs' Logistic Equation of Growth (Krebs, 1978).

Primarily, the rate of growth/death of the bacteria in each of the soils were correlated with the values obtained for each soil parameter in each of 23 different type of soils using Linear and Multiple Regression Analyses at $P \leq 0.01$ and 0.05. Then, all the survival data from all the soils and all the soil parameter data were analyzed together by Principle Component Analysis procedure.

Results and discussion

The release of *Azospirillum* and other PGPR into soils has a long history of unpredictable and often disappointing results. One of the main obstacles has been the often poor establishment and survival of the introduced bacteria in the soil prior to root colonization (Bashan and Levanony, 1990, Michiels et al. 1989, Jagnow, 1987). Survival of PGPR in the soil is a crucial issue for both inoculant users and manufacturers, stemming from the common agrotechnical belief that seed inoculation is impractical in many cases (perennial plants, vegetative propagated plants, trees, or where more than one inoculation per season is required, etc). When the PGPR is applied to the soil, it is hoped that it will survive long enough to find its target plant. Thus, before one considers an expensive field inoculation, the prospect of bacterial survival in the soil must be considered.

The aim of this study was to address the soil phase survival of *Azospirillum* by creating sufficient data which will allow future modeling and prediction of bacterial behavior in any given soil without laborious studies of bacterial survival in every field. To this end, we collected 23 soil types representing different climatological conditions, from tropical to arid zones, and compared them all in identical, commonly used experiments using the most available stains of *A. brasilense*.

One fact emerged immediately and was clear even without the aid of any statistical analysis. *A. brasilense* proved to be a rhizosphere bacteria. It survived well regardless of any of the soil characteristics as long as plants were growing in these soils. In the absence of plants, the survival picture differed significantly. The general survival differences were more geographical than climatological, i.e., in the arid soils of Israel the bacteria poorly survived while it proliferated in the arid soils of Baja California. An example for this tendency is given in Fig. 1 in which the survival of *A. brasilense* in the rhizosphere and soil in three representative soil types is demonstrated.

With the aid of several statistical analyses performed on the bacterial survival data and the soil parameters data, we were able to sort out the major soil factors affecting the survival of *A. brasilense* in these soils. Two factors, the levels of $CaCO_3$ and rough sand were negatively and significantly correlated with bacterial survival (Fig 2).

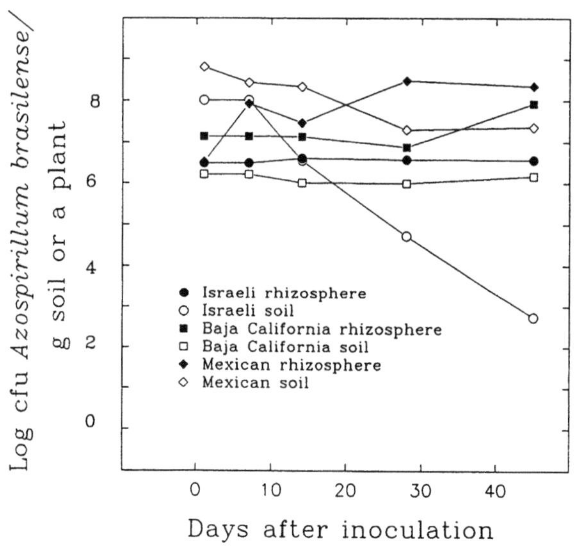

Fig. 1. Survival of *Azospirillum brasilense* in the soil and rhizosphere of 3 soil types obtained from Israel, Baja California Sur, Mexico and Central Mexico.
Each point represents the mean of 3-5 independent samplings from pots, each conducted in triplicates. For simplicity, the standard deviations were not drawn and are: Israeli soil; rhizosphere (R) ±1.4851, soil (S) ±1.5275. Baja California soil; R ±1.39024, S ±1.5149. Central Mexico soil; R ±1.24098, S ±1.5127.

No other single parameter was responsible by itself for survival. However, Principle Component Analysis revealed that when several factors were grouped together (clay, nitrogen, organic matter and water holding capacity), they positively affected survival. On the other hand, levels of $CaCO_3$ and rough sand together with the level of fine sand in the soil had a negative influence on the survival of the bacteria. The effects of single soil parameters on the survival and proliferation of soil and rhizosphere bacteria are known (Foster, 1988; van Elsas, 1992; van Elsas and van Overbeek, 1993). The novelty of this study is that only the combined action of several, lessor parameters significantly and positively determined the survival of *Azospirillum* in the soil. Nevertheless, these parameters are unimportant when plants are growing in the same type of soils. What is still unknown are the proportional effects of each parameter on theoverall effect, and whether manipulation of the soil parameters will alter the bacterial survival.

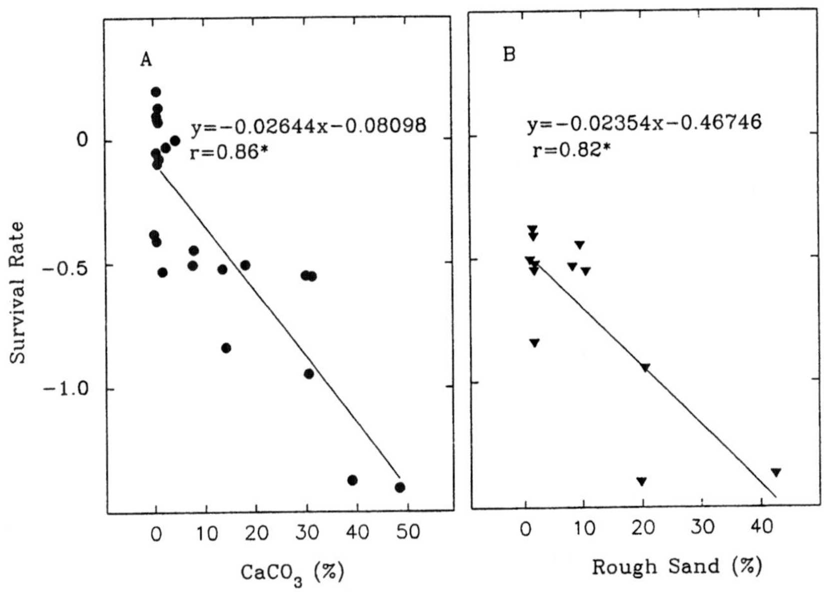

Fig. 2. Linear regression analyses between percentage of $CaCO_3$ (A) and rough sand (B) and the survival rate of *A. brasilense* in 23 and 13 soil types, respectively. Missing points in each sub-figure are overlapped by other printed points. * - Significance of the regression at $P \leq 0.05$.

In conclusion, we propose that: (i) *A. brasilense* is a rhizosphere colonizer which survives poorly in most soils for prolonged periods and, (ii) some major physical and chemical soil parameters may affect the survival of this bacteria in plantless soils.

Acknowledgements

This study was written for the memory of the late Mr. Avner Bashan from Israel who encouraged agricultural research. We thank Ms. Langen Corlay, and Mrs. Thelma Castellanos for excellent technical assistance, Mr. Gustavo Padilla and Ms Yolanda Maya for helpful discussions on soil analyses and soil classification, Mr. Roy Bowers for meticulously clarifying the English and Dr. J. Döbereiner, EMBRAPA, Brazil for donating strain Sp-245. This study was partially supported by Consejo Nacional de Ciencia y Tecnologia (CONACyT), Mexico grant # 225130-5-3541A.

References

Albrecht SL, Gaskins MH, Milam JR, Schank SC, Smith RL (1983) Ecological factors affecting survival and activity of *Azospirillum* in the rhizosphere. In *Azospirillum* II Genetics, physiology, ecology. Edited by W. Klingmüller, Birhäuser Verlag, Basel. pp. 138-148.

Baldani VLD, Alvarez MA de B, Baldani JI, Döbereiner J (1986) Establishment of inoculated *Azospirillum* spp. in the rhizosphere and in roots of field grown wheat and sorghum. Plant Soil 90:35-46.

Baldani VLD, Baldani JI, Döbereiner J (1987) Inoculation of field-grown wheat (*Triticum aestivum*) with *Azospirillum* spp. in Brazil. Biol. Fertil. Soils 4:37-40.

Bashan Y (1986 a) Enhancement of wheat roots colonization and plant development by *Azospirillum brasilense* Cd. following temporary depression of the rhizosphere microflora. Appl. Environ. Microbiol. 51: 1067-1071.

Bashan Y (1986 b) Significance of timing and level of inoculation with rhizosphere bacteria on wheat plants. Soil Biol. Biochem. 18: 297-301.

Bashan Y, Holguin G, Lifshitz R (1993) Isolation and characterization of plant growth-promoting rhizobacteria. In: Methods in plant molecular biology and biotechnology. (Eds.) B.R. Glick. and J.E. Thompson . CRC Press, USA. pp. 331-345.

Bashan Y, Levanony H (1990) Current status of *Azospirillum* inoculation technology: *Azospirillum* as a challenge for agriculture. Can. J. Microbiol. 36: 591-608.

Bashan Y, Mitiku G, Ziv-Vecht O, Levanony H (1991) Estimation of minimal numbers *Azospirillum brasilense* using time-limited liquid enrichment combined enzyme-linked immunosorbent assay. Soil Biol. Biochem. 23: 135-138.

Bashan Y, Levanony H, Ziv-Vecht O (1987) The fate of field-inoculated *Azospirillum brasilense* Cd in wheat rhizosphere during the growing season. Can. J. Microbiol.33: 1074-1079.

Chapman HD, Pratt PF (1984) Analytical methods for soil, plants and water. University of California, Citrus Experimental Station, Riverside, USA. pp. 149-152. (In Spanish)

De Coninck K, Horemans S, Randombage S, Vlassak K (1988) Occurrence and survival of *Azospirillum* spp. in temperate regions. Plant Soil 110:213-218.

Döbereiner J, Baldani VLD (1979) Selective infection of maize roots by streptomycin-resistent *Azospirillum lipoferum* and other bacteria. Can. J. Microbiol. 25:1264-1269.

Döbereiner J, Marriel IE, Nery M (1976) Ecological distribution of *Spirillum lipoferum* Beijerinck. Can. J. Microbiol. 22:1464-1473.

Foster RC (1988) Microenvironments of soil microorganisms. Biol. fertil. Soils 6:189-203.

Germida JJ (1986) Population dynamics of *Azospirillum brasilense* and its bacteriophage in soil. Plant Soil 90:117-128.

Harris JM, Lucas JA, Davey MR, Lethbridge G, Powel KA (1989) Establishment of *Azospirillum* inoculant in the rhizosphere of winter wheat. Soil Biol. Biochem. 21:59-64.

Jackson ML (1976) Soil chemical analysis. Prentice Hall Inc. Englewood Cliffs, NJ. USA, p. 49.

Jagnow G (1987) Inoculation of cereal crops and forage grasses with nitrogen-fixing rhizosphere bacteria: possible causes of success and failure with regard to yield response-a review. Z. Pflanzeneraehr. Dueng. Bodenkde. 150:361-368.

Krebs, C.J. (1978) Ecology: the experimental analysis of distribution and abundance. Harper and row, N.Y., USA. 753 pp.

Levanony H, Bashan Y (1991) Enumeration and identification of rhizosphere bacteria by advanced immuno techniques. In: Plant growth-promoting rhizobacteria-progress and prospects, (Eds.) C. Keel, B. Koller and G. Défago. pp. 231-237. IOBC/WPRS Bulletin, Zürich, Switzerland.

Michiels K, Vanderleyden J, Van Gool A (1989) *Azospirillum*-plant root associations: a review. Biol. Fertil. Soils 8:356-368.

Nayak DN, Ladha JK, Watanabe I (1986) The fate of marker *Azospirillum lipoferum* inoculated into rice and its effect on growth, yield and N_2 fixation of plants studied by acetylene reduction, $^{15}N_2$ feeding and ^{15}N dilution techniques. Biol. Fertil. Soils 2:7-14.

Puente M-E, Bashan Y (1993) Effect of inoculation with *Azospirillum brasilense* strains on the germination and seedlings growth of the giant columnar cardon cactus (*Pachycereus pringlei*). Symbiosis 15: 49-60.

Royce CF, Jr (1980) An introduction to sedimentary analysis. Arizona State University. pp. 180-202.

Sadasivam KV, Negi M, Tilak KVBR (1986) Survival of *Azospirillum brasilense* and *Azotobacter chroococcum* in organic-amended soil-based carriers. Zentralbl. Mikrobiol. 141:567-570

Shawky BT (1989) Studies on the occurrence of asymbiotic nitrogen-fixing *Azospirillum* species in the soils and rhizosphere of some plants in Egypt. Zentralbl. Mikrobiol. 144:581-594.

Smith RL, Schank SC, Milam JR, Baltensperger AA (1984) Responses of *Sorghum* and *Pennisetum* species to the N_2-fixing bacterium *Azospirillum brasilense*. Appl. Environ. Microbiol. 47:1331-1336.

Vandenhove H, Merckx R, van Steenbergen M, Vlassak K (1993) Microcalorimetric characterization, physiological stages and survival ability of *Azospirillum brasilense*. Soil Biol. Biochem. 25:513-519.
van Elsas JD (1992) Environmental pressure imposed on GEMMOS in soil. In The release of Genetically modified microorganisms. (eds.) DES Stewart-Tull and M. Sussman, Plenum Press, NY. pp. 1-14.
van Elsas JD, van Overbeek LS (1993) bacterial responses to soil stimuli. In Starvation in bacteria (ed.) S. Kjelleberg, Plenum Press, N.Y. pp. 55-79.
Volonteri HJ (1983) [Ciencia del suelos. Revista de la Asociación Argentina de la Ciencia del Suelo]. 1: 98-99 (in Spanish).

Chemotaxis in Plant-Associated Bacteria: the Search for the Ecological Niche

Igor B. Zhulin* and Barry L. Taylor
Department of Microbiology and Molecular Genetics
Loma Linda University
Loma Linda CA 92350
U. S. A.

Abstract

Chemotaxis is the movement of organisms towards or away from chemicals. Plant-associated bacteria respond chemotactically to certain compounds released by plants. The chemotactic response provides bacteria with a competitive advantage and allows them to occupy specific ecological niches. In this brief review, we bring together data on various aspects of chemotaxis in some plant-associated bacteria, with special attention to their ecology.

Role of chemotaxis in establishing *Rhizobium*-legume symbiosis

The *Rhizobium*-legume symbiosis culminates in the formation of nitrogen-fixing root nodules. Thus, rhizobia experience two different modes of life cycle: free-living cells and bacteroids (cells inside nodules). Free-living cells are motile by means of flagella and capable of chemotaxis towards compounds excreted by plant roots. Chemotaxis appears to help a free-living rhizobial cell to achieve its ultimate state, to become a bacteroid. Non-motile and non-chemotactic mutants of different *Rhizobium* species were shown to be less competitive in nodulation than the wild type strains (Hunter and Fahring, 1980; Ames and Bergman, 1981; Mellor *et al.*, 1987; Catlow *et al.*, 1990).

*Permanent address: Institute of Biochemistry and Physiology of Plants and Microorganisms, Russian Academy of Sciences, Saratov, 410015 Russia

NATO ASI Series, Vol. G 37
Azospirillum VI and Related Microorganisms
Edited by I. Fendrik et al.

It appears that motility and chemotaxis facilitate the initial contact of rhizobia to the host root surface, increase the efficiency of nodule initiation, and increase the rate of nodule development (Caetano-Anolles *et al.*, 1988*b*; Parco *et al.*, 1994).
Several studies indicated that root exudates of leguminous plants attract rhizobia (Currier and Strobel, 1976; Gitte et al., 1978; Gaworzewska and Carlile, 1982; Barbour *et al.*, 1991). Rhizobia are strongly attracted to a few sites on the root surface (Gulash et al., 1984; Malek, 1989; Caetano-Anolles *et al.*, 1992). Presumably, these are the sites of metabolite excretion. Among metabolites excreted in reasonable amounts by plant roots, organic acids are the best attractants for rhizobia, although amino acids and sugars also act as attractants. The spectrum of attractants may vary not just on the species level, but also from strain to strain (Bowra and Dilworth, 1981; Burg *et al.*, 1982; Gaworzewska and Carlile, 1982; Gotz *et al.*, 1982; Parke et al., 1985; Aguilar *et al.*, 1988; Armitage *et al.*, 1988; Malek, 1989; Kape *et al.*, 1991; Robinson and Bauer, 1993). Two species of fast-growing rhizobia, *R. leguminosarum* and *R. meliloti*, have a similar spectrum of attractants. However, slow-growing *Bradyrhizobium japonicum* was shown to have a different spectrum, not being attracted by sugars or by majority of amino acids (Barbour *et al.*, 1991). Aromatic and hydroaromatic acids, another class of nutrients for rhizobia, were also shown to be strong attractants. *B. japonicum* were attracted by shikimate, quinate, protocatechuate and some other aromatic acids at threshold concentrations as low as 10^{-6} M, whilst *R. leguminosarum* responded to these compounds at concentrations of 2 orders of magnitude higher.
The ability of bacteria to respond chemotactically to potential nutrients has obvious survival advantages. Some compounds excreted by plant roots may not be of nutritional value but have a biological significance for the mode of growth and/or life cycle. Currier and Strobel claimed that a 60 KDa glycoprotein excreted by trefoil roots attracted a homologous strain of *Rhizobium spp*. but not other rhizobial strains at concentration of 10^{-16} M (Currier and Strobel, 1977). This is the only report in the chemotaxis literature of bacterial chemotaxis caused by proteins. This observation still requires an explanation of how such a large molecule could permeate the outer membrane and cause the chemotactic response at such negligible concentration.
Several plant root phenolic compounds are known to stimulate expression of the rhizobial nodulation genes (*nod*) that are involved in nodule initiation and development. The majority of these compounds appear to be flavonoids and isoflavonoids. The attempts to reveal the role of *nod*-inducing compounds as potential chemoeffectors for rhizobia demonstrated that some of these compounds cause attraction of rhizobia at very low concentrations. Luteolin, a flavone that acts

as a specific inducer of *nodABC* gene expression in *R. meliloti*, attracted the bacteria at 10^{-8} M (Caetano-Anolles *et al.*, 1988*a*). Three other principal *nod* gene inducers secreted by alfalfa roots, 4',7-dihydroxyflavone, 4',7-dihydroxyflavanone, and 4,4'-dihydroxy-2-methoxychalcone caused very similar chemotactic responses at concentrations as low as 10^{-10} M (Dharmatilake and Bauer, 1992). Two other flavonoids, naringenin and apigenin, were not attractants for *R. meliloti* but they evoked positive chemotaxis in *R. leguminosarum* with peaks at 10^{-5}-10^{-6} M (Aguilar *et al.*, 1988; Armitage *et al.*, 1988). Surprisingly, the inhibitors of *nod*-induction, umbelliferone, acetosyringone (Aguilar *et al.*, 1988), and kaempferol (Armitage *et al.*, 1988) all produced strong positive chemotactic responses in *R. leguminosarum*, although at much higher concentrations. In *B. japonicum*, however, *nod*-inducing flavonoids are not principal attractants. Genistein and some other isoflavonoids produce very weak chemotactic responses in comparison with the action of succinate or glutamate (Barbour *et al.*, 1991). On the other hand, aromatic acids such as hydroxycinnamic and coumaric acids, elicit a strong chemotactic response and some of them also enhance *nod* gene induction in this bacterium (Kape *et al.*, 1991).

The mechanism of the chemotactic response in rhizobia is still unknown despite the fact that several chemotaxis mutants have been isolated and characterized (Bergman *et al.*, 1988). No dedicated chemoreceptors were found during an extensive study of *Rhizobium meliloti* chemotactic machinery (Robinson and Bauer, 1993), so chemotaxis towards plant metabolites seems to be dependent on utilization of the metabolite by the bacteria. It is not clear whether chemotaxis towards flavonoids is based on chemoreception or metabolism. Nevertheless, chemotaxis towards flavonoids and some other aromatic compounds is independent of *nod* gene induction (Armitage *et al.*, 1988; Kape *et al.*, 1991; Dharmatilake and Bauer, 1992).

In conclusion, it is necessary to point out, that rhizobia show positive chemotaxis toward two classes of chemicals excreted by plant roots. Firstly, the bacteria are attracted by nutrients which are released by roots in considerable amounts. Organic acids, such as succinate, as well as some aromatic acids appear to be the best attractants for rhizobia. Secondly, rhizobial cells show chemotaxis towards flavonoids and some other phenolic compounds which do not serve as nutrients but play in important biological role. These compounds are excreted by legume plant roots in small amounts. However, phenolics might be principal attractants since they elicit the chemotactic response at extremely low concentrations.

Role of chemotaxis in associative symbiosis of *Azospirillum* with plants

Bacteria of the *Azospirillum* genus have been isolated from the rhizosphere of a wide variety of plants in all geographical regions. The bacteria were shown to be capable of chemotaxis towards several amino acids, sugars and organic acids (Okon *et al.*, 1980; Barak *et al.*, 1983). It appears that almost all common metabolites that are typical of plant root exudates serve as attractants for azospirilla (Zhulin *et al.*, 1988). The peak concentration for these attractants is about 10^{-3} M. Apparently the best growth substrates, such as malate, succinate, gluconate and fructose, are strong attractants and poor substrates, such as amino acids, are weak attractants. Organic acids seem to be the best attractants for *Azospirillum* as they are for *Rhizobium*. Chemotaxis in *Azospirillum* is strain-specific (Reinhold *et al.*, 1985) as it is in *Rhizobium*. The chemotaxis systems in azospirilla and rhizobia may be quite similar (Table 1).

Beside the chemotactic response, azospirilla are capable of positive chemokinesis, i.e. the cells increase their speed when moving towards attractants (Zhulin and Armitage, 1993).

A recent study demonstrated that *Azospirillum* has a highly sensitive chemotactic response towards aromatic acids that are characteristic of the rhizosphere, such as benzoate, protocatechuate and hydroxybenzoate. Threshold levels of these attractants ranged from 10^{-6} to 10^{-11} M (Lopez-de-Victoria and Lovell, 1993). It is not clear however what mechanism could be involved in sensing of these compounds at such low concentrations since aromatic acids penetrate freely across the cytoplasmic membrane and do not require a receptor.

The mechanism of the chemotactic response in *Azospirillum* is under current investigation. No dedicated chemoreceptors have been found in *A. brasilense* and it was suggested that the main chemosensory pathway in this bacterium is based on metabolism of chemoeffectors (Zhulin and Armitage, 1993). Our recent work confirmed this proposal by demonstrating that sensing of the proton motive force is probably involved in signal transduction during *Azospirillum* taxis (Zhulin *et al.*, 1995).

One of the most remarkable features of *Azospirillum* is its ability to respond chemotactically to low oxygen concentrations (Barak *et al.*, 1982). The aerotactic response is so strong that it could mask taxis to other chemicals (Okon *et al.*, 1980).

Table 1. Comparison of chemotaxis in *A. brasilense, R. meliloti* and *E. coli*

Chemotaxis feature	*A. brasilense*	*R. meliloti*	*E. coli*
Strong attractants	organic acids	organic acids	amino acids
Weak attractants	amino acids	amino acids	organic acids
Repellents	not found	not found	some amino acids
Taxis to phenolics	positive	positive	negative
Strain-specificity	yes	yes	no
Dedicated chemoreceptors	not found	not found	exist

Since microaerobic conditions are typical of the rhizosphere, aerotaxis may play an important role in attracting azospirilla to plant roots. Azospirilla are also capable of redox taxis, i.e. movement towards optimal redox potential (Grishanin *et al.*, 1991). We also found that azospirilla responded negatively to low and high osmolarity and showed positive chemotaxis to optimal osmolarity (Zhulin and Taylor, unpublished data). These responses might be important in the rhizosphere where bacteria experience different concentrations of oxygen, ions and other chemicals that affect redox potential and osmolarity.

Experiments *in situ* provided valuable information on *Azospirillum* behaviour in natural environments. In the soil, *A. brasilense* cells are capable of moving up to 30 cm from the inoculation point towards the plant roots (Bashan and Levanony, 1987). This movement was proved to be an active process since a non-motile mutant did not move towards plant roots (Bashan and Holguin, 1994). Wild type cells had a competitive advantage in the rhizosphere in comparison with non-motile (Bashan and Holguin, 1994) and non-chemotactic (Kimmel *et al.*, 1990) mutants. Studies by Bashan (Bashan and Levanony, 1987; Bashan and Holguin, 1994) demonstrated that bacterial migration within the plant root system is a prerequisite in the plant-bacterium recognition and this migration is a consequence of bacterial chemotaxis.

Role of *Agrobacterium* chemotaxis in host-pathogen interactions

Agrobacterium tumefaciens is the etiologic agent of crown gall tumor disease in plants. The *Agrobacterium*-plant interaction is a unique example of interkingdom DNA transfer. For a recent review on the *Agrobacterium* Ti-plasmid and crown gall tumorigenesis see Ankenbauer and Nester (1993).

A. tumefaciens can transfer DNA from its Ti- (Tumor Inducing) plasmid to the plant chromosome. Expression of T-DNA in the plant cell causes a neoplastic transformation resulting in the production of a crown gall. Tumor cells produce some unusual metabolites, opines, that are of little value for a plant but can be used as substrates by *A. tumefaciens*. Thus, the bacterium creates an ecological niche for itself by means of "genetic engineering". In order to reach the habitat and to occupy the ecological niche, the bacteria employ chemotaxis towards compounds exuded and excreted by plant roots. Sugars, major components of plant root exudates, are strong attractants for *Agrobacterium* (Loake *et al.*, 1988). ChvE, a periplasmic sugar-binding protein similar to the chemoreceptors for sugars in *Escherichia coli*, was shown to be involved in signal transduction during chemotaxis of *A. tumefaciens* to sugars (Cangelosi *et al.*, 1990). Chemotaxis towards sugars in *A. tumefaciens* appears to be Ti-plasmid-independent, since both Ti-plasmid-harbouring cells and cells without the plasmid showed the same response (Loake *et al.*, 1988).

The most intriguing, yet controversial, chemotactic feature of *A. tumefaciens* is their ability to respond to phenolic compounds released by wounded plant roots. Acetosyringone and some other wound-specific phenolics were shown to induce *vir* genes in *Agrobacterium*. The *vir* region is located on the Ti-plasmid and it is responsible for bacterial virulence. VirA and VirG proteins are members of a family of bacterial two-component regulatory signal transduction systems, and they have been reported to mediate a chemotactic response toward acetosyringone (Shaw *et al.*, 1988). VirA is homologous to CheA, a central chemotaxis regulator in *E. coli*. Chemotaxis of *A. tumefaciens* toward acetosyringone requires phosphorylation of VirG by VirA in the same manner as chemotaxis in *E. coli* requires phosphorylation of CheY by CheA (Palmer and Shaw, 1992). Bacteria lacking the Ti-plasmid as well as *virA* and *virG* mutants are unable to chemotaxis toward acetosyringone (Shaw *et*

al., 1988). However, some strains of *A. tumefaciens* carrying the Ti-plasmid do not show chemotaxis towards acetosyringone (Parke *et al.*, 1987)
Attraction to sugars exuded by both wounded and non-wounded plants lead both types of *Agrobacterium* cells (with and without the Ti-plasmid) to the rhizosphere. If the plant is non-wounded, microbes colonize it without eliciting disease. In this case, *Agrobacterium* can be considered a plant commensal organism. A wounded plant releases phenolics that attract only Ti-plasmid-harbouring cells, causing crown gall disease. This type of *Agrobacterium*-plant relationship is parasitic. Thus, taxis towards two different types of chemicals may be involved in a search for two different ecological niches for *A. tumefaciens*.

Conclusing remarks

Three representatives of different genera of plant-associated bacteria that occupy different ecological niches demonstrate complicated behaviour in response to chemicals excreted by plants. Symbiotic rhizobia, associative azospirilla, and pathogenic agrobacteria, are all capable of chemotaxis towards attractants released by plant roots. Chemotaxis towards specific flavonoids and other aromatic compounds might be a first step in recognition and selection of a partner in highly specific *Rhizobium*-legume symbiosis. A wide variety of attractants for azospirilla, along with a simple chemosensing system based on metabolism of chemoeffectors, may explain the ability of these protocooperative bacteria to recognize a wide variety of plants. Two different chemosensing pathways in *Agrobacterium* are involved in commensal and parasitic modes of plant-bacteria interactions. Thus, chemotactic behaviour plays an important role in the life of bacteria being responsible for an active search of the habitat and occupation of the ecological niche.

References

Aguilar JM, Ashby AM, Richards AJ, Loake GA, Watson MD, Shaw CH (1988) Chemotaxis of *Rhizobium leguminosarum* biovar phaseoli towards flavonoid inducers of the symbiotic nodulation genes. J Gen Microbiol 134:2741-2746

Ames P, Bergman K (1981) Competitive advantage provided by bacterial motility in the formation of nodules by *Rhizobium meliloti*. J Bacteriol 148:728-729

Ankenbauer RG, Nester EW (1993) The *Agrobacterium* Ti-plasmid and crown gall tumorigenesis: a model for signal transduction in host-pathogen interactions *In* Kurjan J, Taylor BL (eds) Signal Transduction. Prokaryotic and Simple Eukaryotic Systems pp 67-104 Academic Press San Diego

Armitage JP, Gallagher A., Johnston AWB (1988) Comparison of the chemotactic behaviour of *Rhizobium leguminosarum* with and without the nodulation plasmid. Mol Microbiol 2:743-748

Barak R, Nur I, Okon Y (1983) Detection of chemotaxis in *Azospirillum brasilense.* J Appl Bacteriol 53:399-403

Barak R, Nur I, Okon Y, Henis Y (1982) Aerotactic response of *Azospirillum brasilense.* J Bacteriol 152:643-649

Barbour WM, Hattermann DR, Stacey G (1991) Chemotaxis of *Bradyrhizobium japonicum* to soybean exudates. Appl Environ Microbiol 57:2635-2639

Bashan Y, Holguin G (1994) Root-to-root travel of the beneficial bacterium *Azospirillum brasilense.* Appl Environ Microbiol 60:2120-2131

Bashan Y, Levanony H (1987) Horizontal and vertical movement of *Azospirillum brazilense* Cd in the soil and along the rhizosphere of wheat and weeds in controlled and field environments. J Gen Microbiol 133:3473-3480

Bergman K, Gulash-Hoffee M, Hovestadt RE, Larosilliere RC, Ronco II PG, Su L (1988) Physiology of behavioral mutants of *Rhizobium meliloti*: evidence for a dual chemotaxis pathway. J Bacteriol 170:3249-3254

Bowra BJ, Dilworth MJ (1981) Motility and chemotaxis towards sugars in *Rhizobium leguminosarum.* J Gen. Microbiol 126:231-235

Burg, D, Guillaume J, Tailliez R (1982) Chemotaxis by *Rhizobium meliloti.* Arch Microbiol 133:162-163

Caetano-Anolles G, Crist-Estes DK, Bauer WD (1988*a*) Chemotaxis of *Rhizobium meliloti* to the plant flavone luteolin requires functional nodulation genes. J Bacteriol 170:3164-3169

Caetano-Anolles G, Wall LG, De Micheli AT, Macchi EM, Bauer WD, Favelukes G (1988*b*) Role of motility and chemotaxis in efficiency of nodulation by *Rhizobium meliloti.* Plant Physiol 86:1228-1235

Caetano-Anolles G, Wrobel-Boerner E, Bauer WD (1992) Growth and movement of spot inoculated *Rhizobium meliloti* on the root surface of alfalfa. Plant Physiol 98:1181-1189

Cangelosi GA, Ankenbauer RG, Nester EW (1990) Sugars induce the *Agrobacterium* virulence genes through a periplasmic binding protein and a transmembrane signal protein. Proc Natl Acad Sci USA 87:6708-6712

Catlow H, Glenn YAR, Dilworth MJ (1990) Does rhizobial motility affects its ability to colonize along the legume root? Soil Biol Biochem 22:573-575

Currier WW, Strobel GA (1976) Chemotaxis of *Rhizobium spp.* to plant root exudates. Plant Physiol 57:820-823

Currier WW, Strobel GA (1977) Chemotaxis of *Rhizobium spp.* to a glycoprotein produced by birdsfoot trefoil roots. Science 196:434-436

Dharmatilake AJ, Bauer WD (1992) Chemotaxis of *Rhizobium meliloti* towards nodulation gene-inducing compounds from alfalfa roots. Appl Environ Microbiol 58:1153-1158

Gaworzewska ET, Carlile MJ (1982) Positive chemotaxis of *Rhizobium leguminosarum* and other bacteria towards root exudates from legumes and other plants. J Gen Microbiol 128:1179-1188

Gitte RR, Rai PV, Patil RB (1978) Chemotaxis of *Rhizobium* spp. towards root exudate of *Cicer arietinum* L. Plant Soil 50:553-566

Gotz R, Limmer N, Ober K, Schmitt R (1982) Motility and chemotaxis in two strains of *Rhizobium* with complex flagella. J Gen Microbiol 128:789-798

Grishanin R, Chalmina II, Zhulin IB (1991) Behaviour of *Azospirillum brasilense* in a spatial gradient of oxygen and in a "redox" gradient of an artificial electron acceptor. J Gen Microbiol 137:2781-2785

Gulash M, Ames P, Larosiliere RC, Bergman K (1984) Rhizobia are attracted to localized sites on legume roots. Appl Environ Microbiol 48:149-152

Hunter WJ, Fahring CJ (1980) Movement by *Rhizobium* and nodulation of legumes. Soil Biol Biochem 12:537-542

Kape R, Parniske M, Werner D (1991) Chemotaxis and *nod* gene activity of *Bradyrhizobium japonicum* in response to hydroxycinnamic acids and isoflavonoids. Appl Environ Microbiol 57:316-319

Kimmel S, Reinhold-Hurek B, Fendrik I, Niemann EG (1990) Contribution of chemotaxis and aerotaxis to the establishment of *Azospirillum* in the rhizosphere. Symbiosis 9:195-197

Loake GJ, Ashby AM, Shaw CH (1988) Attraction of *Agrobacterium tumefaciens* C58C1 towards sugars involves a highly sensitive chemotaxis system. J Gen Microbiol 134:1427-1432

Lopez-de-Victoria G, Lovell CR (1993) Chemotaxis of *Azospirilllum* species to aromatic compounds. Appl Environ Microbiol 59:2951-2955

Malek W (1989) Chemotaxis of *Rhizobium meliloti* strain L5-30. Arch Microbiol 152:611-612

Mellor HY, Glenn AR, Arwas R, Dilworth MJ (1987) Symbiotic and competitive properties of motility mutants of *Rhizobium trifolii* TA1. Arch Microbiol 148:34-39

Okon Y, Cakmakci L, Nur I, Chet I (1980) Aerotaxis and chemotaxis in *Azospirillum brasilense*: a note. Microb Ecol 6:277-280

Palmer ACV, Shaw CH (1992) The role of VirA and VirG phosphorylation in chemotaxis towards acetosyringone by *Agrobacterium tumefaciens*. J Gen Microbiol 138:2509-2514

Parco SZ, Dilworth MJ, Glenn AR (1994) Motility and the distribution of introduced root nodule bacteria on the root system of legumes. Soil Biol Biochem 26: 297-300

Parke D, Ornston LN, Nester EW (1987) Chemotaxis to plant phenolic inducers of virulence genes is constitutively expressed in the absence of the Ti plasmid in *Agrobacterium tumefaciens*. J Bacteriol 169:5336-5338

Parke D, Rivelli M, Ornston LN (1985) Chemotaxis to aromatic and hydroaromatic acids: comparison of *Bradyrhizobium japonicum* and *Rhizobium trifolii*. J Bacteriol 163:417-422

Reinhold B, Hurek T, Fendrik I (1985) Strain-specific chemotaxis of *Azospirillum* spp. J Bacteriol 162:190-195

Robinson J, Bauer WD (1993) Relationships between C_4 dicarboxylic acid transport and chemotaxis in *Rhizobium meliloti*. J Bacteriol 175:2284-2291

Shaw CH, Ashby AM, Brown AP, Royal C, Loake GJ, Shaw CH (1988) *virA* and *virG* are the Ti-plasmid functions required for chemotaxis of *Agrobacterium tumefaciens* toward acetosyringone. Mol Microbiol 2: 413-418

Zhulin IB, Armitage JP (1993) Motility, chemokinesis, and methylation-independent chemotaxis in *Azospirillum brasilense*. J Bacteriol 175:952-958

Zhulin IB, Sarmiento LE, Taylor BL (1995) Changes in membrane potential upon chemotactic stimulation of *Azospirillum brasilense*. This volume

Zhulin IB, Tretyakova SE, Ignatov VV (1988) Chemotaxis of *Azospirillum brasilense* towards compounds typical of plant root exudates. Folia Microbiol 33:277-280

[illegible] M., Vercell[illegible] (1993) [illegible] [illegible] and [illegible] Elsevier [illegible] 3-319
[illegible] S. [illegible] (19[illegible]) [illegible]
[illegible]

Occurrence, Isolation and Identification of *Azospirillum* Strains in Greece.

I. Kefalogianni, F. Flouri, and C. Balis.
Laboratory of General and Agricultural Microbiology
Agricultural University of Athens
Iera Odos 75, 118 55 Athens
Greece.

Abstract

To study the occurrence and distribution of *Azospirillum* in diverse areas of Greece, soil and plant-root samples were obtained from the most cultivated varieties of wheat, barley and oat. The nitrogen fixation of the samples was measured by the acetylene reduction assay. Nitrogen fixing bacteria, adapted to Greek conditions, were isolated from the rhizosphere, the surface and the interior of root hairs. Using biochemical tests and nutritional characteristics the isolated strains were identified. Screening 53 isolates and comparing with the properties, as described in the literature for the genus *Azospirillum*, 51 isolates were identified as *Azospirillum brasilense* and only 2 as *Azospirillum lipoferum*.

Introduction

Bacteria of the genus *Azospirillum* are very common soil and root inhabitants and they have been isolated from a great diversity of plant species and soils in tropical, temperate and cold regions (Michiels *et al.*, 1989; Döbereiner *et al.*, 1976) Especially in temperate soils, they are present at low levels (Penot *et al.*, 1992; Tyler *et al.*, 1979).

The inoculation of cereals with *Azospirillum* may benefit plants by several mechanisms (Okon and Kapulnik, 1986). In many studies however, results appeared to be controversial and highly variable. Several investigators admit that, as the most soils already contain a natural population of *Azospirillum,* a significant response of the plants to inoculation would only occur, if the inoculated strain could compete successfully with the native population to dominate active sites in the root or rhizosphere. Therefore, large screening programms are currently being undertaken to isolate indigenous plant growth promoting rhizosphere bacteria, which show strongly root colonization (Boddey and Döbereiner, 1988).

Considering the above, this study was designed to investigate the occurrence and distribution of *Azospirillum* in Greece and to isolate strains from the indigenous populations, which are better adapted to Greek soil-climate conditions..

Materials and Methods

<u>Isolation of nitrogen-fixing microorganisms</u>

NATO ASI Series, Vol. G 37
Azospirillum VI and Related Microorganisms
Edited by I. Fendrik et al.

Clusters of cereals were collected from diverse geographic regions of Greece. Excess soil was gently detached by hand. The upper portion of the plants was cut off, leaving about 2-5 cm with leaf sheath (Barraquio *et al.*, 1986). Then, a 50 g. (f.w.) of roots and adherent soil were placed in 250 ml vessels, and tested for acetylene reduction activity (ARA). Ethylene produced by acetylene reduction was analyzed with a Perkin-Elmer F-11 gas chromatograph fitted with a flame ionization detector, and a Porapak R 80-100 mesh stainless steal column (2mm x 2m). From each ARA positive sample inocula of washed roots (WR), sterilized roots (SR) and soil suspension (SS) were transfered to Nfb semi-solid medium (Döbereiner, 1980; Zafar, 1986; Garvin and Lindemann, 1986), incubated at 33°C for 4-6 days and tested for ARA. Samples were accepted as ARA positive, when the produced C_2H_4 exceeded 10 nmol/sample/h (Döbereiner *et. al.*, 1976; Thompson *et al.*, 1984; Tyler *et al.*, 1979). From ARA positive suspensions a loopful was streaked on RC medium (Rodriguez-Caceres, 1982). Single colonies were further purified and tested for acetylene reduction.

Identification

For each isolate the following biochemical tests were performed: oxidase, catalase, gelatin hydrolysis, indode production, urease, citrate utilization, malonate utilization and denitrification. The auxanographic method was used to detect utilization of sucrose, mannitol, sorbitol, maltose, glucose, arabinose, acetate, fumarate, lactate, malate, succinate and propionate. The morphology of the cells was observed by light microscopy. Two type strains of American Type Culture Collection, Sp.7 for *A. brasilense* and Sp.59 for *A. lipoferum*, were used as references for comparison, to evaluate the results.

Results

All samples examined for acetylene reduction were positive, but the nitrogenase activity ranged a lot, from 0.5 to 193.3 nmoles of ethylene produced/h/sample. (Table 1). The isolated strains were positive for oxidase, catalase, urease and denitrification, but negative for gelatin hydrolysis and indole production. Organic acids salts (except malonate) were metabolized by all strains. There were no utilization of sucrose and sorbitol. Only a few strains were able to utilize arabinose. Glucose, was assimilated from the stains SS and WR isolated from the variety of wheat Yecora (Domokos). Some strains were able to assimilate acetate. All the above are summarized in Tables 2.1, 2.2. Microscopically, all isolates were plump, slightly curved or straight rods, Gram negative, with refractile granules.

Table 1. Rates of acetylene reduction for the rhizosphere of cereals in Greece.

Plant species	Variety	Location	Growth stage	Activity (nmol eth./h/sample)
T. turgidum durum	Mexicali	Kopais	milky grain	55.7-83.1
		Skourta	flowering	23-37.9
		Domokos	earl grain filling	2-12.2
		Salonica	late grain filling	28.3-193.3
		Kileler	late grain filling	2-8.9
	Capeiti	Skourta	flowering	15.5-20.3
		Salonica	late grain filling	66-142.9
	Appulo	Salonica	late grain filling	5.2-8.9
	Vitron	Skourta	flowering	71.7-85.1
T. aestivum	Yecora	Domokos	late grain filling	55.7-72
		Salonica	late grain filling	0.5-0.9
		Gonni	late grain filling	0.9-2
	Vergina	Domokos	earl grain filling	0.9-12.5
		Salonica	late grain filling	3.1-16.5
	Siete-Cerros	Salonica	late grain filling	2-12.0
	Generoso	Salonica	late grain filling	0-0.8
	G-07783	Salonica	late grain filling	6.1-8.9
H. vulgare	local	Skourta	earl grain filling	12.2-16.9
	Igri	Domokos	late grain filling	31.1-31.7
	Rivale	Salonica	late grain filling	0.9-1.4
	Carina	Salonica	late grain filling	1.2-4.3
	Georgy	Salonica	late grain filling	0.5-0.9
	local	Kileler	maturity	2-3.3
Avena sativa	local	Skourta	heading	10-15.5
	Kasandra	Salonica	earl grain filling	0.9-0.9

Table 2.1. Results of the biochemical tests concerning strains isolated from the rhizosphere of barley (*Hordeum vulgare*) and oat (*Avena sativa*).

Isolate										sugars						organic acids							
				oxid	catal	gelat	indol	ureas	denitr	arab	malto	sacch	gluc	mannit	sorbit	malon	citr	aceti	lact	prop	fumar	malic	succ
H. vulgare	Igri	Domokos	SS	+	+	-	-	+	+	-	-	-	-	-	-	-	+	-	+	+	+	+	+
			SS_1	+	+	-	-	+	+	-	-	-	-	-	-	-	+	-	+	+	+	+	+
			SS_2	+	+	-	-	+	+	-	-	-	-	-	-	-	+	-	+	+	+	+	+
			WR	+	+	-	-	+	+	-	-	-	-	-	-	-	+	+	+	+	+	+	+
	Rivale	Salonica	SS_1	±	+	-	-	+	+	-	-	-	-	-	-	-	+	-	+	+	+	+	+
			SS_2	±	+	-	-	+	+	+	-	-	-	-	-	-	+	-	+	+	+	+	+
			WR	+	+	-	-	+	+	-	-	-	-	-	-	-	+	-	+	+	+	+	+
			SR	+	+	-	-	+	+	-	-	-	-	-	-	-	+	-	+	+	+	+	+
	Carina	Salonica	SS	+	+	-	-	+	+	+	-	-	-	-	-	-	+	+	+	+	+	+	+
			WR	+	+	-	-	+	+	-	-	-	+	+	-	+	+	+	+	+	+	+	+
			WR_1	+	+	-	-	+	+	-	-	-	-	-	-	-	+	+	+	+	+	+	+
	Georgy	Salonica	SS_1	+	+	-	-	+	+	-	-	-	-	-	-	-	+	+	+	+	+	+	+
			SS_2	+	+	-	-	+	+	+	-	-	-	-	-	-	+	+	+	+	+	+	+
			WR	+	+	-	-	+	+	+	-	-	-	-	-	-	+	+	+	+	+	+	+
	local	Kileler	SS	+	+	-	-	+	+	-	-	-	-	-	-	-	+	+	+	+	+	+	+
			WR	+	+	-	-	+	+	-	-	-	-	-	-	-	+	-	+	+	+	+	+
A. sativa	local	Skourta	SS	+	+	-	-	+	+	-	-	-	-	-	-	-	+	-	+	+	+	+	+
			WR	+	+	-	-	+	+	+	-	-	-	-	-	-	+	-	+	+	+	+	+
	Kasandra	Salonica	SS	+	+	-	-	+	+	-	-	-	-	-	-	-	+	-	+	+	+	+	+
			WR	+	+	-	-	+	-	-	-	-	-	-	-	-	+	-	+	+	+	+	+
A. brasilense	Sp.7	A.T.C.C.		+	+	-	-	+	+	+	-	-	-	-	-	-	+	+	+	+	+	+	+
A. lipoferum	Sp.59	A.T.C.C.		+	+	-	-	+	+	-	-	-	+	+	-	-	+	-	+	+	+	+	+

Table 2.2. Results of the biochemical tests concerning strains isolated from the rhizosphere of wheat (*Triticum turgidum durum* and *Triticum aestivum*).

Isolate				oxid	catal	gelat	indol	ureas	denitr	sugars: arab	malto	sacch	gluc	mannit	sorb	organic acids: malon	citr	acet	lact	prop	fumar	malic	succ
T. turgidum	Mexicali	Kopais	SS	+	+	-	-	+	+	-	-	-	-	-	-	-	+	-	+	+	+	+	+
			WR	+	+	-	-	+	+	-	-	-	-	-	-	-	+	-	+	+	+	+	+
			SR	+	+	-	-	+	+	-	-	-	-	-	-	-	+	-	+	+	+	+	+
			$SR_{2.1}$	+	+	-	-	+	+	-	-	-	-	-	-	-	+	-	+	+	+	+	+
			$SR_{2.2}$	+	+	-	-	+	+	-	-	-	-	-	-	-	+	+	+	+	+	+	+
		Domokos	SS	+	+	-	-	+	+	-	-	-	-	-	-	-	+	-	+	+	+	+	+
			WR	+	+	-	-	+	+	-	-	-	-	-	-	-	+	-	+	+	+	+	+
		Salonica	SS	+	+	-	-	+	+	-	-	-	-	-	-	-	+	-	+	+	+	+	+
			WR	+	+	-	-	+	+	-	-	-	-	-	-	-	+	-	+	+	+	+	+
		Kileler	SS	+	+	-	-	+	+	-	-	-	-	-	-	-	+	-	+	+	+	+	+
			WR	+	+	-	-	+	+	-	-	-	-	-	-	-	+	-	+	+	+	+	+
	Capeiti	Skourta	SS	+	+	-	-	+	+	-	-	-	+	+	-	+	+	-	+	+	+	+	+
			SS_1	+	+	-	-	+	+	-	-	-	+	-	-	+	+	-	+	+	+	+	+
			SS_2	+	+	-	-	+	+	-	+	-	+	+	-	+	+	+	+	+	+	+	+
			SS_3	+	+	-	-	+	+	-	+	-	+	-	-	+	+	+	+	+	+	+	+
		Salonica	SS	+	+	-	-	+	+	-	-	-	-	-	-	-	+	-	+	+	+	+	+
			WR	+	+	-	-	+	+	-	-	-	-	-	-	-	+	-	+	+	+	+	+
	Appulo	Salonica	SS	+	+	-	-	+	+	-	-	-	-	-	-	-	+	-	+	+	+	+	+
			WR	+	+	-	-	+	+	-	-	-	-	-	-	-	+	-	+	+	+	+	+
	Vitron	Skourta	SS	+	+	-	-	+	+	-	-	-	-	-	-	-	+	-	+	+	+	+	+
			WR	+	+	-	-	+	+	-	-	-	-	-	-	-	+	-	+	+	+	+	+
T. aestivum	Yecora	Domokos	SS	+	+	-	-	+	+	-	-	-	+	-	-	-	+	-	+	+	+	+	+
			WR	+	+	-	-	+	+	-	-	-	+	-	-	-	+	-	+	+	+	+	+
		Salonica	SS	+	+	-	-	+	+	-	-	-	-	-	-	-	+	-	+	+	+	+	+
			WR	+	+	-	-	+	+	-	-	-	-	-	-	-	+	+	+	+	+	+	+
		Gonni	SS	+	+	-	-	+	+	-	-	-	-	-	-	-	+	+	+	+	+	+	+
			WR	+	+	-	-	+	+	-	-	-	-	-	-	-	+	-	+	+	+	+	+
	Vergina	Domokos	SS_1	+	+	-	-	+	+	-	-	-	-	-	-	-	+	-	+	+	+	+	+
			SS_2	+	+	-	-	+	+	-	-	-	-	-	-	-	+	+	+	+	+	+	+
			WR	+	+	-	-	+	+	-	-	-	-	-	-	-	+	-	+	+	+	+	+
		Salonica	SS	+	+	-	-	+	+	+	-	-	-	-	-	-	+	-	+	+	+	+	+
			WR	+	+	-	-	+	+	-	-	-	-	-	-	-	+	-	+	+	+	+	+
	Siete-Cerros	Salonica	SS	+	+	-	-	+	+	-	-	-	-	-	-	-	+	+	+	+	+	+	+
			WR	+	+	-	-	+	+	-	-	-	-	-	-	-	+	-	+	+	+	+	+
	Generoso	Salonica	SS	+	+	-	-	+	+	-	-	-	-	-	-	-	+	-	+	+	+	+	+
			WR	+	+	-	-	+	+	+	-	-	-	-	-	-	+	+	+	+	+	+	+
	G-07783	Salonica	SS	+	+	-	-	+	+	-	-	-	-	-	-	-	+	-	+	+	+	+	+
			WR	+	+	-	-	+	+	-	-	-	-	-	-	-	+	+	+	+	+	+	+

Discussion

The variation in nitrogenase activity is not unexpected, since the samples derived from different geographic regions, plant species, varieties and growth stages. It is well documented that nitrogenase activity ranges during the day, or during the crop's growth period (Wani *et al.*, 1988; Pedersen *et al.*, 1978; Van Berkum and Bohlool, 1980). Maximum activity was recorded the afternoon, at the stage of late flowering-early grain filling (Day *et al.*, 1975). In Greece, the occurrence of nitrogen fixing bacteria in the rhizosphere of cereals was more frequent, in comparison with the evidences from other scientists' reports

(Tyler *et al.*, 1979; de Coninck *et al.*, 1988). During the purification procedures some of the isolated strains developed higher acetylene reduction, in mixed cultures. In pure cultures, they lost their activity entirely. A possible explanation is that, the nitrogen fixing bacteria acting in root-culture may not have found the necessary nutrients when transferred to fresh media, (Lindberg and Granhall, 1984; Garvin and Lindemann, 1986). The properties of the isolated strains, were compared with those of the other nitrogen fixing bacteria. According to Bergey's Manual of Systematic Bacteriology (Krieg and Döbereiner, 1984), 51 strains were identified as *Azospirillum brasilense* and only 2, able to utilize glucose, were identified as *A. lipoferum*. Our first purpose, to isolate native populations of *Azospirillum* was achieved. Further studies are underway to verify their use as seed inoculants.

References

Barraquio WL, Daroy MLG, Tirol AC, Ladha JK and Watanabe I, (1986) Laboratory acetylene reduction assay for relative measurement of N_2-fixing activities associated with field-grown wetland rice plants. Plant and Soil 90:359-372.

Boddey RM and Döbereiner J, (1988) Nitrogen fixation associated with grasses and cereals. Recent results and perspectives for future research. Plant and Soil 108:53-65

Coninck de K, Horemans S, Randombage S and Vlassak K, (1988) Occurrence and survival of *Azospirillum* spp. in temperate regions. Plant and Soil 110:213-218.

Day JM and Döbereiner J, (1975) Physiological aspects of N_2-fixation by a *Spirillum* from *Digitaria* roots. Soil Biology and Biochemistry 8:45-50.

Döbereiner J, (1980) Forage grasses and grain crops *in* Methods for Evaluating Biological Nitrogen Fixation. Bergersen FG (ed) pp. 535-555, John Wiley and Sons, New York

Döbereiner J, Marriel IE and Nery M, (1976) Ecological distribution of *Spirillum lipoferum* Beijerinck. Canadian Journal of Microbiology 22:464-1473.

Garvin S, and Lindemann W, (1986) Isolation, characterization and inoculation of N_2-fixing bacteria from dicotyledonous plants. Canadian Journal of Microbiology 32:912-916.

krieg NR and Döbereiner J, (1984) Genus *Azospirillum in* Bergey's Manual of Systematic Bacteriology. Holt JG and Krieg NR, (eds) 9th edition, pp. 94-104, Williams and Wilkins, Baltimore.

Lindberg T and Granhall U, (1984) Isolation and characterization of dinitrogen-fixing bacteria from the rhizosphere of temperate cereals and forage grasses. Applied and Environmental Microbiology 48:683-689.

Michiels K, Vanderleyden J and Van Gool A, (1989) *Azospirillum*-plant root associations:a review. Biology and Fertility of Soils, 8:356-368.

Okon Y and Kapulnik Y, (1986) Development and function of *Azospirillum*-inoculated roots. Plant and Soil 90:3-16.

Pedersen WL, Chakrabarty K, Klucas RV and Vidaver AK, (1978) Nitrogen fixation (acetylene reduction) associated with roots of winter wheat and sorghum in Nebraska. Applied and Environmental Microbiology 35(1):129-135.

Penot I, Berges N, Guinguene C and Fages J, (1992) Characterization of *Azospirillum* associated with maize (*Zea mays*) in France, using biochemical tests and plasmid profiles. Canadian Journal of Microbiology 38:798-803.

Rodrigues Caceres EA, (1982) Improved medium for isolation of *Azospirillum* spp. Applied and Environmental Microbiology 44:990-991.

Thompson JA, Gemell LG, Roughley RJ and Evans J, (1984) Nitrogenase activity associated with pasture grasses in northern New South Wales. Soil Biology and Biochemistry 16(3):217-222.

Tyler ME, Milam JR, Smith RL, Schank SC and Zuberer DA (1979) Isolation of *Azospirillum* from diverse geographic regions. Canadian Journal of Microbiology 25:693-697.

Van Berkum P and Bohlool BB, (1980) Evaluation of nitrogen fixation by bacteria in association with roots of tropical grasses. Microbiological Reviews 44(3):491-517.

Wani SP, Chandrapalaih S, Zambre MA and Lee KK, (1988) Association between N_2-fixing bacteria and pearl millet plants:Responses mechanisms and persistence. Plant and Soil 110:289-302.

Zafar Y, Ashrat M and Malik KA, (1986) Nitrogen fixation associated with roots of kallar grass (*Leptochloa fusca* L. Kunth.). Plant and Soil 90:93-105.

(Vio... et al. 1978; de Cock et al. 1988). During the purification process some of the isolated strains [illegible] lignin [illegible] [illegible] [illegible] [illegible]. They lost their [illegible] activity. A [illegible] [illegible] [illegible] [illegible] [illegible] [illegible] [illegible] [illegible] [illegible] [illegible] [illegible] [illegible] (Holt and [illegible] [illegible]) [illegible] [illegible] [illegible] [illegible] [illegible] [illegible] [illegible] Bergey's Manual of Systematic [illegible] [illegible] [illegible] [illegible] [illegible] were identified as [illegible] [illegible] [illegible] [illegible]. A [illegible] [illegible] [illegible] [illegible] [illegible] [illegible] [illegible]

[illegible]

[illegible]

Influence of the Carbon Substrate on the Composition of the Exocellular Polysaccharides by *Azospirillum brasilense.*

Francesca Marini°, Silvia Speranza° and Maddalena del Gallo
Dept. Food Technology,
Environmental and Microbiological Sciences,
University of Molise,
I-86100 Campobasso.

Abstract

In the present study exocellular-polysaccharide (PS), capsular and low and high MW exo-polysaccharides produced by *Azospirillum brasiliense* were analysed and the influence of growth medium on the composition of all fractions was evaluated.
HPLC analyses showed that the capsular-PS contained significant amounts of glucose, galactose, rhamnose, arabinose and fucose. The approximate molar ratio of these sugars were 2:4:11:2:6. In some preparations glucuronic acid and acetyl moieties were detected.
The qualitative composition of the high MW exo-PS (EPS_i) was similar to CPS but ratios were different. In particular, glucose and galactose were present in a greater amount and glucuronic acid was never detected. The low MW exo-PS (EPS_s) fraction contained a higher galactose and arabinose content in comparison with the capsular fraction and no rhamnose.

Introduction

Azospirillum spp. exhibit a marked pleomorphism under different growth conditions which not only reflects in cell morphology but also in cell surface biochemical changes (Becking, 1985). Detailed studies to understand the origin of these variability and their effects on root colonisation and plant association are still lacking and the precise nature of the interaction between azospirilla and plant root still remains to be elucidate.
Initial binding of bacterial cells to plant surfaces is frequently characterised by an early reversible and occasionally rather unspecific phase, often followed by an

----

° ENEA-Casaccia, Biotechnology and Agriculture Sector, Via Anguillarese 301, I-00060 Rome

NATO ASI Series, Vol. G 37
Azospirillum VI and Related Microorganisms
Edited by I. Fendrik et al.

irreversible attachment step which, in many cases, is tightly specific to the interacting cell types. Bacterial capsular polysaccharides (CPS) like membrane lipopolysaccharides (LPS) and exo-polysaccharides (EPS) are key determinants of many pathogenic and symbiotic associations between bacteria and plants (Halverson and Stacey, 1986).

Several bacterial cell-surface components have been implicated in the binding of the crown-gall pathogen *Agrobacterium tumefacies* to plant tissue: LPS, β-1-2-glucans, two outer-membrane proteins - not related to flagella or pili (Mathysse, 1987) - and cellulose fibrils which are synthesised by *A. tumefaciens* after initial adsorption and which anchor the bacteria to the plant tissue (Mathysse, 1983).

In *Rhizobium meliloti*, the role of exopolysaccharides in nodule development has been well established and their chemical structure and the genes necessary for their production have been studied in detail (Diaz *et al.*, 1989): oligosaccharide molecules and cell wall associated components are know to be involved in the interaction with the host legume (Gray *et al.*, 1992).

It is very likely that in *Azospirillum*-plant associations bacterial cell-wall associated components such as PS, LPS, glycoproteins, fimbriae and flagella are involved. Exocellular PS, in particular, may possibly play a role in the association, either by molecular interaction with plant lectins, either physically, by sticking the bacteria one to each other and to the plant, allowing a massive colonisation of the root surface. *Azospirillum brasilense*, in fact, is known for forming large bacterial aggregates both in liquid medium (Sadasivan and Neyra, 1985) and on root surface (Bashan and Levanony, 1990). Sugar and fatty acid composition of LPS from several *A. brasilense* and *A. lipoferum* strains has been determined (Choma *et al.*, 1987) and lectin binding studies have suggested the presence of several sugars, glucose and galactose in particular, in the capsule of *A. brasilense* cell (Yagoda-Shagam *et al.*, 1988). Tn5 induced *A. brasilense* mutants were isolated which did not stain calcofluor white (CW), a fluorescent dye specifically binding 1-2 and 1-4 β-linked PS (Michiels *et al.*, 1988, Michiels *et al.*, 1991). These mutants are supposed to be affected in the production of a surface polysaccharide and could be very useful in assessing the role of various surface PS components in processes such as plant-interaction and cyst-formation.

Umali-Garcia *et al.* (1980) and Gafny *et al.*, (1986) found that attachment of *A. brasilense* to pearl millet and maize roots, respectively, was stimulated by lectin-like substances present in the root exudate. Vanderleyden *et al.* (see this volume) observed that *Azospirillum* attached polarly to the plant surface by glycosylated proteins of the polar flagellum.

However, the molecular mechanism of plant-*Azospirillum* interaction has not been clarified in all single step. In particular, the role played by the polysaccharidic fraction, either isolated or attached to different proteins (glycoprotein fraction), has not been fully elucidated. This attachment mechanism is characterised by two distinct phases: in a first phase CPS and flagella are mostly involved (Michiels *et al.*, 1991, Croes *et al.*, 1993, Vanderleyden *et al.*, this volume); in a second step, firm irreversible anchoring is detected and CW-binding PS and fibrils, produced at that moment, seem to be involved. Mutants which do not interact with calcofluor, do not anchor plant roots (Michiels *et al.*, 1991) These mutants are also affected in flocculation (Michiels *et al.*, 1990) and fibrillar material production.
Scanning and transmission microscopy observations have shown that two main forms of exocellular-PS exist on the outer surface. One form appears to be more dense and tightly bound to the cell and is referred to as the capsular-PS (CPS), the second form looks lighter, extending away from the cell and is referred to as exo-PS (EPS) (Del Gallo *et al.*, 1989).
Still unknown is the role played by the plant on the attraction of the bacterium and on the influence of signal molecules which induce changes on its morphology and stimulate the association.

Aims of the present study were to evaluate the influence of the growth medium - carbon sources which could be present in root exudates, in particular - on the composition of CPS and EPS fractions (low molecular weight, insoluble in ethanol EPS_s, and high molecular weight, ethanol-soluble EPS_i).

Materials and methods

Bacteria and growth conditions

Azospirillum brasilense Cd (ATCC 29710) was grown in 2 l flasks containing 1 l of minimal medium (Martinez-Drets *et al.*, 1984) supplemented with one of the following carbon sources: arabinose 10 g x l^{-1}, galactose 10 g x l^{-1}, fructose 10 g x l^{-1}, Na-malate 8.5 g x l^{-1}, Na-gluconate 7 g x l^{-1}.

Polysaccharides isolation and purification

Bacterial cultures were grown aerobically at 30 C in a rotary shaker for 4 days to the late stationary phase. The bacteria were removed by centrifugation, 8000 rpm, and the different PS fractions were purified as described (Del Gallo and Haegi, 1990).

Chromatographic sugar-analysis

Monosaccharidic composition of the different PS fractions, was analysed by HPLC using as a Beckman 156 Refractive Index Detector and 163 Variable Wavelength Detector. Identification of neutral sugar was performed on a Bio-Rad Aminex HPX-87C column, while polar sugar and organic acid were detected by a Bio-Rad Aminex HPX87H column, following the procedure previously described (Del Gallo and Haegi, 1990). For quantitative analyses the method of the "2 point absolute titration curve" described on the Shimadzu Chromatopac C-R3A instruction manual was utilised. HPLC-grade sugars and organic acids (Aldrich) were used as standards.

Colorimetric sugar quantification

A preliminary quantification of the polysaccharides present in the different fractions was performed by the Anthrone reagent using glucose as a standard (Dische, 1962). Total protein content in both EPS and CPS fractions was estimated by the Lowry's method (Lowry *et al.*, 1951). LPS contamination of the different fraction was estimated by the presence of 2-keto-3-deoxyoctonate, by its reaction with thiobarbituric acid in the Karkhanis modification (Karkhanis *et al.*, 1978).

Results and discussion

Preliminary analyses of CPS and EPS of *A. brasilense* Cd (Haegi and Del Gallo, 1991) revealed that the quantitative composition of both fractions increased with culture ageing and reached a peak after 4 days. In the present work we examined the composition of PS (capsular- and exo- polysaccharides) of 4-days-old cultures of *A. brasilense* Cd grown in different carbon sources. EPS were analysed after separation into low and high molecular weight fractions, EPS_s and EPS_i, soluble or insoluble in ethanol.

Yield of total PS from *A. brasilense* varied approximately from 4.17% to 504.83% of the total dry weight, depending on the different carbon source in which the bacterium was grown (table 1). When cells were grown in the presence of organic acids, however, these differences were less marked, even though the amount of PS produced were much lower.

Table 1. Total PS (% CPS and EPS), produced by *Azospirillum brasilense*, grown in different carbon sources and cell dry weight.

C source	CPS %	EPS %	total PS (mg/l)	PS %cell d.w.	cell d.w. (mg/l)
fructose	0.51	99.49	3395.33	279.59	1211
arabinose	0.45	99.54	2676.03	299.60	890
galactose	1.28	98.72	2847.70	504.83	560
Na-gluconate	36.73	63.27	29.11	7.68	380
Na-malate	32.87	67.13	18.46	4.17	440

The amount of proteins in the EPS fractions varied from 3.93% to 351.1%, while in the CPS varied from 30.81% to 94.49% (Table 2). Proteins seemed to reflect the same pattern of the PS. The higher % of proteins present in the EPS fraction was mostly due to a lower production of EPS in Na-malate and Na-gluconate conditions.

Table 2. Proteins present in both CPS and EPS fractions expressed both in % of the respective fraction and in µg/ml.

C source	% proteins CPS	% proteins EPS	proteins µg ml/CPS	proteins µg ml/EPS
fructose	84.07	8.32	14.66	280.94
arabinose	94.49	3.93	11.55	104.80
galactose	63.11	4.06	22.97	114.05
Na-gluconate	45.36	351.00	4.85	64.61
Na-malate	30.81	246.30	1.87	30.54

2-ketodeoxyoctonate, one of the main component of *Azospirillum* cell wall (Choma *et al*, 1987) was detected at levels varying between 0,15% to 1,46%, showing a low sample contamination by membrane LPS.

Quantification of PS by Anthrone (Figure 1) showed different production patterns: *A.brasilense* Cd produced more EPS when grown on a medium supplied with sugars as carbon sources in comparison with a medium supplied with organic acids. In particular this bacterium showed consistent variations in the ratios of high and low molecular weight PS when grown under different conditions (Figure 1, right).

The monosaccharide composition of CPS is shown in the figure 2. CPS contained significant amounts of glucose, galactose, rhamnose, arabinose and fucose. The approximate molar ratio of these sugars was 2:4:11:2:6. In some preparations -

cultures grown in fructose, arabinose or galactose - glucuronic acid and acetyl moieties were detected.

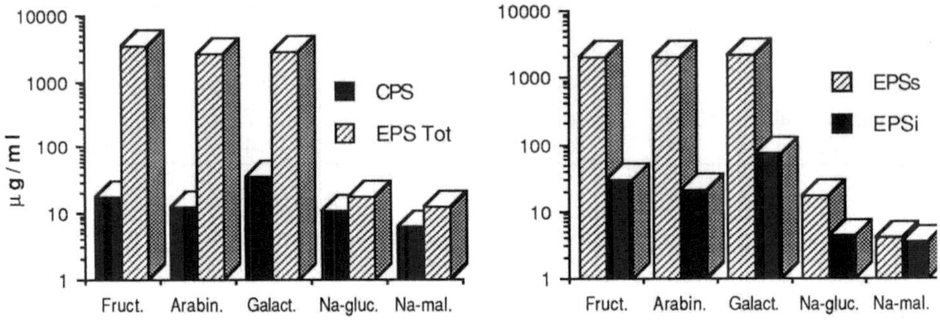

Figure 1. On the left: total CPS and EPS (µg/ml) produced by *A. brasilense* Cd grown in different carbon sources. On the right, the amount of PS present in both low and high MW fractions, EPSs and EPSi, produced at the same growth conditions (Fruct., fructose; Arabin., arabinose; Galact., galactose; Na-gluc., Na-gluconate; Na-mal., Na-malate).

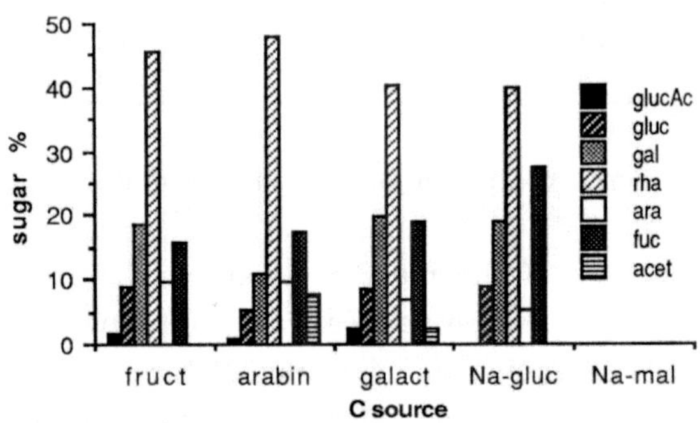

Figure 2 Sugar composition (gluAc, glucuronic acid; glu, glucose; gal, galactose; rha, rhamnose; ara, arabinose; fuc, fucose; acet, acetate) in % of *A.brasilense* Cd grown in different carbon sources. CPS of Na-malate grown culture were not detected (fru, fructose; ara, arabinose; gal, galactose; Na-glu, Na-gluconate; Na-mal, Na-malate).

Qualitative composition of EPS_i was similar to CPS but ratios were different (Figure 3, right). In particular, glucose and galactose were present in a greater amount and no glucuronic acid was detected. The EPS_S fraction contained a higher galactose and arabinose amount and no rhamnose (Figure 3, left).

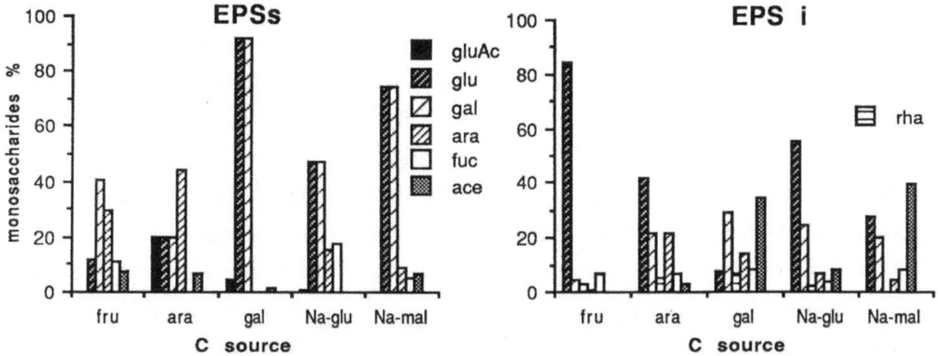

Figure 3. Sugar composition (gluAc, glucuronic acid; glu, glucose; gal, galactose; rha, rhamnose; ara, arabinose; fuc, fucose; acet, acetate) in % of the EPS s of *A. brasilense* Cd strain grown in different carbon sources (fru, fructose; ara, arabinose; gal, galactose; Na-glu, Na-gluconate; Na-mal, Na-malate).

Conclusions

Costerton *et al.* (1979) have coined the term glycocalyx to describe all types of polysaccharide-containig structures, lying outside the integral elements of the bacterial outer membrane. Polysaccharide capsules are found in a wide range of bacteria: in mollicutes (Razin, 1973), in Gram-positive and Gram-negative eubacteria (Sutherland, 1982) and in cyanobacteria (Rippka *et al.* 1979). Capsule usually consists of heavily hydrated gels of heteropolysaccharides with repeating unit structures that contain negatively-charged groups.

Several structural studies on EPS produced by *Azospirillum* have been conducted (Konnova *et al.*, 1990; Konnova *et al.*, 1992; Del Gallo and Haegi, 1990; De Troch, 1992). The differences between the PS of the two strains *A. brasilense*, Sp7 and Cd, though very close phylogenetically (Fani *et al.*, 1991; Gündisch*et al.*, 1993), is not surprising, since antibodies raised against Cd - which mostly recognise EPS - do not cross react with Sp7 (Levanony and Bashan, 1989). Konnova *et al.* (1990, 1992) found EPS associated with lipids which can only be removed by chloroform extraction.

Several studies indicated differences in PS composition of *Lactobacillus casei* and *Rhizobium trifolii* due to different carbon sources (Wicken *et al*, 1983;

Zevenhuizen, 1986). The influence of the growth medium on chemical composition of exopolisaccharides of *Pseudomonas aeruginosa* was recently described by Marty *et al.* (1992). EPS isolated from *A. brasilense* Sp7 and Sp245 were different when isolated from liquid or solid medium (P. De Troch, 1992, 1993). In batch culture, medium composition inevitably changes during growth with, in many cases, unpredictable effects on cell physiology. More than in any other bacterial polymer, apparently, the PS production (the fraction which is contact with the external environment), is markedly influenced by environmental conditions. Ratio of carbon to nitrogen in the medium and the growth phase seem to be the factors with the greatest influence.

In order to establish a possible influence of the carbon sources on *Azospirillum* PS composition, we isolated both CPS and EPS fractions produced by strain Cd grown in mineral medium supplied with several carbon sources present in the root exudates. CPS composition was similar in the different conditions. While EPS, the outer fraction, changed with the different carbon sources. Moreover, *A.brasilense* Cd produced more EPS when grown on a medium supplied with sugar as carbon source in comparision with a medium supplied with organic acids. Zaady *et al.* 1993 studied the different behaviour of *A.brasilense* Cd when grown with either malate or fructose as the carbon sources. Inoculation with fructose-grown bacteria resulted in a highly significant increase in root surface area and root and foliage dry weight of maize seedilings, as compare to plants inoculated with malate-grown bacteria. *Azospirillum* is known to move chemotactically towards gradients of sugar concentration that exist in cereal root exudates and in maize roots that contain D-fructose. As showed in this study fructose stimulates more EPS production than malate as carbon sources. This fact can explains the highest adsorption obtained with cells grown on D-fructose. We can speculate that in the associative process the plant can influence the bacterial EPS composition through the root exudation. This can be important for the specificity of the interaction.

Acknowledgements

The authors wish to thank the ENEA for giving the opportunity of working in the Casaccia Laboratories and the CNR (P.F. RAISA-Italian National Council for Research) for founding this research.

References

Bashan Y, Levanony H (1990) Currents status of *Azospirillum* inoculation technology: *Azospirillum* as a challenge for agriculture. Can J Microbiol 36: 591-608

Becking J H (1985) Pleomorphism in *Azospirillum*, In Azospirillum III: Genetics, Physiology, Ecology. Klingmüller W (*ed*) Springer-Verlag Berlin Heidelberg *pp* 243-262

Choma A, Russa R, Mayer H, Lorkiewicz Z (1987) Chemical analysis of *Azospirillum* lipopolysaccharides. Arch Microbiol 146: 341-345

Costerton JW, Irvin RT, Chen KJ (1979) The bacterial glycocalyx in nature and disease. Ann Rev Microbiol 35: 299-324

Croes C L, Moens S, Van Bastelaere E, Vanderleyden J, Michiels KW (1993) Polar flagellum mediates *Azospirillum brasilense* adsorption to wheat roots. J Gen Microbiol 139: 2261-2269

De Troch P, Philip-Hollingsworth S, Orgambide G, Dazzo FB, Vanderleyden J (1992) Analysis of extracellular polysaccharides isolated from *Azospirillum brasilense* wild type and mutant strains. Symbiosis 13: 229-241

De Troch P (1993) Dissertationes de agricultura: Bacterial surface polysaccharides in relation to plant interaction: a genetic and chemical study of *Azospirillum brasilense*. *pp*. 279

Del Gallo M, Negi M, Neyra C (1989) Calcofluor- and lectin-binding exocellular polysaccharides of *Azospirillum brasilense* and *Azospirillum lipoferum*. J Bacteriol 171: 3504-3510

Del Gallo M, Haegi A (1990) Characterization and quantification of exocellular polysaccharides in *Azospirillum brasilense* and *Azospirillum lipoferum*. Symbiosis 9: 155-161

Diaz C L, Melchers L S, Hooijkaans P J J, Lugtenberg B J J, Kijne J W (1989) Root lectin as a determinant of host-plant specificity in the *Rhizobium*-legume symbiosis. Nature 338: 579-581

Dische Z (1962) General colour reactions. Methods Charbohydr Chem 1:477-479

Fani R, Bazzicalupo M, Gallori E, Giovannetti L, Ventura S, Polsinelli M (1991) Restriction fragment lenght polymorphism of *Azospirillum* strains. FEMS Microbiol Lett 83: 225-230

Gafny R, Okon Y, Kapulnik Y, Fischer M (1986) Adsorption of *Azospirillum brasilense* to corn roots. Soil Biol Biochem 18: 69-75

Gray JX, deMaagd R, Rolfe BG, Johnston AWB, Lugtenberg JJ (1992) The role of the *Rhizobium* cell surface during symbiosis. In Molecular signals in plant-microbe communications Verma DPS (*ed*)CRC press *pp* 360-376

Gündisch C, Kirchhof G, Baur M, Bode W, Hartmann A, (1993) Identification of *Azospirillum* strains by RFLP and pulsed field gel electrophoresis. Microbial Releases 2: 41-45

Haegi A, Del Gallo M (1991) *Azospirillum*-plant interaction: a biochemical approach. In "Nitrogen Fixation". M. Polsinelli, R. Materassi, M. Vincenzini, (*eds*). Kluwer Acad Publishers *pp* 147-153

Halverson LJ, Stacey G (1986) Plant Microbe Interactions. Microbiol Rev 50: 193

Karkhanis YD, Zeltner JY, Jackson JJ, Carlo DJ (1978). A new and improved microassay to determine 2-keto-3-deoxyoctonate in lipopolysaccharide of gram-negative bacteria. Analytical Biochemistry 85: 595-601

Konnova SA, Baberdina IV, Makarov OE, Skvortsov IM, Ignatov VV(1990) Polysaccharide-lipid complex degradation, fractionation and monosaccharide composition of polysaccharides. Mikrobiol Zhurn 52: 40-46

Konnova SA, Makarov OE, Skvortsov IM, Ignatov VV (1992) Exopolysaccharides of bacteria *Azospirillum brasilense* Sp 245 and Sp 107. Mikrobiol Zhurn 54: 31-42

Levanony H, Bashan Y (1989) Localization of specific antigens of *Azospirillum brasilense* Cd in its exopolysaccharide by immuno-gold staining. Curr Microbiol 18: 145-149

Lowry OH, Rosenbrough NJ, Farr RL, Randal RJ (1951) Protein measurement with the Folin phenol reagent. J Biol Chem 193: 265-275

Martinez-Drets G, Del Gallo M, Burpee C, Burris RH (1984) Catabolism of carbohydrates and organic acids and expression of nitrogenase by Azospirilla. J Bacteriol 159: 80-85

Marty N, Dournes J L, Chabanon G, Montrozier H (1992) Influence of nutrient media on the chemical composition of the exopolysaccharide from mucoid and non-mucoid *Pseudomonas aeruginosa*. FEMS Microbiol Lett 98: 35-44

Matthysse AG (1983) Role of bacterial cellulose fibrils in infection. J Bacteriol 154: 906-915

Matthysse AG (1987) Characterization of non attaching mutants of *Agrobacterium tumefaciens*. J Bacteriol 169: 313-323

Michiels K, Vanderleyden J, Van Gool A, Signer E R (1988) Isolations and characterization of *Azospirillum brasilense* loci that correct *Rhizobium meliloti exo*B and *exo*C mutations. J Bacteriol 170: 5401-5404

Michiels K, Vanderleiden J, Van Gool A (1989) *Azospirillum*-plant root associations: A review. Biol Fertil Soils 8: 356-368

Michiels K W, Verreth C, Vanderleyden J (1990) *Azospirillum lipoferum* and *Azospirillum brasilense* surface polysaccharide mutants that are affected in flocculation. J Appl Bacteriol 69: 705-711

Michiels K,Croes CL, Vanderleyden J (1991) Two different modes of attachment of *Azospirillum brasilense* Sp7 to wheat roots. J Gen Microbiol 137: 2241-2246

Razin S (1973) The physiology of mycoplasmas. Adv Microbial Physiol 10: 1-80.

Rippka R, Deruelles J, Waterbury JB, Herdman M, Stanier RY (1979) Generic assignments, strain histories and properties of pure cultures of cyanobacteria. J Gen Microbiol 111: 1-61

Sadasivan L, CA Neyra (1985) Flocculation in *Azospirillum brasilense* and *Azospirillum lipoferum*: exopolysaccharides and cyst formation. J Bacteriol 163: 716-723

Sutherland IW (1982) Biosynthesis of microbial exopolysaccharides. Adv Microbial Physiol 23: 79-150

Umali-Garcia M, Hubbel DH, Gaskins MH, Dazzo FB. (1980) Association of *Azospirillum* with grass roots. Appl Environ Microbiol 39: 219-226

Vanderleyden J, Costacurta A, Desair J, De Troch P, Dobbelaere S, Keijers V, Michiels K, Milcamps, Moens S, My Bekri A, Pedersen D, Van Bastelaere E,

Vande Broek A, Van Dommeln A *Azospirillum*-Cereals: a Partnership with Future (this volume)

Wicken AJ, Ayres A, Campbell LK, Knox K W (1983) Effect of growth conditions on production of Rhamnose-containing cell wall and capsular polysaccharides by strains of *Lactobacillus casei* subsp. *rhamnosus*. JBacteriol 153: 84-92

Yagoda-Shagam J, Barton LL, Reed WP, Chiovetti R(1988). Fluorescein isothiocyanate-labeled lectin analysis of the surface of the nitrogen-fixing bacterium *Azospirillum brasilense* by flow cytometry. Appl Environ Microbiol.54: 1831-1837

Zaady E, Perevolotsky A, Okon Y (1993) Promotion of plant growth by inoculum with aggregated and single cell suspensions of *Azospirillum brasilense* Cd. Soil Biol Biochem 25: 819-823.

Zevenhuizen LPTM (1986) Selective synthesis of polysacchrides by *Rhizobium trifolii* strain TA-1. FEMS Microbiol Lett 35: 43-47

Identification of *Azospirillum* by Oligonucleotide Probes after isolation from soil and Sorghum rhizoplan contaminated or not by the Parasitic plant:*Striga*

Mahbubul Kabir and Denis Faure
Laboratoire d'Ecologie Microbienne du Sol
URA C.N.R.S. 1450, Université Claude Bernard Lyon-I
43 Bd. du 11 Novembre 1918, Villeurbanne Cedex ,France

Abstract

A total of 154 isolates were identified as *Azospirillum* species, after isolation from the soils and the rhizoplan, using three species-specific probes of *Azospirillum* ssp based on 16S-rRNA gene. Among these isolates, 151 strains were found positive signals after colony hybridization with Aba probe (specific to *A. brasilense/amazonense* ssp) and only 3 isolates showed positive signals when treated with Al probe(specific to *A. lipoferum*). No signal was recorded with the Ai probe (specific to *A. irakense*). Differential biochemical characteristics indicate that, all of the 151 strains are unable to use sucrose as a sole source of C and their colonies on potato agar are large pink and raised margin. This results confirmed that they belonged to *A. brasilense* species. While rest of the 3 isolates considered as *A. lipoferum* group, by their ability to use glucose and by their requirement for biotin.

Key Words: *Azospirillum*, Colony hybridization, Isolation, Oligonucletide probe, Parasitic plant *Striga*, Sorghum rhizoplan.

Introduction

Azospirillum is a well known nitrogen fixing soil bacterium which is usually found in association with roots of grasses and cereals (Döbereiner and Day 1975, Laksmi-kumari et al 1976). These rhizospheric nitrogen fixing bacteria, which have a plant growth promoting capacity on cereals (effect PGPR) are interesting candidates for agronomic application.The parasitic plant: *Striga* have the same plant host of the rhizosphere of cereals. The angiosperm *Striga* spp are obligate parasitic weeds of tropical cereals (Sorghum, Rice, Maize...).

In this study, we described a simple method of isolation of *Azospirillum* from the soil and rhizoplan of sorghum plant, contaminated or not by the parasitic plant, *Striga* and then identification by colony hybridization technique using different species-specific oligonucletide probes of *Azopirillum* ssp based on 16S-rRNA gene.

NATO ASI Series, Vol. G 37
Azospirillum VI and Related Microorganisms
Edited by I. Fendrik et al.

Materials and Methods

Soil and cultivar

The soil was sampled from two sites located in the east of Mali. The main characteristics of soils sampled from the two sites are described in Table 1. In this experiment, we used a sorghum cultivar *Tiémarifing* (sensible to *Striga*).

Oligonucleotide probes and Bacterial strains

We have used 3 species-specific probes (Al, specific for *A. lipoferum*, Aba specific for *A. brasilense/amazonense* and Ai, which is specific to *A. irakense*) for *Azospirillum* ssp (data not shown). Several strains of *Azospirillum* ssp used in this study are listed in Table 2.

Table 1. The selected chemical and biological properties of the studied soils.

Main characteristics	Soil contaminated (SC)	Soil not contaminated (SNC)
pH-H2O	6.5	6.4
-KCl	6.0	6.0
Organic matter (g. kg.$^{-1}$)	9.6	9.0
Numbers of C.F.U. on *NA* (x10^7)	5.2 ± 1.8	1.78 ± 1.4
Numbers of C.F.U. on *NFB* (x10^7)	2.5 ± 1.3	0.4 ± 0.13
Numbers of *AODC* (x10^9)	3.1 ± 0.7	2.4 ± 0.24
Numbers of *IF* (x10^5)	4.04 ± 2.6	2.54 ± 0.8
IF%AODC	0.013	0.010

Note: Values are mean± standard deviation. All results presented here are expressed as g. g^{-1} of dry soil. C.F.U=Colony Forming Unit, AODC=Acridine Orange Direct Count and IF=Immunofluorescence direct count.

Media

Two major media were employed for isolation: Nitrogen free medium (Nfb) was used according to Baldani et al 1980 and Congo Red medium was applied as described by Caceres, 1982. In the both media the pH was adjusted to 7.0 with KOH. Besides, rich culture medium like TY, (Difco) and PDA, (Bio-merieux, France) were used in this experiment. PDA was applied as to differentiate the *Azospirillum* ssp as described by Magalhaes et al. 1983.

Table 2. List of the *Azospirillum* strains studied for different tests.

Azospirillum ssp	Origin	Reference or Source
A. lipoferum		
4B	Rice, France	Bally et al, .1983
4T	Rice, France	Bally et al, .1983
BR17	Maize, Brazil	Tarrand et al., 1978
Sp59b	Wheat, Brazil	Tarrand et al., 1978
CRT 1	Maize, France	Fage et al., 1988
MRB 16	Rice, Bangladesh	Rahman M
TVV 3	Rice, Vietnam	Van Tran Van
TVV 7	Rice, Vietnam	Van Tran Van
A. brasilense		
Sp7	Digitaria, Brazil	Tarrand et al., 1978
Cd	Cynodon dactilon, USA	Döbernier J
Sp 245	Wheat, Brazil	Baldani et al., 1983
A95	Rice, France	Rinaudo G
NO40	Rice, Ezypte	Omar N
A. amazonense		
Y1	Digitaria, Brazil	Magalhaes et al., 1983
Y2	Hyperrenia, Brasil	Magalhaes et al., 1983
Y6	Pennisetum purpureum, Brazil	Magalhaes et al., 1983
A. irakense		
KBC1	Rice, Irak	Khammas et al., 1989
KaC5	Rice, Irak	Khammas et al., 1989
A. halopraeferens		
7108	Kallargrass,Pakistan	LMG
7110	Kallargrass,Pakistan	LMG
7113	Kallargrass,Pakistan	LMG

LMG=Laboratorium Microbiologie Rijksuniversiteit, Gent, Belgium

Isolation procedures

A simple model of microcosm was prepared as described in Fig. 1. This model was prepared with or without a sorghum seed, with a prior surface sterilization step according to Thomas-Bauzon et al, 1982. After 12 days of incubation, samples of soil (5g humid soil) and roots (successively washed with sterile water) of sorghum plants (roots of five plants) were taken for homogenization in 50 ml sterile distilled water in a Waring blender (Eberbec Corp) for 2 minutes and then a small volume (2ml) of liquid samples (soil and roots) were transferred into tubes, contained 5ml semisolid Nfb malate medium. Then the tubes were placed under microaerophilic condition according to Krieg and Döberiner, 1984 . After enrichment of microaerobic bacteria forming a white, dense, undulating diffuse pellicle under the surface (1 to 4 cm) of the Nfb medium, the bacteria were streaked on to plates of RC medium and incubated at 37°C for 72h (as described by

Caceres, 1982).Then typical small, dense, single colonies stained as scarlet color were subcultured on Nfb liquid medium to check for purity.

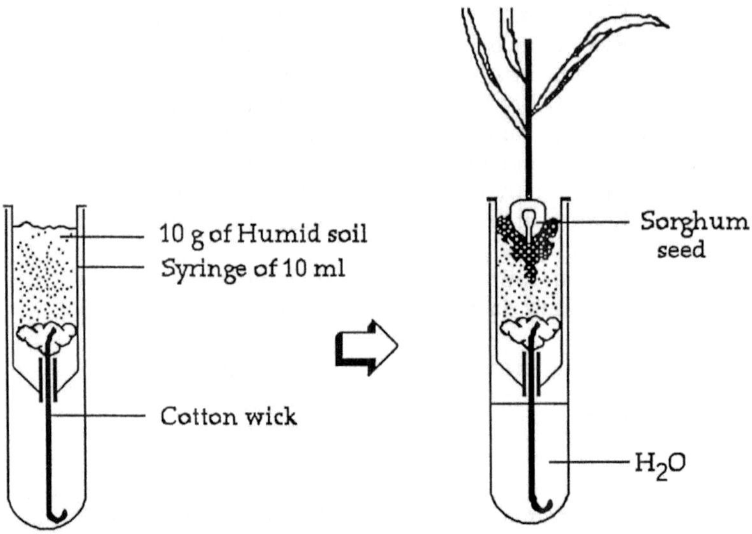

Fig. 1. A simple model of a microcosm.

Colony-hybridization using different oligonucleotide probes

Synthetic DNA oligonucleotide probes were 5'-end labelled with 32 p by using polynucleotidekinase and (y-32p) ATP as described by Sambrook et al. (1989).

Bacterial colonies were grown directly on the nylon bond membrane filter, lysed according to the protocol described by Maniatis et al. 1982, and baked at 80°C for 2h in a vacuum oven.

The baked membrane filters were hybridized according to the protocol by Maniatis et al. 1982, followed by exposure and development of X-ray film (after incubation for 12h at - 80°C) to localize the individual hybridized signals.

Biochemical tests

For differentiation of the *Azopirillum* ssp, dinitrogen growth on various carbon substrates was assayed in semi-solid Nfb medium, modified by replacing malate with glucose, sucrose and glycerol (5g/L). Biotin requirement were also checked for certain isolates as described by Tarrand et al., 1978.

Results and Discussion

Isolation and Identification by using different specific probes

A total of 175 microaerobic, scarlet colored bacteria, were isolated(Table 3), all of which had some typical morphological characters, resembling known *Azospirillum* strains as described by Caceres, 1982.

Table 3. Colony hybridization of the different isolates (obtained after modified enrichment method, Krieg and Döberiner, 1984), using different oligonucleotide probes.

Soil or Rizoplan	Total colony on RC medium	Scarlet colonies tested with probes	Positive colonies after hybridization with Probes		
			Al probe	**Aba probe**	**Ai probe**
SC	77	38	0	34	0
SCR	141	75	1	64	0
SNC	75	34	2	28	0
SNCR	83	28	0	25	0

Note: SC or SNC=Soil contaminated or not by *Striga*. SCR or SCNR=Sorghum rhizoplan, cultivated in the studied soils. RC=Red congo. Al probe=Specific with *A. lipoferum*, Aba probe=*A. brasilense/amazonense*, and Ai probe = *A. irakense*

All 175 strains were tested with probes in three distinct colony hybridization experiments (Table 3). A total of 151 positive signals were obtained when treated with the **Aba** probe. On the other hand, only 3 strains showed positive signals with **Al** probe. No signal was found with the **Ai** probe .

The **Aba** probe has one limitation, it cannot differentiate between the *A. brasilense* and *A. amazonense*. Therefore, to differentiate between the species, some particular biochemical tests were carried out to distinguish the *Azospirillum* ssp (Table 4).

Biochemical characterization

The *Azospirillum ssp.* differ when they grow on PDA (Table 4). All 151 colonies were found to be small, white, round and smooth and after 1 to 2 weeks they progressively became dry and wrinkled with varying shades of pink resembling known *A. brasilense.* None of the strains possessed the typical white, raised color of *A. amazonense* as described by Magalhaes et al., 1987. The 3 isolates which were positive with Al probe became pink colored on PDA confirming that they are similar to known *A. lipoferum.*

The use of various carbon sources for N_2 dependant growth is summarized in Table 4. No positive growth was recorded, when sucrose was a sole source of carbon. Thus this confirmed that all the isolates which had positive signals with **Aba** probe belonged to the *A. brasilense* species. Besides they grew with glycerol and no growth was found with D-glucose.The 3 strains differ from other strains by their requirement for biotin and can utilize D-glucose as a sole source of carbon. These results confirmed that they belong to *A. lipoferum* species.

Table 4. Differential characteristics of the new isolates of *Azospirillum* ssp in comparison with known *Azospirillum* ssp.

Characteristics	*A. lipoferum*	*A. brasilense*	*A. amazonense*	Isolates hybridized with **Probe Al**	Isolates hybridized with **Probe Aba**
Motility	+	+	+	+	+
Colony type on RC medium	Scarlet	Scarlet	Pink	Scarlet	Scarlet
Colony type on PDA medium	Pink raised	pink raised	White flat raised	Pink raised	[a]Pink raised
Biotin requirement	+	-	-	+	-
Utilisation of Carbon:					
Malate	+	+	+	+	[a]+
D-Glucose	+	-	-	+	-
Glycerol	+	+	-	+	[a]+
Sucrose	-	-	+	-	-

Note: Symbols: +, positive for all strains; -, negative for all strains; Nd, not determined. [a] Explanation of signs: more than 90% of the strains.

In conclusion, the results proved the significance of the method for identification of *Azospirillum* ssp using different oligonucleotide probes. Our results show that the biochemical tests correlated to the results obtained by colony hybridization. In this study it is clearly observed that more than 90% scarlet color bacteria identified as *Azospirillum ssp.*. Among these isolates only one species, *Azospirillum brasilense* is dominantly (151 isolates) found in the both soils as well as in the rhizoplan. Between the 3 *A.lipoferum* strains, 2 were isolated from the SNC soil and the rest was found from the rhizoplan ofthe SC soil.

References

Bally R., Thomas-Bauzon D., Heulin T., Balandreau J., Richard C. and De Ley J. 1983. Determination of the most frequent N_2-fixing bacteria in a rice rhizosphere.Can. J. Microbiol. 29: 881-887.

Baldani V. L. D. and Döbereiner J. 1980. Host-plant specificity in the infection of cereals with *Azospirillum* ssp. *Soil Biol. Biochem.* 12: 433-439.

Döbereiner J. and Day J. M. 1976. Associative symbioses in tropical grasses: characterization of microorganisms and dinitrogen fixing sites. *in*: Proceedings of the Ist intern. Symp. on Nitrogen Fixation. 2: 518-538.

Enrique A. R. CACERES, 1982. Improved Medium For Isolation of *Azospirillum* spp. *Appl. Env and Microbiol.* 990-991.

Khammas K. M., Ageron E. Grimont P. A. D. and Kaiser P. 1989. *Azospirillum irakense* sp. nov., a nitrogen-fixing bacterium associated with rice roots and rhizosphere soil. Res. Microbiol. 679-693.

Krieg N. R. and Döbereiner J. 1984. Genus *Azospirillum* , Tarrand, Krieg and DÖbereiner 1979, in "Bergey's Manual of Systematic Bacteriology", vol. 1 94-104, Williams & Wilkins, Baltimore.

Lakshmi-Kumari M., Kavimandan S. K. and Subba Rao N. S. 1976. Occurrence of nitrogen fixing *Spirillum* in roots of rice, sorghum, maize and other plants. *Ind. J. Exp. Biol.*, 14: 638-639

Magalhaes F. M., Baldini J. I., Souto S. M., Kuykendall J. R. and Döbereiner J. 1983. A new acid tolerant *Azospirillum* species. *An. Acad. Bras. Cien.*,55, 417-430

Maniatis T. E., Fritsh E. F. and Sambrook J. 1982. Molecular cloning-A laboratory manual. Cold Spring Harbour Laboratory, Cold Spring Harbour, NewYork.

Sambrook J., Fritsh E. F. and Maniatis T. E. 1989. Molecular cloning-A laboratory manual. Cold Spring Harbour Laboratory, Cold Spring Harbour, NewYork.

Thomas-Bauzon D., Weinhard P., Viiecourt P. and Balandreau J. 1982. The spermospher model. I. Its use in growing, counting, and isolating N_2-fixing bacteria from the rhizospher of rice. *Can. J. Microbiol.* 28: 922-928

Tarrand J. J., Krieg N. R. and Döbereiner J. 1978. A taxinomic study of the *Spirillum lipoferum* group, with description of a new genus. *Azospirillum lipoferum* comb. nov. and *Azospirillum brasilense* sp. nov. *Can. J. Microbiol.* 39: 219-226.

Ecological Aspects of Azorhizobium caulinodans in the Rhizosphere of Rice Seedlings

Catherine Van Nieuwenhove, Lieven Van Holm, Raf Vanden Bruel and Karel Vlassak[1] 1

BNF Project

Institute of Fundamental Studies

Hantana Road

Kandy

Sri Lanka

Abstract

The nitrogenase activity of *A. caulinodans*, as a free living organism, in nitrogen free semi liquid culture medium under normal atmospheric conditions was measured with ARA, at different times after incubation with acetylene. A significant higher activity was observed compared with the blanks, this activity was stable upto 24h incubation. In liquid medium the AR was measured only under N_2 atmosphere and stable for 12h after acetylene injection. The survival and the nitrogenase activity of *A. caulinodans* on the roots of rice seedlings was tested with 6 rice varieties under sterile conditions in test tubes. A good survival was measured but no nitrogenase activity was found 14 days after transplanting with none of the 6 varieties. Activity was measured 3 days after inoculation. After supplementing the growth medium with different concentrations of sodium-lactate as a carbon source, high AR activities were measured 14 days after transplanting the rice seedlings.

Introduction

Long term productivity of soils, in Low External Input and Sustainable Agricultural Systems (LEISA) has to rely on renewable resources such as soil micro organisms and green manures. Green manuring has been practised in rice cultivation for more than 2000 years in Asian countries. *Sesbania* species have been found to be suitable for rice as green manure because of their ability to grow in heavy soils, with stand waterlogging and flooding and tolerate soil salinity (Becker, 1990). *S. rostrata* in particular fixes more dinitrogen than many other leguminous plants. *Azorhizobium caulinodans* is responsible for the formation of these highly effective N_2-fixing root and stem nodules on *S. rostrata* (Dreyfus et al., 1988). Besides its

[1] Laboratory of Soil Fertility and Soil Biology, Faculty of Agriculture, K.U. Leuven, Kardinaal Mercierlaan 92, 3001 Leuven, Belgium.

NATO ASI Series, Vol. G 37
Azospirillum VI and Related Microorganisms
Edited by I. Fendrik et al.

symbiotic ability, *A. caulinodans* can also fix dinitrogen as a free-living organism when the vitamin nicotinic acid is available (Elmerich et al., 1983, Gebhardt et al., 1984). After incorporation of the easily decomposable biomass of *S. rostrata*, *A. caulinodans* is released in the rice rhizosphere and can fix dinitrogen in association with the rice roots (Ladha et al., 1989). However, *A. caulinodans* is not indigenous to Sri Lanka and it also doesn't belong to the normal rhizobacterial population of rice. Basic information on the ecology of the bacterium in the rhizosphere and the rhizoplane of rice has to be collected.

Results and discussion

a. OD measurements of the full grown culture

The development of a fast method to standardize inocula was envisaged. The cell density of a full grown *A.c.* culture was measured with plate counts (drop plate method, Somasegaran and Hoben, 1985) on YEB medium (5g/l pepton, 1g/l yeast extract, 5g/l beef extract, 5g/l sucrose, 0.5g/l $MgSO_4.7H_2O$) and OD_{540} measurements of dilution series (Fig. 1). To test the reliability of the method, CFU and OD was measured of several samples. Environmental factors as temperature, light and shaking speed influenced the OD of the culture.

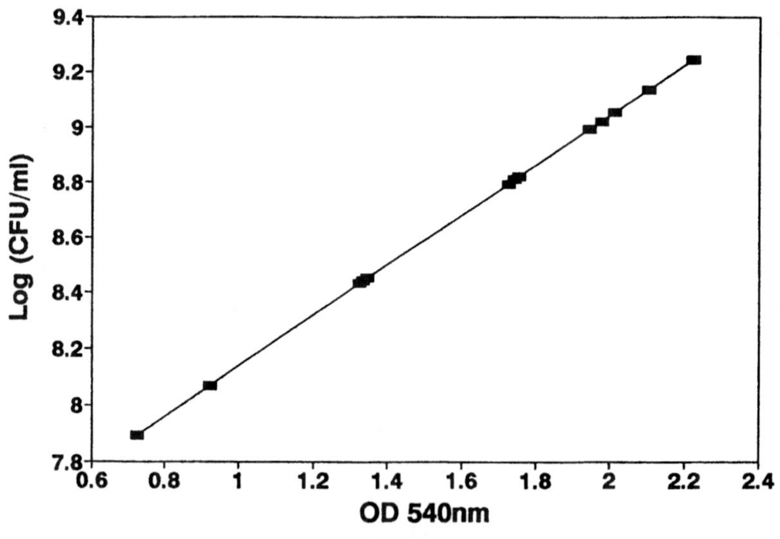

Fig. 1: Linear regression of optical density versus plate count

b. ARA of A.C., as a free living organism

Diazotrophic bacteria are sensitive for O_2 (Michiels, 1989). Nitrogenase is rapidly inactivated by O_2. In semi liquid medium, m.o. can migrate towards the place where the O_2 tension is optimal for fixing N_2. *A.c.* was subcultured in semi liquid N-free LSO medium (Elmerich, 1982). The ARA was measured after 3 days. Fig. 2 shows the C_2H_2 produced in semi liquid LSO medium after incubation for 4, 6, 8, 12 and 24h with C_2H_2. The reduction of C_2H_2 starts without lag period and stays stable till 24h after incubation. In liquid medium, *A.c.* was subcultured under atmospheric and dinitrogen conditions. We only observed ARA when *A.c.* was grown under N_2 atmosphere (97%/3% N_2/O_2), this confirms De Bruijns findings (1989). The production of C_2H_2 remained stable upto 12h after incubation but then stops abruptly (Fig. 3).

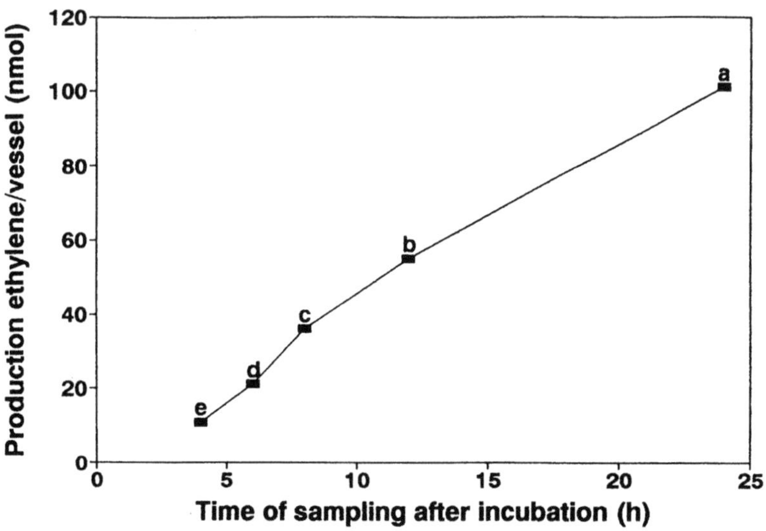

Fig. 2: Cumulative acetylene production of *A.c.* in semi liquid LSO medium. The letters show the significant differences at 5% probability level.

c. Survival and ARA of *A.c.* in association with roots of rice seedlings

The rice varieties Deveredderi, Rathu heenati, Herath banda, Bg 94-1, Bw 267-3 and Ld 181-5 were tested. The rice seeds were sterilized after decortication and allowed to germinate on nutrient agar plates. Three old sterile seedlings were transplanted in test tubes (250mm by

20mm). Nitrogen-free semi-liquid nutrient medium for rice according (yoshida, 1976) was used as substrate. The inoculation was done at transplanting with 1ml *A. caulinodans* liquid culture. The rice varieties did not show ARA after 14 days of growth. These results are contradictory with the results of Ladha (1987, 1989). Although the maintenance of a high bacterial population (10^{10}cfu/g dry root) on the rice rhizoplane suggest a saprophytic growth of *A.c.* Since external factors such as temperature, light, pH, O_2 tension, availability of nutrients may all have influenced the nitrogenase activity of the bacteria, priority was given to the hypothesis of exhaustion of exudates. After selecting the rice variety Ld 181-5, known as high yielding and pest resistant, the acetylene reduction activity was measured at different time intervals after inoculation. Only after 3 days of rice growth a significant production of ethylene was observed (Table 1). Thus the exhaustion of suitable carbon sources in the medium can be the reason for the drop of AR in later stages under the mentioned conditions.

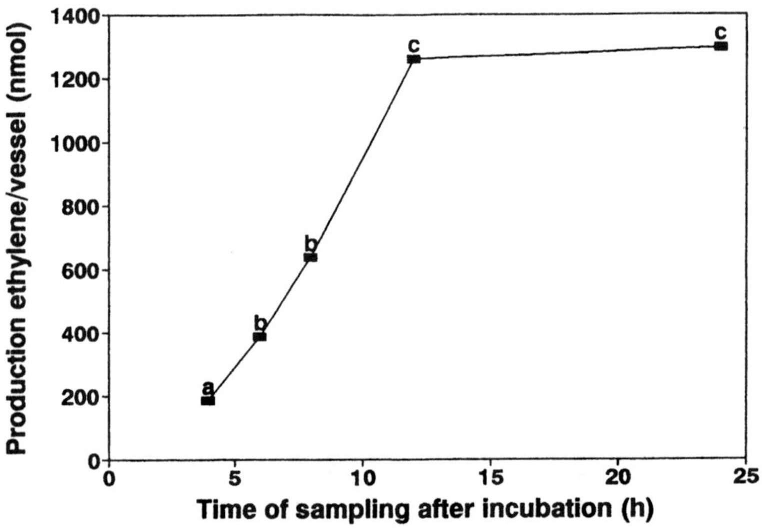

Fig. 3: Cumulative AR activity of *A.c.* in liquid LSO medium under N_2 atmosphere. The letters show the significant differences at 5% probability level.

Table 1: Optimal time for measuring ARA of rice seedlings

		ARA (μmol/24 h)	Length (cm)	DW shoots (mg)	DW roots (mg)	CFU/g dry roots
3 days	I	111[a]	7.4[d]	3[d]	9	1.9e+10
	B	88[b]	7.9[d]	2[d]	7	
6 days	I	92[b]	12.1[c]	5[c]	7	3.1e+10
	B	92[b]	12.4[c]	5[c]	6	
9 days	I	92[b]	15.8[b]	6[b]	7	2.7e+10
	B	88[b]	16.1[b]	6[b]	7	
14 days	I	88[b]	18.3[a]	9[a]	8	2.0e+10
	B	88[b]	19.1[a]	10[a]	8	

I=Inoculated seedlings B=Not inoculated seedlings

d. Energy requirement of A.c.

Dinitrogen fixation is an energy-expensive process. When *A. caulinodans* is used as an inoculant for rice, its energy-requirements have to be met to allow rice to benefit from the bacteria-plant association. Rice seedlings of the variety Ld 181-5 were grown under sterile conditions in test tubes in semi-solid Yoshida-plant nutrition medium. After supplementing the medium with different concentration of sodium-lactate (0.25%, 0.1%, 0.05% and 0.025%) as a carbon source, high AR activities were measured 14 days after transplanting the rice seedlings. The results were 1177nmol/plant for the medium supplemented with 0.25% sodium-lactate, 899nmol/plant with 0.1%, 235nmol/plant with 0.05% and 155nmol/plant with 0.025%. Ladha et al. (1989) reported an ethylene production of 368nmol/plant when inoculated wityh *A.c.* and 365nmol/plant when inoculated with *Azospirillum lipoferum,* in similar conditions but without supplementing the rice growing medium with an extra carbon source. The high ethylene production with the 2 highest sodium-lactate concentrations reflects the need of *A. caulinodans* for a high energy supply. These results confirm the findings of Vande Broek et al. (1993) with the *Azospirillum brasilense*-wheat root association.

Conclusions

The use of OD measurements is restricted to studies on the kinetics of bacterial growth and can not be used for standardization of inocula. The nitrogenase activity of *A.c.* in free living conditions was measured. The experiments give no evidence for varietal specificity. High AR was measured when the rice growing medium was supplemented with an extra carbon source. An active diazotrophic association between rice seedlings and *A.c.*, with a high dinitrogen fixing capacity, is not yet found. The possibility is still valid, if the extremely large population on the rhizoplane of rice seedlings could fix N_2 and release it to the plants.

Acknowledgement

The authors gratefully acknowledge the Belgian Administration for Development Cooperation (BADC) for the financial assistance, intervention number 14672/11, the Sri Lanka Government for its contribution to the BNF project at the IFS and the members of the BNF project for their support.

References

Becker M (1990) Potential use of the stem-nodulating legumes *Sesbania rostrata* and *Aeschynomene afrespera* as green manure for lowland rice (Oryza sativa L.). PhD. Thesis, IRRI, Philippines

De Bruijn FJ (1989) The unusual symbiosis between the diazotrophic stem-nodulating bacterium *Azorhizobium caulinodans* and its host, the tropical legume *Sesbania rostrata*. In: Nestor E, Kosuge T (eds) (1989) Plant-Microbe Interactions. McGraw-Hill New York

Dreyfus B, Garcia GL and Gillis M (1988) Characterization of *Azorhizobium caulinodans* gen. nov., sp. nov., a stem-nodulating nitrogen-fixing bacterium isolated from *Sesbania rostrata*. International Journal of Systematic Bacteriology 38,89-98

Elmerich C, Dreyfus BL, Reysset G, Aubert JP (1982) Genetic analysis of nitrogen fixation in a tropical fast-growing *Rhizobium*. EMBO Journal 1,4,499-503

Elmerich C, Dreyfus B and Aubert JP (1983) Nicotinic acid requirement and degradation by *Sesbania rhizobium* strain ORS571. FEMS Microbiology Letters 19,281-284

Gebhardt C, Turner GL, Gibson AH, Dreyfus BL and Bergersen FJ (1984) Nitrogen-fixing growth in continuous culture of a strain of *Rhizobium* sp isolated from stem nodules on *Sesbania rostrata*. Journal of General Microbiology 130,843-848

Michiels K, Vanderleyden J, Van Gool A (1989) *Azospirillum*-plant root associations: A review. Biol.Fertil.Soils 8,356-368

Ladha JK, Tirol-padre A, Punzalan GC, Watanabe I (1987) Nitrogen-fixing (C_2H_2-reducing) activity and plant growth characters of 16 wetland rice varieties. Soil Sci. Plant Nutr. 33,187-200

Ladha JK, Garcia M, Miyan S, Tirol-Padre A, Watanabe I (1989) Survival of *Azorhizobium caulinodans* in the soil and rhizosphere of wetland rice under *Sesbania rostrata*-rice rotation. Appl.Environ.Microbiol. 55,454-460

Somasegaran P, Hoben HJ (eds) (1985) Methods in Legume *Rhizobium* Technology. NifTal Hawaii

Vande Broek A, Michiels J, Van Gool A, Vanderleyden J (1993) Spatial-Temporal Colonization Patterns of *Azospirillum brasilense* on the Wheat Root Surface and Expression of the Bacterial nifH Gene during Association. MPMI 6,5,592-600

Yoshida S, Forno D, Cock J, Gomez K (1976) Laboratory manual for physiological studies of rice. IRRI Philippinnes

Associative, symbiotic N_2-fixers and scavenger strains, affected by Cu^{2+} and Zn^{2+} *in vitro*.

Borbála Biró[1]; Bayoumi Hamuda H. E. A. F.[2] and Mihály Kecskés[2]
[1] Research Institute for Soil Science and Agricultural Chemistry of the Hungarian Academy of Sciences, H-1022, Budapest, Herman O. ut 15., Hungary.

Abstract

An investigation was conducted *in vitro* to establish the Cu^{2+} and Zn^{2+} (sulphate and chloride) sensitivity of various bacteria (N_2-fixing *Rhizobium, Azospirillum* and fluorescent *Pseudomonas* sp.), which were candidate strains of biofertilizers for some important agricultural crops (e. g. *Triticum aestivum, Zea mays* and *Vicia faba*). The growth rate (optical density) of the microbes was measured in modified liquid media. Some of the isolates proved to be non N_2-fixing fluorescent-putida type pseudomonads, which were coisolates with the *Azospirillum* strains. These bacteria were the least sensitive group to heavy metals, however, not much variation in metal-sensitivity was found among them. For the N_2-fixers, the sensitivity was rather strain- than species-dependent. In the case of *Rhizobium* strains Cu^{2+} was more toxic than Zn^{2+}, whereas the opposite was found for the associative N_2-fixing *Azospirillums* and N-scavenger pseudomonads. The sensitivity order of the strains was highly influenced by the *in vitro* techniques. The modified liquid method, however, was useful when discussing the toxicity and selection of the strains for further *in vivo* examinations.

key words: heavy-metal tolerance, *in vitro* techniques, *Rhizobium leguminosarum* bv. *viceae*, *Azospirillum* sp., fluorescent pseudomonads, N-scavengers.

Introduction

The members of the genus *Azospirillum* are closely associated with the roots of many agriculturally important crops (Bashan and Levanony, 1990). Positive responses to their inoculation are reported both in tropical and temperate regions (Neyra and Döbereiner, 1977 and others). Although there are also opposite opinions (e.g. Albrecht et *al.* 1981), inoculation with metal-tolerant strains might be especially important in contaminated soils. Research in this area, however, has been largely neglected and only experiments on the effect of agrochemicals have been reported (Haahtela et al. 1988, Gadkari, 1987). The beneficial effect of plant-growth-promoting (PGPR) pseudomonads has already been established (Bayoumi Hamuda, 1992). Selected strains of these are also applicable as biofertilizers against plant damaging fungi and for PGPR activity (Kloepper et *al.*, 1980).

[2] *Department of Microbiology, University of Agricultural Sciences, Gödöllő, Páter K. ut 1, H-2100,*

NATO ASI Series, Vol. G 37
Azospirillum VI and Related Microorganisms
Edited by I. Fendrik et al.

On the basis of these findings, the main purpose of our investigation was to compare the differences in metal sensitivities of various N_2-fixing strains (associative azospirilla and symbiotic rhizobia), as well as some N-scavenger pseudomonads, so as to identify the possibilities for *in vitro* selection and finding potential candidates for further soil and plant inoculation.

Materials and Methods

Rhizobium leguminosarum bv. *viceae* strains were isolated from the root-nodules of *Vicia faba* L.; type strains (Sp 7-ATCC 29145, Sp 59-ATCC 29707) and Hungarian *Azospirillum* isolates (Km 5, Be 7) originating from the rhizosphere of wheat (*Triticum aestivum* L.) and maize (*Zea mays* L.), and two N-scavenger *Pseudomonas* strains (Hurek et al. 1988) were selected as test organisms. The effect of Cu^{2+} and Zn^{2+} heavy metals (in sulphate and chloride forms) was examined at 25, 50, 100, 200 and 400 µM concentration in liquid yeast-extract-mannitol (YEM), nitrogen-free maleic acid (Nfb) and nutrient media (Döbereiner and Day, 1976; Vincent, 1981). Due to the formation of precipitates, inorganic additives were omitted (Bayoumi Hamuda, 1992). Tubes, containing 5.0 ml medium, with or without the addition of heavy metal, were inoculated with 100 µl bacterial suspension to provide a density of 10^6 CFU/ml and incubated in a rotary shaker (150 rpm) at 28°C for 48 h in three replicates. Sensitivity order of bacteria was also checked on agar plates, where metal-containing paper discs were put on the densely seeded surfaces. Growth rate (optical density) was measured spectrophotometrically: OD_{550} for the *Rhizobium* and *Azospirillum* strains, and OD_{640} for the *Pseudomonas* strains in the liquid method. Data are shown as percentage of controls (inoculated, without metals) and the least significant difference ($LSD_{5\%}$) was calculated after an analysis of variance.

Results

Effect of Cu^{2+} and Zn^{2+} (sulphates) on the growth of investigated bacterial groups

Effect of $CuSO_4$ on the growth of the various groups of soil and rhizosphere bacteria is shown in Table 1. Among the three groups of microbes, like to earlier results (Biró and Kecskés, 1984) the pseudomonads were significantly less sensitive. *Azospirillum brasilense* Sp 7, on the other hand was highly affected by low metal concentrations, but this strain was also the best survival at the level of 200 µM Cu^{2+} ions. Hungarian *Azospirillum* isolates were affected as the authentic type strains, although strain Km 5, which is suggested for *in vivo* experiments was outstandingly tolerant. There was an especially high diversity of the *Rhizobium leguminosarum* bv. *viceae* strains at the various Cu^{2+} concentrations among the three groups investigated (cell reduction amounted to 51 % and 78 % in comparison to the metal-free control tubes). Lowest observed effect concentrations (LOECs, Chaudry et *al.* 1993) were 25-50 μM for rhizobia, 50-100 μM for azospirilla and 200 μM for pseudomonads. Strains *R. leguminosarum* Bükköny 75/4 and *Azospirillum brasilense* Sp 7 proved to be the most sensitive to Cu^{2+}, already 25 μM dose of this metal (with sulphate anion) significantly reduced the cell-number.

Table 1.
Effect of $CuSO_4$ on the growth of some *Rhizobium leguminosarum* bv. *viceae*, *Azospirillum* sp., *A. lipoferum* Sp 59, *A. brasilense* Sp. 7 and *Pseudomonas* sp. in liquid media, after 48 h incubation at 28 °C in a rotary shaker. (Optical Density, as % of the metal-free control)

strains	**doses of heavy metal (μM/ml)**				
	25	**50**	**100**	**200**	**400**
Rhizobium leguminosarum					
HB -3841	90.9 *a*	43.7 *a*	31.2 *a*	18.7 *a*	6.2 *a*
Lóbab Z	105.7 *a*	59.7 *b*	42.7 *b*	24.8 *a*	10.9 *b*
Bükköny 75/4	43.2 *b*	28.8 *c*	21.6 *c*	14.4 *b*	3.4 *a*
mean	**76.6**	**40.7**	**28.5**	**18.3**	**6.8**
Azospirillum					
Sp 59	100.1 *a*	101.3 *a*	86.7 *a*	20.6 *a*	10.4 *a*
Sp 7	67.5 *b*	72.3 *b*	59.5 *b*	50.0 *b*	11.0 *a*
Be 7	109.5 *a*	69.8 *b*	47.1 *b*	30.3 *a*	15.4 *b*
Km 5	111.0 *a*	98.0 *a*	92.6 *a*	37.7 *a*	11.6 *a*
mean	**97.0**	**85.3**	**73.9**	**34.6**	**11.8**
Pseudomonas sp.					
Km8	101.5 *a*	102.0 *a*	94.0 *a*	63.0 *a*	9.6 *a*
Ke 1	95.1 *a*	94.0 *a*	93.1 *a*	51.7 *a*	13.2 *b*
mean	**98.3**	**98.0**	**93.5**	**57.3**	**11.4**
LSD 5%	**15.3**	**12.5**	**10.3**	**9.1**	**4.2**

a, b: values with the same letter within the columns are not significantly different at $P_{0.005\%}$.

In Table 2 the *in vitro* effect of $ZnSO_4$ can be seen on the growth of the three groups of microbes investigated. In comparison to the Cu^{2+} tolerance, the rhizobia were less affected by Zn, whereas the opposite was found for the associative N_2-fixers and *Pseudomonas* strains. *R. leguminosarum* Lóbab Z, and *Azospirillum* sp. Km 5 had extremely high tolerance to Zn^{2+} concentrations (lowest observed inhibiting doses were 100 μM and 50 μM, respectively). Other hungarian *Azospirillum* strain (Be 7), however, proved to be especially sensitive to 25 μM Zn^{2+}. Average cell reduction of the investigated bacterial representatives ranged from 67 % (*Azospirillum brasilense* Sp 7) to 13 % (*R. leguminosarum.* Lóbab Z). Therefore, high variation existed even when a single method is used to test the metal sensitivity.

Table 2.
Effect of $ZnSO_4$ on the growth of some *Rhizobium leguminosarum*, *Azospirillum* sp., *A. lipoferum* Sp 59, *A. brasilense* Sp 7, and *Pseudomonas* sp. in liquid media, after 48 h incubation at 28 °C in a rotary shaker. (Optical Density, as % of the metal-free control)

strains	doses of heavy metal (µM/ml)				
	25	**50**	**100**	**200**	**400**
Rhizobium leguminosarum					
HB -3841	112.5 *a*	68.7 *a*	37.5 *a*	12.5 *a*	6.2 *a*
Lóbab Z	118.9 *a*	109.0 *b*	87.2 *b*	70.9 *b*	49.0 *b*
Bükköny 75/4	103.0 *b*	87.1 *c*	50.0 *c*	33.1 *c*	17.5 *c*
mean	**111.5**	**88.3**	**58.2**	**38.8**	24.2
Azospirillum					
Sp 59	72.8 *a*	54.6 *a*	14.2 *a*	11.7 *a*	10.2 *a*
Sp 7	91.5 *b*	57.4 *a*	17.0 *a*	15.5 *a*	8.5 *a*
Be 7	78.5 *a*	62.6 *a*	28.1 *b*	8.3 *a*	6.4 *a*
Km 5	108.4 *c*	75.5 *b*	43.2 *c*	25.2 *a*	7.9 *a*
mean	**87.8**	**70.0**	**33.1**	**17.7**	**9.2**
Pseudomonas sp.					
Km 8	147.4 *a*	69.1 *a*	15.1 *a*	9.5 *a*	3.2 *a*
Ke 1	110.9 *b*	83.1 *b*	28.2 *a*	15.0 *a*	6.8 *a*
mean	**119.2**	**76.1**	**22.1**	**12.2**	**5.0**
LSD 5%	**13.2**	**14.4**	**14.5**	**16.8**	**12.7**

a, *b*: values with the same letter within the columns are not significantly different at $P_{0.005\%}$.

Table 3.
Sensitivity of *Azospirillum* strains to 100 µM Cu^{2+} and Zn^{2+} chloride salts using liquid and agar methods (OD_{550} as % of control, and inhibition zone in mm, respectively).

strains	growth response with heavy metals							
	Zn^{2+}		Zn^{2+}		Cu^{2+}		Cu^{2+}	
	liquid (%)		agar (mm)		liquid (%)		agar (mm)	
A. lipoferum Sp 59	55.0	(2)[a]	2.5	(2)[a]	73.3	(2)[a]	1.8	(1)[a]
A. basilense Sp 7	48.3	(3)	3.0	(2)	83.3	(1)	2.9	(2)
A. sp. Be 7	40.0	(4)	2.1	(1)	45.3	(3)	3.1	(2)
Km 5	61.5	(1)	4.2	(3)	85.0	(1)	2.5	(2)

[a] sensitivity order of bacteria at the $P_{0.005\%}$ level of significance

Solid and liquid methods influencing the sensitivity order of bacteria.

According to Table 3 the sensitivity of the N_2-fixing *Azospirillum* strains was highly influenced by the *in vitro* methodology and no positive correlation (figure is not shown) could be found between the results found in YEM broth and agar plates. In both cases, however, Zn^{2+} was more deleterious than Cu^{2+} and strain Km 5, originating from the histosphere of maize, had the best survival not only in the case of sulphates, but also with chloride salts.

Discussion

Different ecological factors affect the growth of various microbial populations in the soil and in the rhizosphere; therefore, the effect of sewage sludge with heavy metal contamination, must also be taken into consideration. Because microorganisms have such a rich spectrum of mechanisms to provide heavy-metal tolerance (Wood and Wang, 1983) there is a real need to establish the physiological and genetical changes influenced by the application of various metals. The first step consist of *in vitro* experiments for various purposes. It was realized, however, that due to complex formation, the toxicity of Zn and Cu *in vitro* may differ in the presence of agar (Märtensson, 1992). To overcome this problem metal-chelating buffers have been used to study metal availability and toxicity to bacteria (Chaudry et *al.*, 1993). In these particular cases the level of free metal ions was established, but the minimum inhibitory concentrations (MIC) of metals depended on the organic and inorganic additives. Although it was recommended that media components should only be added where they are absolutely necessary for the growth of microorganisms (Angle et *al.*, 1992), it is still difficult to compare metal sensitivities with such varying media (Jagnow et al. 1979). Despite attempts to prevent metal binding, this process still occurs even in the most well-defined media. One particular method, however, applied throughout a series of examinations can be suitable for screening the sensitivity order of various strains. Taking this into consideration, the broth culture method was preferred in this study, and because of precipitations between the studied chemicals and medium components, the inorganic salts were omitted (Bayoumi Hamuda, 1992). The growth rates obtained with this procedure were useful in selecting strains for *in vivo* examinations. Although care should be taken when extrapolating *in vitro* results to *in vivo* circumstances, it is necessary to obtain data about the limiting concentrations of the metals and the sensitivity of various groups of microorganisms.

Acknowledgements

This research was in part supported by a grant (OTKA 2O37/1991) from the Hungarian State Research Foundation.

References

Albrecht, S. L.; Okon, Y.; Lonnquist, J. and R. H. Burris (1981): Nitrogen fixation by corn- *Azospirillum* association in a temperate climate. *Crop Science*, **21**: 301-306.

Angle, J. S.; McGrath, S. P. and Chaudry, A. M. (1992): Effects of media components on toxicity of Cd to rhizobia. *Water, Air and Soil Pollution*, **64**: 627-633.

Bayoumi Hamuda, H.E.A.F. (1992): Factors influencing the optimalization of *Rhizobium leguminosarum* and *Vicia faba* symbiosis. *Ph.D. Dissertation and thesis*. University of Agricultural Sciences, Gödöllő, Hungary.

Bashan, Y. and H. I. Levanony (1990): Current status of *Azospirillum* inoculation technology: *Azospirillum* as a challenge for agriculture. Can. J. Microbiol. **36**: 591-608.

Biró, B. and M. Kecskés (1984): Herbicide sensitivity of *Coronilla Rhizobium* and *Pseudomonas* rhizobacterium strains. *Acta Microbiol. Acad. Sci. Hung.*, **31:** 302-303.

Chaudry A. M.; McGrath S. P.; Giller K. E.; Angle J. S. and R. L. Chaney (1993): Screening of isolates and strains of *Rhizobium leguminosarum* bv. *trifolii* for heavy metal resistance using buffered media. *Environm. Toxicol. Chem.* **12**: 1643-1651.

Döbereiner, J. and J. M. Day (1976): Associative symbioses in tropical grasses: characteriza-tion of microorganisms and dinitrogen-fixing sites. *In*: *Proc. 1st Int. Symp. on N_2-fixat.* (Eds. Newton W. E., Nyman C. J.) Washington Univ. Press, Pullmann, W.A.

Gadkari, D.(1987): Influence of the herbicides Arelon, Goltix and Stomp on growth and nitrogenase activity of *Azospirillum lipoferum. Zentralbl. Mikrobiol.*, **142**: 587-594.

Haahtela, K.; Kilpi, S. and K. Kari (1988): Effects of phenoxy acid herbicides and lyphosate on nitrogenase activity (acetylene reduction) in root associated *Azospirillum, Entero-bacter* and *Klebsiella. FEMS Microbiol. Ecol.*, **53**: 123-127.

Hurek, T.; Reinhold, B.; Grimm, I.; Fendrik, I. and E. G. Niemann (1988): Occurence of effective nitrogen-scavenging bacteria in the rhizosphere of kallar grass. *Pl. Soil*, **110**: 339-348.

Jagnow, G.; Heinemeyer, O. and S. Draeger (1979): Choice of liquid, semisolid or soil suspension media: an important factor modifying the effect of pesticides on the nitrogenase (C_2H_2) activity of *Clostridium pasteurianum, Azotobacter chroococcum* and *Spirillum lipoferum* Beijerinck. *Ecotoxicol. Environ. Saf.* **3**: 152-158.

Kloepper, J.W.; Leong, J.; Teintze M. and M. N. Schroot (1980): Enhanced plant growth by siderophores produced by plant growth promoting rhizobacteria. *Nature*, **286**: 885-886.

Märtensson, A. M. (1992): Effects of agrochemicals and heavy metals on fast-growing rhizobia and their symbiosis with small-seeded legumes. *Soil Biol. Biochem.*, **24:** 435-445.

Neyra, C. A.; Döbereiner, J. (1977): Nitrogen fixation in grasses. *Adv. Agron.*, **29**: 1-38.

Vincent, J. M. (1981): The Prokaryotes. A Handbook on habitats, isolation and identification of bacteria, (eds.: M. P. Syars, H. Stolp, H. G. Truper, A. Balows and H. G. Schlegel). Springer-Verleg, Berlin, pp. 818-841.

Wood, J. M. and H. K. Wang (1983): Microbial resistance to heavy metals. *Environm. Sci. Technol.* **17**: 582-590.

SECTION VI:

FIELD EXPERIMENTS AND APPLICATIONS

Inoculation effects of *Azospirillum* on sunflowers (*Helianthus annus*) under different fertilization and irrigation regimes.

Robin Itzigsohn, Zahera Abbass, Shlomo Sarig[1] and Y. Okon
Department of Plant Pathology & Microbiology
Faculty of Agriculture,
The Hebrew University of Jerusalem
Rehovot 76100, Israel

Abstract

In greenhouse experiments with different levels of NPK application, the only significant differences between *Azospirillum* inoculated and non-inoculated sunflowers (*Helianthus annus)* were both in the root surface area (RSA) and shoot weight. However, when different irrigation results were combined, *Azospirillum* led to a significant increase in plant height after 18 and 28 d and in shoot weight as compared to the control. The effects of *Azotobacter paspali, Azospirillum brasilense* and *Azospirillum lipoferum* inoculation on sunflower growth and seed yield were investigated in field trials. No significant differences were detected between the treatments. There was, nevertheless a trend towards an increase in germination percentage with *A. brasilense* Cd. A possible trend also exists in the 1000 weight of the seeds. A combination of these two trends could conceivably lead to an increase in yield at the field level.

Introduction

The growth promoting effects of the genus *Azospirillum* on plants of diverse species has been well documented (Okon 1985). Research indicates that the agronomic and climatic conditions are critical in achieving growth promotion in the field (Okon and Labandera-Gonzalez, 1994). Sarig et al. (1988) observed that field grown sorghum performed better under sub optimal nutrient and moisture conditions when inoculated with *Azospirillum*, and in reviewing the literature it becomes evident that this is a general phenomenon in the *Azospirillum*-plant association (Okon and Labandera-Gonzalez 1994). In order to assess the useful applicability of an inoculant, as many factors as possible affecting the interaction between the plant and bacterium should be determined. These include the nutrient levels, cultivars used, inocula genera, species or strains and their levels need to reach the threshold of effect.

Present Address:
[1]Department of Life Sciences, Bar Ilan University, Ramat Gan, Israel

NATO ASI Series, Vol. G 37
Azospirillum VI and Related Microorganisms
Edited by I. Fendrik et al.

The following series of experiments attempted to determine some of the distinct parameters affecting growth of sunflowers (*Helianthus annus*) under different conditions. The sunflower has been observed to be a good model system for studying the effects of the plant growth promoting rhizobacterium, *Azotobacter paspali* (Abbas and Okon 1993a,b). Such a system could well be adapted to investigate the *Azospirillum*--sunflower interaction, as it is cultivated in Israel and elsewhere as a non-irrigated and non-fertilized crop.

Materials and Methods

Organisms and growth conditions

Greenhouse experiments

These experiments were assayed with the sunflower cultivar "Odil" (Pioneer France Mais, Aussonne, France) under varying NPK, fixed PK and varying N, and irrigation regimes. This cultivar, bred for arid regions, and therefore suitable for summer greenhouse conditions was assayed with both *Azospirillum brasilense* and "Azogreen", (Pioneer France Mais, Aussonne, France) an inoculant consisting of micro-encapsulated *A. lipoferum* strain CRT-1 bacterial cells dehydrated in a polymer matrix (Fages 1994).
Azospirillum brasilense Cd (ATCC-29729) and *Azospirillum lipoferum* CRT-1 (Fages 1992) were grown on malate minimal medium (Okon *et al.* 1976) supplemented with 0.25% NH_4Cl and 0.01% yeast extract at pH 6.8. *Azotobacter paspali* (ATCC 23367) was grown on Nil medium (Robson 1986). Cells were harvested after 20 h at 30°C at the late logarithmic growth phase by centrifugation (1000 x g) and were resuspended in 30 mM potassium phosphate buffer, pH 7. Non-sterile pots of 5 L capacity were filled with sifted Rehovot loam-sand. Six seedlings were placed in small depressions in each pot and thinned to 4 after emergence. Inoculum was applied at the time of sowing at a rate of 1 mL plant^{-1} with appropriate *Azospirillum* cfu mL^{-1}, followed by 5 mL tap water. Fertilization was carried out with a modified Hoagland nutrient solution: 100% NPK: 8 mM N, 1mM P and 1 mM K. In the experiment with variable N, PK levels were maintained as above. Pots were arranged in a completely randomized design and all experiments were analyzed by running a 2-way ANOVA programme using the Duncan's Multiple Range Test (p=0.05).

Field trials

These experiments were carried out over a period of two growing seasons using two cultivars of sunflowers: an Israeli edible variety and one French oil variety. The region is hot in the summer with no rainfall during most of the growing season. The soil is a Groumosol Typic

Chromoxerert. In order to maintain weed free fields the germination inhibitor Treflan (2 l ha^{-1}) was applied, at the depth at which seeds were sown and the water potential was near field capacity at the planting date.

Year 1

Seeds of the edible cultivar cv. DY-3 (Hazera Seed Co., Gadera, Israel), were inoculated as follows: with *Azospirillum brasilense* Cd. or *A. lipoferum* CRT-1 or *A. paspali*. The inoculant was prepared by inoculating sterile peat to a final concentration of bacteria of 10^9cfu.g^{-1}. Eight g of freshly prepared peat inoculant was mixed with 5 l water and applied at the rate of 125 ml m^{-1}. The seeds were sown mechanically at a depth of 4-5 cm and were inoculated with the peat slurry two weeks after emergence. The controls received sterile peat.

Year 2

Seeds of the oil variety "Odil" (Pioneer France Mais Co.) were sown in plots of 10 M long at a rate of 1 seed per 20 cm with 1 M between rows. After emergence plants were thinned to 40 cm apart. At the time of sowing, seeds were placed in depressions and were inoculated as follows: with *Azospirillum brasilense* Cd. Approximately 9 ml of 5 x 10^{-7} cfu.ml^{-1} per seed were applied in liquid form in a 300 mM phosphate buffer suspension. Nine mL of buffer was applied to seed depression with both "Azogreen" (Fages, 1990) and control. Additionally, 150 ml of tap water was supplied to all treatments before covering the seed. The sole treatment after emergence was hand weeding and covering of the heads with paper bags after pollination in order to prevent losses due to birds' feeding.

Analyses

Root surface area (RSA): The titration method was used for measurement of root surface area (Carley and Watson 1966). Roots were washed from their growth medium in tap water, air dried, and dipped in a 3 M solution of HCl. A solution of 0.1 M NaOH with 0.5 mL phenolpthalein was used for titration. The volume in mL of NaOH titrated was proportional to the root surface area to which the HCl was adsorbed.

Leaf surface area (LSA): The delta T Area Measurement (T Devices Ltd.) was used to measure the leaf area in the field and is expressed in cm^2.

Shoot weight: was determined during the growing season by evaluating randomly thinned shoots at different intervals.

Results and Discussion

Greenhouse experiments

In experiments with different levels of NPK application, the only significant differences between *Azospirillum* inoculated and non-inoculated sunflowers were both in the root surface area (RSA) and shoot weight. A significant increase was observed at the 100% NPK level, but not at the 50 and 10% levels (Table 1). When the different fertilization results were combined, *Azospirillum* led to a significant increase in plant height after 18 and 28 d, in RSA and shoot weight as compared to control (Table 2).

TABLE 1: Effect of different levels of NPK with or without *Azospirillum brasilense* Cd inoculation on yield of greenhouse grown sunflowers, cv. Emil.

Treatment	*Azospirillum* Cd	Plant ht. (18d) cm	Plant ht (28d) cm	RSA (mL NaOH)	Root wt. (g)	Shoot wt. (g)
100% NPK	+	8.9a	12.8a	11.8a	0.58a	1.6a
	-	7.5a	11.1a	7.1b	0.48a	1.0b
50% NPK	+	7.4a	10.6a	7.7a	0.32a	1.1a
	-	6.5a	8.7a	6.3a	0.17a	0.8a
10% NPK	+	7.1a	9.8a	5.6a	1.0 a	1.0a
	-	6.0a	7.4a	3.8a	1.0 a	0.9a

Numbers followed by the same letter are not significantly different from each other for each level of fertilization at p=0.05.

TABLE 2: Effect of *Azospirillum brasilense* Cd inoculation on yield of greenhouse grown sunflowers, cv. Emil.

Treatment	*Azospirillum* Cd	Plant ht. (18d) cm	Plant ht (28d) cm	RSA (mL NaOH)	Root wt. (g)	Shoot wt. (g)
	+	7.85a	11.09a	7.93a	0.64a	1.21a
	-	6.69b	9.07b	6.22b	0.63a	0.95b

Numbers followed by the same letter are not significantly different from each other at p=0.05.

In the second experiment with fixed levels of PK and varying levels of N, a significant increase was observed in all of the parameters checked at the level of 40% N, when inoculated with *Azospirillum* (Table 3). When the different N fertilization results were combined, *Azospirillum* was shown to cause a significant increase in plant height after 18 and 28 d, in RSA, root and shoot weight as compared to the control (Table 4). The different irrigation regimes were shown to have different effects dependent upon inoculation. When irrigated to field capacity, significant differences were observed in both the RSA and shoot weight when inoculated with *Azospirillum* (Table 5). Although there was a trend to higher values under inoculation, a significant increase was seen only in shoot weight. When the different irrigation results were combined, *Azospirillum* led to a significant increase in plant height after 18 and 28 d and in shoot weight as compared to the control (Table 6).

TABLE 3: Effect of different levels of N with fixed PK with or without *Azospirillum brasilense* Cd inoculation on yield of greenhouse grown sunflowers, cv. Emil.

Treatment	*Azospirillum* Cd	Plant ht. (18d) cm	Plant ht (28d) cm	RSA (mL NaOH)	Root wt. (g)	Shoot wt. (g)
100% N	+	9.0a	12.8a	11.8a	0.59a	1.6a
	-	7.5a	11.1a	7.1b	0.49a	1.0b
80% N	+	8.0a	19.1a	8.1a	0.68a	1.8a
	-	6.7a	15.8a	5.7a	0.51a	1.3a
40% N	+	10.3a	23.7a	8.1a	0.81a	2.1a
	-	7.0b	17.6b	4.2b	0.41b	1.2b
0% N	+	8.3a	17.9a	5.3a	0.58a	1.5a
	-	6.9a	17.0a	4.2a	0.42a	1.2a

Numbers followed by the same letter are not significantly different from each other for each level of fertilization at p=0.05.

TABLE 4: Effect of different levels of N with fixed PK with or without *Azospirillum brasilense* Cd inoculation on yield of greenhouse grown sunflowers, cv. Emil

Treatment	*Azospirillum* Cd	Plant ht. (18d) cm	Plant ht (28d) cm	RSA (mL NaOH)	Root wt. (g)	Shoot wt. (g)
	+	8.87a	18.39a	8.33a	0.66a	1.7a
	-	7.02b	15.38b	5.30b	0.46b	1.2b

Numbers followed by the same letter are not significantly different from each other at p=0.05.

TABLE 5: Effects of different irrigtion regimes with or without *Azospirillum brasilense* Cd inoculation on yield of greenhouse grown sunflowers, cv. Emil

Irrigation	*Azospirillum* Cd	Plant ht. (18d) cm	Plant ht (28d) cm	RSA (mL NaOH)	Root wt. (g)	Shoot wt. (g)
Field capacity	+	9.0a	12.8a	11.8a	0.59a	1.6a
	-	7.5a	11.1a	7.1b	0.49a	1.0b
Wilting point	+	5.8a	10.5a	5.3a	0.42a	1.1a
	-	3.9a	8.3a	4.5a	0.30a	0.7b

Numbers followed by the same letter are not significantly different from each other for each level of irrigation at p=0.05.

TABLE 6: Effects of different irrigation regimes with or without *Azospirillum brasilense* Cd. inoculation on yield of greenhouse grown sunflowers, cv. Emil

Azospirillum Cd	Plant ht. (18d) cm	Plant ht (28d) cm	RSA (mL NaOH)	Root wt. (g)	Shoot wt. (g)
+	7.38a	11.66a	8.18a	0.51a	1.35a
-	5.74b	9.71b	6.22a	0.40a	0.86b

Numbers followed by the same letter are not significantly different from each other for each level of irrigation at p=0.05.

The results from the above greenhouse experiments are in accordance with the previous results for *Azospirillum* inoculation where the highest positive effects are observed under intermediate levels of NPK, N under fixed PK fertilization and moisture levels in the soil. This is true for sunflowers (Abbass and Okon 1993 a,b) and for other plants (Okon and Labandera-Gonzalez 1994).

Field trials

The effects of *A. paspali*, *A. brasilense* and *A. lipoferum* inoculation on plant growth and seed yield with sunflowers cv. DY-3 are summarized in Table 7 and figures 1,2 and 3. The plants were grown over a period of 19 weeks during which plant height, weight, head diameter and weight were measured at weeks 3, 7, 10, 14 and at harvest. At three weeks plants showed a significant increase in dry weight with all three inocula relative to the control. With *A. paspali* causing an increase of 51%, *A. brasilense* Cd. 39% and *A. lipoferum* 36%. No differences, however, were observed between the different bacteria (Table 7). At seven weeks from

TABLE 7: Comparison between different growth parameters of field grown DY-3 sunflower inoculated by *Azobacter paspali* (A.p), *Azospirillum brasilense* (Cd.) and *Azospirillum lipoferum* (CRT-1).

	3 weeks	7 weeks			10 weeks	
Treatment	Shoot wt g 10 plants^{-1}	3rd leaf area (cm^2 plants^{-1})	7th leaf area (cm^2 plants^{-1})	stem diam. (cm)	Plant ht (cm)	6th leaf area (cm^2 plants^{-1})
Control	6.1b	144a	172a	2.3a	119b	366a
A.p.	9.2a	130a	169a	2.3a	131a	316a
Cd	8.5a	144a	168a	2.4a	140a	309a
CRT-1	8.3a	160a	162a	2.2a	139a	409a

Numbers followed by the same letter are not significantly different at p=0.05.

emergence, no differences were seen between bacteria or relative to the control in the 6th and 7th leaves, stalk diameter at the second internode, head, leaf and stalk weight or whole shoots (sum of the 3 former parameters) (Table 7, Fig. 1). Added to the fact that there is a large inherent variance between plants in this cultivar leading to uneven ontogenic development, the heads are still small (approximately 4 cm), consequently decreasing the possibility of observing significant differences. The head diameters as measured at 7, 10 and 14 weeks are presented in Fig. 2. Although at 7 and 10 weeks no significant differences were seen between the treatments, however, at 10 weeks increases of 26%, 22% and 17% were observed for CRT-1, *A. paspali* and Cd., respectively. At 14 weeks the trend was similar and *A. paspali* caused a significant increase relative to the control, but not to the other inocula. Significant increases in the order of 18%, 17% and 10% for Cd., CRT-1 and *A. paspali*, respectively relative to the control, were observed in plant height at 10 weeks, however, not between inoculation treatments. The 6th leaf area as not found to be significantly different regardless of the treatment (Table 7). The plants were harvested 19 weeks from sowing. The heads were weighted, the seeds separated and the empty heads and seeds weighed again (Fig. 3). No significant differences above the controls could be found between the empty and full heads, although in all cases the weights of inoculated plants were slightly higher. The weight of seeds was higher (11-19%) than the control in all treatments but statistically significant only with strain CRT-1 (Fig. 3).

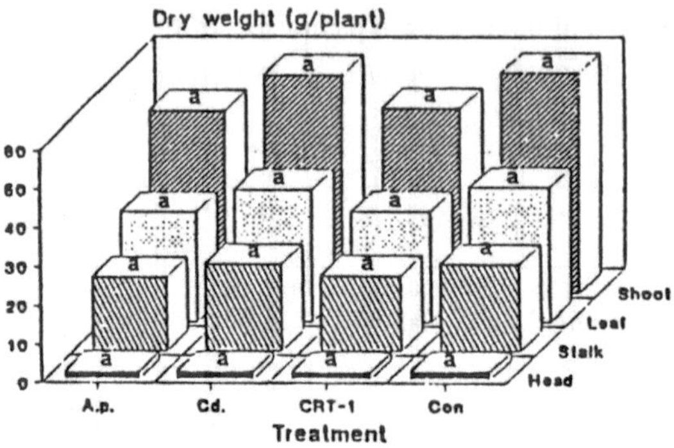

Figure 1: Comparison between the dry weight of whole plants inoculated by *Azotobacter paspali* (A.p.) *Azospirillum brasilense* (Cd.) and *Azospirillum lipoferum* (CRT-1) on field grown DY-3 sunflower harvested at 7 weeks. Significant differences between treatments are expressed by different letters where p=0.05 using Duncans Multiple Range Test.

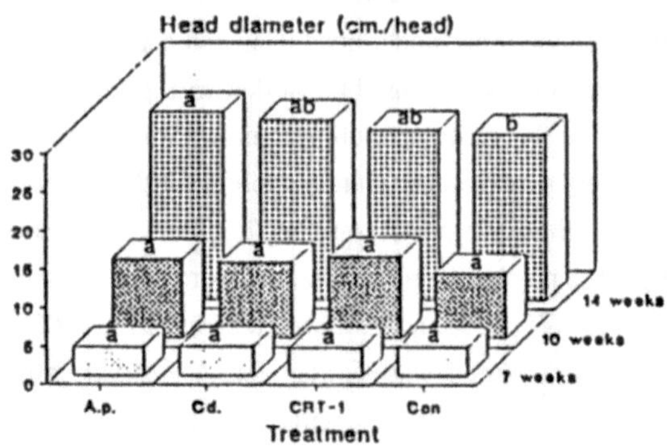

Figure 2: Comparison between the head diameter of whole plants inoculated by *Azotobacter paspali* (A.p.). *Azospirillum brasilense* (Cd.) and *Azospirillum lipoferum* (CRT-1) on field grown DY-3 sunflower harvested at 7 weeks. Significant differences between treatments are expressed by different letters where p=0.05 using Duncans Multiple Range Test.

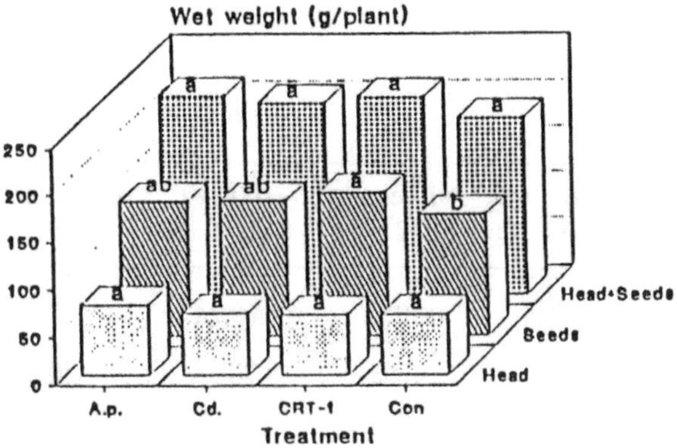

Figure 3: Comparison between field dried head weights of sunflower cv. DY-3 at harvest inoculated by *Azotobacter paspali* (A.p.). *Azospirillum brasilense* (Cd.) and *Azospirillum lipoferum* (CRT-1) at harvest. Significant differences between treatments are expressed by different letters where p=0.05 using Duncans Multiple Range Test.

In the field experiments with the cultivar DY-3, a general tendency of growth promotion was observed. However, this was inconsistent throughout the growing season. Whereas at 3 weeks the shoot weight with all three bacteria was found to be significant, this did not follow through at 7 weeks. The head weight was found to be significantly higher only after 14 weeks when inoculated with *A. paspali*. The parameters of 6th and 7th leaves and stalk diameter were not found to have any relevance as regards a response to inoculation.

In the field trial with sunflower cv. Odil, no significant differences were detected between the treatments (Tables 8 and 9).

TABLE 8: Application of *Azospirillum brasilense* Cd. and "Azogreen" on several growth parameters of sunflower cv. Odil grown under field conditions.

Treatment	Shoot wt. (g)	Germination/ 50 seeds	Plant ht. 4 weeks (cm)	Plant Ht 6 weeks (cm)
A. brasilense Cd	149a	32.2a	34.3a	78.0a
"Azogreen"	152a	23.9a	34.5a	74.4a
Control	137a	24.3a	31.1a	74.8a

Numbers followed by the same letter are not significantly different at p=0.05.

TABLE 9: Application of *Azospirillum brasilense* Cd. and "Azogreen" on several final yield parameters of sunflower cv. Odil, grown under field conditions.

Treatment	Head diam. (cm)	Head wt. (g)	Full seed wt (g)	1000 wt	Full seed %
A. brasilense Cd	24.1a	176a	106a	69.2a	82.7a
"Azogreen"	24.3a	183a	109a	70.9a	80.2a
Control	24.5a	182a	110a	68.1a	82.4a

Numbers followed by the same letter are not significantly different at p=0.05.

There was, nevertheless a trend towards an increase in germination percentage with *A. brasilense* Cd. relative to both Azogreen and the control. A possible trend also exists in 1000 weight of the seeds (Table 9). A combination of these two trends could conceivably lead to an increase in yield at the field level. There are several possible reasons for this lack of significant results. Soon after sowing, climatic conditions became very hot and dry with temperatures exceeding the 40°C mark. Despite the fact that the seeds were placed in moist soil it is possible that there was not enough moisture for the bacteria to reach the elongating root. Additionally, Azogreen which comes in a formulation of desiccated alginate beads, needs a minimum level of humidity to release its encapsulated contents. It is therefore likely that in both cases the bacterial numbers reaching the root were not at the threshold levels usually required to cause a promotion of plant growth. To support this assertion it was shown that the tested germination percentage of the supplied seeds was 96%, as compared to 75% in the field.

The above experiments emphasize the fact that both the nutrient concentrations and the level of soil moisture are critical in determining whether or not *Azospirillum* promotes plant growth. These must be assayed for each climatic zone and soil type and the desired crop.

Acknowledgement

This work was supported by a grant from Pioneer France Mais, Aussonne, France.

References

Abbass Z and Okon Y (1993a) Physiological properties of *Azotobacter paspali* in culture and the rhizosphere. Soil Biol Biochem 25:1061-1063.

Abbass Z and Okon Y (1993b) The effect of *Azotobacter paspali* growth promotion in the rhizosphere. Soil Biol Biochem 25:1075-1083.
Carley HE and Watson RD (1966) A new gravimetric method for estimating the root surface area. Soil Sci 102:289-292.
Fages J (1990) An optimized process for manufacturing an *Azospirillum* inoculant for crops. App Microbiol Biotechnol 32:473-478.
Fages J (1992) An industrial view of *Azospirillum* inoculants: formulation and application technology. Symbiosis 13:15-26.
Fages J (1994) Azospirillum inoculants and field experiments. In: *Azospirillum*/Plant Associations , Y. Okon (ed.)., pp. 87-110, CRC Press, Boca Raton.
Fallik E, Sarig S and Okon Y (1994). Morphology and physiology of plant roots associated with *Azospirillum* . In: *Azospirillum* /Plant Associations, Okon Y (ed.). pp. 77-86, CRC Press, Boca Raton
Okon Y (1985) *Azospirillum* as a potential inoculant for agriculture. Trends in Biotechnol 3:223-228.
Okon Y, Albrecht SL and Burris RH (1976) Factors affecting growth and nitrogen fixation of *Spirillum lipoferum* . J Bacteriol 127:1248-1254.
Okon Y and Labandera-Gonzalez CA (1994) Agronomic application of *Azospirillum* : an evaluation of 20 years world-wide field inoculation. Soil Biol Biochem (in press).
Robson RL (1986) Nitrogen fixation in strains of *Azotobacter chroococcum* bearing deletions of a cluster of genes coding for nitrogenase. Arch Microbiol 146:74-97.
Sarig S, Blum A and Okon Y (1988) Improvement of the water status and yield of field-grown grain sorghum (*Sorghum bicolor*) by inoculation with *Azospirillum brasilense*. J Agric. Sc. Cambridge 110:271-277.

Inoculation of genetically modified strains of *Azospirillum*: monitoring of population dynamic.

Corrado Fogher, Massimo Delledonne and Lorenzo Frigerio
Istituto di Botanica e Genetica vegetale
Università Cattolica S. Cuore
Via E. Parmense, 84
29100 Piacenza
Italy

Abstract

Azospirillum brasilense has been tested as inoculum for both cereals and grasses, in a wide range of climatic and agronomical conditions. Several reports show a positive effect on grain yield and protein content in vegetative organs. Breeding programs to obtain plant materials with an improved capacity of association with these bacteria can be planned. In particular, for maize it seems possible to exploit the variability in the response to inoculum observed in germplasm collections. For the development of breeding programs we need to easily monitor the population dynamic in the rhizosphere to select for enhanced metabolic exchange in maize-*Azospirillum* association. Fusion of the promoter region of the constitutive gene *nod*G of *A. brasilense* Sp7 with the reporter gene *lac*Z and the transformation of the chimeric construct in the *Azospirillum* strain Cd, widely used in field inoculum experiment on plants, give a reliable and rapid method for the identification of the inoculated bacterial cells. The standard colony plate count, on minimal or nutrient agar media supplemented with x-gal and antibiotic, allows to verify the exact bacterial cell number of the inoculated *Azospirillum* strain in the rhizosphere, rhizoplane and root cortex of the inoculated plants.

Introduction

Bacteria of the genus *Azospirillum* are diazotrophs associated with the roots of graminaceous plants. Cereals and grasses inoculated with *Azospirillum* show an increased development of the root apparatus, with a greater proton flux in roots (Bashan, 1990). *Azospirillum* has been tested as inoculum for both cereals and grasses, in a wide range of climatic and agronomical conditions (Kapulnik and Okon, 1983; Favilli *et al.* 1987; Bertolini *et al.* 1990). Several reports show a positive effect on grain yield and protein content in vegetative organs (Favilli *et al.* 1987; Kapulnik *et al.* 1981; Sarig *et al.* 1988) but often there are inconsistent and unpredictable yield results, probably because several factors influence the size of *Azospirillum* population and their potential contribution to plant growth, in dependence also of the plant genotype.

Breeding programs aimed to obtain plant materials with an improved capacity of association with these bacteria can be planned and, in particular for maize, it seems possible to exploit the variability in the response to inoculum observed in germplasm collections (Von Bülow and

NATO ASI Series, Vol. G 37
Azospirillum VI and Related Microorganisms
Edited by I. Fendrik et al.

Döbereiner, 1975; Ela *et al.* 1982; Fogher *et al.* 1993). Cultivars exhibit different phenotypic expression of nitrogen fixation supportive traits (*nis*) and examples for maize are reported in table I.

Table I. Average nitrogenase activity (μmol C_2H_4/plant/h) of maize genotypes associated with two *Azospirillum* strains.

Genotype	Average activity	
	A. brasilense Cd	*A. lipoferum 59*
Va93	116.2	-
Mo 17	95.8	102.7
Pool 17*	1276.7	-
Va 93 x Pool 17	75.2	140.1
Mo 17 x Pool 17	103.3	97.6

L.S.D. (0.05): 343.8; * Pool 17 is a CIMMYT population (see Fogher *et al.* 1993 for more details).

This shows a genetic role of the host plant and that the trait is recessive. To transfer the *nis* trait from responsive populations to inbred lines used in hybrid constitutions, in other words, for the development of breeding programs, we need to easily monitor the population dynamic in the rhizosphere to select for enhanced metabolic exchange in maize-*Azospirillum* association. Preparation of plant material is essentially based on the extraction of progenies in segregating offsprings derived from crosses and backcrosses between standard inbreeds and exotic populations known for the affinity with *Azospirillum* (Tab. I). Differences in leaves aspect, between duplicated N-fertilized and N-free inoculated plots, plant (figure 1) and ear size, root development (figure 2), ear insertion height and resistance to common parasites have been used as selection criteria in every generation.

At the end of the process, single crosses between lines provide the seed for combining ability evaluation of the selected genotypes; these materials are also tested for affinity to *Azospirillum* in several ways as, for example, controlling the presence of bacteria on the surface of washed roots (figure 3) and inside sterilized roots (figure 4).

Figure 1. Height of control (C) and inoculated (T) experimental hybrid plants.

Figure 2. Roots development in experimental material, control (left) and inoculated (right).

Figure 3. Scanning electron microscopy of a washed maize root.

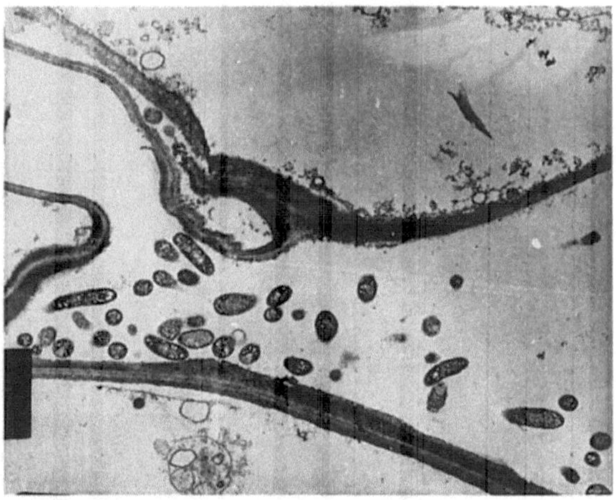

Figure 4. Transmission electron microscopy of a section of an inoculated maize root.

Previously proposed methods for the identification and enumeration of a specific species or inoculated strain (Shank *et al.* 1979; De-Polli *et al.* 1980; Rennie, 1980; Bashan and Levanony, 1985) give valid contributions but are not suitable in a breeding program. To simplify our breeding work we have genetically modified two *Azospirillum* strains (Sp7 and Cd) introducing a chimeric gene that allows to easily identify and quantify the inoculated strain.

Results and discussion

The chimeric reporter gene has been prepared starting from the cloned *A. brasilense* Sp7 *nod*G gene (Fogher *et al.* 1985; Delledonne *et al.* 1990). A fragment of about 500 bp covering the region upstream of the ATG start codon until 37 bp after the ATG to the *Nco*I site (figure 5) as been cloned in the plasmid pQF50 (Farinha and Kropinski, 1990) cut with *Bam*HI and filled in with Klenow (Sambrook *et al.* 1989). In the *Pst*I site of the vector DNA has been inserted also the Kanamycin resistance box (KanBlock, Pharmacia). The constructs were tested in *E. coli* and the selected, ß-galactoside active, plasmid (pIGP 13) has been transferred by conjugation, through a triparental mating using as helper the plasmid pRK2073 (Tp^R), to *A. brasilense* strains Sp7 and Cd.

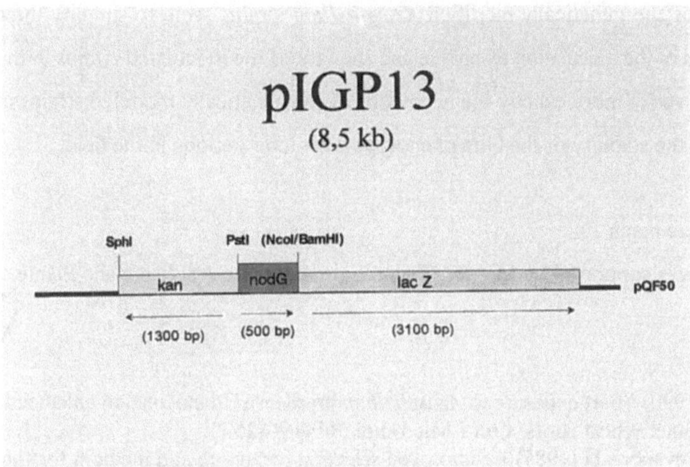

Figure 5. Physical map of the plasmid pIGP 13.

The transconjugants Sp7 (pIGP 13) and Cd (pIGP 13) containing the *nod*G-*lac*Z fusion were phenotypically blue on NB and NFB media containing 40 µg/ml X-gal, with or without antibiotics (Km, Ap). The presence of the plasmid pIGP 13 was confirmed by Southern blot analysis using as probe the plasmid pQF50, the transconjugants result to be stable after 72 hours growth on rich media without selection pressure.

The genetically modified strains of *Azospirillum* were used in a preliminary test, carried out in a confined greenhouse area, to check the validity of the system for monitoring the population dynamic in the rhizosphere and to study the colonization of the roots. Twenty-five maize plants were grown in 50-L pots and inoculated, at the stage of three leaves, with 50 ml of liquid culture of the modified strains or sterile media (control) and then irrigated to help the dispersion of the bacteria.

Enumeration of the bacteria was done at 40 and 65 days from the inoculation, using one gram of roots and adjacent soil or sterilized roots. The dilutions were sprayed on rich (NB) and minimal (NFB) media with and without antibiotic (Km 50 mg/l, Ap 50 mg/l) and with X-gal. Plates were controlled after 2, 4 and 7 days at 30°C. The number of bacteria, showed as blue colony, of both strains was comparable and ranging from 2.5×10^4 to 29.5×10^4. The variation inside the same pot and between pots was of the same amount and no blue colonies were found in the control pots. The blue colonies were recognized as the inoculated strains doing experiment of colony blot hybridization using as probe the plasmid pQF50. The blue colonies are easily visible even in presence of fungi after 7 days at 30 °C.

The use of the genetically modified *Azospirillum* strains seem to simplify the control of the effectiveness of the inoculation technique and the fate of the inoculated strains in the rhizosphere. We need to verify more exactly the behaviour of the genetically modified strains in the field and particularly the stability of the blue phenotype over long periods in the field.

Acknowledgements
This work was supported by M.A.F., Tecnologie Avanzate Applicate alle Piante.

References

Bashan Y (1990) Short exposure to *Azospirillum* brasilense Cd inoculation enhanced proton efflux of intact wheat roots. Can J Microbiol 36:419-425.

Bashan Y, Levanony H (1985) An improved selection technique and medium for the isolation and enumeration of *Azospirillum brasilense*. Can J Microbiol 31:947-952.

Bertolini M, Bressan M, Snidaro M, Fogher C, Marocco A (1990) Inoculo con *Azospirillum* e concimazione azotata in mais. Inform Agr 46:51-53.

Delledonne M, Porcari R, Fogher C. Nucleotide sequence of the *nod*G gene of *Azospirillum brasilense*. Nucl Acids Res 18:6435.

De-Polli H, Bohlool BB, Döbereiner J (1980) Serological differentiation of *Azospirillum* species belonging to different host-plant specificity groups. Arch Microbiol 126:217-222.

Ela SW, Anderson HA, Brill WJ (1982). Screening and selection of maize to enhance associative bacterial nitrogen fixation. Plant Physiol 70:1564-1567.

Farinha MA, Kropinski AM (1990) Construction of broad-host-range plasmid vectors for easy visible selection and analysis of promoters. J Bacteriol 172:3496-3499.

Favilli F, Balloni W, Cappellini A, Granchi L, Savoini G (1987) Esperienze pluriennali di batterizzazione in campo con *Azospirillum* spp. di culture cerealicole. Ann Microbiol 37:169-181.

Fogher C, Delledonne M, Frigerio L, Lorenzoni C (1993) Breeding for associative nitrogen fixation in maize. Eucarpia Proc XVIth Conf Maize Sect p. 198-202.

Fogher C, Dusha I, Barbot P, Elmerich C (1985) Heterologous hybridization of *Azospirillum* DNA to *Rhizobium nod* and *fix* genes. FEMS Microbiol Lett 30:245-249.

Kapulnik Y, Okon Y (1983) Benefits of Azospirillum inoculation on wheat: effects on root development, mineral uptake, nitrogen fixation and crop yield. Experientia Supp 48:163-170.

Kapulnik Y, Okon Y, Kigel J, Nur I, Henis Y (1981) Effects of temperature, nitrogen fertilization and plant age on nitrogen fixation by *Setaria italica* inoculated with *Azospirillum brasilense* (strain cd). Plant Physiol 68:340-343.

Rennie RJ (1980) Dinitrogen-fixing bacteria: computer-assisted identification of soil isolates. Can J Microbiol 26:1275-1283.

Sambrook J, Fritsch EF, Maniatis T (1989) Molecular cloning. A laboratory manual. Cold Spring Harbor Lab Press New York.

Sarig S, Blum A, Okon Y (1988) Improvement of the water status and yield of field-grown grain sorghum (*Sorghum bicolor*) by inoculation with *Azospirillum brasilense*. J Agric Sci 110:271-277.

Schank SC, Smith RL, Weiser GC, Zuberer DA, Bouton JH, Quesenberry KH, Tyler ME, Milam JR, Littell RC (1979) Fluorescent antibody technique to identify *Azospirillum brasilense* associated with roots of grasses. Soil Biol Biochem 11:287-295.

Von Bülow JFW, Döbereiner J (1975) Potential for nitrogen fixation in maize genotypes in Brazil. Proc Nat Acad Sci 72:2389-2393.

Growth stimulation in graminees, legumes, and crucifers by *Pseudomonas*-Bacteria

G. Höflich and W. Wiehe
Zentrum für Agrarlandschafts- und Landnutzungsforschung
Institut für Ökophysiologie der Primärproduktion
Eberswalder Str. 84
D-15374 Müncheberg
Germany

Abstract

The fluorescent Pseudomonas strain PsIA12, selected out of 105 bacterial isolates from the wheat rhizosphere, promoted the development of young plants of wheat, maize, alfalfa, pea, rapeseed, oil raddish and mustard in greenhouse experiments.Seed inoculation with this strain also repeatedly stimulated the plant growth in field experiments on loamy sand and sandy loam.The Pseudomonas strain PsIA12 produces phytohormones (cytocinin,auxin)and siderophores, is able to solve Ca-phosphate and to metabolize pectin and cellulose. This strain showes antagonistic activities against soil borne plant root pathogenes and stimulates the nodulation of alfalfa and pea roots.The long-term survival and establishment of this strain in the rhizosphere of various crops was studied by an antibiotic resistant mutant.
Electron microscopic investigations demonstrated the colonization not only of the rhizoplane but also of the root cap mucigel, the intercellululare spaces of the living cortex and the ruptures of the main root cortex by the emerging lateral roots.

Introduction

Rhizosphere microorganisms are able to promote plant growth by mobilization of nutrients and production of phytohormones or by phytosanitary effects. Due to a high genetical variability of microorganisms it is possible to isolate plant-growth-promoting microorganisms from the rhizosphere. Positive results are reported from many countries with symbiotic *Rhizobium* bacteria on legumes. In tropical climates it was possible to promote the growth of gramineae by inoculation with *Azospirillum* spp. and *Azotobacter chroococcum*.The aim of our research was to prove, that it is also possible to promote the growth of different crops by associative bacteria under temperate climatic conditions. The following questions were investigated:

- From which bacterial genera is it possible to select associative plant-growth-promoting strains?
- Are reproducible plant-growth-promoting effects restricted to special crops or soils?
- Which are the main physiological characteristics and which colonization mode of introduced bacteria causes the plant-growth-promotion?

NATO ASI Series, Vol. G 37
Azospirillum VI and Related Microorganisms
Edited by I. Fendrik et al.

Results and discussion

105 bacterial strains of different genera were isolated from the rhizosphere of winter wheat. Strains of *Pseudomonas, Xanthomonas, Azotobacter, Flavobacterium* and *Bacillus* promoted the development of young wheat plants in first, orientating greenhouse experiments (Tab.1). Isolates of *Agrobacterium, Streptomyces* and coyneform bacteria showed no positive effects. With the exception of *Azotobacter* and *Bacillus,* all investigated genera included strains with deleterious effects on the developing wheat plants.

Table 1. Percentage of growth stimulating bacterial strains among different bacterial genera and groups; pot experiment with wheat.

bacterial group	effective strains (%) single stimulation(1)		reproducible stimulation(1)	
	shoot	root	shoot	root
Pseudomonas	18	45	8	12
Xanthomonas	12	0	0	0
Azotobacter	100	100	0	0
Agrobacterium	0	0	0	0
Flavobacterium	23	32	0	0
coryneforme bacteria	0	14	0	0
Streptomyces	0	10	0	0
Bacillus	100	100	0	0

(1) growth stimulation is indicated by extra dry matter yield of wheat > 5 % compared to non-inoculated plants

Reproducible growth stimulation on wheat were only found with four *Pseudomonas* strains. Especially the fluorescent *Pseudomonas* strain PsIA12 promoted reproducibly shoot and root growth of wheat, rape, oil raddish, mustard, pea, lupin and alfalfa in greenhouse experiments (Tab 2.).

Table 2. Influence of *Pseudomonas fluorescens* strain PsIA12 on the growth of different crops (greenhouse experiments with loamy sand; non-inoculated control = 100%)

crop	No. of experiments	dry matter[1] shoot	root	root-length[2]	No. of laterals[2]
winter wheat	6	109+	134+	-	-
rape	3	105+	102	90	167+
oil radish	3	118+	104	119	233+
mustard	2	110+	108	107	-
pea	3	112+	110	121+	129
lupin	2	121+	112	160+	218+
alfalfa	3	117+	106	117	113

[1] 4 weeks old plants
[2] 2 weeks old plants
\+ significant at $p<0.05$, LSD test
\- not determined

Seed inoculation with this strain also repeatedly stimulated the plant growth of maize, oil radish, mustard, *Phacelia* and pea in field experiments on loamy sand (Tab.3).

Table 3. Growth promotion in different crops by *Pseudomonas fluorescens* strain PsIA12; field experiments on loamy sand (1985-1993).

crop	No. of experiments	shoot dry matter control $t \cdot ha^{-1}$	extra yield $t \cdot ha^{-1}$	%
maize	4	15.89	1.01+	6
rape	2	1.95	0.38	19
oil radish	2	2.25	0.40+	18
mustard	2	2.33	0.74+	30
Phacelia	1	3.00	1.12+	37
pea	1	3.36	0.92+	27

\+ significant at $p<0.05$, LSD test

Co-inoculation with strain PsIA12 and *Rhizobium* bacteria improved the effect of a single *Rhizobium* inoculation in alfalfa, broad bean and pea on loamy sand, sandy loam and chernozem (Höflich *et al.* 1992). The degree of plant-growth-promotion is influenced by soil and weather factors and by the crop cultivars (Höflich 1993).

Physiological investigations of the metabolism of the *Pseudomonas* strain PsIA12 showed no nitrogenase activity, but the production of phytohormones (cytokinin, auxin) and siderophores. It is able to solu-bilize Ca-phoshate and to metabolize pectate and cellulose. This strain shows *in vitro* antagonistic activities against soil borne plant root pathogenes (*Fusarium* spp., *Rhizoctonia solani, Helminthesporum sativum* and *Geaumannomyces graminis*).

The nodulation of alfalfa and pea roots and the colonization with AM fungi was stimulated; the infection with root pathogens was reduced in inoculated young pea plants (Höflich *et al.* 1992). The inoculation also stimulated root length and branching (Tab. 2).

The survival of an antibiotic-resistant mutant (120ppm rifampicin) of strain PsIA12 was investigated in greenhouse experiments. PsIA12 was able to survive in the rhizosphere of wheat, maize, rape, oil raddish, pea, and alfalfa (Tab. 4).

Table 4. Rhizosphere colonization (root segments 0-3cm) in different crops by *Pseudomonas fluorescens* strain PsIA12 (inoculation: 10^5-10^6 cfu · seed^{-1}); greenhouse experiment with loamy sand.

crop	rhizosphere colonization (10^3 cfu·g root^{-1}) weeks after inoculation				
	1	2	4	6	8
rape	808	64	86	109	10
oil radish	560	110	35	99	8
pea	122	333	212	2	2
alfalfa	535	250	265	11	1
wheat	1080	400	1380	114	26
maize	-	195	17	-	-

- not determined

In field experiments with pea, lupin, and maize the strain was recorded in the rhizosphere of these plants during the whole vegetation time until seed maturity (Wiehe and Höflich 1994).

Electron microscopic investigations demonstrated the colonization not only of the rhizoplane but also of the mucigel of the root cap, the intercellular spaces of the living cortex and the sites of the emerging lateral root in different plants (Wiehe and Höflich 1994).

The cause of the growth stimulation by strain PsIA12 is obviously not N_2-fixation, but possibly a stimulated enlargement of the nutrient absorbing root area by microbially produced phytohormones, an improved nutrient uptake, a protection against soil-borne root pathogens or a better stress-resistance.

The importance of microbial physiological characteristics may vary in dependence on ecological factors. The production of phytoeffective microbial metabolites depends on plant assimilation. Therefore, attention is payed to the principles of plant production, like crop rotation, seed bed preparation, control of soil-pH. Basic fertilizing can stimulate the development of young plants and therefore enhance the effectivity of the bacterial inoculation. Plant damage caused by extreme growth conditions mostly can not be compensated by inoculation.

Conclusions

- Fluorescent pseudomonads play a prominent role in plant- growth-promotion under temperate climatic conditions. It is possible to select effective strain (eg strain PsIA12).
- The importance of individual physiological principles of plant-growth-promotion (N2-fixation, P-mobilization, antagonism, phytohormone production) may be varied by multiple interactions between inoculated bacteria, crops and ecological factors.
- Growth stimulation by the *Pseudomonas* strain PsIA12 with gramineae (wheat, maize), crucifers (rape, mustard, oil raddish), and legumes (pea, broad been, alfalfa) may be caused by
  * different phytoeffective physiological characteristics
  * mode of rhizosphere colonization, eg growing root parts (root cap, root hair zone, cortex intercellular spaces, ruptures of emerging laterals)and production of microbial metabolites directly on the sites of their demand
  * survival of inocula in the rhizosphere of different crops during vegetation time
  * stimulation of other phytoeffective microorganisms (rhizobia, AM fungi)
- the level of plant-specific differentiated growth-stimulation depends on soil, weather factors, and infestation with pathogenes

References

Höflich G (1993) Effect of variety and environmental factors on the phyto-effectivity of bacterial inoculations in pea. Zentralbl Mikrobiol 148: 315-324.

Höflich G, Glante F, Liste HH, Weise I, Ruppel S, Scholz-Seidel C (1992.): Phytoeffective combination effects of symbiotic and associative microorganisms on legumes. Symbiosis 14: 427-438.

Wiehe W, Höflich H (1994) Survival of plant-growth promoting rhizosphere bacteria in the rhizosphere of inoculated and non-inoculated plants under field conditions. in preparation.

Wiehe W, Hecht-Buchholz Ch, Höflich G (1994) Electron microscopic investigations on root colonization of *Lupinus albus* and *Pisum sativum* with two associative plant growth promoting rhizobacteria, *Pseudomonas fluorescens* and *Rhizobium leguminosarum* bv.*trifolii*. Sybiosis 17: xx-yy.

Experimental System for Creating New Type of N_2-Fixing Plants

Sz. S. Varga, É. Preininger, P. Korányi, I. Gyurján
Department of Plant Anatomy,
Eötvös Loránd University,
1088 Budapest, Puskin utca 11-13.,
Hungary

Abstract

Artificial associations were established between nitrogen-fixing *Azomonas* or *Azotobacter* cells and plant tissues based on the induced carbon and energy dependency of diazotrophs on plant metabolic activity. Plant regeneration was achieved from mixed callus-bacterium associations. Light and electron microscopy was used to show the location of bacteria in intercellular spaces of callus and regenerated plant tissues. The nitrogen-fixing ability of the partnership was proved by acetylene reduction assay.

Introduction

Tissue culture techniques have been developed to study natural associations and the possibility of creating new associations between plant tissue cultures and diazotrophs (Carlson and Chaleff, 1974; Child, 1975; Gusev et al., 1986; Scowcroft and Gibson, 1975; Vasil et al., 1979; You and Zhou, 1989). The establishment of artificial symbioses between N_2-fixing prokaryotes and plants is an attainable solution to the transfer of diazotrophy to new hosts. Associations were established between nitrogen-fixing *Azotobacter* cells and carrot tissues based on the induced carbon and energy dependency of diazotrophs on plant metabolic activity (Korányi et al., 1993; Varga et al, 1994). Plants regenerated from callus also contained bacteria in the intercellular spaces (Varga et al, 1994). It was defined an experimental system to investigate the extension nitrogen fixing capability to non-leguminous crop species by using aerob diazotrophs (Varga, 1993).

NATO ASI Series, Vol. G 37
Azospirillum VI and Related Microorganisms
Edited by I. Fendrik et al.

Materials and methods

Azotobacter vinelandii (CCM 289 and DSM 87), *A. paspali* (DSM 2283), *A. beijerinckii* (DSM 378), *A. zettuovii* (CRS-H6), *Azomonas insignis* (CRS-HK5), *A. agilis* (DSM 89), *A. macrocytogenes* (DSM 721) and carrot cells were sustained as described in Varga et al. (1994). Mature embryo derived wheat calluses and sugar-cane calluses from surface sterilized leaves were obtained and grown on MS media (Murashige and Skoog, 1962) supplemented with 3 mgl^{-1} 2,4-dichlorophenoxy acetic acid (2,4-D). Plant regeneration was obtained by somatic embryogenesis in the case of carrot, wheat and sugar-cane cultures on hormone free MS media. *Nicotiana tabacum* calluses were grown on a MS medium supplemented with 1 mgl^{-1} α-naphthaleneacetic acid (NAA), 0.5 mgl^{-1} kinetin. To induce organogenesis the tobacco calluses were transferred to MS medium with 0.5 mgl^{-1} indole-3yl-acetic acid (IAA), 1 mgl^{-1} kinetin. *Populus tremula* shoot cultures were grown aseptically on MS medium with 1 mgl^{-1} IAA and 6-benzylaminopurine (BAP). Callus cultures were derived and maintained on MS media supplemented with 0.5 mgl^{-1} 2,4-D. Plant regeneration were obtained by using hormones of shoot-culture media. Tomato calluses were obtained on MS media contained 5 mgl^{-1} 2,4-D and 0.1 mgl^{-1} kinetin. Organogenesis in the case of tomato calluses were obtained on MS medium with 2 mgl^{-1} NAA and 1 mgl^{-1} BAP. To generate the diazotroph-plant tissue association, the undifferentiated plant cells were washed in a sucrose-free medium and inoculated with cells of bacteria. The mixed cultures were incubated on solidified media containing 3% carbon source cannot utilizable by the bacterial partner, but utilizable by plant cells (lactose or maltose or galactose) as the sole energy source. The "low-sucrose media" contain 0.1-0.5% sucrose in addition to main carbon source to increase the bacterial growth activity. Calluses associated with bacteria could be maintained by subcultivation every 1.5-3 months. Plant materials were incubated at 25±2 °C under a 14-h photoperiod (Tungsram, Tungsraflor-36, 250-300 μmol m^{-2} s^{-1}). The pH of all media was 7.2 before autoclaving. After regeneration, the plantlets were planted in soil in a green-house. Fixation and electron microscopy of plant materials were similar as described earlier (Varga et al., 1994). The bacterial numbers were estimated by plating tenfold serial dilutions. Nitrogenase-activity was estimated by the acetylene-reduction technique (Hardy et al., 1968).

Results

Several independent lines of evidence demonstrated that diazotrophs depended on metabolites derived from plant tissues for growth in the absence of a carbon source that could be utilized by the bacteria. During the mutual cultivation of plant cells and bacteria belonging to *Azotobacteraceae* both not to intensive callus growth and multiplication of the bacteria were observed on lactose containing media. However, it was surprising that in the case of poplar the cultures grew intensively. On media containing maltose most of the plant culture grown similarly to cultures grown on saccharose containing media. On "low-sucrose media", the bacteria were usually observed on the surface of the medium and the calluses, and did not overgrow the plant tissues. The number of bacteria was estimated in the mixed calluses growing on N-containing media and higher numbers of bacterial cells were found in cultures living on the "low-sucrose medium" than on the medium containing carbon source for plant cells only. The higher number of bacteria in the regenerating region of the mixed calluses could result more efficient diazotrophs incorporation into regenerated plant tissues. Electron micrographs of mixed calluses showed the presence of bacterial cells in the intercellular regions of the growing tissues. The forced association in the regenerated plants were shown by estimating the number of bacteria (10^3-10^7 bactria/g fresh weight) in the surface-sterilized leaves and by the acetylene-reduction assay (Tab. 1).

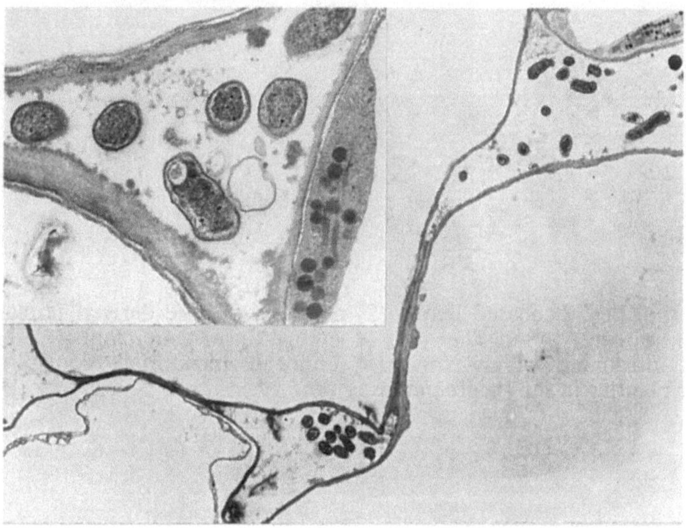

Fig. 1. Transmission electron micrographs of *A. insignis* cells among the petiole cells of *Daucus* plant regenerated on maltose containing media.

Improved growth (Fig. 3) by the associations compared with controls showed that the bacteria contributed significantly to plant development. Whole plant, roots, leaves and surface sterilized leaves all showed nitrogen fixing activity (Tab. 1). The rate of nitrogenase activity measured may be dependent upon the used bacterial and plant species (and strains), upon the number of associated bacteria, upon the plant products (e.g. carbon sources) utilizable by microsymbiont, temperature etc. The mixed culture derived plant containing no bacteria did not show acetylene—reduction activity similarly to the control plants. Light and electron micrographs of leaves of regenerated plants showed the presence of bacterial cells in the intercellular regions of the inner tissues (Figs. 1,2). The bacteria were often observed in the intercellular spaces of cortical parenchyma cells and were rarely in the xylem or pith region. Bacteria did not invade living cells.

Fig. 2, 3. In the TTC-incubated leaves (2) of mixed culture derived poplar plant *A. vinelandii* cells appeared among the cells of lamina. Sugar-cane plants (3) regenerated from calluses inoculated with *A. agilis* (9) and the appropriate control plants (C) 2 months after planting in soil (in greenhouse).

Discussion

The associations between cells of plants and bacteria belonging to *Azotobacteraceae* was established as a result of by the growing dependency of the bacterial cells on metabolic products of the plant cells in cultures. The preference of mutual growth for stabilizing the callus-bacteria associations could be obtained during cultivation on media with reduced

Tab. 1. Nitrogenase activity of the mixed bacteria-plant cells derived plants. Plants harboring no bacteria did not show acetylene reduction. Plants were grown in soil (in pot) at least for 2 months.
* after surface sterilization

Materials	Aceylene reduction (nmol/g FW/day)
Daucus X	
A. zettuovii	
whole plant	172
whole plant	120
leaves	61
A. paspali	
whole plant	240
leaves*	600
leaves	57
leaves*	17
A. vinelandii	
whole plant	168
leaf*	760
leaf*	530
leaf	40
A. agilis	
whole plant	96
leaf	278
leaf	87
leaf*	56
leaf	58
A. insignis	
whole plant	86
whole plant	90
leaf*	110
leaf*	65
leaf	240
leaf	350
root	100
A. macrocytogenes	
leaf	45
leaf	26
Tobacco X	
A. paspali	
leaves	800
A. vinelandii	
leaves	7330
leaves	1170
leaves	250
leaf*	7250
leaf*	1130
A. insignis	
whole plant	21
sugar-cane X	
A. agilis	
leaf	1240
leaf	379

Materials	Aceylene reduction (nmol/g FW/day)
sugar-cane X	
leaf	320
root	1030
A. insignis	
leaf	47
leaf	26
A. vinelandii	
leaf	54
leaf	3120
whole plant	778
whole plant	112
whole plant	184
Populus tremula X	
A. beijerinckii	
leaf	490
leaf	153
A. macrocytogenes	
leaf	150
leaf	25
A. vinelandii	
leaf	83
Tomato X	
A. beijerinckii	
leaf	33900
leaf	286
root	15900
root	154
A. agilis	
root	180
leaf	484
leaf	341
leaf	593
leaf	3410
A. vinelandii	
leaf	47
leaf	57
leaf	472
root	133
root	57
A. insignis	
leaf	29
leaf	406
root	133
root	2870
root	6000
wheat X	
A. insignis	
leaf	1630

bound-nitrogen content, or on media containing fixed nitrogen source, which can be utilised only by the plant cells. The usefulness of this co-culture system was shown by plant regeneration, in contrast to the previous associations achieved between *Azotobacter* and carrot calluses (Carlson and Chaleff, 1974) and by the fact that this intercellular association could be maintained in whole plants. The observed endophytic diazotrophs population indicates the possibility of multiplication of the bacteria inside the regenerated plants. The N_2-fixing activity in roots and leaves suggested that this system is a successful nitrogen-fixing association. Presumably, plant cells received reduced N [e.g. NH_3, amino acids, vitamins (Gonzalez-Lopez et al., 1983;), growth substances(Brown and Burlingham, 1968)] from endophytic bacterial cells which, in turn, depended upon the plant cells to satisfy their requirement for carbon and energy. These results suggest that N_2-fixing ability may be extended to non-leguminous plants by non invasive, aerobic diazotrophic bacteria.

REFERENCES

Brown, M. E. and Burlingham, S. K. (1968) Production of plant growth substances by Azotobacter chroococcum. J. Gen. Microbiol. 24: 369-377.

Carlson, P. S. and Chaleff, R. S. (1974) Forced association between higher plant and bacterial cells in vitro. Nature 252: 393-395.

Child, J. J. (1975) Nitrogen fixation by a Rhizobium sp. in association with non-leguminous plant cell cultures. Nature 253: 350-351.

Gonzalez-Lopez, J., Salmeron, V., Moreno, J., and Ramos-Cormenzana, A. (1983) Amino acids and vitamins produced by Azotobacter vinelandii ATCC 121837 in chemically-defined media and dialysed soil media. Soil Biol. Biochem. 15: 711-713.

Gusev, M. V., Korzhenevskaya, T. G., Pyvovarova, L. V., Baulina, O. I., and Butenko, R. G. (1986) Introduction of a nitrogen-fixing cyanobacterium into tobacco shoot regenerates. Planta 167: 1-8.

Hardy, R. W. F., Holsten, R. D., Jackson, E. K., and Burns, R. C. (1968) The acetylene-ethylene assay for N_2-fixation - laboratory and field evaluation. Plant Physiol. 43: 1185-1207.

Korányi P.; Varga Sz. S.; Preininger É.; Gyurján I. (1993) Symbiotic Integration of Diazotrophic Bacteria with Plant Cells in Artificial Cytobioses. *In*: Endocytobiology V. 465-472. Tübingen University Press, Tübingen

Murashige, T., and Skoog, F. (1962) A revised medium for rapid growth and bio assays with tobacco tissue cultures. Physiol. Plant. 15: 473-497.

Scowcroft, W. R. and Gibson, A. H. (1975) Nitrogen fixation by Rhizobium associated with tobacco and cowpea cell cultures. Nature 253: 513-523.

Varga Sz. S. (1993) Process for the development of novel type of plants with nitrogen-fixing capacity also in their leaves.; PCT patent: PCT/HU 93/00020,

Varga Sz. S.; Korányi P.; Preininger É.; Gyurján I. (1994) Artificial associations between *Daucus* and nitrogen-fixing Azotobacter cells in vitro. Physiol. Plant. 90: 786-790.

Vasil, V., Vasil, I. K., Zuberer, D. A., Hubbel, D. H. (1979) The biology of Azospirillum-sugarcane association. I. Establishment of the association. Z. Pflanzenphysiol. 95: 141-147.

You, C., and Zhou, F. (1988) Non-nodular endorhizospheric nitrogen fixation in wetland rice. Can. J. Microbiol. 35: 403-408.

Survival of Associative Nitrogen Fixers in Rhizoplane as a Criterion for Estimation of Their Effect on Inoculated Plants

Belimov A.A., Kunakova A.M., Alekseyeva E.G.,
Martsinkevitch O.N., Kojemyakov A.P.
Research Institute for Agricultural Microbiology
Podbelsky Sh.3, Pushkin-8
Saint-Petersburg, 189620
Russia

Abstract

The population dynamics of introduced associative nitrogen fixers belonging to various genera (*Azospirillum, Arthrobacter, Flavobacterium, Agrobacterium*) in the rhizoplane of homologous and nonhomologous plants with respect to the origin of their initial isolation (rice, wheat, ryegrass, oat, barley, Pe-tsai cabbage) was studied in pot experiments. The overall pictures of survival of the studied strains on roots of plants were similar in appearance, even though the population size and dynamics of the introduced bacteria depended on the plant genotype. No correlation was established between survival of bacteria and their effect on plant development. It was concluded that the effect of inoculation can not predicted using the data on the population size of the introduced bacteria.

Introduction

It has been known that interactions of *Azospirillum* with inoculated plants can vary significantly depending on genotypes of plants and bacteria (Bashan and Levanony, 1990; Bouton *et al.*, 1979; Kesava Rao *et al.*, 1990). In some instances the maximum effect of inoculation was produced by homologous strains, i.e. those which have been initially isolated from the roots of plant species used for subsequent inoculation (Boddey *et al.*, 1986; Reynders and Vlassak, 1982; Wani *et al.*, 1985). On the other hand, the same strains are able to form successful associations with a wide variety of plants (Bashan *et al.*, 1989; Vassyuk, 1989). Investigation of the survival of *Azospirillum* on roots showed that in most instances there was no correlation between the number of introduced bacteria and their effect on plants

NATO ASI Series, Vol. G 37
Azospirillum VI and Related Microorganisms
Edited by I. Fendrik et al.

(Bashan *et al.*, 1989; Boddey *et al.*, 1986; Kesava Rao *et al.*, 1985). The aim of the present paper is to compare some of the diazotrophs belonging to various genera in relation to their ability to colonize the rhizoplane of homologous and nonhomologous plants and to improve the development of the inoculated plants.

Materials and methods

The names of the studied strains and the origin of their initial isolation: *Azospirillum lipoferum* 137 (rhizosphere of rice cv.Dubovsky-129 grown in grey dessert soil), *Azospirillum lipoferum* 407 (cultivated flower garden chernozem), *Azospirillum brasilense* 406 (virgin grey soil), *Azospirillum brasilense* 504 (rhizosphere of oat cv.Narumsky-943 grown in sandy grey soil), *Arthrobacter mysorens* 7 (rhizosphere of annual ryegrass grown in chernozem), *Agrobacterium radiobacter* 10 (rhizosphere of wheat grown in chernozem), *Flavobacterium* sp.L30 (rhizosphere of rice cv.Krasnodarsky-424 grown in derno-podzolic soil). The following plants were used: lowland rice cv. Dubovsky-129 and cv. Krasnodarsky-424; annual ryegrass cv. Izorsky; Pe-tsai cabbage cv. Khibinskaya; spring barley cv. Temp; oat cv. Narumsky-943; spring wheat cv. Atlant; winter wheat purchased from the Agricultural Chemistry Station of Zibo (China).

In pot experiments the plants (6 plants per pot) were grown in pots (5 pots for harvest measurement; 3 pots for counting of bacteria) containing 2.5 kg non-sterile derno-podzolic soil (humus 0.9-1.5%; total N 0.096-0.126%; available P 5-10 mg P_2O_5/100 g; available K 7-12 mg K_2O/ 100g; ph=6.0-6.2). Soil was additionally fertilized with (mg/kg): $(NH_4)_2SO_4$ - 68; K_2HPO_4 - 250; KCL - 125, $MgCl_2$ - 60; H_3BO_3 - 3; $MnSO_4$ - 3; $ZnSO_4$ - 3; Na_2MoO_4 - 1.5. Inoculum cultures of bacteria were grown in a liquid DN medium (Rodriguez-Caceres, 1982) supplemented with 1 g/l yeast extract and sucrose and cultivated using a shaker at 28°C for 3 days. Inoculations were carried out in the amount of 10^7-10^8 cells per seed. The soil moisture was maintained at 60% WHC (rice was submerged 28 days after planting). Nitrogen content in plants was determined by the Kjeldahl method using the "Kjeltec-auto" analyser. The spontaneous antibiotic-resistant mutants of the strains were used: *Azospirillum* strains were resistant to 40 mg/l rifampin; the other strains were resistant to 40 mg/l rifampin and 500 mg/l streptomycin simultaneously. The roots were washed and homogenized in sterile water. The bacteria in the homogenate were assumed to be the bacteria located in the rhizoplane. The homogenates were serially diluted and plated on solid DN medium (5 replications) amended with proper amount of antibiotics.

Results and discussion

The strain *A. lipoferum* 137 possessed lower viability in the rhizoplane of homologous rice cv. Dubovsky-129 than in that of cv. Krasnodarsky-424 and Pe-tsai cabbage during the whole period of the experiment (Fig. 1,A). The high survival rate of *A. mysorens* 7 was observed in the rhizoplane of homologous plant genus (ryegrass). Nevertheless, at an early stage of plant development the maximum number of this strain was detected on the roots of winter wheat from the ACS of Zibo and rice cv. Krasnodarsky-424 (Fig. 1,B). The strain *A. radiobacter* 10 was able to colonize the root surface of both wheat cultivars very actively (Fig. 1,D). However, the population of *A. radiobacter* 10 introduced in the rhizoplane of wheat had no significant advantage over that introduced in the rhizoplane of rice cv. Krasnodarsky-424 as well as in the rhizoplane of ryegrass and Pe-tsai cabbage at the beginning or end of the experiment respectively. The strain *Flavobacterium* sp. L30 colonized the roots of homologous rice cv. Krasnodarsky-424 more actively as compared with cv. Dubovsky-129 (Fig. 1,C). There is no escape from the conclusion that the interaction between *Flavobacterium* sp. L30 and homologous cultivar has higher degree of relatioship, because: (a) the number of remaining three strains in the rhizoplane of cv. Krasnodarsky-424 was also higher; (b) the population of *Flavobacterium* sp. L30 on the roots of rice plants decreased most rapidly and disappeared by 45th day after inoculation.

The reliable effect (by 15-20%) on dry weight of both rice cultivars was achieved by inoculation with *Flavobacterium* sp. L30 only (Table 1). It was a result of an increase in the weight of shoots, other than roots (data are not shown). The dry weight and nitrogen accumulation in ryegrass plants inoculated with *Flavobacterium* sp L30 were higher by 16-19% as compared with the control treatment but differences were not significant. Three of the studied strains increased significantly the weight of Pe-tsai cabbage and the maximum positive yield response as well as the nitrogen accumulation were observed when the plants

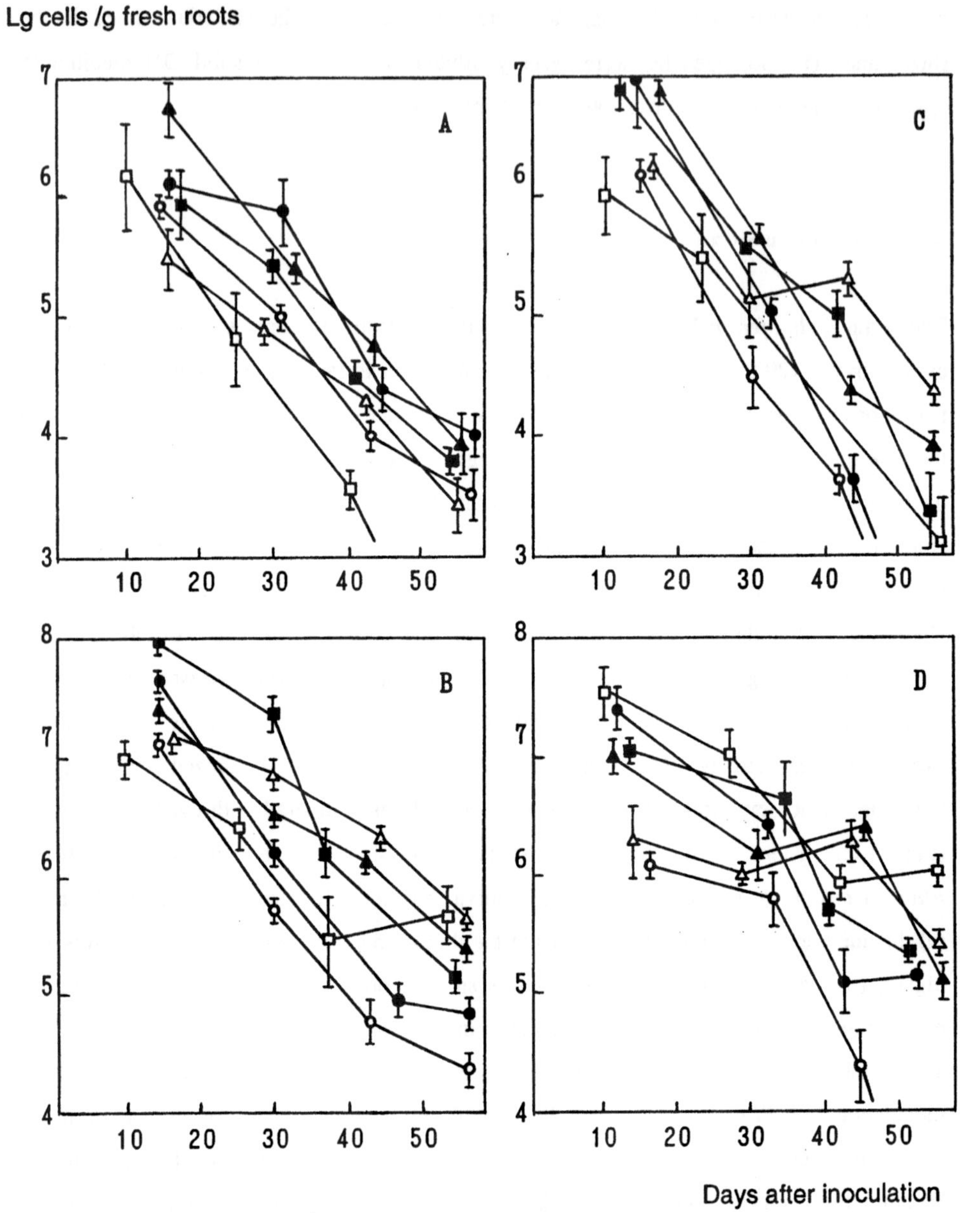

Fig.1. Survival of introduced bacteria in rhizoplane of inoculated plants. Strains A - *A. lipoferum* 137; B - *A. mysorens* 7; C - *Flavobacterium* sp. L30; D - *A. radiobacter* 10. Plants: ○ rice cv. Dubovsky-129; ● rice cv. Krasnodarsky-424; Δ ryegrass; □ wheat cv. Atlant; ■ wheat from ACS of Zibo; ▲ Pe-tsai cabbage. In this and Figure 2, vertical bars represent the confidence intervals at P=0.05.

Table 1. Effect of nitrogen fixing bacteria on yield (g/pot) and nitrogen accumulation (mg/pot) in inoculated plants.

Treatments	Rice v. Dubovsky-129		Rice cv: Krasnodarsky-424	
	Dry weight of plants	Total N in plants	Dry weight of plants	Total N in plants
Uninoculated control	4.8	132	3.4	109
A. lipoferum 137	4.4	138	3.7	106
A.mysorens 7	4.8	154	3.9	126
A. radiobacter 10	4.9	139	3.7	123
Flavobac. sp.L30	5.5*	146	4.1*	119
$LSD_{0.05}$	0.6	23	0.6	23

	Ryegrass		Pe-tsai cabbage		Wheat cv Atlant	
	Dry weight of plants	Total N in plants	Dry weight of plants	Total N in plants	Grain yield	Total N in grain
Uninoculated control	6.4	91	5.3	216	4.1	144
A. lipoferum 137	6.5	102	5.6	235	4.9*	170*
A.mysorens 7	6.7	94	6.3*	248	4.5	154
A. radiobacter 10	7.2	100	6.6*	244	4.6	161
Flavobac. sp.L30	7.4	108	6.7*	275*	4.8	167
$LSD_{0.05}$	1.1	18	0.9	45	0.75	25

	Oat		
	Grain yield	Dry weight of roots	Total N in plants
Uninoculated control	8.2	1.4	347
A. lipoferum 137	8.9	2.0*	372
A.mysorens 7	9.0	1.8*	379
A. radiobacter 10	9.0	1.9*	399*
Flavobac. sp.L30	8.8	1.5	383
$LSD_{0.05}$	0.83	0.37	40

The reliable influence on wheat cv. Atlant was obtained after inoculation with *A. lipoferum* 137 only. The inoculation with were inoculated with *Flavobacterium* sp. L30 increased the grain yield of wheat by 18%, however the difference was not significant because of wide scatter of the data. Unfortunately, the response of winter wheat from ACS of Zibo was not studied.

In the second experiment the survival of various *Azospirillum* strains in rhizoplane of oat and barley was studied. It was established that both strains belonging to species *A. lipoferum* colonized actively the rhizoplane of oat (Fig. 2,A). On the contrary, the strains *A. brasilense* were not capable of inhabiting the roots of oat for a long time despite the fact that one of them (*A. brasilense* 504) was initially isolated from the rhizosphere of the present oat cultivar. It was interesting to observe that *A. brasilense* 406 colonized very actively the root surface of barley during the whole period of plant vegetation (Fig. 2,B). In either case, *A. brasilense* 504 was the least effective root colonizer. The effect of inoculation of oat was most conspicuous in the increase of the root system (Table 1). Maximum accumulation of nitrogen in the oat plants was observed after inoculation with *A. brasilense* 406 which possessed low viability. The response of barley to inoculation with all of the studied strains was not significant (data are not shown).

There is some evidence that the survival of the introduced *Azospirillum* on the roots can vary significantly depending on the plant genotype (Baldani and Döbereiner, 1980) and the homologous strains are the most effective root colonizers (Pereira *et al.*, 1988). In contrast, no specificity in *Azospirillum* survival rate on the roots of various plants was detected (Harris *et al.*,1989; Bashan *et al.*, 1989). In our experiment performed on *Azospirillum* strains having different origin of isolation the population dynamics depended significantly on the plant genotype. It is important that both strains belonging to *A. lipoferum* colonized the rhizoplane of oat more actively as compared with *A. brasilense*. Our results support the data of the previous experiments showing that these species prefer different plant genera (Bashan and Döbereiner, 1980) and *A. lipoferum* inhabits the root surface more actively than that of *A. brasilense* (Vassyuk *et al.*, 1989).

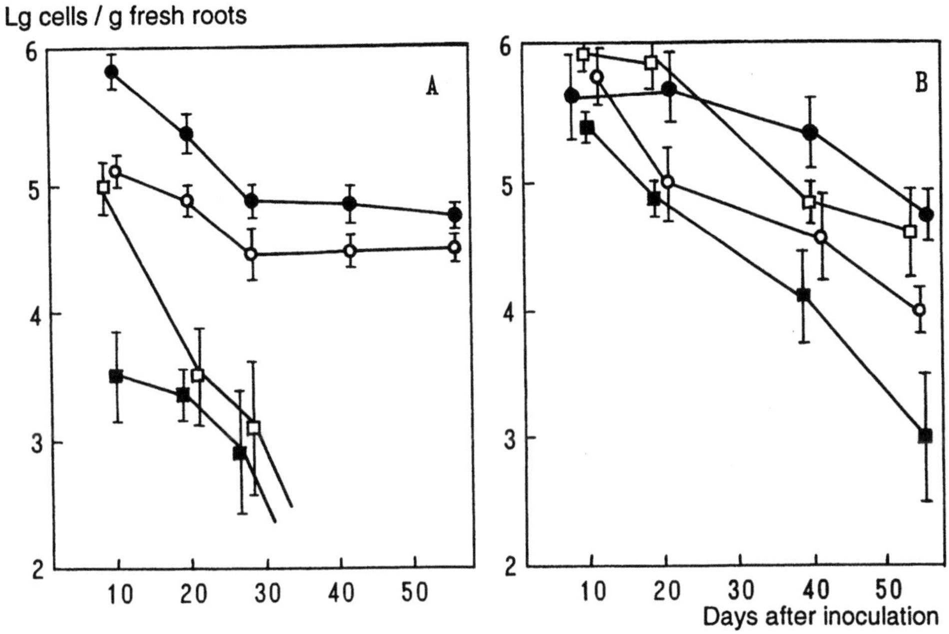

Fig. 2. Survival of *Azospirillum* strains in rhizoplane of oat cv. Narumsky-943 (A) and barley cv. Temp (B). Strains: ○ *A. lipoferum* 137; ● *A. lipoferum* 407; □ *A. brasilense* 406; ■ A. brasilense 504.

Although the population size of the studied diazotrophs varied depending on the genotype of the inoculated plants, the influence of plants on the bacteria survival was not specific. The following evidences may be given: (a) the difference in population size and dynamics of the introduced strains among the plant genera and cultivars in many instances was not significant; (b) the ratio of the population densities among the strains in the rhizoplane of various plants remained approximately identical, as an example, the number of *A. mysorens* 7 and *A. radiobacter* 10 was always higher as compared with the other bacteria; (c) the effect of homologous plant species on the survival of the introduced bacteria was not reliable. The response of the plants to inoculation with the studied diazotrophs varied. Nevertheless, the inoculation with Flavobacterium sp. L30 was as a rule most successful. No relationship between the effect of inoculation and the survival of introduced strains or the origin of their initial isolation were established. Thus, the survival rate of the present strains was determined essentially by the bacterial genotype, and yet the response of plants to inoculation was not be predicted using the data on the population size of the introduced bacteria.

References

Baldani VLD, Döbereiner J (1980) Host-plant specificity in the infection of cereals with *Azospirillum* spp. Soil Biol. Biochem. 12: 433-439

Bashan Y, Levanony H (1990) Current status of *Azospirillum* inoculation technology: *Azospirillum* as a challenge for agriculture. Can J Microbiol 36: 591-608

Bashan Y, Ream Y, Levanony H, Sade A (1989) Nonspecific responses in plant growth, yield and root colonization of noncereal crop plants to inoculation with *Azospirillum brasilense* cd. Can J Botany 67: 1317-1324

Boddey RM, Baldani VLD, Baldani JI, Döbereiner J (1986) Effect of inoculation of *Azospirillum* spp. on nitrogen accumulation by field-grown wheat. Plant Soil 95: 109-121

Bouton JH, Smith RL, Shank SC, Burton GW, Tyler ME, Littell RC, Galaher NR, Quesenberry KH (1979) Response of pearl millet inbrids and hybrids to inoculation with *Azospirillum brasilense*. Crop Sci. 19: 12-16

Harris JM, Lucas JA, Davey MR, Lethbridge G (1989) Establishment of *Azospirillum* inoculant in the rhizosphere of winter wheat. Soil Biol. Biochem. 21: 59-64

Kesava Rao PS, Arunachalam V, Tilak KVBR (1990) Genotype-dependent response to *Azospirillum* treatment in yield and nitrogenase activity in *Brassica juncea* L. Curr. Sci. 59:607-609

Pereira JAR, Cavalcante VA, Baldani JI, Döbereiner J (1988) Field inoculation of sorghum and rice with *Azospirillum* spp. and *Herbaspirillum seropedicae*. Plant Soil. 110: 269-274

Reynders L, Vlassak K (1982) Use of *A. brasilense* as biofertilizer in intensive wheat cropping. Plant Soil 66: 217-223

Rodriguez-Caceres E (1982) Improved medium for isolation of *Azospirillum* spp. Appl. Environ. Microbiol. 44: 990-991

Vassyuk LF (1989) Application of nitrogen fixers inhabiting roots of nonleguminous plants. Biological Nitrogen in Agriculture SU. Mishustin EN (eds) Nauka Moscow (in Russian)

Vassyuk LF, Borovkov AV, Khalchitsky AE, Ionkova SV, Chmeleva ZV (1989) Bacteria of the genus *Azospirillum* and their effect on productivity of nonhomologous plants. Microbiology 58: 642-648

Wani SP, Chandrapalaiah S, Dart PJ (1985) Response of pearl millet cultivars to inoculation with nitrogen-fixing bacteria. Expl. Agric. 21: 175-182

Adhesion of Rhizosphere Bacteria to Roots of Maize and Wheat

R.Pastorelli, A.Gori and F.Favilli
Dipartimento di Scienze e Tecnologie Alimentari e Microbiologiche,
Sez. di Microbiologia Applicata, University of Florence,
Piazzale delle Cascine, 27 - 50144 Firenze, Italy.

Abstract

Several bacterial strains were isolated from the rhizosphere of important agronomical crops and classified in the genera *Enterobacter*, *Pseudomonas*, *Azospirillum*, *Azotobacter* and *Bacillus* on the basis of their morphological and biochemical characters. Investigations on the efficiency of colonization and adhesion of these bacteria to roots of maize and wheat by scanning electron microscopy have been carried out.

Keywords: rhizosphere, adhesion, root surface, SEM.

Introduction

The ability of rhizobacteria to colonize roots is an essential requisite for the success of the competition of the bacterium in the rhizosphere. The rhizobacteria dispose of various properties that are important to promote competitiveness, one of these is the capacity to adhere to root surface. Living firmly anchored to root, bacteria can benefit more quickly from root exudates (Michiels *et al.*, 1989).

In this work we have investigated the capacity to adhere to maize and wheat roots of several bacterial strains isolated from the rhizosphere of important agronomical crops.

NATO ASI Series, Vol. G 37
Azospirillum VI and Related Microorganisms
Edited by I. Fendrik et al.

Materials and Methods

Bacterial strains used are reported in Table 1, the species assignation was performed following Holt (1984). Plant inoculation was carried out as described by Bazzicalupo *et al.* (1985). Plants were maintained in controlled environment growth cabinet (16^h light, 21°C; 8^h dark, 17°C) for three days.

For scanning electron microscopy, samples of washed inoculated-roots were fixed for 18^h at 4°C with 2.5% glutaraldehyde in 0.1 M sodium phosphate buffer (pH 7.0), washed three times, postfixed for 2^h at 4°C with 1% osmium tetroxide and dehydrated in a grade series of ethanol. The specimens were dried to critical point (Critical Point Dryer 020 Balzers) mounted on stubs, coated with gold and examined with a Philips 505 scanning electron microscope (Haahtela *et al.*, 1988, de Freitas and Germida, 1990).

Table 1. Bacterial strains used.

Strain	Source	Species
EnunAp2	barley	*Azospirillum lipoferum*
EnunAp3	barley	*Azospirillum lipoferum*
EnunEt5	wheat	*Enterobacter cloacae*
EnunEt7	wheat	*Enterobacter sakazakii*
EnunPs17	wheat	*Pseudomonas* sp.
EnunAz2	wheat	*Azotobacter chroococcum*
EnunB12	wheat	*Bacillus subtilis*
EnunB20	wheat	*Bacillus megatherium*

Results

SEM observations showed that all the bacterial strains tested were able to colonize and adhere to root surface of maize and wheat, although the colonization level varied along each root. Figure 1A shows how most of the adhering bacteria were associated on the surface of epidermal cells in the root-hair regions, while the strains EnunB12 and EnunPs17 seemed to be connected to root surface and each other by fibrillar material (Fig. 1B). No bacterial colonization was observed near root meristem. Strains EnunEt5 and EnunPs17 were observed to be attached to the cell surface of maize and wheat roots by a single pole of the bacterial cell while the other strains along their full extent (Fig. 1C).

The majority of the bacterial cells occurred, on root surface, mainly in single units or small aggregations, with the exception of the strain EnunB12 which formed extensive aggregations on the surface of roots (Fig. 1D).

Azospirillum and *Bacillus* strains were also able to adhere to root-hair surface (Fig. 1E). SEM observations provided evidence for internal colonization of *Azospirillum* (Fig. 1F).

Conclusions

Bacteria tested prefer to colonize root-hair regions in which there is more release of root exudates; the modality of adhesion to root surface depends on bacterial strain and for some seems to be supported by fibrillar material anchoring.

Internal root colonization has been observed only for *Azospirillum* strains. It is known that *Azospirillum* strains are able to external and internal colonization of roots. The way of *Azospirillum* penetration into intercellular spaces is still controversial matter and not fully understood (Bashan and Levanony, 1990), but in our case SEM observations suggested the penetration of bacteria in root injuries.

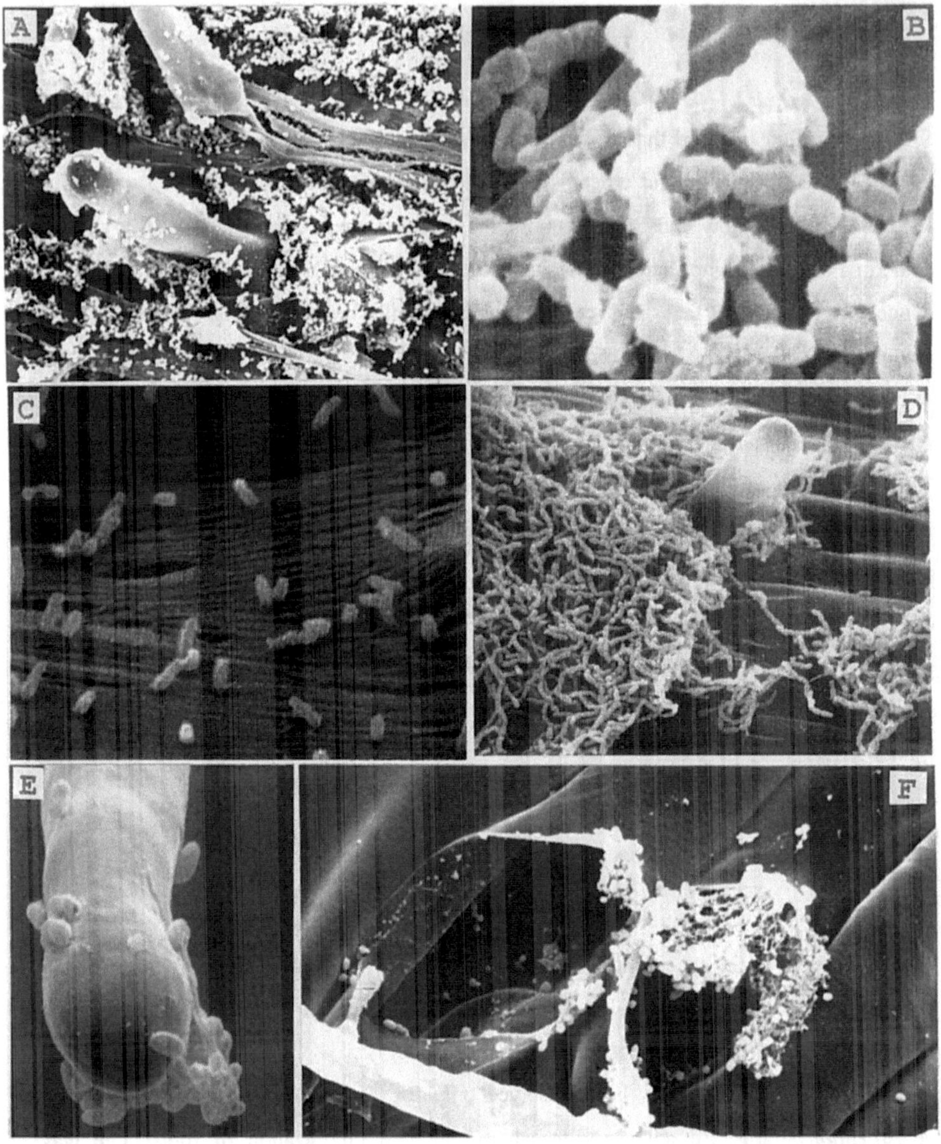

Figure 1. Scanning electron micrographs: (A) EnunAp3 adhered to maize root, X655; (B) EnunB12 anchored to root surface by fibrillar materal, X5000; (C) EnunEt5 anchored to maize root by a single pole of the bacterial cell, X2500, (D) EnunB12 formed aggregations on maize root surface, X600; (E) EnunAp2 adhered to root hair surface, X2620; (F) internal colonization of maize root by EnunAp3, X1250.

Acknowledgement

Work carried out according to a contract of Ministry for Universities and Scientific and Technological Research to Enichem Agricoltura S.p.A., referring to the National Research Program in Advanced Biotechnologies.

References

Bashan Y and Levanony H (1990) Current status of *Azospirillum* inoculation technology: *Azospirillum* as a challenge for agriculture. Can J Microbiol 36:591-608

Bazzicalupo M, Cresta E and Favilli F (1985) An in vitro assay for evaluating the *Azospirillum* wheat association. In Azospirillum III: Genetics, Phisiology, Ecology. W.Klingmuller, ed. Springer-Verlag, pp. 139-146

de Freitas JR and Germida JJ (1990) A root tissue culture system to study winter wheat-rhizobacteria interactions. Appl Microbiol Biotechnol 33:589-595

Haahtela K, Laakso T, Nurmiaho-Lassila EL and Korhonen TK (1988) Effects of inoculation of *Poa pratensis* and *Triticum aestivum* with root-associated, N_2-fixing *Klebsiella, Enterobacter* and *Azospirillum*. Plant and Soil 106:239-248

Holt JG (1984) Bergey's manual of systematic bacteriology. NR Krieg, ed. The Williams and Wilkins Co. Baltimore.

Michiels K, Vanderleyden J and Van Gool A (1989) *Azospirillum*-plant root associations: a review. Biol Fertil Soils 8:356-363

Occurrence of Dinitrogen-fixing Bacteria and Acetylene Reduction Activity in a Rhizosphere of Five Selected Plants from Tropics

Stefan Russel, Chuka Chidolue Ifiorah*
Department of Agricultural Microbiology
Warsaw Agricultural University
02-528 Warsaw
ul. Rakowiecka 26/30
Poland

Abstract

The aim of the papers is to characterize the occurence and distribution of dinitrogen-fixing bacteria including species of *Azospirillum*, *Azotobacter* and *Clostridum* in rhizosphere of the following plant species: *Delonicsregia*, *Jecopina corsinea*, *Lagestroemia indica*, *Lawsonia inermis* and *Zae mays*.
The samples of roots and rhizosphere soil were taken from Botanical Garden and University farm, University of Nigeria Nsukka Campus. The plant roots were studied for nitrogenase activity by acetylene reduction method. The effect of selected plants on bacteria was also determined.
Plate counts and tube dilution methods revealed the stimulation of development of bacteria in the rhizosphere of the investigated plants. The strongest stimulative effect of root systems on *Azotobacter sp*. was observed in the rhizosphere of *Zea mays*. Development of *Clostridium sp*. in the rhizosphere of selected plants was very irregular and dufficult to proper estimation.
All investigated root samples showed nitrogenase activity.

Introduction

Diazotrophic rhizocoenoses have been described for higher plants in tropical, temperate, terrestial and aquatic environment with a variety of bacteria including species of *Azospirillum*, *Azotobacter*, *Bacillus*, *Beijerinckia* and *Clostridium* (Knowles, 1977; Neyra and Döbereiner, 1977).
The widespread distribution of dinitrogen-fixing associative symbioses has let to interest in determining their relative importance to, and potential in, agricultural systems (Rennie, 1981).
The objective of these studies were to characterize the occurrence and distribution of nitrogen-fixing bacteria in rhizosphere of the following plants: *Delonics regia*, *Jecopina corsinea*,

---

*Department of Microbiology, University of Nigeria, Nsukka Campus, Nigeria

NATO ASI Series, Vol. G 37
Azospirillum VI and Related Microorganisms
Edited by I. Fendrik et al.

Lagestroemia indica, Lawsonia inermis and *Zea mays*. Additionally attempts to determine the plants root nitrogenase activity were also made.

Materials and methods

The roots samples were taken of the following plants: *Delonics regia, Jecopina corsinea, Lagestroemia indica, Lawsonia inermis* and *Zea mays*. The first four plants have been kindly supplied by the Botanical Garden and the last one by the University Farm, University of Nigeria, Nsukka Campus.

The plants were gently up-rooted with the attached soil particles still in place. Collected samples were carried in separate plastic bags to the laboratory for analysis.

The bacteria number determination were done by plate or tube dilution methods. For the counting number of *Azospirillum*, *Azotobacter* and *Clostridium* spp. media preparation of Rodriquez (1982), Becking (1981) and Abd-el-Malek (1971), respectively were employed.

The effect of rhizosphere was expressed by the ratio: R:S, where: R - number of bacteria in rhizosphere soil and, S - number of bacteria in nonrhizosphere soil.

The nitrogenase activity of plant roots was measured by acetylene reduction method (van Berkum and Bohlool, 1980).

Results and Discussion

The rhizosphere soil is a highly favourable site for the proliferation and activity of numerous microorganisms (Rovira and Davey, 1975). The bacterial density in rhizosphere is affected by roots exudates and is higher than in nonrhizosphere soil. The roots influence on microorganisms measured by plate method is often expressed as a rhizosphere effect.

In Table 1 the rhizosphere effect in root areas of five selected plants measured as R:S ratio is shown. In this case the rhizosphere effect was defined as the ratio of total number of microorganisms per 1g of rhizosphere soil to total number of microorganisms of 1g of adjacent nonrhizosphere soil. The amounts of microorganisms in rhizosphere of investigated plants were higher than in soil out of rhizosphere. These differences depended on date of analysis and species of plants. The highest effect of rhizosphere was observed in the case of *Lawsonia ineris* (from 14.7 to 19.9).

Figures 1, 2 and 3 show the dynamic of development of dinitrogen-fixing bacteria belonging to *Azospirillum*, *Azotobacter* and *Clostridium* spp. in rhizosphere of investigated plants. The rhizosphere region is a good environment for dinitrogen-fixing bacteria (Neyra and Döbereiner,

1977). In present work the samples of rhizosphere soil from five plants were collected monthly from April 15, and afterwards were analysed for total number of *Azospirillum*, *Azotobacter* and *Clostridium* spp.

Table 1. Effect of rhizosphere for bacteria in investigated soil samples

Selected plants	R:S and dates of analysis				
	April 15	May 13	June 15	July 14	Aug. 14
Delonics regia	10.3	13.2	11.8	8.2	8.1
Jecopina corsinea	6.1	10.8	14.6	7.0	7.9
Lagestroemia indica	8.7	7.4	7.9	4.8	2.9
Lawsonia inermis	14.7	19.9	17.4	15.1	15.8
Zea mays	7.1	4.6	8.9	7.2	3.1

It appears (Fig. 1) that the strongest stimulative effect of root system on *Azotobacter* sp. was observed in rhizosphere of *Zea mays* but the weakest in rhizosphere of *Delonics regia*. The total number of *Azospirillum* cells was growing up in all studied rhizosphere systems up to third term of analysis and then droped a little bit down (Fig. 2). The lowest number of *Azospirillum* cells was

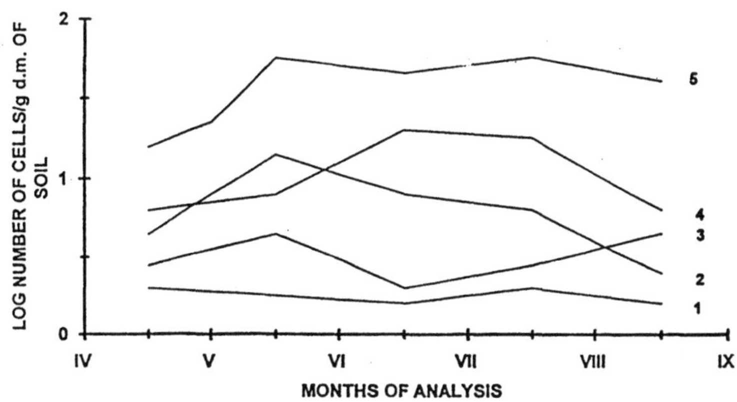

Fig. 1 Total number of *Azotobacter* sp. in rhizosphere of selected plants: 1. *Delonics regia*, 2. *Jecopina corsinea*, 3. *Lagestroemia indica*, 4. *Lawsonia inermis*, 5. *Zea mays*

seen in soil sample taken from *Delonics regia* rhizosphere but the highest in rhizosphere of

Jecopina corsinea and *Lagestroemia indica*. Development of *Clostridium* sp. in rhizosphere of selected plants was very irregular and difficult to proper estimation (Fig. 3).

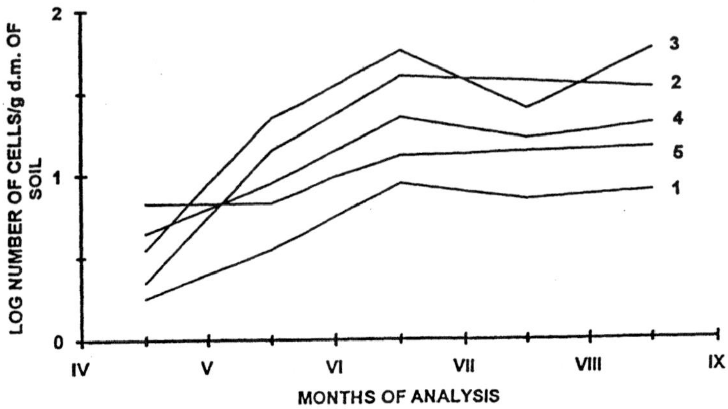

Fig. 2 Total number of *Azospirillum* sp. in rhizosphere of selected plants: 1. *Delonics regia*, 2. *Jecopina corsinea*, 3. *Lagestroemia indica*, 4. *Lawsonia inermis*, 5. *Zea mays*

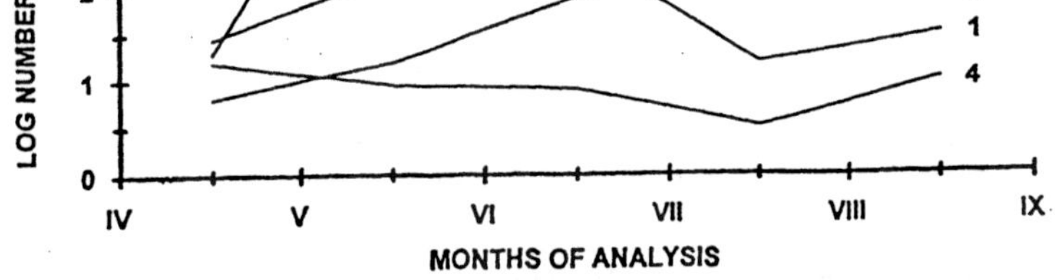

Fig. 3. Total number of *Clostridium* sp. in rhizosphere of selected plants: 1. *Delonics regia*, 2. *Jecopinia corsinea*, 3. *Lagestroemia indica*, 4. *Lawsonia inermis*, 5. *Zea mays*

The presence of nitrogen-fixing microorganisms in the rhizosphere, rhizoplane, and root cortical tissue of plant have been reported (Umali-Garcia, Hubbell, Gaskins and Dazzo, 1981; Partiquin and Döbereiner 1978). The activity of dinitrogen-fixers within and around root may be associated with requirements for plant specifity. Host-plant specificity in associative symbioses was first proposed by Döbereiner (1966) for an association between grass *Paspalum notatum* and *Azotobacter paspali*. *Azotobacter paspali* has not been found in association with any other grass species (Baldani and Döbereiner 1980).

Table 2. Acetylene reduction by roots of selected plants (Nmol ethylene per 3h and g of d.m. of roots)

Selected plants	Dates of analysis				
	April 15	May 13	June 15	July 14	Aug. 14
Delonics regia	3.2	4.6	3.1	2.8	2.3
Jecopina corsinea	3.8	5.9	5.7	7.3	5.6
Lagestroemia indica	3.1	3.9	3.8	5.8	4.1
Lawsonia inermis	2.9	3.7	5.8	5.5	5.1
Zea mays	8.4	7.1	5.9	6.2	5.4

In our experiments by means of the acetylene reduction technique, dinitrogen-fixation has been shown to occur on the roots of all investigated plants (Table 2). However, it is difficult to establish the correlation between nitrogenase activity and the number of interested us dinitrogen-fixers.

Conclusions

1. The number of bacteria in rhizosphere of studied plants is higher then in nonrhizosphere adjacent soil.
2. The investigated rhizosphere regions are favourable sites for development dinitrogen-fixing bacteria belonging to the following genera: *Azospirillum*, *Azotobacter* and *Clostridium*.
3. It is difficult to establish in our investigation the correlation between activity of nitrogenase on root pieces of investigated plants and the number of *Azospirillum*, *Azotobacter* cells in respective rhizospheres.

References

Abd-El-Malek A. (1971) Isolation of anaerobic nitrogen fixers from soil. Plant and Soil 35:432-434

Baldani V.L.D., Döbereiner J. (1980) Host-plant specificity in the infection of cereals with *Azospirillum* spp. Soil Biol Biochem. 12:433-439

Becking J.J. (1981) The family *Azotobacteriacea*. In: The procayotes. Handbook on habitats, isolation and identification of bacteria. Vol. 1. Springer Verlag.

Berkum P., Van Bohlool B.B. (1980) Evaluation of nitrogen fixation by bacteria in association with roots of tropical grasses, Microbiol. Rev. 44:491-517

Döbereiner J. (1966) *Azotobacter paspali* sp. uma bacteria fixadora de nitrogenio na rhizosfera de *Paspalum notatum*. Pesqu. Agropecu. Bas. 1:357-365

Knowles R. (1977) The significance of a symbiotic nitrogen fixation by bacteria. In: A treatise on dinitrogen fixation. J.Wiley and Sons, New York

Neyra C.A., Döbereiner J. (l977) Nitrogen fixation in grasses. Adv. Agronom. 29:1-38

Patriquin D., Döbereiner J. (1978) Bacteria in the endorhizosphere of maize in Brazil. Can. J. Microbiol. 24:734-742

Rennie R.J. (1981) Potential use of induced mutations in improve symbioses of crop plants with N_2-fixing bacteria. In: Induced mutations a tool in plant breeding. International Atomic Energy Agency, Vienna. 293-321

Rodriquez E.A. (1982) Improved medium for isolation of *Azospirillum* species. Applied Environ. Microbiol. 44:990-991

Rovira A.D., Davey C.B. (1975) Biology of the rhizosphere. In: Plant root and its environment. Univ. Press Virginia.

Umali-Garcia M., Hubbell D.H., Gaskins M.H., Dazzo B. (1980) Association of *Azospirillum* with grass roots. Applied Environ. Microbiol. 39:219-226

Umali-Garcia M., Hubbell D.H., Gaskins M.H., Dazzo B. (1981) Absorption and mode of entry of *Azospirillum brasilense* to grass roots. In: Associative N_2 fixation. Vol.I, CRC Press, Boca Raton

Dominant N_2-Fixing Microflora in a Paddy Soil and Effect of Rice Inoculation with *Azospirillum* Bacteria

Nadia Kaloianova and Ognian Kostov

Soil Microbiology Department

N. Poushkarov Institute of Soil Science and Agroecology

Shosse Bankya 7

1080 Sofia

Bulgaria

Abstract

The composition of N_2 fixing non rhizosphere and rhizosphere microflora was studied. Field experiments with *Azospirillum brasilense sp* 7 and *Azospirillum lipoferum sp* 3624 were carried out. The highest yield response was established using composted sawdust as carrier containing *Azospirillum lipoferum* strain. The data suggest that except yield, the N use efficiency and microbial criteria responsible for the quality of paddy soil (biomass C and CO_2 production) should be used as complex criterion for the selection of promising *Azospirillum*-plant systems as well as other diazotrophic bacterial inoculants. *Azospirillum* inoculation could be used as a criterion for further rice breeding activity.

Introduction

Bulgarian field experiments related to assessment of *Azospirillum* inoculation generally resulted in increasing of total yield up to 20% (Kaloianova *et al.* 1989; Kostov *et al.* 1992) or equivalent yields in inoculated variants as compared to heavy dressed control soil (Kaloianova *et al.* 1989). The equivalent yields in inoculated variants can be considered as a possible potential to reduce chemical nitrogen fertilization (Favilli *et al.* 1993; Kostov *et al.* 1992). In our previous research in pot experiments, we established the positive response of cultivar Krasnodar 424 to *Azospirillum* inoculation which resulted in higher top and root biomass compared to the control (Kaloianova *et al.* 1993). The aim of this study was to find out the composition of the

NATO ASI Series, Vol. G 37
Azospirillum VI and Related Microorganisms
Edited by I. Fendrik et al.

N_2 fixing (C_2H_2 reduction) bacterial population of a paddy soil and to assess the effect of rice inoculation with *Azospirillum* strains in 4 successive years with different methods of inoculation.

Materials and methods

The composition of N_2 fixing non rhizosphere and rhizosphere microflora was studied as recommended by Thomas-Bauzon *et al.* (1982). An alluvial soil {Total C-1.23%; Total N-0.083%; pH (H_2O)-6.9} where rice is generally grown in Bulgaria was used both with spermosphere model and in field experiments. The field experiments were carried out during the period of 1986-1990. The inoculations were done using *Azospirillum brasilense sp* 7 and *Azospirillum lipoferum sp* 3624. The inoculum was introduced as follows: 1. as suspension on rice seeds before sowing; 2. double inoculation (the first time as a suspension on seeds before sowing and in the beginning of the tillering); 3. 3 weeks seedlings were soaked for 24 hours and than transplanted by hand on the field (1987); 4. the seeds were treated with composted sawdust carrier containing *Azospirillum* bacteria. The N (0; 90 and 180 kg/ha) was applied as NH_4SO_4. The P and K in doses of 160 kg/ha and 90 kg/ha were applied as superphosphate and K_2SO_4 respectively. The cultivar was Krasnodar 424. Three field replicates (each replicate of $100m^2$) were used. The reported yield was harvested from 4 plots of $20m^2$.

In 1986 the dynamics (during the plant growth stages) of soil microbial biomass C, CO_2 and Eh were also measured.

Results and discussion

In the spermosphere model positive ethylene production in all replicates of rhizosphere and non rhizosphere soil were established in dilutions of 10^7 and 10^4 respectively. Finally 60 strains were isolated from which 30% from non rhizosphere and 15% from rhizosphere soil gave a positive response to ethylene production. Thirteen N_2 fixing isolates were considered as belonging to families: (i) *Enterobacteriaceae* (g. *Citobacter*-3 isolates; g. *Enterobacter*-2 isolates; g.

Klebsiella-3 isolates); (ii) *Spirillaceae* (g. *Azospirillum*-2 isolates); ;(iii) *Bacillaceae* (*Bacillus polymyxa*-2 isolates). Isolates from g. *Pseudomonas* and g. *Azotobacter* were not detected. Our data show a higher density of N_2 fixing microflora, higher N_2 fixing activity and lower bacterial diversity compared to the results of Tomas-Bouzon (1982). This suggest that the inoculation of rice with *Azospirillum* bacteria could have positive response. The yield data during the experimental years are given in Table 1.

Table 1. Effect of inoculation on rice yield and N use efficiency.

Treat ments*	1986		1987		1989		1990	
	kg/ha[a]	kg/kg[b]	kg/ha[a]	kg/kg[b]	kg/ha[a]	kg/kg[b]	kg/ha[a]	kg/kg[b]
1.	6000							
2.	6500							
3.	5600							
4.	6810	75.6						
5.	7100	78.8						
6.	7400	82.2						
7.	7000	38.8	5200	28.8	5430	30.1	7060	39.2
8.	7830	43.5	-	-	-	-	-	-
9.	8030	44.6	6100	33.8	6000	33.3	-	-
10.			7500	41.6	6270	34.8	-	-
11.			-	-	6680	37.1	8560	47.5
12.			5350	29.7	-	-	-	-
13.			5900	32.7	-	-	-	-
LSD p=0.05	930		710		490		440	

*Treatments: 1. $N_0P_{16}K_9$; 2. $N_0P_{16}K_9$ + *Az. brasilense sp.* 7; 3. $N_0P_{16}K_9$ + *Az. lipoferum sp.* 3624. 4. $N_9P_{16}K_9$; 5. $N_9P_{16}K_9$ + *Az. brasilense sp.* 7; 6. $N_9P_{16}K_9$ + *Az. lipoferum sp.* 3624; 7. $N_{18}P_{16}K_9$; 8. $N_{18}P_{16}K_9$ + *Az. brasilense sp.* 7; 9. $N_{18}P_{16}K_9$ + *Az. lipoferum sp.* 3624; 10. $N_{18}P_{16}K_9$ + *Az. lipoferum sp.* 3624 (double inoculated); 11. $N_{18}P_{16}K_9$ + *Az. lipoferum sp.* 3624 (sawdust composted carrier); 12. $N_{18}P_{16}K_9$ (seedlings); 13. $N_{18}P_{16}K_9$ + *Az. lipoferum sp* 3624 (seedlings inoculation); [a] kg grain/ha; [b] kg grain/kg N applied.

It can be seen that the size of yields varies significantly during the years. At moderate N dose (90 kg/ha) the inoculation did not resulted in significant increase of the yield but resulted in the highest N use efficiency (75.6 kg grain/kg N applied at control soil and 82.2 kg grain/kg N applied at *Azospirillum lipoferum* treatment). The highest N dose (180 kg N/ha) stimulate the inoculation effect probably due to a low N mineralization potential of this soil. In 1990 the highest yield was established using composted sawdust as carrier containing *Azospirillum lipoferum* strain. In this year N use efficiency was also higher as compared to 1987 and 1989.

The lower yields and N use efficiency in 1987 and 1989 could be due to climatic factors (windy weather and dropping of the temperature during the blossoming stage of rice in this region). These data suggest that inoculation could change the duration of the rice plant stages and consequently the growth technology has to be adjusted to the inoculation effect. Unfortunately, all cultivars used in Bulgarian agriculture are selected without considering microbial factors, such as inoculation. *Azospirillum* inoculation could be used as a criterion for further rice breeding activity. Dynamics of biomass C, CO_2 and Eh during the plant stages of rice (data not shown), indicate an accumulation of root biomass which resulted in increasing of biomass C, CO_2 and reducing of Eh in inoculated treatments. These results suggest that inoculation could have a positive effect on the quality of the soil (Doran et al. 1994). Thus, the inoculation results in an additional ecological value-quality of the soil. Except yield, the N use efficiency and microbial criteria responsible for the quality of the paddy soil (microbial C, CO_2 and Eh) should be used as a complex criterion for the selection of promising *Azospirillum*-plant system as well as other diazotrophic bacterial inoculants.

References

Doran JW, Sarrantonio M, Janke R (1994) Strategies to promote soil quality and health. In: *Management in Susatainable Farming Systems* (CD Pankhurst, BM Doube, VVSR Gupta and PR Grace, Eds.) pp 230-247. CSIRO, East Melbourn, Victoria, 3002, Australia.

Favilli R, Pastorelli R, Gori A (1993) Response of sugar beet to *Azospirillum* bacterization in field experiments. Agr Med 123: 281-285

Kaloianova N, Dimitrova A, Kostov O (1989) Effect of *Azospirillum* inoculation on the yield of maize. Agrochem Soil Sci 39: 395-399

Kaloianova N, Dimitrova A, (1993) Effect of inoculation with *Azospirillum* bacteria on growth and yield of rice. Soil Sci Agrochem 26: 62-66

Kostov O, Dimitrova A & Kaloianova N (1992) Associative nitrogen fixation. Bull Blg Cent Inst Sci Technolog Inform ND-1489: 1-19

Thomas-Bouzon D, Weinard P, Villecurt P, Balandreau J (1982) The spermosphere model. I. It use in growing, counting and isolating N_2 fixing bacteria from the rhyzosphere of rice. Can J Microbiol 28: 922-928

Isolation and Characterization of *Azospirillum* Strains from Soil and Rice Plants in North Vietnam

Nguyen ngoc Dung[1], Maria Kraut and Walter Klingmüller
Department of Genetics
University of Bayreuth
Universitätsstrasse 30
95447 Bayreuth
Germany

Abstract

From rice field soils and rice roots, 51 bacterial strains were isolated showing the characteristics of *Azospirillum*. They could be distributed into six groups due to the utilization of sugars, the requirement for biotin, and the formation of acid from glucose. Most of the isolates showed the phenotype of *Azospirillum lipoferum*. Only four isolates were similar to *Azospirillum brasilense*. Ten isolates differed phenotypically from the *Azospirillum* species described so far. Several isolates were selected for the genotypical characterization by ribotyping and hybridization with species specific oligonucleotide probes. These analyses confirmed the phenotypic classification of three isolates as *A. brasilense*. The majority of the isolates seemed to be more related to *A. lipoferum*. Two isolates may represent a new species.

Key Words

Azospirillum, characterization, ribotyping, restriction fragment length polymorphism (RFLP), oligonucleotide probes, ribosomal DNA

Introduction

Bacteria of the genus *Azospirillum* are associated with roots of cereals, such as rice and wheat, and grasses from various geographical origins. Strains of *Azospirillum* are capable of N_2 fixation and influence growth and nutrient uptake of the host plants. Until now, five species were described: *A. brasilense*, *A. lipoferum* (Tarrand et al. 1978), *A. amazonense* (Magalhães

---

[1] Present address: Institute of Biotechnology, National Centre of Naturescience and Technology, Nghia Do, Tu Liem, Hanoi, Vietnam

NATO ASI Series, Vol. G 37
Azospirillum VI and Related Microorganisms
Edited by I. Fendrik et al.

et al. 1983, *A. halopraeferens* (Reinhold et al. 1987), and *A. irakense* (Khammas et al. 1989). They were differentiated on the basis of physiological, morphological and molecular properties. In Vietnam, rice is the most important food. Since Vietnam is not able to produce enough inorganic fertilizers, we are interested to develop biological fertilizers to increase the rice production. Therefore, we isolated a number of bacteria showing associative N_2 fixation with rice seedlings. Here we present the data of their phenotypical and phylogenetical characterization.

Materials and Methods

Bacterial strains, media and growth conditions

A. amazonense Y1 (DSM 2787, Magalhães et al. 1983), *A. brasilense* Sp7 (DSM 1690, Tarrand et al. 1978), *A. halopraeferens* Au4 (DSM 3675, Reinhold et al. 1987), *A. irakense* KBC1 (CIP 10311, Khammas et al. 1989), and *A. lipoferum* Sp59b (DSM 1691, Tarrand et al. 1978) were used as references in the genotypical characterization. DNA from *Escherichia coli* DH10b (Grant et al. 1990) and *Saccharomyces cerevisiae* was included in the hybridization with oligonucleotides as negative control. The isolates were obtained from rice field soils as well as sterile and unsterile roots from rice according to the procedures decribed by Döbereiner et al. (1976), Thomas-Bauzon et al. (1982), and Cacares (1982). Routinely, the strains and isolates were grown in LB-medium (Miller 1972). For phenotypical characterization, the isolates were cultivated in MAZ-medium (Albrecht and Okon 1980) supplemented with carbon sources, RC-medium (Cacares 1982), BMS-medium (Krieg and Döbereiner 1984), and Nfb-medium (Döbereiner et al. 1976). Associative fixation of N_2 with rice seedlings was tested in the spermosphere model (Thomas-Bauzon et al. 1982).

DNA isolation and hybridization

The genomic DNA of isolates and reference strains was isolated according to Masterson et al. (1985). For the isolation of DNA from *S. cerevisiae* the cells were broken in a French Press (American Instrument Company, USA) with 147 MPa. The extraction of DNA was performed as described by Marmur (1961). The DNA was blotted directly (slot blots) or after restriction and electrophoresis onto nylon membranes (Qiabrane, Quiagen, Germany). The slot blot membranes were hybridized with species specific oligonucleotide probes derived from 23S rDNA sequences of *Azospirillum* ssp. and an eubacterial oligonucleotide probe (Kirchhof and Hartmann 1992, G. Kirchhof and A. Hartmann pers. communication). The oligonucleotides were labelled with Digoxigenin and the membranes hybridized according to the instructures of the manufacturers (Boehringer, Germany).

To obtain a probe for the 16S rRNA genes, the 16S rDNA of *A. brasilense* Sp7 was amplified using primers deduced from the conserved ends of the genes (positions 8-27 and 1540-1528, S. Selenska-Pobell pers. communication). For the identification of fragments carrying 23S rRNA genes, a 4 kbp *XbaI/Hin*dIII fragment from pKK3535 (Brosius et al. 1981) was used as gene probe. The fragment contained the 23S and 5S rRNA genes of the *E. coli rrnB* operon and a small part of the pBR322 *tet* gen. The gene probe for *nifH* was an amplification product of the gene from *A. brasilense* Sp7 using primers at the conserved positions 25-45 and 429-449 (G. Romanowski pers. communication). Hybridizations with the Digoxigenin labelled probes were performed as described by Church and Gilbert (1984).

Results and Discussion

Isolation and phenotypical characterization

From rice field soils and rice roots 51 strains were isolated which were mobile vibroids, able to fix N_2, formed scarlet-red colonies on congo red medium and grew microaerobically as a thin pellicle a few millimeters below the surface of the semi-solid medium. All isolates showed associative N_2 fixation with rice seedlings.

Phenotypically, they could be distributed into six groups (Tab. 1). The members of the groups I and II exhibited the same characteristics as the type strains of *A. brasilense* and *A. lipoferum*, respectively, and seemed to belong to these species. Some isolates (groups III to VI) differed to some extent from all species described so far. It was unlikely that these isolates belong to the species *A. amazonense* and *A. halopraeferens* since they formed pink colonies on potato medium and did not require a higher salt concentration for growth. The physiological differences between these isolates and *A. irakense* or *A. lipoferum* were not significant. Therefore, these isolates possibly were members of one of these species. Because physiological characteristics are not sufficient for the classification of the isolates we made attemps to characterize them with phylogenetical methods.

Phylogenetical characterization

By using gene probes for 16S (Fig. 1) and 23S rDNA as well as for the *nifH* gene (not shown), the RFLPs of *Eco*RI (Fig. 1) and *Sal*I (not shown) digested genomic DNA of 16 isolates were studied. Each pair of the isolates QO 1-1a and QO 1-1b, Mat 1-4 and Mat 2-1b as well as Arm

Table 1: Phenotypical characteristics of the isolates and *Azospirillum* species

Characteristic	Groups of the isolates I[f]	II[g]	III[h]	IV[i]	V[j]	VI[k]	A.a.[a]	A.b.[b]	A.h.[c]	A.i.[d]	A.l.[e]
Growth with											
glucose	-	+	+	+	+	+	+	-	-	+	+
mannitol	-	+	+	-	-	-	-	-	+	-	+
α-ketoglutarate	-	+	+	-	+	+	-	-	+	±	+
sucrose	-	-	-	+	+	-	+	-	-	+	-
Biotin-requirement	-	+	-	-	-	+	-	-	+	-	+
Formation of acid with glucose	-	+	+	+	+	+	-	-	-	-	+
Colonies on											
potato medium[l]	pink	pink	pink	pink	pink	pink	white	pink	-[m]	n.d.	pink
congored medium[n]	red[o]	red[o]	red[o]	red[o]	red[o]	red[o]	n.d.	red[o]	-[m]	n.d.	red[o]

The data for the *Azospirillum* species are from Tarrand *et al.* 1978, Magalhães *et al.* 1983, Krieg and Döbereiner 1984, Drets *et al.* 1985, Reinhold *et al.* 1987 and Khammas *et al.* 1989.

a *A. amazonense* Y1
b *A. brasilense* Sp7
c *A. halopraeferens* Au4
d *A. irakense* KBC1
e *A. lipoferum* Sp59b
f DA 2-3-2, QO 1-1a, QO 1-2, SS B
g Lun 2-6-1, Lun 2-7-1, Lun 2-16-1, Mat 1-3, Mat 1-4, Mat 2-1a, Mat 2-1b, DA 5-1, DA 2-3-1, DA 3-1, DA 4-2 DA 4-3a, DA 4-3b, DA 5-2, DA 5-3, DA 6-3a, DA 6-3b, GL 1-2, GL 2-2, TB 80, SS I-1, SS II-1, SS 1, SS 2-1, SS 2-2, SS A-2, SS A-3, SS b, SS III-3, SS IV-1,SS IV-2, SS 3-2, SS 3-4, SS4-1, SS 4-2, SS c
h DA 9-3a, DA 9-3b, DA 10-2
i DA 10-1
j Arm 2-2, DA 6-2-2, GL 1-1
k BB 1-1, S III-1, SS IV-3
l BMS medium (Krieg and Döbereiner 1984)
m no growth
n RC medium (Rodriguez-Caceres 1982)
o scarlet-red
\+ good growth
\- no growth
± depends on strain
n.d. not determined

2-2 and GL 1-1 showed identical RFLPs with all probes and enzymes. All other strains could be differentiated by at least one RFLP. From the RFLPs of the 16S and 23S rDNA fragments produced by *Eco*RI digestion, a dendrogram was derived (Fig. 2). The major part of the examined strains were more closely related to *A. lipoferum* Sp59b than to other *Azospirillum* species although some of them showed other physiological characteristics as described for *A. lipoferum*.

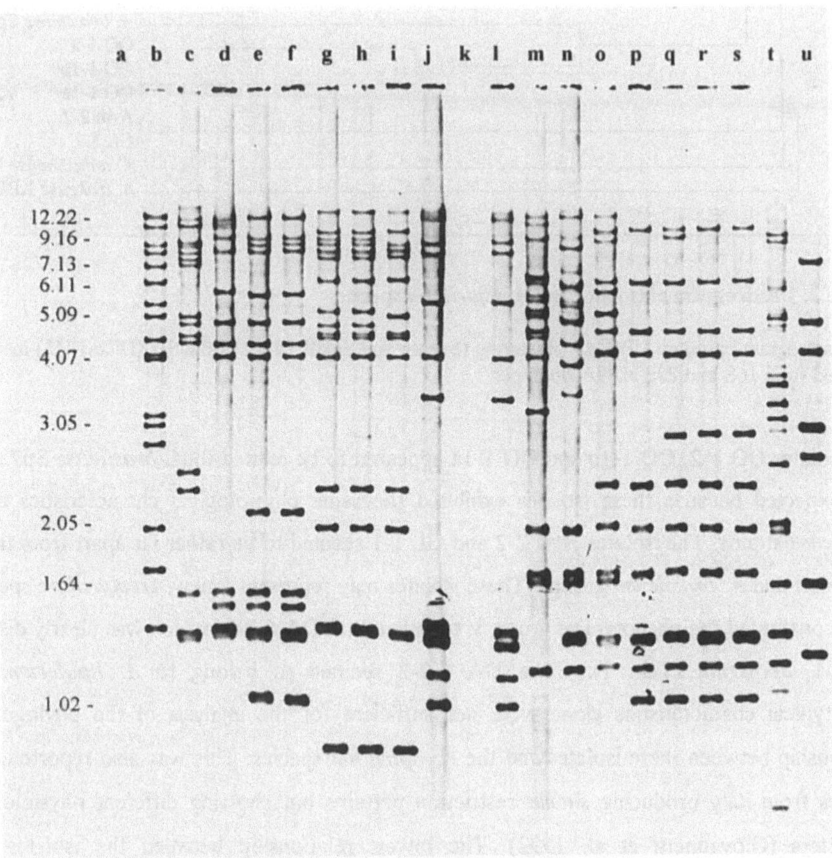

Figure 1: Ribotypes of genomic DNA from isolates and *Azospirillum* species

The genomic DNA of *A. lipoferum* Sp59b (**b**), *A. brasilense* Sp7 (**c**), QO 1-2 (**d**), QO 1-1b (**e**), QO 1-1a (**f**), Mat 1-4 (**g**), Mat 2-1b (**h**), Mat 2-1a (**i**), Arm 2-2 (**j**), GL 1-1 (**l**), SS III-3 (**m**), DA 2-3-1 (**n**), DA 6-2-2 (**o**), DA 6-3a (**p**), DA 9-3b (**q**), DA 10-1 (**r**), DA 10-2 (**s**), *A. amazonense* Y1 (**t**), *A. irakense* KBC1 (**u**) was digested with *Eco*RI and hybridized with a probe for 16S rRNA genes. A kilo base ladder (**a, k**) was included as size standard. Sizes are given in kbp.

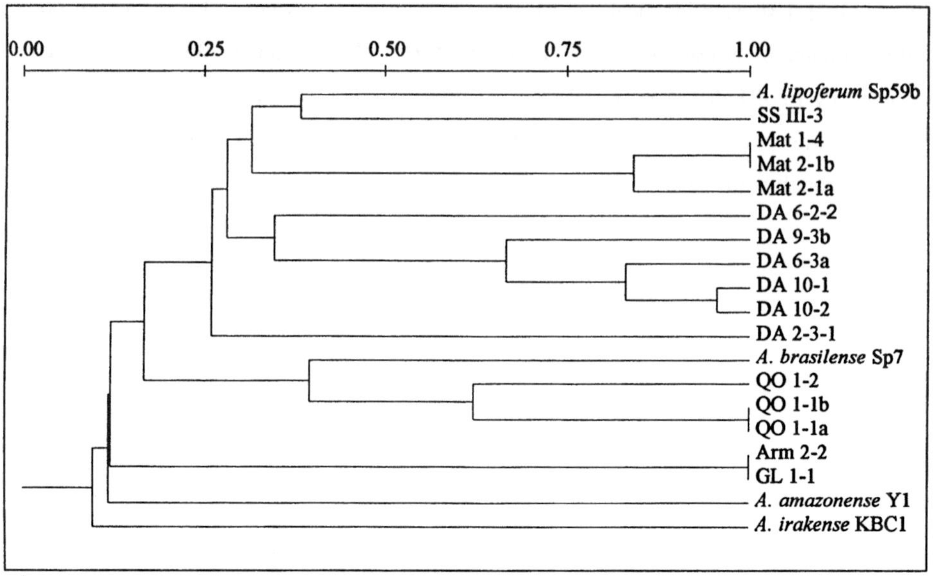

Figure 2: Dendrogram of isolates and *Azospirillum* species

The dendrogram based on UPGMA clustering (Sneath and Sokal 1973) of the SD (Dice 1945) matrices obtained from 16S and 23S rDNA ribotypes.

The isolates QO 1-2, QO 1-1b and QO 1-1a appeared to be related to *A. brasilense* Sp7. This was expected because these isolates exhibited the same physiological characteristics as *A. brasilense* strains. The isolates Arm 2-2 and GL 1-1 seemed to be rather far apart from the *A. lipoferum* and *A. brasilense* groups. These isolates may represent a new *Azospirillum* species. They constituted the phenotypical group V together with DA 6-2-2, which was clearly distinct from *A. lipoferum* (Tab. 1). Since DA 6-2-2 seemed to belong to *A. lipoferum*, the phenotypical characteristics alone were not sufficient for the analysis of the phylogenetic relationship between these isolates and the *Azospirillum* species. This was also reported with isolates from italy producing similar restriction patterns but showing different physiological characters (Giovannetti et al. 1992). The lowest relationship between the isolates and *Azospirillum* species was found with *A. amazonense* Y1 and *A. irakense* KBC1.

To improve the phylogenetic characterization of the isolates, the genomic DNA of 17 isolates were hybridized with species specific deoxyoligonucleotide probes (Fig. 3). The DNA of the isolates QO 1-1a, QO 1-1b and QO 1-2 hybridized with the probe for *A. brasilense*. This was in agreement with the results of the phenotypical characterization and the ribotype dendrogram. The DNA of the other isolates did not hybridize with the probes for *A. amazonense*, *A. halopraeferens* and *A. irakense*. This indicated, that these isolates belong to *A. lipoferum* or a new *Azospirillum* species. We could not decide which of the two possibilities was right, since in our hands the probe for *A. lipoferum* was not specific enough for the identification of this species.

At the moment we can say that some of the isolates are *A. brasilense* and none of them belong to *A. amazonense*, *A. halopraeferens* or *A. irakense*. We suppose that the other isolates are *A. lipoferum* strains. Some of them, especially Arm 2-2 and GL 1-1, may represent an *Azospirillum* species not described so far.

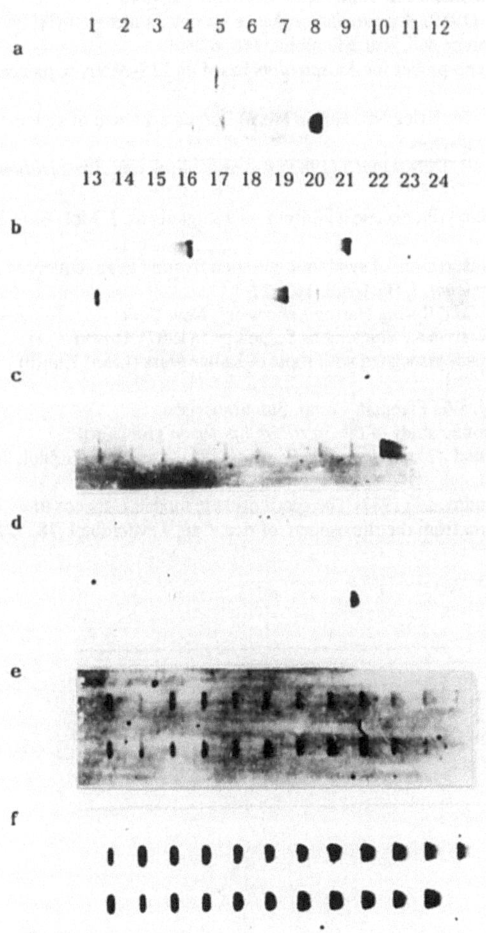

Figure 3: Hybridization with species specific oligonucleotide probes

The genomic DNA was blotted onto nylon membranes and hybridized with oligonucleotide probes derived from variable sequences of the 23S rDNA of:

a *A. amazonense*
b *A. brasilense*
c *A. halopraeferens*
d *A. irakense*
e *A. lipoferum*
f Eubacteria

The slots were loaded with DNA of the following strains:

1 SS III-3 | 2 SS B
3 GL 1-1 | 4 QO 1-2
5 DA 6-3a | 6 Mat 2-1a
7 Mat 2-1b | 8 DA 9-3a
9 QO 1-1b | 10 DA 10-2
11 DA 9-3b | 12 DA 10-1
13 QO 1-1a | 14 DA 6-2-2
15 Arm 2-2 | 16 Mat 1-4
17 DA 2-3-1
18 *A. lipoferum* Sp59b
19 *A. brasilense* Sp7
20 *A. amazonense* Y1
21 *A. irakense* KBC1
22 *A. halopraeferens* Au4
23 *E. coli*
24 *S. cerevisiae*

Acknowledgements: We thank Dr. A. Hartmann and Dr. G. Kirchhof for their kind gift of bacterial strains and especially of oligonucleotide probes prior to publication. The stay of D.n. Nguyen in Bayreuth was supported by the Deutscher Akademischer Austauschdienst (DAAD).

References:

Albrecht SL, OkonY (1980) Culture of *Azospirillum*. In: San Pietro A (ed.): Methods in Enzymology 69, Acad. Press, New York

Brosius J, Ullrich A, Raker MA, Gray A, Dull TJ, Gutell RR, Noller HF (1981) Construction and fine mapping of recombinant plasmids containing the *rrn* ribosomal RNA operon of *E.coli*. Plasmid **6**: 112

Cacares EAR (1982) Improved medium for isolation of *Azospirillum* spp. Appl. Environ. Microbiol. **44**: 990

Church BM, Gilbert W (1984) Genomic sequencing. Proc. Natl. Acad. Sci. USA **81**: 1991

Dice LR (1945) Measures of the amount of ecological association between species. Ecology **26**: 297

Döbereiner J, Mariel I.E., Nery M (1976) Ecological distribution of *Spirillum lipoferum* Beijerinck. Can. J. Microbiol **22**: 1464

Drets GM, Fabiano E, Cardona A (1985) Carbohydrate catabolism in *Azospirillum amazonense*. Appl. Environ. Microbiol. **50**: 183

Giovannetti L, Fedi S, Gori A, Mantaini P (1992) Identification of *Azospirillum* strains at the genome level with total restriction pattern analysis. Syst. Appl. Microbiol. **15**: 37

Grant SGN, Jessee J, Bloom FR, Hanahan D (1990) Differential plasmid rescue from transgenic mouse DNAs into *Escherichia coli* methylation-restriction mutants. Proc. Natl. Acad. Sci. USA **87**: 4645

Khammas KM, Ageron E, Grimont PAD, Kaiser P (1989) *Azospirillum irakense* sp. nov., a nitrogen-fixing bacterium associated with rice roots and rhizosphere soil. Res. Microbiol. **140**: 679

Kirchhof G, Hartmann A (1992) Development of gene-probes for *Azospirillum* based on 23S-rRNA sequences. Symbiosis **13**: 27

Krieg NR, Döbereiner J (1984) Genus *Azospirillum*. In: Krieg NR, Holt JG (eds) Bergey's manual of systematic bacteriology. The Williams and Wilkings Co., Baltimore

Magalhães RM, Baldani JI, Souto SM, Kuykendall JR, Döbereiner J (1983) A new acid-tolerant *Azospirillum* species. An. Acad. Brasil. Cienc. **55**: 417

Marmur J (1961) A procedure for the isolation of deoxyribonucleic acid from microorganisms. J. Mol. Biol. **3**: 208

Masterson RV, Prakash RK, Atherly AG (1985) Conservation of symbiotic nitrogen fixation gene sequences in *Rhizobium japonicum* and *Bradyrhizobium joponicum*. J. Bacteriol. **163**: 21

Miller J (1972) Experiments in molecular genetics. Cold Spring Harbor Laboratory, New York

Reinhold B, Hurek T, Fendrik I, Pot B, Gillis M, Kersters K, Thielemans S, De Ley J (1987) *Azospirillum halopraeferens* sp. nov., a nitrogen-fixing organism associated with roots of Kallar grass (Linn) Kunth). Intern. J. Syst. Bacteriol. **37**: 43

Sneath PHA, Sokal RR (1973) Numerical taxonomy. WH Freeman Comp. San Franscisco

Tarrand JJ, Krieg NR, Döbereiner J (1978) A taxonomic study of the *Spirillum lipoferum* groups with *Azospirillum lipoferum* (Beijerinck) comb. nov. and *Azospirillum brasilense* sp. nov.. Can. J. Microbiol. **24**: 967

Thomas-Bauzon D, Weinhard P, Villecourt P, Balandreau J (1982) The spermosphere model. I. Its use in growing, counting and isolating N_2-fixing bacteria from the rhizosphere of rice. Can. J. Microbiol. **28**: 922

Subject Index

(The first page only of the article to which the key word is referred, is indicated).

LIST OF PARTICIPANTS FROM NATO COUNTRIES

ARW NO. : 940558

TITLE : NATO Advanced Research Workshop on *Azospirillum* and Related Microorganisms

Director: I. Fendrik **LOCATION:** Sárvár/ Hungary **DATES:** September, 4-7, 1994

1. KEY SPEAKERS

Belgium

Jos Vanderleyden
F.A. Janssen Memorial Lab. f. Genetics
Catholic University of Leuven
Willem de Croylaan 42
3030 Heverlee - Leuven
Belgium

France

M. de Zamaroczy
Departement des Biotechnologies
Institut Pasteur
25 Rue du Dr Roux
75724 Paris Cedex 15
France

Germany

Istvan Fendrik
Inst. f. Biophysik
Universität Hannover
Herrenhäuser Str. 2
30419 Hannover
Germany

Anton Hartmann
Institut für Bodenökologie
GSF - Neuherberg
Ingolstädter Landstr. 1
85764 Oberschleißheim
Germany

Fritz Schönbeck
Institut für Pflanzenkrankheiten und Pflanzenschutz
Universität Hannover
Herrenhäuser Str. 2
30419 Hannover
Germany

Italy

Maddalena Del Gallo
Dipartimento Agrobiotecnologie
ENEA - Casaccia
Via Anguillarese 301
00060 Rome
Italy

Renato Fani
Dip.to Biologia Animale e Genetica
Università di Firenze
Via Romana 17
50125 Firenze
Italy

United Kingdom

E. Cocking
University of Nottingham
Plant Genetic Manipulation Group
Life Science
NG 72 RD Nottingham
U.K.

Janett Sprent
Dep. of Biol. Sciences
Univ. of Dundee
Dundee, DDI 4HN
U.K.

United States

Carlos Neyra
The State University of New Jersey
New Brunswick - New Jersey 08903-0231
USA

2. OTHER PARTICIPANTS

Belgium

Paul De Troch
F.A.Janssens Lab. of Genetics
K.U. Leuven
Willem de Croylaan 42
3030 Heverlee - Leuven
Belgium

S. Dobbelaere
F.A.Janssens Lab. of Genetics
K.U. Leuven
Willem de Croylaan 42
3030 Heverlee - Leuven
Belgium

Karel Vlassak
Catholic University of Leuven
Kardinaal Mercierlaan 92
3001 Heverlee
Belgium

France

Florence Arsene
Institut Pasteur
25 Rue du D' Roux
75724 Paris Cedex 15
France

René Bally
Universitée Claude Bernard
Lyon I.A.S.B.S.E. U.R.A. CNRS 1450
Ecologie Microbienne du Sol, Bât. 741
43, Boulevard du 11 novembre 1918
69622 Villeurbanne cedex
France

Denis Faure
Université Claude Bernard Lyon I
43, Bd du 11 Novembre 1918
69100 Villeurbanne
France

M. Kabir
Ecologie Microbienne du Sol
URA CNRS 1450, UCBL-1
69622 Villeurbanne
France

Paul Kaiser
Laboratoire de Microbiologie
Institut National Agronomique
Rue Claude Bernard 16
75231 Paris Cedex 05
France

Germany

G. Höflich
Forschungszentrum f. Bodenfruchtbarkeit
Wilhelm-Piek-Str. 72
15374 Müncheberg
Germany

Walter Klingmüller
Lehrstuhl für Genetik
Universit Bayreuth
Universitätsstr. 30
95447 Bayreuth
Germany

Tina Stein
Institut für Biophysik
Universität Hannover
Herrenhäuser Str. 2
30419 Hannover
Germany

Wolfgang Zimmer
Botanisches Institut
Universität Köln
Gyrhofstr. 15
50931 Köln
Germany

Greece

Fotini Flouri
General & Agricult. Microbiol. Lab.
Agricultural University of Athens
Iera Odos 75, Votanicos
118 55 Athens
Greece

Io Kefalogianni
General & Agricult. Microbiol. Lab.
Agricultural University of Athens
Iera Odos 75, Votanicos
118 55 Athens
Greece

Italy

Paola Barbieri
Universita di Milano
Genetica e Biologia dei Microorganismi
Via Celoria
20133 Milano
Italy

Franco Favilli
Dipart. Scienze e Tecnologie Alimentari
e Microbiche e C.S. dei Microorganismi
Autotrofi CNR
Università di Firenze
P.le delle Cascine 27
50144 Firenze
Italy

Corrado Fogher
Istituto di Genetica
Università Cattolica del Sacro Cuore
Via Emilia Parmense 84
I-29100 Piacenza
Italy

Alessandro Gori
Dip.to Scienze Tecnol. Alim. Microb.
Piazzale delle Cascine, 27
50144 Firenze
Italy

Francesca Marini
Dipartimento Agrobiotecnologie
ENEA - Casaccia
Via Anguillarese 301
00060 Rome
Italy

Roberta Pastorelli
Dipartimento di Scienze e Tecnologie Alimentari e
Microbiologie
Piazzale delle Cascine 27
50144 Firenze
Italia

Silvia Tabacchioni
Dipartimento Agrobiotecnologie
ENEA - Casaccia
Via Anguillarese 301
00060 Rome
Italy

Carlo Viti
Dipartimento di Scienze e Tecnologie Alimentari e
Microbiologie
Piazzale delle Cascine 27
50144 Firenze
Italy

United States

Igor Zhulin
Loma Linda University
School of Medicine
Department of Microbiology
Loma Linda, California 92350
USA

LIST OF PARTICIPANTS FROM NON-NATO COUNTRIES

1. KEY SPEAKERS

Brazil

Johanna Döbereiner
EMPRAPA - CNPBS
km47, Seropédica
23851 Rio de Janeiro
Brazil

Hungary

Istvan Gyurjan
Eötvös Lorand University
Department of Plant Science
H- 1088 Budapest
Puskin U. 11-13
Hungary

Tibor Sik
Dept. Biotechnolog
Univ Agricult Sci
H-2103 Gödöllö
Hungary

Israel

Yaakov Okon
Dept. of Plant Pathology and Microbiology
Faculty of Agriculture
The Hebrew University of Jerusalem
Rohovot
Israel

Mexico

Yoav Bashan
Dept. of Microbiology
Centro de Investigaciones Biologicas de B.C.S.
P.O. Box 128
23000 La Paz, B.C.S.
Mexico

Russia

V. V. Ignatov
Institute of Biochemistry and Physiology
of Plants and Microorganisms
Prosp. Enthusiastov 13
Saratov 410015
Russia

Igor Skvortsov
Institute of Biochemistry and Physiology
of Plants and Microorganisms
Prosp. Enthusiastov 13
Saratov 410015
Russia

2. OTHER PARTICIPANT

Albania

Kristao Sini
Soil Research Institute
Tirana
Albania

Bulgaria

Ognian Kostov
Soil Science Inst.
7 Shosse Bankya
1080 Sofia
Bulgaria

Hungary

Majid Afsharian
University of Agricultural Sciences
Department of Biotechnology
H 2103 Gödöllö
Hungary

Andor Balint
Univ Agricult Sciences
Dept Plant Genetics
H- 2103 Gödöllö
Hungary

Borbala Biro
Institute of Soil Sci & Agrochem
Hungarian Acad Sci
H-1022 Budapest
Hermann O. U. 15.
Hungary

Gabor Gyulai
Univ Agricult Sciences
Dept Plant Genetics
H- 2103 Gödöllö
Hungary

Márton Jolánkai
Hungarian Academy of Sciences
Department of Natural Sciences
Nádor-u. 7.
1051 Budapest
Hungary

Eva Kàrpati
Dept. Biotechnology
Univ. Agricult. Sci. Gödöllö
2103 Gödöllö
Hungary

Pal Koranyi
Eötvös Lorand University
Department of Plant Anatomy
Puskin utca 11-13
1088 Budapest
Hungary

Judit Kovács
"Joliot-Curie"
National Res. Inst.
for Radiobiology and Radiohygiene
1088 Budapest
Hungary

Eva Preininger
Eotvos Lorand University
Dept of Plant Anatomy
Puskin U. 11-13
1088 Budapest
Hungary

Szilard Sandor Varga
Eotvos Lorand University
Dept of Plant Anatomy
Puskin U. 11-13
1088 Budapest
Hungary

Valeria Varga
University of Agric. Sciences Gödöllö
Dep. of Biotechnology
2103 Gödöllö
Hungary

Poland

St. Muszynski
Warsaw Agricultural University
Department of Plant Breeding
ul. Nowoursyynowska
02-766 Warsaw
Poland

S. Russel
Lehrstuhl f. Mikrobiologie
Akademie der Landwirtschaft
ul. Rakoviecka 26/30
02-524 Warschau
Poland

Russia

Ludmila P. Antoniuk
Academy of Sciences
Institute of Biochemistry and
Physiology of Plants and Microorganisms
pr. Entuzlastov 13
Saratov, 410015
Russia

Andrew Belimov
All Russian Research Institute for Agricultural
Microbiology
Pushkin 8, Podbelsky Sh., 3
189620 Saint -Petersburg
Russia

M.I. Chumakov
Institute of Biochemistry and Physiology
of Plants and Microorganisms
Russia Academy of Sciences
Prosp. Entuziastov, 13
Saratov, 410015
Russia

Alexander Iosipenko
Institute of Biochemistry & Physiology of
Plants & Microorganisms
Entuziastov Avenue 13
410015 Saratov
Russia

Elena I. Katzy
Institute of Biochemistry and Physiology of
Plants and Microorganims
pr. Entuziastov 13
410015 Saratov
USSR

Larissa Mantora
Inst.of Biochem. and Physiol.of Plants and
Microorg.
Lab. of Physical Chem. of Cell Struct.
IBPPM
13 Entuziastov Prosp.
410015 Saratov
Russia

3. OBSERVER

Argentinia

Carlos Bellone
Facultad de Agronomia y Zootecnia
Universidad Nacional de Tucmán
Av. Roca 1900
4000 San Miguel de Tucmán
Argentina

Camizo de Bellone
Facultad de Agronomia y Zootecnia
Universidad Nacional de Tucmán
Av. Roca 1900
4000 San Miguel de Tucmán
Argentina

Raul Pedraza
Facultad de Agronomia y Zootecnia
Universidad Nacional de Tucmán
Av. Roca 1900
4000 San Miguel de Tucmán
Argentina

India

Anil Kumar Tripathi
School of Biotechnology
Faculty of Science Banaras Hindu University
221005 Varanasi
India

Italy

Jindrich Shejba
J.S. Konsulent
Via Dei Glicini 11
00012 Guidonia (Roma)
Italia

Mexico

Beatriz Baca
Miguel Angel Mazcarva Esparza
Dept de Investigaciones Microbiologias
Apdo Postal 1622
Universidad Autónoma de Puebla
Puebla
Mexico

Uruguay

Gloria Martinez-Drets
Division Bioquimica
Inst. Invest. Biol. Clemente Estable
Av. Italia 3318
Montevideo
Uruguay

NATO ASI Series G

NATO ASI Series G

Vol. 21: **Health Related Effects of Phyllosilicates.**
Edited by J. Bignon. 462 pages. 1990.

Vol. 22: **Evolutionary Biogeography of the Marine Algae of the North Atlantic.**
Edited by D. J. Garbary and G. R. South. 439 pages. 1990.

Vol. 23: **Metal Speciation in the Environment.**
Edited by J. A. C . Broekaert, Ş. Güçer, and F. Adams. 655 pages. 1990.

Vol. 24: **Population Biology of Passerine Birds. An Integrated Approach.**
Edited by J. Blondel, A. Gosler, J.-D. Lebreton, and R . McCleery.
513 pages. 1990.

Vol. 25: **Protozoa and Their Role in Marine Processes.**
Edited by P. C. Reid, C. M . Turley, and P. H. Burkill. 516 pages. 1991.

Vol. 26: **Decision Support Systems.**
Edited by D. P Loucks and J. R. da Costa. 592 pages. 1991.

Vol. 27: **Particle Analysis in Oceanography.** Edited by S. Demers. 428 pages. 1991.

Vol. 28: **Seasonal Snowpacks. Processes of Compositional Change.**
Edited by T. D. Davies, M . Tranter, and H . G. Jones. 484 pages. 1991.

Vol. 29: **Water Resources Engineering Risk Assessment.**
Edited by. J. Ganoulis. 551 pages. 1991.

Vol. 30: **Nitrate Contamination. Exposure, Consequence, and Control.**
Edited by I. Bogárdi and R. D. Kuzelka. 532 pages. 1991.

Vol. 31: **Industrial Air Pollution. Assessment and Control.**
Edited by A. Müezzinoğlu and M. L. Williams. 245 pages. 1992.

Vol. 32: **Migration and Fate of Pollutants in Soils and Subsoils.**
Edited by D. Petruzzelli and F. G. Helfferich. 527 pages. 1993.

Vol. 33: **Bivalve Filter Feeders in Estuarine and Coastal Ecosystem Processes.**
Edited by R. F. Dame. 584 pages. 1993.

Vol. 34: **Non-Thermal Plasma Techniques for Pollution Control.**
Edited by B. M. Penetrante and S. E. Schultheis.
Part A: Overview, Fundamentals and Supporting Technologies. 429 pages. 1993.
Part B: Electron Beam and Electrical Discharge Processing. 433 pages. 1993.

Vol. 35: **Microbial Mats. Structure, Development and Environmental Significance.**
Edited by L. J. Stal and P. Caumette. 481 pages. 1994.

Vol. 36: **Air Pollutants and the Leaf Cuticle.**
Edited by K. E. Percy, J. N. Cape, R. Jagels and C. J. Simpson. 405 pages. 1994

Vol. 37: **Azospirillum VI and Related Microorganisms.**
Edited by I. Fendrik, M. del Gallo, J. Vanderleyden, and M. Zamaroczy.
588 pages. 1995.

Vol. 38: **Molecular Ecology of Aquatic Microbes.**
Edited by I. Joint. 423 pages. 1995.

MIX
Papier aus verantwortungsvollen Quellen
Paper from responsible sources
FSC® C105338

If you have any concerns about our products,
you can contact us on
ProductSafety@springernature.com

In case Publisher is established outside the EU,
the EU authorized representative is:
Springer Nature Customer Service Center GmbH
Europaplatz 3, 69115 Heidelberg, Germany

Printed by Libri Plureos GmbH
in Hamburg, Germany